T0255830

Biodiversity in Dead Wood

Fossils document the existence of trees and wood-associated organisms from almost 400 million years ago, and today there are between 400 000 and 1 million wood-inhabiting species in the world. This is the first book to synthesize the natural history and conservation needs of wood-inhabiting organisms.

Presenting a comprehensive introduction to biodiversity in decaying wood, the book studies the rich diversity of fungi, insects and vertebrates that depend upon dead wood. It describes the functional diversity of these organisms and their specific habitat requirements in terms of host trees, decay phases, tree dimensions, microhabitats and the surrounding environment. Recognizing the threats posed by timber extraction and insensitive forest management, the authors also present management options for protecting and maintaining the diversity of these species in forests as well as in agricultural landscapes and urban parks.

JOGEIR N. STOKLAND is a researcher at the Norwegian Forest and Landscape Institute and an Associate Professor at the University of Oslo, Norway. He has conducted research on forest biodiversity, dead wood dynamics and species diversity in decaying wood for more than 20 years. His expertise covers both entomology and mycology.

JUHA SIITONEN is a researcher at the Finnish Forest Research Institute. He has conducted research on the effects of forest management on dead wood and saproxylic species, including beetles and polypores, for more than 20 years. He is a member of the Finnish Beetle Working Group, and has been involved in the Red List assessments of Finnish fauna.

BENGT GUNNAR JONSSON is a Professor of Plant Ecology at Mid Sweden University. His research focuses on forest history and dynamics and its role in maintaining forest biodiversity. He has played an active role in several national conservation projects initiated by the Swedish Environmental Protection Agency and the Swedish Forest Agency.

ECOLOGY, BIODIVERSITY AND CONSERVATION

The world's biological diversity faces unprecedented threats. The urgent challenge facing the concerned biologist is to understand ecological processes well enough to maintain their functioning in the face of the pressures resulting from human population growth. Those concerned with the conservation of biodiversity and with restoration also need to be acquainted with the political, social, historical, economic and legal frameworks within which ecological and conservation practice must be developed. The new Ecology, Biodiversity and Conservation series will present balanced, comprehensive, up-to-date and critical reviews of selected topics within the sciences of ecology and conservation biology, both botanical and zoological, and both 'pure' and 'applied'. It is aimed at advanced final-year undergraduates, graduate students, researchers and university teachers, as well as ecologists and conservationists in industry, government and the voluntary sector. The series encompasses a wide range of approaches and scales (spatial, temporal and taxonomic), including quantitative, theoretical, population, community, ecosystem, landscape, historical, experimental, behavioural and evolutionary studies. The emphasis is on science related to the real world of plants and animals rather than on purely theoretical abstractions and mathematical models. Books in this series will, wherever possible, consider issues from a broad perspective. Some books will challenge existing paradigms and present new ecological concepts, empirical or theoretical models, and testable hypotheses. Other books will explore new approaches and present syntheses on topics of ecological importance.

Ecology and Control of Introduced Plants
Judith H. Myers and Dawn Bazely

Invertebrate Conservation and Agricultural Ecosystems
T. R. New

Risks and Decisions for Conservation and Environmental Management
Mark Burgman

Ecology of Populations
Esa Ranta, Per Lundberg and Veijo Kaitala

Nonequilibrium Ecology
Klaus Rohde

The Ecology of Phytoplankton
C. S. Reynolds

Systematic Conservation Planning
Chris Margules and Sahotra Sarkar

Large-Scale Landscape Experiments: Lessons from Tumut
David B. Lindenmayer

Assessing the Conservation Value of Freshwaters: An International Perspective
Philip J. Boon and Catherine M. Pringle

Insect Species Conservation
T. R. New

Bird Conservation and Agriculture
Jeremy D. Wilson, Andrew D. Evans, and Philip V. Grice

Cave Biology: Life in Darkness
Aldemaro Romero

Biodiversity in Environmental Assessment: Enhancing Ecosystem Services for Human Well-being
Roel Slootweg, Asha Rajvanshi, Vinod B. Mathur and Arend Kolhoff

Mapping Species Distributions: Spatial Inference and Prediction
Janet Franklin

Decline and Recovery of the Island Fox: A Case Study for Population Recovery
Timothy J. Coonan, Catherin A. Schwemm and David K. Garcelon

Ecosystem Functioning
Kurt Jax

Spatio-Temporal Heterogeneity: Concepts and Analyses
Pierre R. L. Dutilleul

Parasites in Ecological Communities: From Interactions to Ecosystems
Melanie J. Hatcher and Alison M. Dunn

Zoo Conservation Biology
John E. Fa, Stephan M. Funk and Donnamarie O'Connell

Marine Protected Areas: A Multidisciplinary Approach
Joachim Claudet

Biodiversity in Dead Wood

JOGEIR N. STOKLAND
Norwegian Forest and Landscape Institute and University of Oslo, Norway

JUHA SIITONEN
Finnish Forest Research Institute, Vantaa, Finland

BENGT GUNNAR JONSSON
Mid Sweden University, Sundsvall, Sweden

CAMBRIDGE
UNIVERSITY PRESS

University Printing House, Cambridge CB2 8BS, United Kingdom

One Liberty Plaza, 20th Floor, New York, NY 10006, USA

477 Williamstown Road, Port Melbourne, VIC 3207, Australia

314-321, 3rd Floor, Plot 3, Splendor Forum, Jasola District Centre, New Delhi - 110025, India

79 Anson Road, #06-04/06, Singapore 079906

Cambridge University Press is part of the University of Cambridge.

It furthers the University's mission by disseminating knowledge in the pursuit of education, learning and research at the highest international levels of excellence.

www.cambridge.org
Information on this title: www.cambridge.org/9780521717038

First published 2012
Reprinted 2013

A catalogue record for this publication is available from the British Library

Library of Congress Cataloging in Publication data
Stokland, Jogeir N.
Biodiversity in dead wood / Jogeir N. Stokland, Juha Siitonen, Bengt Gunnar Jonsson.
p. cm. – (Ecology, biodiversity, and conservation)
Includes bibliographical references and index.
ISBN 978-0-521-88873-8 (hardback) – ISBN 978-0-521-71703-8 (paperback)
1. Forest biodiversity. 2. Forest litter–Biodegradation. 3. Wood–Deterioration.
4. Forest ecology. 5. Wood-decaying fungi. 6. Saproxylic insects.
I. Siitonen, Juha. II. Jonsson, Bengt Gunnar. III. Title.
QK46.5.D58S88 2012
577.34–dc23 2012000301

ISBN 978-0-521-88873-8 Hardback
ISBN 978-0-521-71703-8 Paperback

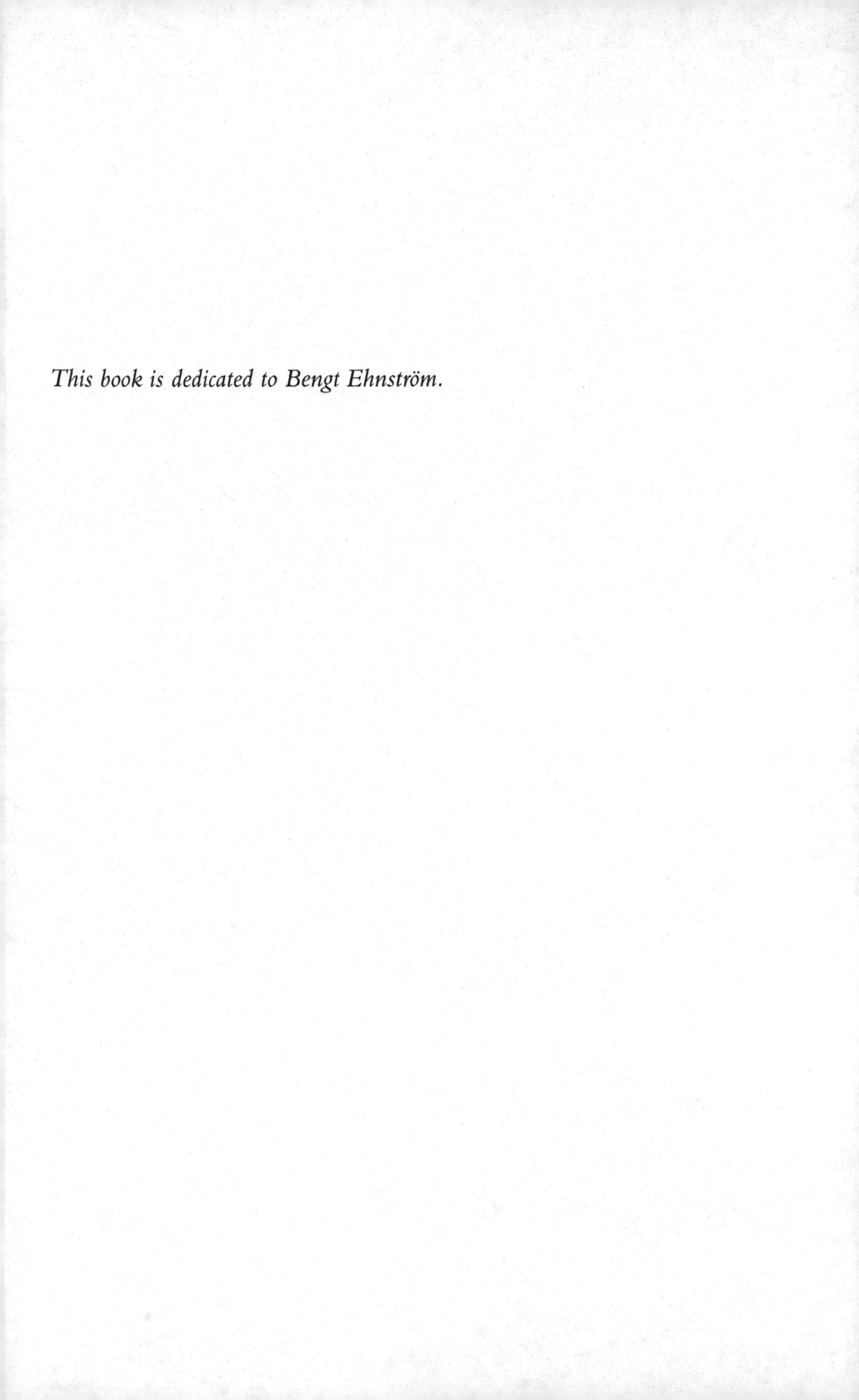

This book is dedicated to Bengt Ehnström.

Contents

Preface

The last few decades have witnessed a rapidly increasing interest in the importance of dead and decaying trees for biodiversity. During their decomposition, dead trees offer habitats for thousands of species. This diversity has been studied by researchers interested in particular organism groups, such as cavity-nesting birds, wood-decaying fungi or saproxylic invertebrates. A holistic overview of the species communities inhabiting trees after their death has been lacking, and our aim is to provide such an overview here.

The scope of the book is global, but we admit that it has a strong north European bias. There are two reasons for this. Firstly, much of the research and many of the scientific publications about species living in dead wood originate from northern Europe, although during the last decade an increasing number of papers dealing with saproxylic organisms have also been published in North America, Australia, Japan and elsewhere. Secondly, our own studies have taken place in Fennoscandia, and our empirical knowledge is mainly derived from the boreal and temperate parts of Europe. We admit that we only have superficial first-hand experience of tropical forests and the temperate and evergreen forests of other continents.

We have written this book with several kinds of reader in mind: biologists and students of biology with an interest in forest ecology and biodiversity, forest and park managers, nature conservation managers, and people interested in nature and natural sciences. This readership includes people with very different background knowledge. Thus, it is likely that the book will cover both familiar and unfamiliar topics for most of our intended readers.

Much of this book is about fungi and insects. Expert mycologists and entomologists might find their own fields of expertise rather superficially treated in some sections, as we have not reviewed everything of potential relevance to each topic. Instead, we have tried to write about mycology for entomologists and about entomology for mycologists.

Similarly, our aim has been to write about ecology directed towards forest and park managers, and about forest dynamics and management directed towards people with a background in ecology. Hopefully, this will make the text more accessible to readers without expertise in any particular discipline.

We have tried to keep the amount of specialized terminology to a minimum and to explain terms and concepts when we first use them. We have used vernacular names for species and higher taxonomic groups wherever these exist, and have provided the scientific names in parenthesis. However, for most individual species we have used the Latin names only, since most wood-inhabiting species lack established vernacular names.

We have made every effort to keep the various topics updated with the most recent and most relevant publications. In many cases we have also highlighted important studies that are several decades old but still represent valuable knowledge. Throughout the book we have made numerous references to the primary literature so that the interested reader can access this for further details. Our intention has been to cite publications that, in combination, provide up-to-date coverage of each topic. However, in some cases, we may still have overlooked important references. This should be borne in mind by the reader.

When writing the individual chapters, we have been given many valuable pieces of information. We would particularly like to thank the following people for reviewing different chapters: Keith Alexander, Peter Baldrian, Manfred Binder, Mattias Edman, Michael S. Engel, Shawn Fraver, Jacob Heilmann-Clausen, David Hibbett, Jyrki Muona, Björn Nordén, Thomas Ranius and Graham Rotheray. Any potential mistakes remain our own responsibility. We also thank all the photographers who have kindly allowed us to use their splendid photos to illustrate this book.

Finally, we would like to pay tribute to the Swedish entomologist and naturalist Bengt Ehnström, to whom we have dedicated this book. He has an impressive knowledge of biodiversity in decaying wood and seems to recognize virtually every insect species as a personal friend. Bengt's warm personality and everlasting willingness to share his knowledge as a field guide, a speaker and a writer has been, and will remain, a great inspiration for innumerable people with an interest in nature and the life found in dead trees.

1 · *Introduction*

Jogeir N. Stokland, Juha Siitonen and Bengt Gunnar Jonsson

This book is about life in dead trees. All over the world one can find a fascinating diversity of life forms in decaying wood – first and foremost a wide variety of fungi and insects. These organisms carry out the hidden but highly important work of wood decomposition.

A fundamental question frequently revisited in this book is: 'Why is the species diversity of wood-inhabiting organisms so tremendously high?' In most chapters we approach this question indirectly by highlighting the key properties of dead wood, along with the environmental factors and processes that bring about the diversity we can observe. We also discuss species richness explicitly in Chapter 11. There are at least two good reasons for addressing the biodiversity in dead wood. One is that the diversity of wood-inhabiting organisms is a multifaceted and interesting phenomenon that deserves attention for its own sake. Another reason is that this diversity is being seriously threatened due both to the loss and fragmentation of forests and because of the greatly reduced amount of dead wood in managed forests and other woodlands. Thus, we need to understand the role of dead wood for biodiversity in order to manage and maintain it while efficiently utilizing forest resources.

From the outset, the subject of this book could be presented in several different ways. One type of book could be directed at academic biologists and would discuss the subject in the context of ecological and evolutionary theories. Another type of book could focus on biodiversity-oriented management of dead wood and the associated species in forests, agricultural landscapes and urban green areas. We have chosen the middle ground, with an emphasis on describing the diversity and the underlying ecological factors, but we have also added chapters related to management. This choice is underpinned by our belief that a deeper knowledge about wood-inhabiting species has a strong applied potential, and that it is useful for people with a broad interest in forests and nature conservation.

1.1 Biodiversity in decaying wood

Most people are completely unaware of the diversity of life that exists inside decaying trees. In fact, even the majority of biologists have a relatively limited knowledge of this diversity. Thus, we shall briefly introduce the variety of organisms that, in different ways, depend upon decaying wood.

As we have already mentioned, wood-inhabiting species primarily consist of fungi and insects. Among the fungi we find several groups that are dominated by or contain wood-inhabiting species. The most important wood-decaying species belong to the basidiomycetes (Basidiomycota), including polypore fungi or bracket fungi (a polyphyletic group with representatives in Hymenochaetales, Polyporales, Gloeophyllales and others), and corticioid fungi (another polyphyletic group represented in Hymenochaetales, Corticiales, Russulales and others). Furthermore, we find many other basidiomycete groups that are dominated by wood-decaying species such as jelly fungi (Dacrymycetales), and several different families and genera of agaric fungi (Agaricales). Large numbers of wood-inhabiting species are also found in the other main phylum of fungi, the sac fungi or ascomycetes (Ascomycota), including yeasts (Saccharomycotina) and many other groups. Even if the names of these taxa may not be informative to a non-specialist reader, the appearance of these fungi is often attractive both in shape and colour (see Figure 1.1 and book cover) and they have very interesting ways of living. Most of these fungi are wood decomposers but several of them have entirely different ecological roles.

Among insects, there are several groups where a significant proportion of the species live in decaying wood (Figure 1.2). These include four key orders that comprise the majority of wood-inhabiting insects: beetles (Coleoptera), gnats and flies (Diptera), wasps, bees and ants (Hymenoptera) and termites (Isoptera; nowadays placed in Dictyoptera). In addition, several other insect orders contain wood-inhabiting species, such as moths (Lepidoptera), bugs (Hemiptera), thrips (Thysanoptera), snakeflies (Raphidioptera) and zorapterans (Zoraptera). However, this does not complete the list of saproxylic invertebrates; for instance, mites (Acari) are well represented in decaying wood. This is a hyperdiverse group of small invertebrates belonging to the arachnids (Arachnida). The number of wood-inhabiting mite species may be as large as in the above-mentioned major insect orders, but their ecology and habitats are generally much less well known. Other invertebrate taxa such as pseudoscorpions (Pseudoscorpionida)

Figure 1.1. Wood-inhabiting fungi representing different taxonomic groups: (a) *Fomitopsis pinicola* (photo John Munt); (b) *Pleurotus ostreatus* (© Jens H. Petersen/MycoKey); (c) *Phlebia tremellosa* (photo Atli Arnarson); (d) *Xylaria hypoxylon* (photo Mikel A. Tapia Arriada); (e) *Lachnellula subtilissima* (photo Dragiša Savić); (f) *Bisporella citrina* (photo Dragiša Savić).

Figure 1.2. Representatives of different insect orders with numerous saproxylic species: (a) "the hoverfly" *Volucella inflata* (photo Dragiša Savić); (b) the giant woodwasp *Urocerus gigas* (photo Nikola Rahmé); (c) the click beetle *Ampedus quadrisignatus* (photo Nikola Rahmé); (d) the longhorn beetle *Acanthocinus henschi* (photo Nikola Rahmé); (e) the stag beetle *Lucanus cervus* (photographer unknown, see http://www.dreamstime.com/royalty-free-stock-photo-stag-beetle-image10207875).

and nematodes (Nematoda) are also well represented in decaying wood. In marine waters, we find both molluscs and crustaceans that bore into submerged wood. This broad taxonomic diversity is paralleled by a wide range of functional roles including those of detritivores, fungivores, predators, scavengers, parasitoids, and various types of symbiosis (commensalisms, mutualism).

Among the vertebrates there are various species with direct associations to wood, such as woodpeckers, a few mammals that eat woody materials, and there is even a group of tropical fish (catfish in the genus *Panaque*) that appear to have a specialized diet of wood.

There are also many species that live in dead and decaying trees but do not use them for their nourishment. A large number of both vertebrate and invertebrate species use snags, logs and cavities in living trees for breeding and other purposes.

1.2 Saproxylic species: defining the concept

In the previous section we briefly introduced various groups of species living in decaying wood. The term *saproxylic* has become well established to denote species that are dependent on dead woody material at some stage of their life cycle. It is derived from the Greek words *sapros* and *xylon*, meaning 'decayed' and 'wood', respectively. This term represents the essence of biodiversity in dead wood and we shall explore its conceptual content, especially since various authors have used it in somewhat different ways.

A term akin to saproxylic was first used by Silvestri (1913) when he described the insect order Zoraptera as new to science. Silvestri called the invertebrates living specifically in decaying wood 'saproxylophiles', in contrast with insects living in soil, dung or carcasses. Dajoz (1966) picked up and used the term saproxylic for insects living in decaying wood. Later he extended the term to include species occurring in recently dead wood (Dajoz, 2000).

It is usual to refer to Speight (1989) for a definition of saproxylic species. Speight defined saproxylic invertebrates as:

> species of invertebrates that are dependent, during some part of their life cycle, upon the dead or dying wood of moribund or dead trees (standing or fallen), or upon wood-inhabiting fungi, or upon the presence of other saproxylics.

The publication by Speight dealt mainly with saproxylic invertebrates, but he also briefly mentioned saproxylic vertebrates and fungi. A strict use of Speight's definition would exclude species confined to the bark, but typical usage includes such species as well. There are several publications that discuss alternative definitions of saproxylic species. Here we only mention Alexander (2008), who pointed out that a definition connecting saproxylic species only to dead or moribund trees may be too restrictive, because hollow trees are often healthy, or at least not moribund.

It is also relevant to mention the term *xylobiont*, which is frequently used in the German literature. Schmidl and Bussler (2004) provided the following definition for xylobiontic beetles:

> species that reproduce and spend obligatorily most of their lifespan in any kind of wood and in any kind of decay stage, including fungi living on wood.

Thus, the meaning is close to the definition by Speight, but it also includes species living in healthy trees.

Note that the definitions above mainly concern animals. In these definitions, the wood-inhabiting fungi simply represent a habitat or medium for the animals like the wood itself. In the mycological literature it is unusual to characterize fungi as saproxylic, although it is possible to find some recent references to saproxylic fungi. Mycologists talk instead about wood-inhabiting fungi or wood-decaying fungi.

Still another related term is *epixylic*, meaning 'on wood'. This term is used for moss and lichen species that prefer to grow on the surface of dead wood. Our definition of saproxylic species in the next paragraph includes epixylic species as a functional subcategory.

In this book we adopt a broad ecological approach, and it therefore becomes essential to include fungi among the saproxylics. We use the term saproxylic based on the following definition:

> any species that depends, during some part of its life cycle, upon wounded or decaying woody material from living, weakened or dead trees.

In this context, woody material refers not only to wood, but also bark and sap (from inner bark, sapwood, or flowing from wounds) at any stage of decay. Thus, we include species living in wounds, dead branches or cavities of otherwise healthy trees. On the other hand, we do not include piercing and sucking insects (such as aphids or scale insects) that dwell on bark and feed on sap from healthy trees. Neither do we include endophytic fungi living inside living trees, unless they are active during the decomposition of the tree when it is dead. At this point we should stress that, irrespective of where one draws the line between saproxylic and non-saproxylic species, the distinction will remain somewhat arbitrary in the sense that species on either side of the boundary will be quite similar.

1.3 Structure of the book

We have written the book so that each chapter has a distinct focus and can be read separately from the others. However, many topics treated in separate chapters are closely related to each other, and we have made numerous cross-references to link such topics. When all the chapters are viewed together, they form four parts that cover different aspects of biodiversity: functional diversity (Chapters 2–4), structural diversity (Chapters 5–9), compositional diversity (Chapters 10 and 11), and finally biodiversity conservation and management (Chapters 12–17).

In the first part we describe how different organisms are functionally associated with decaying wood and with other organisms living in this habitat. The first chapter treats the different ways in which wood-inhabiting species decompose or digest bark and wood. Next we describe how the species living in decaying wood make up a food web with a nutritional link to decaying wood. In the final chapter of this part we describe species that have a spatial link to dead wood but do not depend upon wood as a food or energy source.

The second part, on structural diversity, highlights how different types of decaying wood support various species assemblages. These are treated in separate chapters focusing on host trees, mortality factors and decomposition phases, specific microhabitats, and tree size. The last chapter in this section shows how the surrounding environment has a strong effect on the species composition inside decaying wood.

In the part dealing with compositional diversity, we focus on species diversity itself, i.e. the identity and variety of the species inhabiting decaying wood. First we describe how woody plants and wood-inhabiting organisms originated almost 400 million years ago and evolved until the present. Then we explore the saproxylic diversity that we know today and try to quantify the species richness in various organism groups.

The last part of the book represents a distinct shift in focus of interest. While we describe saproxylic species under natural conditions at a detailed (substrate) level in the first three parts, we here describe how species diversity is maintained at the landscape scale. We also adopt a conservation and management perspective and explore the ways in which we could modify land-use practices in order to maintain the diversity of saproxylic organisms in forests and other woodland types.

1.4 Knowledge, disciplines and perspectives

The knowledge about saproxylic organisms is scattered across thousands of scientific papers and hundreds of books from quite different and distinct disciplines. In addition, there are innumerable publications in national journals – mainly of a taxonomic and faunistic nature – about the species living in dead wood. Naturally, we have not been able to access all these information sources. Nevertheless, we have tried to present a broad treatment of the biology and natural history of saproxylic species. Three things have become apparent while examining various information sources.

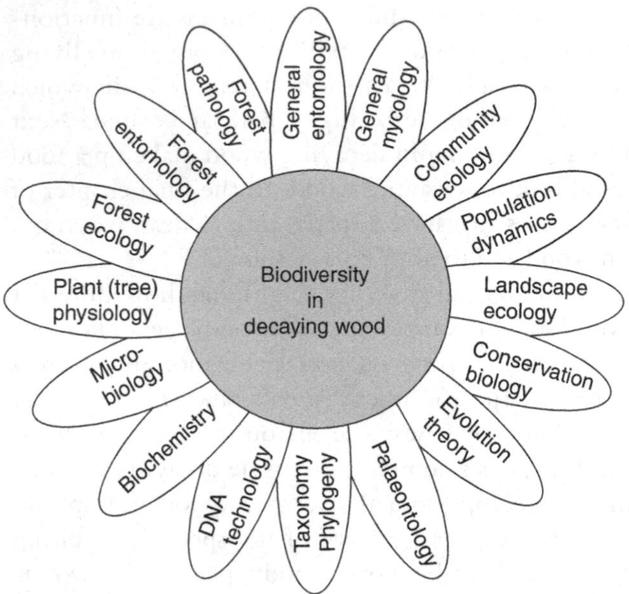

Figure 1.3. Different research disciplines that together form the basis for understanding the biodiversity in decaying wood.

First, a great deal is known about saproxylic organisms and it is quite demanding to obtain an overview of this knowledge. The main reason is that it has been developed in very different disciplines (Figure 1.3). These can be quite narrow in scope such as, for instance, the functioning of cellulose- and lignin-degrading enzymes, forest pathology and forest entomology, and tree physiology. In other cases the relevant pieces of information must be extracted from much wider disciplines such as ecology, general entomology, mycology, palaeontology, taxonomy and phylogeny, where the specific characteristics of saproxylic organisms are treated superficially or indirectly. But there are also publications that focus specifically on the diversity of wood-inhabiting species. These publications typically include keywords such as 'saproxylic', 'wood-decaying', 'wood-inhabiting', or they may contain the keywords 'woody debris', 'dead wood' or 'decaying wood' in combination with a particular taxonomic group. A classical and much cited work about the ecology of dead wood in temperate forests is the review by Harmon et al. (1986). Key books on wood-inhabiting fungi include those by Rayner and Boddy (1988) and Boddy et al. (2008). Similarly, for insects, one can find good overviews in Dajoz (2000) and Lieutier et al. (2004). Useful reviews on the effects of forest management on saproxylic species include those of Siitonen (2001), Grove (2002a, 2002b) and Jonsson et al. (2005). Some larger works even make the cross-over between the fungal–insect division, such as the special volume of *Ecological Bulletins* edited by Jonsson and Kruys (2001), which is devoted to the ecology of coarse woody debris in boreal forests, as well as a book about the 'afterlife of trees' by Bobiec et al. (2005).

The second thing that becomes apparent is that species living in dead wood are viewed quite differently in different disciplines. In most knowledge fields there is a neutral attitude towards saproxylic species. They are simply research objects. But in some disciplines (forest management, forest pathology, forest entomology, arboriculture) there is a strong presumption that saproxylic species are generally unwanted and should be controlled or eliminated. Here one finds terms such as pest species, disease and tree or forest damage – a vocabulary that is commonly used and appropriate when the economic value of forest resources is the primary interest. In other disciplines such as conservation biology, the attitude towards saproxylic species is positive. Here the focus is on species with declining population trends, which is considered undesirable and should be counteracted.

A third finding is that the amount of knowledge varies substantially from one organism group to another. For example, the understanding of cellulose- and especially lignin-degrading enzymes is much better for basidiomycetes than for ascomycetes. And among the basidiomycetes it is much better for initial decomposers than for decomposers that occur later in the decay succession. Similarly, the understanding is generally better for early-successional species than for late-successional species, for wood consumers as compared with predators and parasitic species, for pathogenic species rather than non-pathogenic species, etc. Even in studies of saproxylic species as such, there are big differences. Saproxylic beetles have been much more intensively studied than saproxylic gnats, flies, wasps and mites, and similarly, basidiomycete fungi are much better studied than ascomycetes.

The implication of this uneven spread of knowledge is that for some topics or organism groups we can present well-established and detailed facts, whereas in other cases the treatment is necessarily rather superficial. We have tended to be selective when a lot of knowledge exists, but present most of what we have come across on topics where little is known.

2 · *Wood decomposition*

Jogeir N. Stokland

When you sit beside a campfire you can easily feel the energy that is tied up in woody material. As the wood burns, it is transformed to carbon dioxide, water vapour and minerals – the elements that the tree tied up through photosynthesis when it was alive and growing. The combustion of wood in the campfire takes only a few hours. In temperate and boreal forest ecosystems the equivalent degradation of a tree typically takes 50–100 years and is carried out by numerous wood decomposers working at a much lower temperature.

This chapter deals with the activity of these decomposers – how they degrade and recycle dead wood in forest ecosystems all over the globe. Fungi are the principal decomposers in terrestrial ecosystems, and especially among the basidiomycetes we find many effective wood-decaying species. Also a large number of invertebrates, such as beetles and termites, take part in the process of wood decomposition. Before we explore this fundamental ecosystem process, we shall describe some key aspects of wood structure.

2.1 Structural wood components

Wood is made up of three structural components: cellulose, hemicellulose and lignin. The chemical composition, synthesis and degradation of these economically important wood constituents have been important research topics for more than 50 years – and they still are. As a result, we have a good understanding of their biochemical properties. It is beyond the scope of this book to go into great detail about these specialized topics, which are regularly reviewed in books and scientific journals (see Buswell, 1991; Markham and Bazin, 1991; Jeffries, 1994; Schwarze et al., 2000b; Vicuña, 2000; Martínez et al., 2005; Baldrian, 2008).

2.1.1 Cellulose

Cellulose has a straightforward molecular structure that is composed of glucose units joined together in long carbohydrate chains. Such chains typically contain several thousand glucose units per molecule. Every second glucose unit is rotated 180° with respect to its neighbours, causing pairs of adjacent glucose units to be a repeating unit. These pairs of 180°-twisted glucose units are called cellobiose. The rotation causes cellulose molecules to be highly symmetrical, and it facilitates the coupling of adjacent cellulose molecules by numerous chemical bonds. Such cellulose chains associate laterally and form crystalline microfibrils, which are the basic units of cellulose organization. The microfibrils, in turn, associate into larger cellulose fibres or are bound together with hemicellulose and lignin to form a mechanically very strong material. Cellulose is usually the dominant constituent in wood and accounts for 40–50% of the dry wood weight of both coniferous and broadleaved trees. Cellulose is a very common biochemical substance and is the building material for all photosynthetic plants, including marine and freshwater algae. Thus, cellulose has been synthesized on our planet for a long time before woody plants evolved (see Chapter 10).

2.1.2 Hemicellulose

Hemicellulose is a carbohydrate like cellulose, but has a more complex molecular structure. It consists of several sugar types that are linked together in shorter chains than those of cellulose. The main difference compared with cellulose is that hemicellulose has a branched structure, where the lateral chains consist of different sugars than the main chain. Hemicellulose molecules do not aggregate like cellulose, but they can associate with the cellulose microfibrils and co-crystallize together with them. Furthermore, hemicellulose combines with lignin to form a non-crystalline structure in which the cellulose fibrils are embedded.

Hemicellulose occurs as different types in coniferous and broadleaved trees. In conifer wood, the predominant hemicelluloses are glucomannans, where the main chain is made up of glucose and mannose sugars. But we also find another type of hemicellulose in conifer wood that is called xylan. In xylan the main structure is a chain of xylose sugar molecules. In broadleaved wood, the predominant hemicellulose is xylan. In addition, the proportion of hemicelluloses in total dry wood weight differs between the two classes of trees. It tends to be somewhat higher in broadleaved wood (25–40%) than in coniferous wood (25–30%).

2.1.3 Lignin

Lignin has a very complex molecular structure made up of aromatic subunits and not sugars as in cellulose and hemicellulose. There are three types of these subunits, and for simplicity we refer to them as the H-, G- and S-units. They form a heterogeneous three-dimensional network that is linked together in an irregular manner with very strong chemical bonds (covalent C–C and C–O–C bonds). This kind of molecular structure is highly resistant to enzymatic degradation. An effective enzyme is one that can act on a repeating molecular structure and cut off pieces in an iterative manner. Thus, when lignin forms an integrated structure together with hemicellulose molecules, the plants have made a substance that is extremely difficult for decomposing organisms to break up.

Lignin composition, in terms of the ratio between H-, G- and S-units, varies substantially between different plant groups. Conifer lignin is made up mostly of G-units, whereas the lignin of broadleaved trees has a mixture of G- and S-units, with the latter occurring in a greater quantity. Lignin is present in larger amounts (25–35%) in coniferous wood than in broadleaved trees (18–25%). This increased proportion, together with the chemical difference, is probably the reason for the refractory nature of conifer lignin.

2.1.4 Cell structure

Cellulose, hemicellulose and lignin have a very distinct distribution pattern inside and between the wood cells. When the cells grow, they first make a *primary cell wall* composed of cellulose fibres, hemicellulose and pectin (a kind of carbohydrate) (see Figure 2.1). When the cells have reached their full size, they form a *secondary cell wall* inside the primary wall. First, the cell constructs the overall architecture of the secondary cell wall, using cellulose, hemicellulose and some structural proteins. Next, lignin deposition begins at the interface between the primary and secondary cell wall. The lignin formation extends outwards into the space between the cells and inwards throughout the secondary cell wall regions, until the cell wall is completed. The outcome

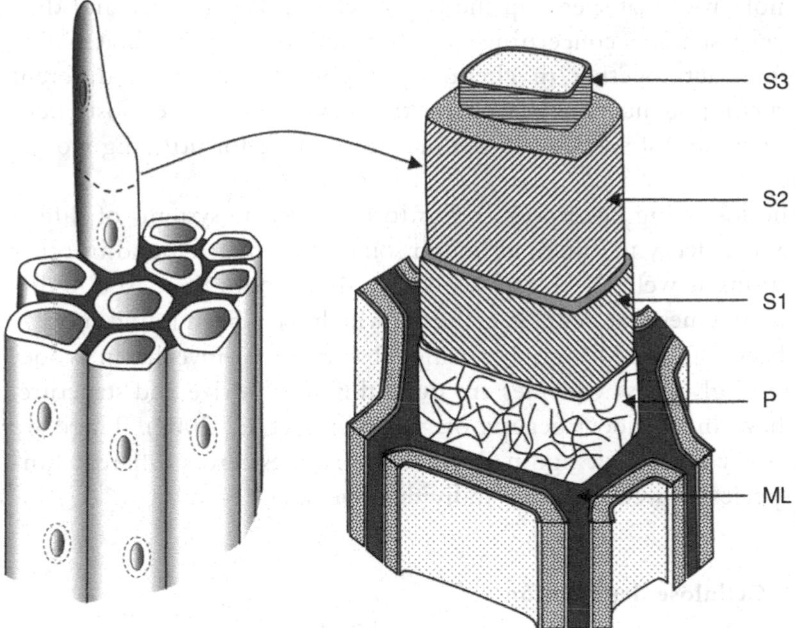

Figure 2.1. A schematic presentation of wood structure showing adjacent tracheid cells, the diameter of each tracheid (left) is approximately 30 μm; wood cell wall layers (right) **S1–S3**: secondary cell wall layers, **P**: primary cell wall, **ML**: middle lamella. Redrawn and modified from Kirk and Cullen (1998).

of this process is a secondary cell wall that is made up of three distinct layers, where lignin dominates in the outer (S1) and inner (S3) layers, and cellulose and hemicellulose dominate in the middle (S2) layer. The space between neighbouring cells, the *middle lamella*, is filled with a lignin-dominated substance and acts as a cement between the cells. This overall cell structure is significant for the wood's resistance to decay and the colonization patterns of fungal hyphae from white-, brown- and soft-rot fungi.

The contents of living wood cells are similar to those of other plant cells and are primarily liquid. When wood cells die, the contents dry up and the interior becomes an empty, typically moist, space that is called the *cell lumen*.

2.2 Enzymatic degradation of wood

Wood-decaying organisms possess a wide range of enzymes that represent an effective toolkit for decomposing woody material. Enzymes are molecules that speed up the rate of chemical reactions, and they are very specific concerning which reactions they facilitate. The enzymes act together in chemical systems operating on different wood components. The organisms that have evolved the most effective enzyme systems are the winners in the race for utilizing woody material.

The following description refers to the enzyme systems of different wood-decaying fungi, although some of them are found in other organisms as well. At present we know about more than 10 000 different enzymes that degrade cellulose which have been found in various bacteria, yeast fungi, and fungi that produce mycelia in wood. Interestingly, these enzymes are quite different in size and structure, and have independent genetic origins (Pérez et al., 2002). There are also many alternative lignin-degrading enzymes across different fungus species, as well as within individual species.

2.2.1 Cellulose degradation

As described above, cellulose has a straightforward chemical structure with a repeating sequence of cellobiose units – the pairs of twisted glucose molecules. Three functionally different types of enzymes are involved in cellulose degradation. The first enzymes that go into action, the *endoglucanases*, attack the long cellulose chains at internal points and split the cellulose into shorter fragments. This initial step produces a large number of terminal ends. The second enzyme type, the *exoglucanases*, operate on these terminal ends and cut off individual cellobiose units. Finally, *beta-glucosidase* enzymes split the cellobiose units into individual glucose molecules. These glucose molecules are sufficiently small to be absorbed and broken down further by the fungus mycelium through internal cell metabolism.

In reality, several equivalent enzymes carry out each of the enzymatic steps described above. In the second step, for example, one exoglucanase enzyme operates on the positively loaded ends of the cellulose fragments, while another type operates on the negatively loaded ends. In a similar way, other equivalent enzymes carry out cellulose degradation under different conditions. In addition to the enzyme-based degradation of cellulose, some fungi have also evolved a machinery for the production of hydroxyl (OH) radicals. Once produced, OH radicals are able to cleave cellulose, and also hemicellulose and lignin (Baldrian and Valášková, 2008).

2.2.2 Hemicellulose degradation

Since hemicellulose is composed of different sugars to cellulose, other enzymes are needed to degrade it. Furthermore, different enzymes degrade xylan (the hemicellulose type found in broadleaved wood), and glucomannan (which predominates in conifer wood).

The complete degradation of both hemicellulose types requires the cooperative action of a variety of enzymes, and there are five steps involved in these processes (Pérez et al., 2002). The first step in xylan degradation is carried out by the enzyme endoxylanase, which attacks internal points in the main chain and produces smaller molecule fragments. Next, three different enzyme types chop off specific chemical bonds that connect the side chains to the main chain of the molecule. Finally, there is an enzyme that works on the same chemical bonds of the main chain as the first enzyme, but this one operates only on smaller fragments and produces xylose. Xylose is the monosaccharide (single sugar molecule) building unit of xylan that corresponds to the glucose molecule in cellulose. The degradation of glucomannan, the conifer hemicellulose, is quite similar to that of xylan, as an initial enzyme breaks up the molecule, others break up specific chemical bonds connecting side chains, and finally two different enzymes chop off glucose and mannose molecules.

2.2.3 Lignin degradation

Due to the complex nature of lignin, researchers have taken many decades to elucidate how it is degraded, and this process is still not fully understood. A breakthrough came in 1983 when researchers discovered that an H_2O_2-dependent enzyme was directly involved in lignin degradation by the white-rot fungus *Phanerochaete chrysosporium* (Glenn et al., 1983; Tien and Kirk, 1983). This enzyme belongs to a class of enzymes called lignin peroxidases. Further research has demonstrated that different white-rot fungi produce at least 10 different types of lignin peroxidase (Cullen, 1997). A key function of these enzymes is to cleave the strong C–C and C–O–C bonds that account for more than 50% of the bonds that link together the lignin subunits. Thus the lignin peroxidases cause extensive breakdown of the lignin structure. Furthermore, these powerful enzymes have several other catalytic functions including cleavage of the aromatic rings that are found in the subunits themselves.

The H_2O_2 that the lignin peroxidases depend upon is produced by another group of enzymes – the oxidases. Several oxidases have been suggested as the producers of H_2O_2 but it seems that only one type, abbreviated as GLOX (glyoxal oxidase), is secreted under the conditions where lignin is degraded (Kirk and Farrell, 1987). The GLOX enzymes respond to the activity of the peroxidase enzymes and appear to switch on H_2O_2 production according to the demand of the peroxidases. On the other hand, the GLOX enzymes use part of the peroxidase products as their substrate for H_2O_2 production. Thus, the two enzymes act in concert and represent an advanced and very powerful coupled enzyme system. Logically enough, GLOX enzymes have been detected in several wood-decaying fungi (Orth et al., 1993).

In addition to lignin peroxidase, there exist at least two other classes of peroxidase enzymes. One is manganese peroxidase, which was discovered at the same time as lignin peroxidase and it is now known to be widely distributed among lignin-degrading fungi. The other is the quite rare 'versatile peroxidase', which has been described from *Pleurotus* and a few other genera of white-rot fungi (Martínez et al., 2005). These two enzyme classes also depend upon H_2O_2 to oxidize their substrate.

Laccases represent a final class of enzymes with the capacity to degrade lignin. Laccases have been known for a long time in plants, fungi, insects and bacteria. They have a variety of functions, including pigment synthesis, fruiting-body morphogenesis, and detoxification (Mayer and Staples, 2002). These enzymes are not as strong as the lignin peroxidases, and their low redox potential only allows the direct oxidation of phenolic lignin units, which often represent less than 10% of the total lignin. Interestingly, the laccases use atmospheric oxygen in the oxidizing process and thereby operate without the co-action of another enzyme system.

In summary, the production of lignin peroxidase, Mn-peroxidase and versatile peroxidase is limited to the white-rot fungi among the basidiomycetes. Most of these fungi rely on Mn-peroxidase in combination with laccase. It seems that lignin peroxidase is produced by no more than four or five genera of fungi, and versatile peroxidase appears to be even rarer (P. Baldrian, personal communication).

2.2.4 Sugar degradation

It is quite unusual to deal with sugar fermentation in a text on wood decomposition. But in this book we use an extended concept of woody material that also includes sap. The reason is that sap represents an important nutritional source for several of the fungi and invertebrates that occur in wounds and the inner bark of recently dead trees.

Sugar fermentation is a process that is completely different from wood degradation. First of all, the raw material is not structural wood components but small sugar molecules – the building blocks of wood cell walls. Secondly, the small sugar molecules occur as soluble molecules in sap. Sap either occurs in the inner bark, as a liquid that transports photosynthetic products, or in the sapwood as a liquid containing sugars from storage deposits in the trunk or the roots. In both cases, the sap does not have sufficient oxygen for aerobic decomposition to take place. Instead, the sugar molecules are decomposed though a biochemical pathway called fermentation. This process takes place without the use of oxygen and it is mostly carried out by yeasts.

Essentially, fermentation is the conversion of a sugar molecule to alcohol and CO_2. This process has been well known for centuries in the production of beer and wine. By mixing yeast and sugar in a water-filled container, one can produce ethanol. The same process occurs in the sap of living or recently dead trees, provided that sugar-fermenting organisms can gain access to the resource. Several yeasts (but not all of them) are capable of sap fermentation. Since ethanol is a volatile substance, a certain proportion of it evaporates. This explains the use of ethanol as an orientation cue by forest insects colonizing wounded or recently dead trees.

Just like the wood decomposers described above, different sugar fungi facilitate alternative enzymatic pathways, although they have not been thoroughly studied from this perspective. At least some fungi degrade the carbon compounds in a way that produces a strong ester odour (Malloch and Blackwell, 1993).

2.3 Fungal decomposition and rot types

Collectively, the wood-decaying fungi are very rich in species numbers and are specialized in many ways. Different ecological specializations are the topics of Chapter 3 (functional specialization) and Chapters 5–9 (habitat and substrate specialization). Here we examine the decomposition activity that brings about different rot types.

Fungal rot types are traditionally divided into three major categories, *white rot*, *brown rot* and *soft rot*, which are recognized from their overall effects on wood coloration and consistency. Brown and white rot have been known for a long time, and by the 1920s scientists understood that fundamentally different biochemical processes caused these rot types. At that time, several studies of rot in forests and wooden buildings revealed that white rot causes the degradation of all the major structural components in wood – cellulose, hemicellulose and lignin, whereas brown rot degrades and removes the carbohydrates only, leaving the lignin virtually unchanged (see Rayner and Boddy, 1988, for a historical perspective). The third rot type – soft rot – was identified later, when Savory (1954) studied wood decay in water-cooling towers. A very good description of the different rot types can be found in the first chapter of Eriksson et al. (1990). This chapter was written by Robert Blanchette and documents both macroscopic and microscopic processes in detail, illustrated with many photographs.

The vocabulary surrounding the topic of wood degradation and rot types can be somewhat confusing until one obtains an overview. We will just mention that the terms 'heart rot', 'top rot', 'root rot' and 'butt rot' refer to the position in the tree where the decomposition takes place. They are not other rot types compared with brown and white rot. Furthermore, alternative terms have been used for both brown rot and white rot, and these will be mentioned below.

2.3.1 White rot

White-rot fungi degrade all the structural wood components – cellulose, hemicellulose and lignin. Thus, they have the full range of enzyme systems outlined above. This does not mean that all white-rot fungi decompose wood in the same way. It is evident that two forms of white rot occur: simultaneous and selective lignin degradation (Otjen and Blanchette, 1986; Rayner and Boddy, 1988). Most white-rot fungi seem to produce just one of these forms, but some species can cause alternative forms under different conditions (Blanchette, 1991).

The mycelium of the simultaneous white-rot fungi grows through the wood by following the space (lumen) inside the dead wood cells. Here the mycelium activates the enzyme systems simultaneously and attacks the cell walls progressively from the inside. This form of white rot degrades rather large sections evenly and leaves a pale and loose wood structure (Figure 2.2a). *Fomes fomentarius*, *Phanerochaete chrysosporium*, *Trametes versicolor* and *Trichaptum* spp. are species that display this kind of white rot (Blanchette, 1984; Martínez et al., 2005).

Figure 2.2. Visual appearance of different rot types: (a) white rot, simultaneous type, of *Betula* caused by *Fomes fomentarius* (photo Tuomo Niemlä); (b) white rot, selective lignin degradation type, showing the honeycomb-like pattern caused by *Phellinus pini* (photo Robert A. Blanchette); (c) brown rot, with a typical cubic appearance (photo Eric Allen, © 2011 Her Majesty the Queen in right of Canada, Natural Resources Canada, Canadian Forest Service); (d) soft rot of wood (photo Robert A. Blanchette).

The selective lignin degraders (sometimes called sequential decayers) first attack the lignin in the secondary cell walls and the middle lamella, and later degrade the cellulose and hemicellulose components. Selective lignin removal often occurs in small pockets arranged longitudinally in the wood (Figure 2.2b). During the decay process, these pockets appear as distinct bleached regions (caused by the cellulose dominance) and have been referred to as 'pocket rot', 'alveolar rot' or 'mottled rot' by different sources. Finally the pockets become empty and leave a distinctive honeycomb-like pattern in the wood. The ability to selectively remove lignin does not exclude the ability to produce simultaneous decay; the two processes may occur side by side caused by the same fungus. Species with this kind of white rot include

Phellinus pini, *Heterobasidion annosum*, *Ganoderma applanatum*, *Pleurotus* spp. and *Xylobolus frustulatus* (Blanchette, 1984; Martínez et al., 2005).

White-rot fungi are responsible for most of the wood decomposition in temperate and tropical forests, as they are the dominant decayers of broadleaved trees (Ryvarden and Gilbertson, 1993). There are, however, several white-rot fungi, such as those in the genus *Trichaptum*, that decompose coniferous wood. The white-rot fungi need a constant supply of oxygen to carry out the decomposition; therefore, they rarely decompose wood submerged in water. Neither do they decompose live sapwood, due to its high water content. Boddy (1994) has established that many white-rot fungi occur as latent spores inside the live sapwood, and as soon as the water content drops in dead cells, the mycelium develops and the decomposition starts.

2.3.2 Brown rot

Brown-rot fungi selectively degrade the cellulose and hemicellulose fractions of the wood and leave the lignin virtually unaltered. Just like white-rot fungi, the brown-rot fungi expand through the wood by occupying the empty lumen inside the wood cells. But, in contrast to the white rots, they are unable to degrade the lignin-rich inner (S3) cell wall layer. This prevents the large cellulose-degrading enzymes from passing through this layer. Thus, the brown rots start their degradation by producing a large amount of oxalic acid and certain molecules (chelators) that can bind and release metal ions. The oxalic acid sequesters the iron that is bound in the wood cell walls and also lowers the pH around the mycelium. Both the iron oxalate and the chelators are small enough to pass through the inner cell wall layer to the cellulose-rich S2 layer (Goodell et al., 1997; Goodell, 2003). Here, a non-enzymatic reaction takes place as a different pH enables the chelators to use the iron from the iron oxalate and makes it available for a powerful chemical reaction called the Fenton reaction. This results in highly reactive hydroxyl radicals that attack and break up cellulose and hemicellulose chains. This degradation results in smaller sugar chains that diffuse into the cell lumen, where they are processed by cellulose-degrading enzymes, as described earlier in this chapter.

Wood decomposed in this manner has a characteristic appearance both in colour and consistency (Figure 2.2c). Due to the brownish colour of lignin, the decayed wood gradually becomes brown or reddish brown. Therefore the rot type is sometimes called 'red rot'. The consistency of brown rot is also characteristic. When the cellulose chains break up, the wood loses its longitudinal strength and the wood shrinks in a manner that produces numerous cracks across the direction of the fibres. This causes a characteristic appearance, and brown rot is sometimes referred to as 'cubic rot'.

Just like white rot, brown rot encompasses a range of decay patterns that reflect different taxonomic identities of the fungi and most probably suggest that alternative enzymes are in operation. At least three forms of brown rot exist, which is evidenced by different abilities to degrade crystalline cellulose, and demonstrated by macroscopic as well as microscopic changes in wood structure (Rayner and Boddy, 1988, pp. 262–263). One of these forms also seems to degrade lignin to some extent.

The majority of the brown-rot fungi occur specifically in the wood of coniferous trees (Gilbertson, 1980) and they are the principal wood decayers in boreal forests (Renvall, 1995). There are, however, some brown-rot fungi that are equally common on both coniferous and broadleaved trees (e.g. *Fomitopsis pinicola*) and others that predominantly or exclusively decompose the wood of broadleaved trees (e.g. *Laetiporus sulphureus*, *Piptoporus betulinus*). The brown-rot fungi are similar to the white-rot fungi in their need for a constant supply of oxygen to carry out the wood degradation. Thus they are only capable of effective wood degradation in terrestrial environments.

2.3.3 Soft rot

Soft-rot fungi primarily degrade the cellulose and hemicellulose in the central layer of the secondary cell wall (Savory, 1954). Some species also decompose the lignin component of the wood (Eriksson et al., 1990). The erosion of the central cell wall layer leaves the wood with a spongy consistency, which gave rise to the name of this rot type. The wood is further characterized by a darkened surface, due to the substantial removal of cellulose. When dry, the decayed wood is brown with cracks across the fibre direction and has a similar appearance to brown-rotted wood (Figure 2.2d).

Soft rot differs from white and brown rot in how the hyphae grow inside the wood and how the enzymes operate from the hyphal surface. Unlike the white- and brown-rot fungi, which grow rapidly through the wood, following the empty space (lumen) in the cell centre, the soft-rot fungi grow slowly inside the cell walls.

The rot starts with a fine penetrating hypha that grows through the innermost S3 layer into the cellulose-rich S2 layer, where it either branches or continues through the cell wall into the adjacent cell. The branching hypha has a very peculiar growth form. First it extends longitudinally following the orientation of the cellulose microfibrils, but it then stops. Next, a cavity is formed around the fine hypha, which increases in diameter as the cavity develops. Then follows a new phase of longitudinal growth before a new cavity section is formed. This growth–expansion process repeats itself several times over. The hyphae growing inside the cell walls regularly branch in a T-shaped manner and the overall outcome is the extensive breakdown of the cell walls (Figure 2.3, see also Eriksson et al., 1990). Another peculiarity of soft rot is that the enzymes act from the hyphal surface and never appear to diffuse into the wood cell wall.

Soft rot is primarily caused by ascomycetes (Eriksson et al., 1990; Worrall et al., 1997). However, it has recently been discovered that some white-rot fungi can exhibit the characteristics of soft rot when growing through the reaction zone formed around wounds (Schwarze and Baum, 2000). In addition, the brown-rot fungus *Fistulina hepatica*

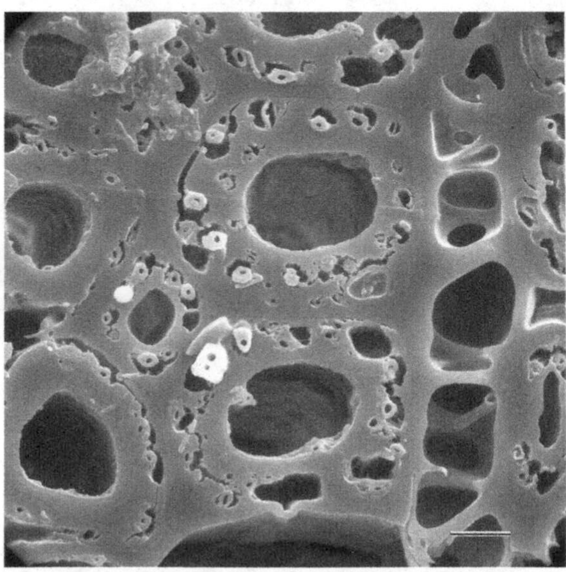

Figure 2.3. Microscopic aspects of soft rot. Cross-section of cells showing extensive soft rot in the S2 cell wall layer (scale bar 10 μm). Reprinted from Eriksson et al. (1990) with permission from Robert A. Blanchette.

uses a soft-rot mode during the initial decay of oak heartwood that is characterized by a high tannin concentration inhibiting the growth of many other wood decomposers (Schwarze et al., 2000a).

Soft-rot fungi are major decay agents in water or high-moisture situations where the activity of white- and brown-rot fungi is inhibited. They also decompose wood buried in the soil, but are outcompeted by basidiomycetes when these become established (Levy, 1987).

2.4 Bacterial wood degradation

Bacteria are ancient organisms that have existed on our planet for billions of years. They are prokaryotes (i.e. lacking a cell nucleus) and tiny in size. In terms of scale, bacterial cells are generally an order of magnitude smaller in size than fungal cells. Most bacteria are unicellular, but filamentous (chain-forming) types exist among the Actinobacteria. Their filamentous growth form fooled early biologists into considering them as primitive fungi. Thus, although these organisms have been commonly referred to as actinomycetes, they are true bacteria, as evidenced by the lack of a cell nucleus. In the more recent literature they are usually called Actinobacteria.

There is a great diversity of bacteria in wood, exhibiting a range of decay mechanisms (Greaves, 1971). One functional group colonizes the sap-conducting cells of wounded or dying trees. These bacteria utilize the cell contents and move easily into neighbouring cells by destroying the pit membranes in the pores that facilitate water movement between adjacent cells (Levy, 1975). Large numbers of bacteria have been found deep in the sapwood of wounded trees. This phenomenon is known as *bacterial wetwood* (Sakamoto and Atsushi, 2002). The organisms causing bacterial wetwood primarily utilize the dissolved sugars in the sapwood and leave the structural cell wall components unaffected (Schmidt and Liese, 1994).

Another group of bacteria degrade both cellulose and lignin. Different types of *erosion bacteria* produce shallow or deep pits in the cell walls (Eriksson et al. 1990, Chapter 1) and may cause extensive destruction after long-lasting attacks. *Tunnelling bacteria* make holes through the cell walls and always leave characteristic secretion bands across the tunnels (Figure 2.4). The secreted mucilage probably aids the initial attachment of the bacteria. Under wet conditions, the mucilage may be essential for enzymes to make contact with the substrate and bind to the cell wall components (Eriksson et al., 1990).

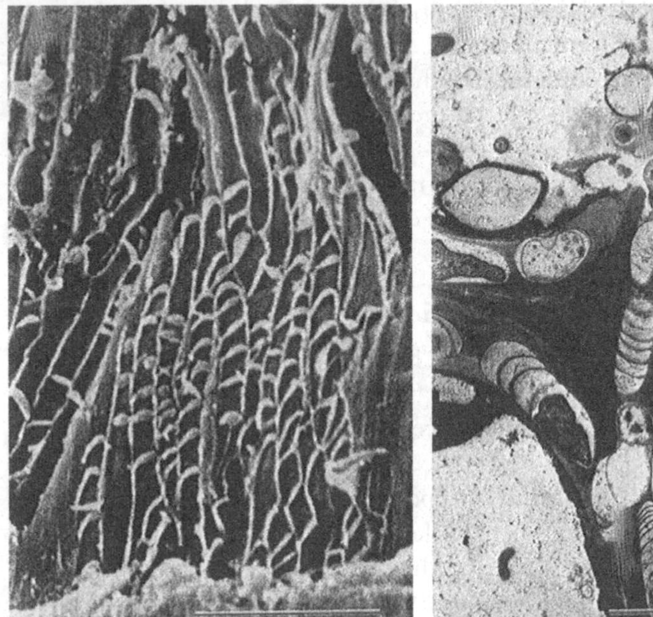

Figure 2.4. Microscopic appearance of wood decay caused by tunnelling bacteria (scale bar 5 μm). Reprinted from Eriksson et al. (1990) with permission from Robert A. Blanchette.

Finally, various bacteria play a significant role as internal symbionts in the gut of wood-boring insects. This occurrence of bacteria is described further in section 2.5.2.

Bacteria are slow wood decomposers compared with the brown- and white-rot fungi. Thus, they are outcompeted and become insignificant where these fungi are active. They do, however, play a significant role under certain conditions. In marine environments, for example, bacteria often appear as the principal wood decomposers. This is clearly demonstrated in archaeological investigations of wood buried in marine sediments or otherwise waterlogged (Kim and Singh, 2000). Both erosion and tunnelling bacteria are frequent decomposers of archaeological woods, and soft-rot fungi also occur together with bacteria in various wet situations. It seems that erosion bacteria are the most tolerant to oxygen deficit, while soft-rot fungi are the least tolerant (Kim and Singh, 2000). Bacteria are responsible for most of the decomposition of wood in freshwater ecosystems, and Actinobacteria can be important decomposers in rivers and lakes (Crawford and Sutherland, 1979).

2.5 Animal degradation of wood

In addition to fungi and bacteria, there are many animals that degrade woody material. The term 'degrade' covers both physical destruction by the grinding action of the mandibles and digestion by means of enzyme activity.

2.5.1 Physical destruction

Throughout the world there are thousands of invertebrate species that feed upon woody material by excavating holes in it or plucking it apart like many termites do. This physical processing reduces the particle size, increases the surface area, and decreases the crystallinity of the cellulose. All these factors are known to increase the susceptibility of cellulose to enzymatic activity (Walker and Wilson, 1991). In addition to the degradation itself, holes made by invertebrates improve aeration to the woody interior and facilitate access for cord-forming fungi, which in turn speeds up fungal decomposition.

2.5.2 Enzymatic digestion

The production of enzyme systems to decompose woody material is widespread among bacteria and fungi but very rare among animals. Nevertheless, there are many invertebrate species that eat and digest woody material. In a comprehensive review, Martin (1991) identified various mechanisms that can account for the digestion of wood by invertebrates:

1. exploitation of the enzymatic capacity of protozoan or bacterial symbionts residing in the digestive tract (Martin considered these as two separate mechanisms);
2. the use of fungal enzymes obtained from the wood, which remain active in the gut after ingestion;
3. the secretion of a complete enzyme system by the insect.

The first mechanism, the exploitation of other organisms residing in the digestive tract, occurs in several systems that have evolved independently of each other. One case is the protozoan symbionts that have been found in the gut of wood roaches (Cryptocercidae) and several genera of lower termites, i.e. those closest to the evolutionary origin of the termites (Cleveland et al., 1934; Martin, 1991; Breznak and

Brune, 1994). The dependence of these insects on protozoa has been recognized since the classic investigations of Cleveland (1924), who demonstrated that the lower termites need cellulose-degrading protozoa for efficient digestion and ultimately for their survival on a diet of wood. The gut symbionts comprise some unique genera of flagellates that seem to be restricted to the lower termites and wood roaches, as they are found virtually nowhere else in nature (Honigberg, 1970). The lower termites and wood roaches are evolutionarily old members of the order Dictyoptera, and it is likely that the symbiotic relationship with protozoans is a correspondingly old association.

Bacteria represent another group of gut symbionts. Gut bacteria contribute to cellulose digestion in a wide range of animals. Both lower and higher termites (Breznak and Brune, 1994) and the American cockroach *Periplaneta americana* (Bignell, 1977; Cruden and Markovetz, 1979) have such symbionts. The rhinoceros beetle *Oryctes nasicornis* has a fermentation chamber filled with bacteria and wood fragments in the swollen hindgut of the U-shaped larvae (Bayon, 1981). This shape is typical for Scarabaeidae larvae and it is likely that all wood-inhabiting Scarabaeidae digest woody material with the aid of bacteria (Dajoz, 2000). Bacteria are also involved in cellulose digestion by the larvae of crane flies (Tipulidae) (Pochon, 1939; Griffiths and Cheshire, 1987) and wood-boring Chironomidae flies (Kaufman et al., 1986). Even in the marine environment, we find invertebrates with cellulose-digesting bacteria. The wood-boring shipworms (Bivalvia: Teredinidae) have a symbiotic relationship with bacteria that provide the host with enzymes necessary for survival on a diet of wood (Waterbury et al., 1983; Sipe et al., 2000). A similar symbiosis has also been demonstrated in deep-sea wood-boring bivalves of the genus *Xylophaga* (Bivalvia: Pholadidae) (Distel and Roberts, 1997).

Yeasts represent yet another group of gut symbionts that might mediate wood digestion. Several saproxylic beetle families (Anobiidae, Bostrichidae, Cerambycidae) have yeasts localized in special pockets of the larval midgut (Dajoz, 2000, and references therein). The exact role of yeast symbionts has been much debated and it is often uncertain whether they take part in wood decomposition or provide nutrients for their host insects. In one beetle family (Passalidae), however, it is reported that the gut yeasts have the ability to degrade xylan, which is a major component of hemicellulose (Suh et al., 2003). It is interesting to observe the rich diversity of yeasts in the gut of wood-inhabiting fungivore beetles. In a study of yeasts in the digestive tract of beetles from several families, Suh et al. (2005) found more than 650 yeasts, of which about 200 species were not known previously.

The second mechanism of wood digestion, the use of ingested fungal enzymes, was discovered in the laboratory of Michael Martin and co-workers – a research group that was particularly interested in digestion by insects and their enzyme systems. First they discovered that the fungus-cultivating termite *Macrotermes natalensis* had active fungal enzymes in its gut (Martin and Martin, 1978). Next, they discovered fungal enzymes in the larval gut of the woodwasp *Sirex cyaneus*, which allowed it to digest woody material. They also established that at least two classes of cellulose-degrading enzymes were not produced by the larvae, but were instead acquired from wood containing mycelia of the corticioid fungus *Amylostereum chailletii* (Kukor and Martin, 1983). They already knew that this woodwasp had a close association with this fungus, as the females inoculate the fungus during oviposition in the sapwood of the host tree (Morgan, 1968). They extended their research to wood-boring beetles and established the presence of fungal enzymes in the larvae of several cerambycid species (Kukor et al., 1988).

The third mechanism, the production by the animal of all the enzymes necessary for cellulose digestion, is difficult to establish unequivocally. In a detailed review, Martin (1991) concluded that only in the case of some roach and termite species is there strong evidence for symbiont-independent cellulose digestion. He questioned the methodology of Lasker and Giese (1956), whose study is regularly cited as a case of self-governed cellulose digestion in the silverfish (Thysanura). Nevertheless, he encouraged researchers to replicate this important study, as Thysanura includes the most primitive known insects, and their mechanisms of cellulose digestion therefore represent important information for understanding the evolution of cellulose digestion in insects. Martin also pointed out that cellulose digestion in wood-boring beetles warrants further study in this respect.

2.6 Ecological aspects

Wood-decaying organisms differ widely in their decomposition effectiveness under differing environmental conditions. In terrestrial environments, the fungi – and especially the basidiomycetes – are the principal wood decomposers, although many ascomycetes might have been overlooked, especially in canopies of broadleaved trees (see

Chapters 5 and 9). Brown-rot fungi are the predominant decomposers of conifer wood in boreal forests, whereas white-rot fungi are more important in temperate and tropical forests. Among the invertebrates, beetles are important decomposers in boreal forests and northern temperate regions (but also in the tropics). In southern temperate and tropical regions, termites replace the beetles as the principal invertebrate decomposers. Termites are especially important in dry tropical environments, while fungi are more important in closed and moist forests.

In aquatic environments, the white- and brown-rot fungi cannot decompose wood due to a lack of oxygen. Instead, different bacteria and soft-rot fungi are the principal decomposers. Among the invertebrates, the aquatic wood decomposers are completely different from those active on land. Both beetles and termites are absent as important wood decomposers. In freshwater, they may be replaced by a small group of wood-boring Diptera of the family Chironomidae, while various molluscs and some crustaceans take on the role of wood consumers in marine waters.

3 · *The saproxylic food web*

Jogeir N. Stokland

Almost every ecology textbook includes a chapter or section on food chains or food webs. Such texts are normally accompanied by an example that starts with a photosynthetic primary producer (the lowest trophic level), followed by a herbivore feeding on the producer, next perhaps some medium-sized predator, and finally a conspicuous top predator (third or fourth trophic level). Subsequently, the example might be expanded with additional organisms at each trophic level to illustrate the concept of a food web. The specific example could be from the marine environment starting with photosynthetic plankton algae and ending with a seal or whale species. Alternatively, the example might depict an African grassland food web, ending up with the powerful cheetahs and lions as top predators.

It is typical that nearly all food web examples have herbivores at the second trophic level. We tend to view decomposer communities as simpler systems composed of two organism groups: decomposers, i.e. bacteria and fungi, and detritivores, i.e. animal consumers of dead matter (Begon et al., 2006). This is far from the truth for the communities associated with decaying wood, which represent complex food webs with several trophic levels above the primary producers (the trees). One can also find all kinds of species interactions, such as predator–prey, competition, parasitism and symbiosis, that are well known from other communities.

An attractive feature of the food web concept is that it cuts through the taxonomic divisions that tend to split descriptions of wood-inhabiting species into texts on fungi (often restricted to basidiomycetes) and invertebrates (often restricted to individual insect groups). In this chapter we explore the full range of species in a community based on energy and nutrients from decaying wood and refer to it as the *saproxylic food web* (Figure 3.1).

The concept of a saproxylic food web is not introduced to replace the wider concept of saproxylic species. It instead highlights a large

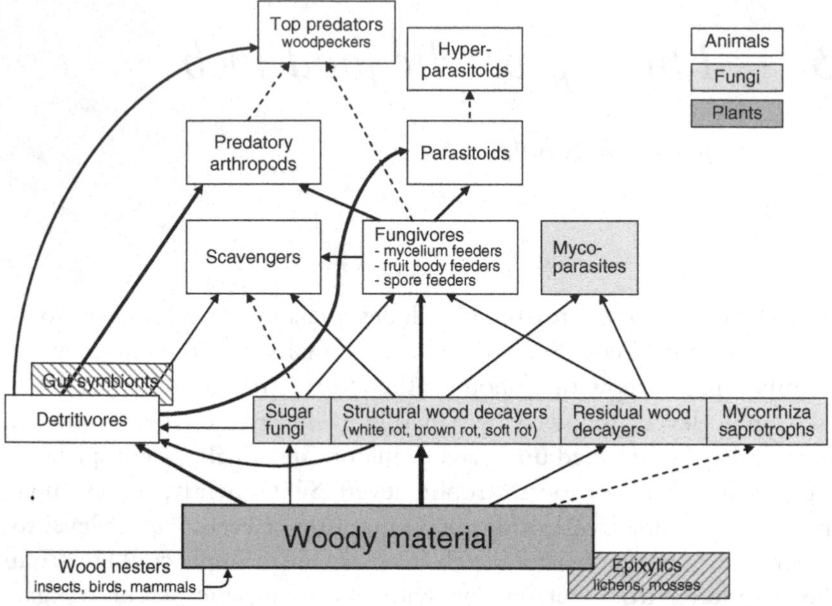

Figure 3.1. The saproxylic food web, with organisms sorted by their functional roles at different trophic levels. Arrows indicate the main nutrition and energy flows and the thickness of the arrows indicates the magnitude of that pathway.

subset of saproxylics and focuses on the feeding interactions between them. Many saproxylic species depend upon woody material in other ways than as a nutrition source; for example, animals that use wood as a breeding site (but find their food elsewhere) as well as mosses and lichens that grow on the surface of dead wood. These saproxylic species are treated separately in Chapter 4.

3.1 Sugar fungi and wood-decaying fungi

Fungi are the most important wood decayers and play a pivotal role by transforming woody material into forms that then become available for many other species living in dead wood. The process of fungal decay starts in wounds and dead branches of live trees. Later, when the tree dies, other fungi gain access to the woody material and immediately speed up the decomposition process. Since woody material contains both simple compounds (in sap and inside live cells) and complex structural compounds (in cell walls) we have divided the fungi into two corresponding groups that are enzymatically very different.

3.1.1 Sugar fungi and staining fungi

The term 'sugar fungus' denotes a fungus that lives on sugars and carbon compounds simpler than cellulose in the initial stages of plant matter decomposition (Garrett, 1951). These fungi are very different from the wood-decaying fungi, as they do not possess enzymes to degrade cellulose or lignin. Instead they utilize the building blocks and energy sources that trees use to construct and maintain woody cells. Garrett used the term sugar fungi to describe the ecology of soil fungi, but later Hudson (1968) used the term with the same meaning for the initial decomposers in woody material. There is a whole suite of fungi that exploit the readily available sugars dissolved in the sap of the inner bark. These fungi colonize wounded and recently dead trees and agree nicely with the definition given by Garrett.

The sugar fungi hold different taxonomic positions, but most of them belong to the ascomycetes or sac fungi. Yeasts are fungi that occur regularly in sap from wounds and in recently dead woody tissue (but also in more decayed wood). The word 'yeast' can be confusing, as it has two separate and distinct meanings. The first refers to a taxonomic group – the true yeasts belonging to the Saccharomycotina among the ascomycetes. The second refers to the morphological type of individual budding cells, in contrast with cells that form long chains (hyphae). Many saproxylic ascomycetes and also some basidiomycetes produce yeasts regularly under certain ecological conditions or as a regular part of their life cycle.

Another group of sugar fungi is made up of the staining (ophiostomatoid) fungi that occur in wounded or recently dead trees. The enzymatic properties of staining fungi enable them to utilize soluble sugars, starches, lipids and proteins (Mathiesen, 1950; Mathiesen-Käärik, 1960a, 1960b; Abraham et al., 1998). They occur abundantly in fresh inner bark, and some grow along the ray parenchyma cells that transport various assimilates from the inner bark to the live sapwood cells. In the sapwood they can also penetrate the pit borders of water-conducting cells in order to benefit from the transport of nutrients from the roots to the canopy. Their enzymatic properties suggest that they may also consume the content of living wood cells.

The staining fungi are best known for causing diseases – in some cases fatal – to different tree species; examples include Dutch elm disease, blue-stain diseases and canker stain diseases (Butin, 1995), and also for causing discoloration to timber. There are, however, many more staining fungi than the pathogenic ones (Gibbs, 1993) and they have been found abundantly in fresh wounds of aspen trees (Hinds, 1972; Hinds and Davidson, 1972), bark beetle galleries (Kirisits, 2004) and in recently dead sapwood of coniferous and broadleaved trees (Seifert, 1993; Butin, 1995). Trichomycetes is yet another taxonomic group of sugar fungi. These are numerous in bark beetle galleries (Kirschner, 2001).

A distinct group of fungi have developed a symbiotic association with ambrosia beetles that bore into living or recently dead trees (Malloch and Blackwell, 1993; Farrell et al., 2001). Many of the ambrosia fungi are closely related to the staining fungi described above. The ambrosia fungi form a mycelial carpet on the wall inside the insect galleries and the insect larvae feed on this mycelium. Just like the staining fungi, ambrosia fungi are unable to degrade structural cell wall components. Therefore, they rapidly deplete the available sugars, and the ambrosia beetles typically need to colonize a new tree after one generation to bring the fungi into contact with fresh resources.

3.1.2 Structural wood decayers

Structural wood decayers are species with the enzymatic ability to degrade more-or-less intact cell wall components (cellulose, hemicellulose and lignin). In Chapter 2 we described the enzymatic activity of wood-decaying fungi and how different species cause brown rot, white rot or soft rot. All these fungi, together with eroding and tunnelling bacteria (see Chapter 2), make up this functional group, which encompasses a large number of species. The structural wood decayers consume nearly all kinds of woody material, but individual species are typically specialized to decay the wood of particular tree species, wood with different properties such as dimension or decay stage, or different vertical and horizontal sections from the roots to the branches and from the bark to the heartwood (see Chapters 5–9).

Structural wood decayers are primarily found among the basidiomycetes. Polypore fungi (of which many belong to Polyporales, according to recent mycological taxonomy) are perhaps the best known, and are important decomposers in forests worldwide. Many species are also found in Hymenochaetales (e.g. the genera *Phellinus*, *Inonotus* and *Trichaptum*), Agaricales (*Armillaria*, *Pleurotus*, *Pholiota*) and Dacrymycetales (*Dacrymyces*, *Calocera*). Also among the ascomycetes we find species that decompose structural wood components, such as in

Xylariales, which contains the genera *Daldinia, Hypoxylon* and *Xylaria* (Worrall et al., 1997; Pointing et al., 2003).

The structural wood decayers provide nutrition for other saproxylic species in two distinct ways. Firstly, by breaking the cell wall material into smaller components, they open the gate for the residual wood decayers described below. Secondly, the fungi themselves represent the main food source for a wide range of invertebrates. In a typical fungal life cycle, the spores colonize woody substrates and develop hyphae that mate and develop a mycelium. This mycelium expands inside the wood and effectively decomposes the wood. After a while, which can take from weeks to years depending on the species, the mycelium develops fruiting bodies on the wood surface. These fruiting bodies produce huge numbers of tiny spores that disperse to new substrates. The mycelium-ridden wood, pure mycelium, fruiting body and spores subsequently become the food for a large number of fungivore insects. In a similar way, different mycoparasites (i.e. fungi parasitizing other fungi) obtain nutrition from the tissues of wood-decaying fungi.

3.1.3 Residual wood decayers

Residual wood decayers are fungi that utilize the decomposition products brought about by the activity of white-rot, brown-rot or soft-rot fungi. They have also been called secondary sugar fungi (Hudson, 1968). However, these fungi do not consume sugar molecules from the sap or recently dead inner bark and sapwood cells. Instead they live on the residuals from partly decomposed cell walls in moderately to strongly decayed wood. Unlike the structural wood decayers, these fungi have been little investigated from a biochemical point of view. Thus, we mainly have indirect evidence suggesting the occurrence of fungi with this nutritional mode. Despite the poor knowledge base, such fungi might still be numerous, as there are many species in well-decayed wood that do not fit neatly into the categories of brown- or white-rot fungi. These may very well belong to the residual wood decomposers.

Information about residual wood decayers can be found scattered throughout the mycological literature, typically presented with some uncertainty concerning the existence of this nutritional mode. In an interesting section under the heading 'Non-decay fungi', Rayner and Boddy (1988, p. 121) listed several fungi that occur in the interaction zones between the mycelia of true wood-decaying fungi: namely species of *Leptodontium*, *Ramichloridium*, *Rhinocladiella*, *Endophragmitella* and *Chaetosphaeria myriocarpa*. They suggest that these fungi could possibly utilize the metabolic products of the wood-decaying fungi (which would qualify them as residual wood decayers). It might also be that some of the successor (i.e. follower) species in the predecessor–successor associations belong to this functional group (Niemelä et al., 1995). Niemelä and co-authors indicated that the primary decayer may alter the decaying wood to make it more favourable for the follower species, whereas Holmer et al. (1997) suggested that primary decayers tend to 'loosen up' the available nutrient resources and allow the follower species to obtain nutrients from the residual material. However, both Niemelä and Holmer suggested this functional mechanism as one of several possible explanations.

In addition, many ascomycetes, agarics and yeasts that occur in well-decayed wood could be residual wood decayers. It is also likely that some of the mycorrhizal fungi that regularly occur in well-decomposed stumps and logs have the enzymatic capacity to degrade remnants of decayed wood. Ectomycorrhizal species have evolved repeatedly from saprotrophic ancestors (Hibbett et al., 2000) and may have retained certain enzymes involved in organic matter degradation (Koide et al., 2008). Both laccase and peroxidase enzymes that contribute to woody cell wall degradation have been found in mycorrhizal species (Chambers et al., 1999; Burke and Cairney, 2002). Chambers and colleagues (Chambers et al., 1999) have shown that *Tylospora fibrillosa*, a mycorrhizal species that occurs commonly in well-decayed wood, has peroxidase-encoding genes, and they also measured the enzymatic activity of these enzymes. It has even been reported that some mycorrhizal species have lignin peroxidase enzymes (Chen et al., 2001), but this claim was later shown to be erroneous (Cairney et al., 2003). Direct interaction between wood-decaying fungi and mycorrhizal fungi is described later in this chapter.

Yet another example suggesting the existence of residual wood decayers comes from a study of the wood-decaying ability of 22 species of aquatic hyphomycetes by Shearer (1992). She found that one group of species caused extensive weight loss and soft-rot cavities, thus being typical soft-rot fungi. Another group of species caused little weight loss and formed no soft-rot cavities. Shearer suggested that these species probably rely on dissolved carbon sources or products from the enzymatic activities of other fungi.

Since there is much uncertainty about the nutritional mode of the fungi described in this section, we definitely need some biochemical studies of their enzymatic activity. A likely outcome of such studies would be that these fungi have various enzyme systems with adaptations that enable them to utilize quite different wood fractions. Another approach would be to study their capacity to decay intact wood and moderately to severely decayed wood conditioned by other wood decomposers. Such experimental studies are needed in order to elucidate the true nature of these fungi.

3.2 Detritivores

The term detritivore (or saprophage) refers to animal species that consume both decaying plant material and the bacteria or fungi that brought about the decay (Begon et al., 2006). This functional group comprises a large number of beetles (Coleoptera), midges and flies (Diptera), termites (Isoptera) and mites (Acari).

The most conspicuous detritivores are those that make distinctive galleries in rather hard bark and wood (Figure 3.2). There could be a case for not calling some of these species detritivores, since the term implies a diet of dead organic matter, whereas several bark- and wood-boring invertebrates undoubtedly attack live trees. But even the aggressive spruce bark beetle *Ips typographus* needs to kill the bark cells and disable the resin defence so that the larvae can develop successfully. Thus, wood-boring insects that attack living trees also function as initial decomposers.

Many detritivores that feed on woody material with a liquid or soft consistency leave very inconspicuous marks, or none at all.

3.2.1 Nutritional value of woody material

If we examine the diet of detritivores more closely, it turns out that their food sources are nutritionally quite different. In an insightful paper entitled ‘Cellulose centered perspective on terrestrial community structure’, Abe and Higashi (1991) pointed out that plant material can be divided into two main categories: cytoplasm (i.e. most of the cell contents) and cell walls. Cytoplasm is rich in protein, lipids and starches. Thus, it is nutritionally well balanced and represents a high-quality food source that is easy to digest. Cell walls contain very little protein and lipids, but are rich in cellulose, hemicellulose and lignin.

Figure 3.2. Larval galleries of (a) the bark beetle *Scolytus ratzeburgi* in inner bark, note the increasing width of larval galleries due to the growth of larvae (photo Dan Damberg); (b) the jewel beetle *Anthaxia zuzannae* in the interface between the inner bark and wood (photo Nikola Rahmé); and (c) the ambrosia beetle *Trypodendron domesticum* in wood (photo Jiří Novák), note the short gallery of individual larvae, as they feed on the ambrosia fungus and not the wood as such.

These cell wall substances constitute *c.* 90% of the total weight in trees (Swift et al., 1979). There is, however, one major challenge associated with this abundant food source – a large enzyme apparatus is needed in order to degrade it.

The tissue type (inner bark, cambium zone, wood), the proportion of living or recently dead cells, and the degree of fungal decay all have an important bearing on the nutritional value of woody material. The inner bark consists of living cells, and this tissue is especially rich in sugars, as it transports photosynthetic products from the canopy. The cambium zone, i.e. the cambium cell layer and the adjacent inner bark and sapwood cell layers, should specifically be mentioned, since many wood borers feed almost exclusively in this zone. The cells in the cambium zone resemble those of the inner bark, but they have an enhanced concentration of proteins. Furthermore, many of these cells have not yet developed lignified secondary cell walls, as the cells are still growing (see Chapter 2 for more details on lignin formation in growing wood cells). The sapwood consists of cells with fully developed cell walls and is composed of living and dead cells. This tissue transports water from the soil to the canopy and this water contains the minerals needed for photosynthesis. The heartwood consists solely of dead cells. In addition, it is impregnated with toxic substances (see Chapter 5) that inhibit the activity of most fungi.

3.2.2 Sap feeders

A tree that is wounded often develops a flow of sap, called sap exudation, that is rapidly colonized by a large number of insects. Many of the adult visitors use the sap as just another sugar-rich nutrient source. But several species are closely associated with sap exudation and their larval development occurs here. The larvae of such sap specialists feed by sucking or filtering this energy-rich soup of sugars, bacteria and yeast cells. Particularly among the Diptera and Coleoptera we find species that are strictly confined to such sap exudations (see Chapter 7), for example the fly *Aulacigaster leucopeza*, several hoverflies in the genus *Brachyopa*, and many sap beetles (Nitidulidae).

In recently dead trees, a sap-like substance develops when the inner bark starts to decay and becomes wet and oily. Normally this takes place during a period that starts soon after death and may continue until 1–2 years after the death of the tree. Different tree species develop this kind of sappy inner bark to a different extent, and aspens and poplars (*Populus* species), in particular, attract many insect species at this stage of decomposition. In this decaying material we find some species that are the same as those that feed on sap exudations, but there are additional species that prefer the sappy inner bark, such as the aspen hoverfly *Hammerschmidtia ferruginea*, several lance flies (*Lonchaea* species) and various fruit flies

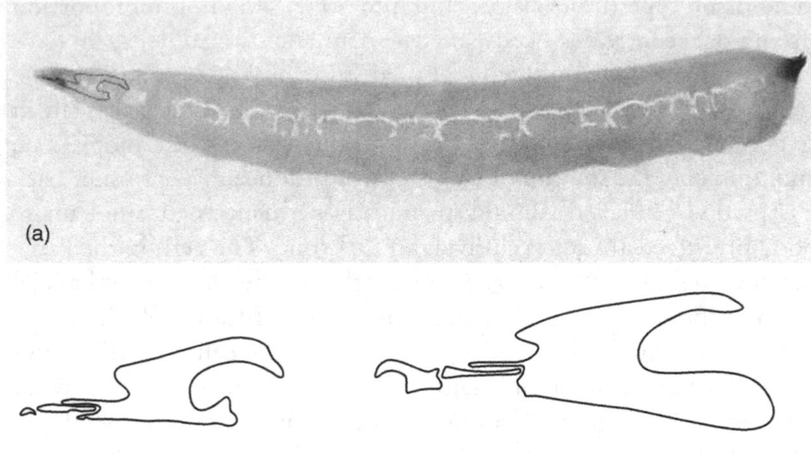

Figure 3.3. The internal head skeleton of *Lonchaea* larvae: (a) larva showing the position of the head skeleton (photo Graham Rotheray); (b) head skeleton of *Lonchaea fraxina*, which feeds in a sucking manner on the semi-liquid material resulting from decomposing inner bark; and (c) head skeleton of *Lonchaea caucasica*, which feeds by scraping on wood surfaces. Line drawings reproduced from MacGowan and Rotheray (2008).

(*Drosophila* species). In the lance flies we find a neat gradation between suckers and scrapers, reflecting fine-tuned adaptations to feeding on woody material with different consistencies (see Figure 3.3).

3.2.3 Inner bark consumers

There is a distinct set of species where the larvae feed almost exclusively in the inner bark. Sometimes these are referred to as phloeophagous species, from 'phloem' (inner bark). The majority of bark beetles (Scolytinae) belong to this category, but many longhorn beetles (Cerambycidae) also feed exclusively on the inner bark. These species typically colonize trees immediately after tree death, when the inner bark still contains sap and nutrient-rich cytoplasm in the cells. Many of the species that feed on the inner bark also consume the cambium zone. Several species of jewel beetles (Buprestidae) and some bark beetles feed almost exclusively on the cambium zone of recently dead trees (Ehnström and Axelsson, 2002). These cambium feeders have a similar diet to those consuming the inner bark, but they are more selective and feed only on the substrate with the highest nutritional quality.

Often the inner bark is rapidly colonized and decomposed by fungi, and for some insects the mycelium might be an important additional nutrition source. Most adult bark beetles bring a variety of fungal spores and bacteria with them from the tree where they developed (Malloch and Blackwell, 1993). The spores germinate and start to grow as soon as the beetles bore into a new tree for mating and egg-laying. Thus, when the eggs hatch, the larvae find themselves in a situation where the fungi have already started to decompose the bark. The fungus-ridden inner bark has a higher nutritional value than inner bark without fungi, and the larval galleries of bark beetles that bring staining fungi with them (e.g. *Ips acuminatus*, *I. sexdentatus*) are typically much shorter than those of other bark beetles. Such bark beetles form a transitional stage towards the situation with ambrosia beetles, where the larvae have a purely fungal diet. The fungi cause a distinct coloration of the adjacent wood and belong to the staining fungi described in Section 3.1.1.

When the inner bark dries up and gradually detaches from the wood surface, another group of detritivores enter the scene. These have well-developed mouthparts that can exert enough force to scrape off or masticate pieces of firm inner bark. Examples are beetles in the genera *Pytho*, *Morpholycus* (see Figure 3.4) and *Pyrochroa*.

3.2.4 Wood consumers

The species that bore into the sapwood or heartwood form another group of detritivores. These are sometimes called xylophagous species, from 'xylem' (meaning wood). Typical representatives are deathwatch beetles (Anobiidae), stag beetles (Lucanidae) and certain scarab beetles (Scarabaeidae). Many longhorn beetles (Cerambycidae) and jewel beetles (Buprestidae) also belong to this category, but they are different species than those feeding on the inner bark. In subtropical and tropical areas we find many wood-boring termites that complement, or even replace, the beetles as principal wood consumers. There are, however, several other species groups that ingest wood. Wood-boring species exist among Hymenoptera (woodwasps), Lepidoptera (the families Sesiidae and Cossidae), Diptera (certain species in the families Tipulidae, Syrphidae and Chironomidae). Even among marine molluscs we find wood-boring representatives such as shipworms (Teredinidae).

Wood consumers differ from the inner bark consumers not only by utilizing another tissue type, but also by ingesting plenty of cellulose and lignin, so they need to handle the enzymatic challenge of digesting this material. Species in this group have developed various associations with the fungi and gut symbionts described in Chapter 2.

Making a strict distinction between bark- and wood-feeding species can be rather arbitrary, as the larvae of many species start to feed in the inner bark or cambium zone and later move into the sapwood as they grow bigger and stronger. Also the digestive ability may change within a species, as demonstrated by Kukor and Martin (1986), who added cellulose-degrading enzymes from a fungus to the food of the longhorn beetle *Saperda calcarata*, which feeds on the inner bark and sapwood of live aspen stems. The beetle is unable to digest cellulose, but was transformed into a cellulose digester by adding active enzymes to the food. The ability to digest cellulose by means of ingested fungal enzymes has been demonstrated in several species of longhorn beetles and woodwasps (Martin, 1991), and this may be widespread among the wood-boring detritivores.

While wood borers are generally cell-wall consumers, with the aid of gut symbionts or fungal enzymes, the larvae of *Cossus*, belonging to the Lepidoptera, are cell-content consumers (see Section 3.2.1). This was clearly demonstrated by Chararas and Koutroumpas (1977), who investigated the enzymes in the digestive tract of these larvae. They found that the *Cossus* gut enzymes degrade small sugar molecules (oligosaccharides), peptides and pectins, and differ profoundly from the gut enzymes of wood-boring beetles such as buprestids and cerambycids. In other words, the *Cossus* larvae, and probably also Sesiidae larvae, act like the herbivorous Lepidoptera caterpillars feeding on the leaves or needles of living trees. They consume the cytoplasm content of live cells, while the cell wall components pass undigested through the gut. This explains why we do not find *Cossus* larvae in dead and decaying wood – there is no cell content left there.

3.2.5 Consumers of fungus-infested wood

Some wood borers have a diet closer to that of the fungivores than to that of other detritivores. The term 'xylomycetophagous' was introduced by Schedl (1958), referring to the ambrosia beetles that bore into wood and cultivate fungi as a food source for their larvae. Literally, the term means 'wood and fungus feeder', and several authors have pointed out that this use of the term is misleading, since the ambrosia beetles feed on a purely fungal diet. When ambrosia beetles bore into the wood they even push out all the wood dust from their tunnels to create good conditions for their fungus culture.

There are, however, several species that feed on a mixture of woody material and fungus mycelia – exactly as the term xylomycetophagous suggests. Such species are found, for instance, in the woodwasps (Siricidae), which depend on the enzymatic activity of associated fungi (of the genera *Stereum* and *Amylostereum*), which the egg-laying females introduce into the wood (see Chapter 2). Wood-boring insects that colonize a trunk some years after the tree has died encounter a situation where the wood has become increasingly interwoven with fungal mycelia. For these species, the mycelium is probably an important nutritional component or even the principal food source. This is underpinned by the fact that many wood-boring insects need pre-rotted wood and predominantly use either brown-rotted or white-rotted wood. Beetles in the families Eucnemidae, Trogossitidae and Melandryidae are such species. Several of these are intimately associated with wood decayed by particular fungus species. The beetle *Peltis grossa*, for example, needs wood decayed by the polypore *Fomitopsis pinicola* (Ehnström and Axelsson, 2002). In fact, there is a gradual transition from species ingesting virtually undecayed wood, through species consuming wood with increasing degrees of decay and fungal mycelia, to exclusively fungivorous species. This transition in diet and food consistency is clearly reflected in morphological adaptations of the mouthparts in the larvae (see Figure 3.4).

3.3 Fungivores

This functional group comprises species that feed more or less exclusively on fungal tissues. Since fungi are the main wood decomposers, it is not surprising that many species have specialized on a fungal diet. The saproxylic fungivores can be grouped into species that feed on the different parts of wood-decaying fungi (typically basidiomycetes): pure mycelia, fruiting bodies or spores. In addition, there are several species that feed on non-basidiomycete fungi such as moulds and ambrosia fungi.

3.3.1 Fruiting-body feeders

Typical fungivores are species whose larvae develop in the fruiting bodies of fungi. Fruiting bodies represent a highly diverse food source with differences in consistency, longevity and chemical composition

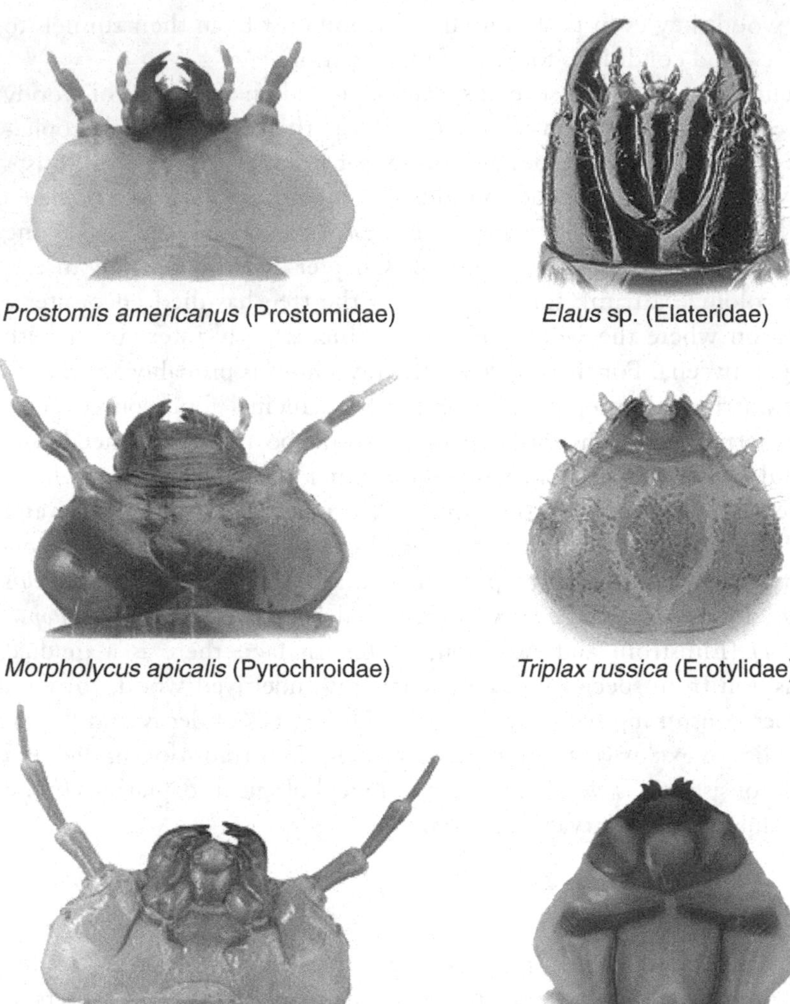

Figure 3.4. Heads of beetle larvae with different feeding habits. The three species on the left-hand side are detritivores from different families that show convergent morphology of mandibles adapted to feed on inner bark or decayed wood. In the case of *Cucujus*, contrasting evidence suggests both a detritivore and a predatory feeding habit for different species. The species on the right-hand side are, from the top: a predator (*Alaus* sp.) with long pointed mandibles, a fungivore species (*Triplax russica*) with small mandibles, which feeds on soft fungal tissue, and a detritivore species (*Hylochares* sp.) belonging to a family that feeds on soft fungus-ridden wood; it probably moves its head from side to side in order to scrape off food particles (all photos Artjom Zaitzev).

across different fungus groups. Furthermore, the fruiting bodies themselves have different qualities during their development from the initial sterile stage, through the fertile spore-producing stage, and finally the different stages of decay or disintegration. Some species feed on the spore-producing layer of live fruiting bodies, but most species utilize dead fruiting bodies. The species that utilize perennial (multi-year) fruiting bodies can spend several generations inside a single fruiting body. In these species, both larvae and adults feed on the fruiting-body tissue, as – for instance – in the black tinder fungus beetle *Bolitophagus reticulatus* (Figure 3.5a). In Chapter 7 we examine various associations with fungal fruiting bodies in detail. Here we shall only mention that typical fruiting-body feeders occur in many insect groups. In the beetle families Ciidae and Erotylidae, nearly all species feed on fungal fruiting bodies, both as larvae and adults, and we also find species with this diet in other beetle families (see Chapter 7). Among the flies, there are many fruiting-body feeders in Mycetophilidae, Platypezidae (Figure 3.5b), Fanniidae and Phoridae (Yakovlev, 1994). Among the moths (Lepidoptera) we also find a family – Tineidae – where a large proportion of the species develop in the fruiting bodies of wood-decaying fungi (Rawlins, 1984).

3.3.2 Spore feeders

Some insects specialize on a diet of fungal spores. In the beetle family Ptiliidae there is a subfamily (Nanosellinae) with species that are tiny in size and slender enough to enter the spore-producing tubes underneath the polypore fruiting bodies. In addition, in the beetle family Staphylinidae there are specialized spore feeders in the genera *Gyrophaena* and *Agaricochara* (Ashe, 1984). These species have modified mouthparts that function as brushes to collect spores. Another adaptation is found in the New Zealand beetle genus *Holopsis* (Corylophidae). Here the larvae of some species have a long snout with a pair of rasping mandibles at the tip, which they insert into the pore tubes of the polypore *Ganoderma* to feed on the spores (Lawrence, 1989). In the Diptera we find yet another amazing adaptation to a diet of spores. The larvae of several species in the family Keroplatidae spin a net underneath the spore-producing surface of bracket fungi to trap the spores as a food source. It has been debated whether Keroplatidae species are predatory, as they also trap small invertebrates in their nets. However, it seems that two different strategies have evolved: predatory species deposit acidic

Figure 3.5. Representatives of species with different functional roles: (a) the fungivorous beetle *Bolitophagus reticulatus*, where both larvae and adults feed on the fungus *Fomes fomentarius* (photo Frank Köhler); (b) galls developed on the fungus *Ganoderma lipsiense* as a result of colonization by the fungivorous gnat *Agathomyia wankowiczii* (photo Laurens Linde); (c) the predatory ant beetle *Thanasimus formicarius*, where both larvae and adults feed on bark beetles (photo Beat Fecker); (d) a female of the parasitoid wasp *Xorides stigmapterus* laying eggs in larvae developing inside wood (photo Loren and Babs Padelford); (e) a nematode trapped by the predatory fungus *Arthrobotrys anchonia* (photo George Barron); and (f) the jelly fungus *Tremella aurantia* (lower left), which is a parasite on the corticioid fungus *Stereum hirsutum*, shown in the upper part of the image (photo Fred Stevens).

droplets on their web, while fungivorous genera such as *Cerotelion* and *Keroplatus* spin less acidic – but hygroscopic – nets under fungi to collect the spores (Evenhuis, 2006).

3.3.3 Mycelium feeders

Many species feed on the mycelium that commonly develops under the loose bark of trees (see Chapter 6), or in well-decayed wood, where the mycelium becomes exposed in cracks when the wood breaks apart. Most of these invertebrates, including the larvae of several beetle and gnat (Diptera) species, ingest all of the mycelial tissue, whereas others are cell-content suckers, including some bugs (Hemiptera) and thrips (Thysanoptera).

A peculiar type of mycelium feeders is represented by the fungus-culturing termites. The termites pluck apart woody material (often partly decomposed) and carry it down into the subterranean nests underneath their mounds. Here they culture some highly specialized fungi (*Termitomyces*) on the organic matter from wood and other dead plants. The fungus produces vegetative nodules that are the principal food of the termites (Mueller et al., 2005).

3.3.4 Ambrosia feeders

A highly advanced group of fungivores are the specialized ambrosia beetles. They make galleries inside wood, where they culture certain fungi that are closely related to the staining fungi described in Section 3.1.1 (Farrell et al., 2001). The ambrosia beetles solely feed upon the fungal mats that cover the walls inside the galleries. Most ambrosia beetles belong to the family Platypodidae, but some bark beetles (e.g. *Xyleborus* spp.) and the beetle species *Hylecoetus dermestoides* (Lymexylidae) are also typical ambrosia feeders.

3.4 Scavengers

Scavengers (also called necrophagous species) are animals that feed upon dead animals. Among the mammals, we are all familiar with hyenas, which eat the carcasses of other animals, but in the microcosmos inside dead wood we also encounter some species with similar feeding habits. In the galleries of bark beetles and other wood borers there are species that have this diet. Scavengers also occur in other microhabitats such as beneath loose bark, in hollow trees, etc.

This functional group comprises relatively few species; there are some typical scavengers among the skin beetles (Dermestidae), the spider beetles (Ptinidae) and soldier flies (Stratiomyidae, e.g. the genera *Zabrachia*, *Pachygaster* and *Neopachygaster*; see Krivosheina, 2006).

3.5 Predators

Predators are species that kill living animals and then eat their prey (Figure 3.5c). In communities living in dead wood, we find many predatory species, especially among beetles and flies, but also among vertebrates and even fungi. The larvae and pupae of detritivorous and fungivorous insects constitute the main food source for predators. This is logical, as the larval stage is a predominant part of the life cycle of saproxylic insects. Furthermore, the larvae are relatively immobile and easy to kill. From the predator's point of view, it is most rewarding to hunt prey that occur abundantly in a small area. Thus, many predators specialize in hunting bark beetle larvae, which are particularly numerous in recently dead trees. At the later decay stages, we find other predators, such as the larvae of click beetles (Elateridae).

3.5.1 Typical predators

Predators exhibit several adaptations that make them effective hunters, such as rapid locomotion and pointed mandibles or mouth-hooks (Figure 3.4). Species that hunt inside the galleries of other insects typically have a cylindrical body shape, like that of the cylindrical bark beetles (Colydiidae). A further specialization is found in a subfamily of the asilid flies, where the larvae have strong tubercules fringing the body. These help to alter the body diameter so that the larvae can move effectively in galleries with different dimensions (Krivosheina, 2006). Under the loose bark there are high concentrations of detritivore larvae, and here we find other predators with a flattened body shape, such as several clown beetles (Histeridae). The number of predators, both in terms of species and individuals, is probably highest in these two microhabitats. However, there are also predators to be found in wounds of living trees, in fungal fruiting bodies, and in wood at all stages of decay – including wood mould in hollow trees.

Typical predators are found in many families of beetles (Cleridae, Rhizophagidae, Elateridae, Histeridae) and flies (Asiliidae, Xylophagidae, Muscidae, Dolichopodidae). Also, in the ancient order Raphidioptera (snakeflies) we find many predator species living under bark (Aspöck, 2002). In all these groups the larvae act as predators, and in most species the adults are also predators.

One might not think of woodpeckers as predators, but they play the role of saproxylic top predators in forests all over the world. With their forceful beak, they rip off bark or peck out pieces of wood. Next they use their exceptionally long and sticky tongue to extract insect larvae from their galleries. In Latin America, we find other vertebrates acting as top predators. Here, several species of ant-eaters tear apart ant nests, termite mounds and rotten trees in their search for insect larvae.

3.5.2 Facultative predators

At least some, and perhaps many, species that have been reported as being predators can survive on a diet of woody material and live as detritivores. Certain species among the lance flies (Lonchaeidae) are a good example. Most lance flies in the genus *Lonchaea* are detritivores (MacGowan and Rotheray, 2008), but the larvae of several *Lonchaea* species occur in galleries of wood-boring beetles (bark beetles, weevils or longhorn beetles). It has been claimed that these larvae are predators on larvae, pupae or even adult insects. In a review of several studies, Ferrar (1987) concluded that *Lonchaea* larvae inhabiting the galleries of other beetles typically live on dead and dying beetle larvae, but have the ability to attack beetle stages that are less well able to protect themselves. However, they have also retained the ability to live as detritivores on decaying woody material. Another example is the big fly genus *Medetera*. Most forest entomologists generally consider *Medetera* species to be strict predators of bark beetle larvae. While many *Medetera* species doubtless predate on bark beetle larvae, there are many species that continue to live under the bark several years after the bark beetles have left. Several of these show clear preferences for specific tree species and perhaps feed on the inner bark tissue (Mats Jonsell, personal communication). In other Diptera families we find cases where closely related species are reported partly as detritivores and partly as predators, e.g. Muscidae (subfamily Azeliinae), Odiniidae and Pallopteridae (Ferrar,

1987). Also among the beetles we find species with this combined diet, e.g. some click beetles (Elateridae) (Koch, 1989).

It is interesting to note that in a laboratory experiment on *Lonchaea corticis* larvae, Taylor (1929) found that the fly larvae with a diet composed of weevil larvae reached maturity much quicker than larvae raised on weevil frass. It seems obvious that if you live on a nutritionally poor diet, you would hardly say no when a fat 'sausage' is offered. On the other hand, decaying inner bark tissue is far more abundant than dead or dying larvae, so combining these two diets is a good survival strategy. Therefore, one should not be puzzled by the fact that some predatory species are not purely carnivorous.

3.6 Predatory fungi

Even among the fungi there are species that act as predators. Several wood-decaying fungi have the ability to trap nematodes as a supplementary nutrient source. This phenomenon has been thoroughly studied by Barron (2003). The agaric *Hohenbuhelia* grows some adhesive appendages on its hyphae that capture nematodes by sticking to them. A similar structure has been found in the closely related oyster mushrooms, *Pleurotus* spp. – when nematodes are present, the hyphae produce secretory appendages with toxic droplets. A nematode that moves through this field of droplets is paralysed within minutes; directional hyphae then grow into the nematode's mouth to digest its body.

In the corticioid genus *Hyphoderma*, some species produce two-celled structures called 'stephanocysts'. For many years their function was unknown, but in the 1990s Tzean and Liou (1993) discovered that their function was to capture nematodes.

The fungus *Arthrobotrys anchonia* has developed an even more sophisticated nematode trap. In this species, a branch of three cells curves round and fuses with itself to form a tiny ring about 0.03 mm in diameter. When a nematode moves into the ring it triggers a response in the fungus and the cells expand rapidly inwards with such power that they hold the nematode, which has no chance of escaping (Figure 3.5e). The *Arthrobotrys* fungi are asexual morphs, and as such they cannot be classified taxonomically due to the absence of fruiting bodies. But, for one species, the sexual state has been observed, and this species was an ascomycete in the genus *Orbilia* (Pfister, 1994). Different *Arthrobotrys* species occur in soil, dung and rotting wood, whereas *Orbilia* species are common on highly decayed wood in temperate and tropical regions. *Arthrobotrys* and other closely related species have developed several types of trapping devices along two lineages: adhesive traps and constricting rings (Yang et al., 2007).

Nematode-trapping fungi occur in at least three unrelated groups of fungi. But these fungi have one feature in common. They occur solely or frequently in decaying wood. Furthermore, they produce cellulose-degrading enzymes – a feature that seems odd for animal predators, since cellulose is never found in animal tissue (Barron, 2003). This led Barron to suggest that the nematodes are dietary supplements for nitrogen uptake, and that cellulose and lignin represent the main energy source. Thus, the predatory habit is merely a secondary capability.

3.7 Parasites

Most of us are familiar with the term 'parasite', which refers to a species that feeds on a live host without killing it. *Parasitoids* are somewhat different, as they typically consume their host almost completely and thereby kill it. Then, one might ask, what distinguishes them from the predators? Well, a predator is a serial killer, while a parasitoid only needs a single host individual to complete its life cycle. *Hyperparasitoids* are parasitoids on other parasitoids. In the next chapter we describe yet another type of parasite – *kleptoparasites*, but such species do not strictly belong to the saproxylic food web, as their food originates from outside woody material.

3.7.1 True parasites

True parasites are very small compared with their host and, with some exceptions, parasites have not been extensively studied among hosts living in dead wood. The main reason is that parasites typically become the subject of research when they occur in economically important species, and most saproxylic species do not belong in this category. We do have, however, a few examples of well-studied host organisms and their associated parasites.

In his classic work, Rühm (1956) documented nematode parasites associated with bark beetles. He found that many nematode species have species-specific associations with their bark beetle host. He also found that some stages of the nematode's life cycle live inside the body of the beetle, whereas other stages live independently in the bark beetle galleries, where they suck or feed on fungal hyphae. It is interesting to note that these nematodes have their enemies. In a study of fungi dispersed by bark beetles, Kirschner (2001) found seven fungus species that were nematode predators. One of these, *Arthrobotrys superba*, was sometimes carried by about 50% of the beetles, and may therefore have a considerable effect on the nematode populations in bark beetle galleries. Also, among the mites, we find several species that are bark beetle parasites (Kielczewski et al., 1983).

3.7.2 Parasitoids

The principal parasitoids in the saproxylic food web are wasps in different families of the large and important insect order Hymenoptera. Diptera also has a large family with a parasitic lifestyle – known as the tachinids. Even among the beetles we find parasitoids in the families Bothrideridae (Eggleton and Belshaw, 1992) and Rhipiphoridae (Švácha, 1994). The wasp and fly species attack their hosts in quite different ways. The female parasitic wasp enters a long egg-laying tube (ovipositor) into the body of the host insect and first injects a paralysing venom that immobilizes the host (Figure 3.5d). She then normally places a single egg into the host or on its body surface. The parasitic flies lack this piercing apparatus and the female instead lays her eggs in the galleries near the host. The newly hatched larva subsequently enters the host without causing any immediate harm and develops inside the growing host larva. Irrespective of the mode of attack, many parasitic wasps and flies play a significant role in regulating the host populations (Kenis and Hilszczanski, 2004).

Individual wasp species typically parasitize a specific developmental stage of the host species. Most species only parasitize the larval stage, others are specialized on pupae, and some species only parasitize the eggs of the host. One would expect that a tiny insect egg would not provide much food for a parasitoid. Remember that parasitoids only utilize one individual of the host – in this case a single egg. It is quite unusual that parasitoids lay eggs in adult insects. However, in the braconid genus *Cosmophorus* the females attack adult bark beetles and hold them tightly around the neck with their mandibles while entering the ovipositor into their body. The tachinid flies seem to be strictly confined to the larval stage of the host, as none are known to parasitize on pupae or eggs (Stireman et al., 2006).

The parasitic wasps with larval development inside the host body typically have a narrower host range than the flies. This is probably related to the fact that these hymenopterans have developed specific poisons to destroy the host's immune system (Strand and Pech, 1995). Those parasitic wasps where the larva is attached on the surface of the host (ectoparasites) generally have a broader host spectrum. In the ichneumonid subfamily Xoridinae, the species are specialized on the larvae of buprestid and longhorn beetles. In the large family Pteromalidae, we find many bark beetle parasitoids. In some of these species it has been demonstrated that the female wasps are specifically attracted to odours associated with bark beetle larvae (Pettersson et al., 2000). In the families Braconidae and Eulophidae, we find specific parasitoids on different fungivorous beetles living inside fungal fruiting bodies (Jonsell et al., 2001). Other parasitoids utilize host species occurring in the dead wood. Members of the ichneumonid genus *Rhyssa* are specialized on woodwasp (Siricidae) larvae. The female *Rhyssa persuasoria* (see Figure 10.8b) is about 10 cm in length, of which about 5 cm is the ovipositor. These females only seek woodwasp larvae that make tunnels several centimetres into the sapwood.

The tachinid flies lack specific adaptations related to the physiological defences of the host, and individual species often have a wide range of host species. It seems that the host range of tachinids is more related to the habitat use of the host (Stireman et al., 2006). Although the mechanism of host location is not well understood, the tachinid females are evidently capable of locating galleries that are inhabited by saproxylic larvae, and lay their eggs into these. Subsequently, the fly larva has a relatively short distance to travel before it finds a host.

The parasitoids occur abundantly as enemies of invertebrates in decaying wood. Again, as for the parasites, it is economically important hosts such as bark beetles, longhorn beetles and buprestid beetles that have been most thoroughly investigated (Kenis and Hilszczanski, 2004). In addition, among fungivorous invertebrates that develop inside fruiting bodies there are many beetles, moths and Diptera that are parasitized by wasps and flies (Komonen et al., 2000; Jonsell et al., 2001). It is likely that species in all the previous feeding groups have some parasitoids associated with them.

Dedicated entomologists who rear saproxylic insects from their immature stages onwards can contribute significantly to improving our knowledge about host–parasitoid associations. They might be disappointed when a parasitic wasp or fly emerges from a larva or pupa instead of a beautiful longhorn beetle or other exciting insect. One reaction to this initial disappointment is to treat the animal that destroyed their specimen as being worthless. Instead they should preserve it or ideally prepare the parasitoid – along with the remains of the host – for identification by an expert on parasitoids (Shaw, 1997).

3.7.3 Hyperparasitoids

Hyperparasitoids are parasitoids on other parasitoids. This form of parasitism has been reported for at least two different genera of Perilampidae wasps. In their study of insects breeding in the fruiting bodies of wood-decaying fungi, Jonsell et al. (2001) reared some *Perilampus* species that were associated with the larvae of fungivorous tineid moths. The *Perilampus* species does not attack the moth larva itself but a parasitic ichneumonid or tachinid larva inside the body of the moth larva. Another example is *Mesopolobus typographi*, which is a hyperparasitoid of *Tomicobia seitneri* which, in turn, is a parasitoid on the bark beetle *Ips typographus* (Ruschka, 1924).

3.8 Mycoparasites

Parasites are not restricted to the animal kingdom. The phenomenon of mycoparasitism (i.e. fungi parasitizing other fungi) is well known among mycologists (Barnett and Binder, 1973; Jeffries, 1995). Mycoparasites are traditionally divided into two categories: biotrophic and necrotrophic.

In biotrophic mycoparasitism, the living host supports the growth of the parasite for an extended period of time and may not appear diseased. The parasites appear highly adapted to this mode of life and several species have special organelles that establish intimate contact between the hyphae of itself and the fungal host (Bauer and Oberwinkler, 1991; Kirschner et al., 1999). Furthermore, such parasites tend to have a narrow host range. Biotrophic mycoparasitism seems to be frequent among certain genera of heterobasidiomycetes (Oberwinkler et al., 1984; Wells, 1994; Zugmaier et al., 1994) and good examples of such relationships are *Tremella aurantia* parasitizing *Stereum hirsutum* (Figure 3.5f), *Tremella encephala* parasitizing *Stereum sanguinolentum*, *Tremella mesentrica* parasitizing different *Peniophora* species, and *Christiansenia pallida* parasitizing *Phanerochaete cremea*.

Necrotrophic mycoparasites are typically aggressive, and their hyphae coil around and frequently penetrate into those of the host fungus. They can also secrete cell-wall-degrading enzymes and release toxins into the local environment to kill the host mycelium. These mycoparasites tend to be non-specific and have a broad range of host fungi. It is important to note that most necrotrophic mycoparasites are capable of indefinite existence as ordinary decomposers, but their growth is often enhanced as they overgrow and kill susceptible fungi (Barnett and Binder, 1973). Thus, there is a clear parallel to the predatory fungi whose main energy source is woody material but who also use live organisms as dietary supplements. It could also be that the killing ability of necrotrophic parasites should be viewed as a competitive rather than a parasitic interaction. Because there are so few detailed studies of interactions between wood-inhabiting fungi, the research of Rayner and co-workers (Rayner et al., 1987; Rayner and Boddy, 1988) is often cited when examples of mycoparasitism are needed. They demonstrated, in laboratory studies, how *Trametes gibbosa* replaced *Bjerkandera adusta* and *B. fumosa*, and how *Lenzites betulinus* replaced *Trametes ochracea* and *T. versicolor.* But *T. gibbosa* and *L. betulinus* are white-rot fungi that rarely seem to interact with these suggested hosts in nature (Niemelä et al., 1995), and it is probably better to consider them as aggressive competitors rather than as parasitic fungi.

3.9 Mycorrhizal fungi

It is well known that several mycorrhizal fungi regularly appear as fruiting bodies on the surface of moderately to severely decayed wood (Renvall, 1995; Nordén et al., 1999; Tedersoo et al., 2003). One reason for their occurrence on woody substrate is simply as a place to attach their fruiting bodies.

Other mycorrhizal fungi occur with well-developed mycelia inside rotten logs and wood fragments buried in the soil (Harvey et al., 1979; Kropp, 1982a). These fungi are instrumental in seed germination and the development of seedlings on or around dead logs – so-called 'nurse logs' (Kropp, 1982b). While these mycorrhizal fungi are capable of nutrient uptake from the forest soil, it is uncertain why they occur in rotten wood. It is likely that at least some mycorrhizal species are involved in wood degradation, as discussed in Section 3.1.3: 'Residual wood decayers'. Another possibility became apparent when Lindahl et al. (1999) demonstrated that the mycorrhizal fungi *Paxillus involutus* and *Suillus variegatus* are able to establish mycelial contact with the wood-decaying fungus *Hypholoma fasciculare* growing on the wood of *Pinus* spp. trees. The mycelia of the mycorrhizal fungi formed dense patches over the *Hypholoma* mycelium and extracted significant amounts of phosphorus from the wood decayer and transferred some of this to their host plant, which was a *Pinus* seedling. This 'short cut' and direct nutrient transport from decaying wood to a new generation of trees is certainly interesting and may have far-reaching implications for boreal forest ecosystem functioning.

3.10 Fungicolous fungi

There are many ways in which fungi can interact with each other, and positive associations between fungal species can result from entirely different mechanisms. Earlier in this chapter we have described residual wood decayers following structural wood decayers, and biotrophic mycoparasites obtaining nutrition directly from the mycelium of a host fungus. Another type of fungal interaction takes place when a fungus utilizes the dead fruiting body or mycelium of another fungus. This interaction is frequent among a variety of hyphomycetes and certain ascomycetes, for example *Cistella hymeniophila* growing on dead *Antrodia serialis* and *A. primaeva* and staining them reddish, and *Hypocrea pulvinata* on dead *Piptoporus betulinus* (Niemelä et al., 1995). In these cases, it is reasonable to assume that the associated species feed on the tissue of the dead host, i.e. a kind of mycosaprotrophism. There are, however, numerous examples of fungi growing on other fungi where the true nature of the association is unclear (Hawksworth, 1981). Until we know the specific nutritional mode of such associated fungi, it is best to classify them as fungicolous, which simply indicates an unknown association with the other fungus.

3.11 Ecological perspectives

Most of this chapter has been devoted to describing the nature of species in different functional groups – which species they are, what they consume, and what feeds on them. In this final section we take another approach and consider some general aspects of the saproxylic food web.

Briefly, this food web consists of four or five trophic levels (Figure 3.1). At the first (bottom) level we find the primary producers, the trees that build up the resource base. The principal actors on the second level are structural wood decayers and several detritivores (many with gut symbionts) that feed on woody material. The sugar fungi and residual wood decomposers also belong to this trophic level. At the third level we find other functional groups, such as fungivores, predators, scavengers and parasites on detritivores. On the fourth level we find parasites of fungivores, and these can have hyperparasites – constituting a fifth trophic level.

When we view the saproxylic species from a food web perspective, a whole suite of interesting questions arise:

- How are the functional groups structured in this food web, and what are the main energy flows in the system?
- How is species diversity correlated with these energy flows?
- What is the functional role and host or resource range of individual species, and what are the dynamic implications of these interactions?

3.11.1 Trophic interactions

In the 1960s and 1970s, most of the ecological literature focused on ecosystem functioning, with an emphasis on trophic levels and energy flow. Recent textbooks have focused more on community structure and species interactions, and here the food web concept plays a key role. In the past two decades, food web analysis has generated much theoretical debate and a lot of empirical and experimental work (for reviews, see Cohen et al., 1990; Pimm et al., 1991; Hall and Raffaelli, 1993; Pimm, 2002; Drossel and McCane, 2003).

A popular question to ask is whether the food web has a bottom-up or top-down control. That is, do species at low trophic levels determine the dynamics of those higher up in the web, or vice versa. This topic has scarcely been investigated for a full-sized food web in dead wood, but there are reasons to believe that it is mostly controlled from below, as the input of dead wood determines the energy base. Furthermore, the saproxylics generally have no, or weak, influence upon the mortality of trees. The exception is some bark beetles confined to coniferous trees, which can cause large-scale mortality among trees (see Chapter 12).

This perspective of trophic interactions is not solely of academic interest. As the top predators are among the first to suffer and become threatened when a particular ecosystem is reduced in spatial extent, so saproxylic species at higher trophic levels may also be particularly vulnerable. In an interesting study of forest fragmentation effects on fungi, fungivores and parasitoids, Komonen et al. (2000) found that the parasitoids were most sensitive and disappeared first in fragmented landscapes. Thus, it is not only the species diversity that is reduced if we extract timber resources too heavily from the forests – but also the complexity of species interactions becomes simpler.

3.11.2 Food web compartments

Studies that focus on trophic interactions tend to emphasize the vertical structure of a food web. But there is also a strong horizontal structure in the saproxylic food web. For simplicity, we have considered woody material as a rather homogeneous resource throughout this chapter in order to keep the focus on the functional roles. But, in reality, this is far from the truth. The saproxylic food web is strongly compartmentalized; that is, it is subdivided into subunits (compartments) within which interactions among species are strong, but between which interactions are weak or non-existent.

One subdivision is between species that occur in the wood of coniferous trees in contrast with those occurring in the wood of broadleaved trees (see Chapter 5). Another distinct subdivision is between species occurring at different decay stages. There is hardly any overlap in species composition between those that co-occur with bark beetles in weakened or recently dead trees, those that live under the bark of slightly to moderately decayed trees, and those that live inside the wood at more advanced stages of decay (see Chapter 6). Furthermore, there is a sharp distinction between species living in specialized microhabitats, e.g. sap exudations of live trees, insect galleries, fungal fruiting bodies, decaying wood, hollow trees, and so on (see Chapter 7). These distinct qualities of woody material are not only fine-scaled criteria with which to classify habitat associations of individual species; their combined effects also divide the saproxylic food web into an array of parallel food web compartments that need to be studied and analysed separately.

3.11.3 Functional roles and species interactions

We have quite a good overview of the functional role of individual species in different organism groups. But we also need to document their host or prey range and the extent to which alternative species are preferred or occur occasionally in their diet.

Recent theoretical and empirical work has brought food web analysis to the forefront of a fundamental question in ecology – the relationship between species diversity and ecosystem stability (McCann, 2000). Back in the 1950s, Odum, Elton and McArthur argued that increasing species diversity enhanced ecosystem stability (see McCann, 2000). In the early 1970s, May (1973) seriously challenged this view, but during the 1990s the diversity–stability relationship was reappraised, based on food web analysis. Nowadays, the key mechanism is perceived to be non-random species interactions across trophic levels, especially the importance of weak interactions that may dampen stronger destabilizing interactions (McCann, 2000).

An important message from this work is that we need not only to sort out how different species in a food web interact with each other, but also to rank or quantify whether these interactions are strong or weak. In other words, we should document whether individual species are monophagous or host-specific on certain species, oligophagous on a narrow set of related species, or polyphagous on a wide range of host species. This challenge leaves ample scope for empirically oriented studies of species interactions between saproxylic organisms.

4 · *Other associations with dead woody material*

Juha Siitonen and Bengt Gunnar Jonsson

Some of the species associated with snags, logs and hollow living trees do not depend on dead wood as a source of nourishment. Instead, they use cavities and other dead-wood microhabitats for various purposes such as nesting, roosting, denning and hibernation. These species may be obligate saproxylics if the availability of dead wood is essential for their survival during some part of their life cycle. Many facultative saproxylic species use dead wood more opportunistically, without being dependent on it. In this chapter, we describe these uses of dead woody material, concentrating on saproxylic species that do not belong to the saproxylic food web described in Chapter 3.

4.1 Vertebrates

4.1.1 Nesting and roosting in cavities

Many forest-dwelling vertebrates utilize holes and cavities in trees. For instance, in Australia, over 300 vertebrate species are known to use cavities (which are generally referred to as hollows in the Australian literature). The list includes 83 mammals (31% of the total terrestrial mammal species in Australia), 114 birds (15%), 79 reptiles (10%) and 27 amphibians (13%) (Gibbons and Lindenmayer, 2002).

Only a small proportion of all the cavity-nesting species can excavate their own holes in the wood; these species are called primary cavity nesters. Secondary cavity users are unable to excavate their own nesting, roosting or denning sites, and rely on holes excavated by the primary species and on naturally occurring cavities. Secondary cavity users can be further divided into those that strictly depend on pre-existing cavities for shelter or breeding sites, on a daily or seasonal basis (obligate cavity users), and species that use cavities opportunistically (facultative cavity users), although the division is somewhat arbitrary

Table 4.1. *Potential benefits and costs associated with cavity-nesting and roosting.*

Benefits	Costs
Protection from predation	Increased risk of predation
Shelter from adverse weather conditions	Low availability of potential nest sites
Thermoregulation	High competition for nest sites
Moisture regulation	Wetting and flooding during rainfall
Less effort in nest construction	High load of ectoparasites in reused sites

(McComb and Lindenmayer, 1999; Gibbons and Lindenmayer, 2002). Nevertheless, for all these species, the availability of cavity trees across the landscape is often a limiting factor for their survival, reproductive output and population density.

There are several ecological and evolutionary factors explaining why so many vertebrate species use cavities (Gibbons and Lindenmayer, 2002). The advantages of cavity use include protection from predators, shelter from adverse weather conditions such as heavy rain and wind, buffering against extreme temperatures and moisture variations, and reduced energy input needed to construct a nest. There are also negative consequences of cavity use, which may include low availability of potential nest sites, and high interspecific and intraspecific competition for nest sites (Gibbons and Lindenmayer, 2002; Newton, 2003). Some predators frequently visit cavities, and it may be difficult to escape from within the confined space. In addition, the importance of mechanisms such as thermoregulation depends on the climatic conditions, and the costs of nest-site competition depend on the availability of cavities. The relative benefits and costs of cavity use (see Table 4.1) constitute selection pressures which can increase or decrease the number of individuals using cavities, and in the long term determine the proportion of species in a regional fauna that specialize in using cavities.

A variety of internal volumes, entrance diameters, depths and positions is needed to meet the ecological requirements of all the cavity-using species. A reduced availability and diversity of cavities can lead to a decline in the density of cavity-dependent species, and in the long run may result in the local extinction of these species.

4.1.2 Formation and availability of tree cavities

Cavities can occur both in living trees and in decaying snags. Hollows in live trees are formed primarily by the action of heart-rot fungi. These fungi are specialized in decaying the dead heartwood of mature living trees. Since heartwood forms the inner core of the trunk and larger branches, and since heart-rot fungi do not grow in the functional sapwood forming the outer wood layer, the trunk may become hollow as a result of decay. The breakage of branches or the action of primary excavators can provide access to the hollow. Saproxylic invertebrate species feeding on the decaying heartwood speed up cavity development (see Chapter 7). Termites are a key agent in hollowing trees both in tropical forests (Apolinário and Martius, 2004) and in savanna woodlands (Werner and Prior, 2007).

Cavities in living trees are more durable, have walls that are more solid, and provide better thermoregulation than cavities in snags (Wiebe, 2001; Paclík and Weidinger, 2007). In contrast to heartwood decay in living trees, when a tree dies it is invaded by sapwood-decaying fungi and it will undergo a rapid decay of the entire tree. This general decay results in easily excavated cavities which, however, deteriorate within a few months or years (Jackson and Jackson, 2004).

Many factors can influence the development of cavities, including tree species, trunk diameter, tree age, and environmental conditions (Gibbons and Lindenmayer, 2002). Certain tree species are more susceptible to becoming hollow and start to produce cavities at younger age than other tree species. This is attributable to the ecological traits of the tree species, such as growth rate, wood density, heartwood formation, decay resistance, branching characteristics and longevity. The propensity of different tree species to form cavities, and the type of cavities, is also closely linked to the specific heart-rot fungus species associated with each tree species (see Chapters 5 and 7). Living conifers usually have a low incidence of cavities as compared with most broadleaved trees, but large conifer snags with soft wood frequently have holes excavated by woodpeckers.

The probability of occurrence of cavities, number of cavities (or entrance holes) per tree, and average size of cavities have been shown to increase with the diameter, age and decreasing crown condition of trees (Lindenmayer et al., 2000; Gibbons and Lindenmayer, 2002; Gibbons et al., 2002; Whitford, 2002; Wormington et al., 2003; Blakely et al., 2008; Fox et al., 2008; Koch et al., 2008a). Large old trees with irregular, senescent crowns are generally the most likely to have cavities. Older trees will have been subject to injuries and consequent infection by fungi for a longer time, they have a higher proportion of heartwood, and may have weakened defences against fungal invasion. For instance, hollows generally do not occur in eucalypts that are under 120 years old, and most eucalypt species reach a 50% probability of containing cavities when they are 180–230 years old (Gibbons and Lindenmayer, 2002; Gibbons et al., 2002; Whitford, 2002; Wormington et al., 2003). Similarly, pedunculate oak (*Quercus robur*) has only about 5% probability of the presence of cavities at an age of 100–200 years, and reaches 50% probability at an age of 200–300 years, depending on the growth rate of the trees (Ranius et al., 2009a).

The number of cavity trees in a stand varies according to region, site type, tree species composition, stand age and disturbance history. All these factors are interrelated, and the individual effects of each factor cannot easily be separated. Estimates of cavity density (average number of cavities per hectare) have been published for many types of forest in different regions. Unfortunately, it is not possible to compare these figures directly, since the minimum size criteria for a cavity have not been consistent between studies, and sometimes no quantitative criteria are given. Boyle et al. (2008) found 62 cavity density estimates through a literature search, but sufficient detail about the minimum size criteria – allowing for comparisons to be made – was provided for only 17 estimates. On the basis of these data (and including entrance size as a covariate in the analysis), cavity density appeared to decrease with latitude, i.e. it was highest in tropical forests and decreased towards the north.

Taking 2 cm as the minimum entrance diameter (this has been used in several studies, since it is close to the smallest diameter that allows small bird and mammal species to enter), the density of cavities in natural temperate forests varies from about 5 (Remm et al., 2006) to 60 per hectare (Carlson et al., 1998). A typical density of cavities usable for vertebrates in natural, mature to old-growth temperate forests in the northern hemisphere appears to be 10–30 per hectare, depending on the region and tree species composition (Bai et al., 2003; Kahler and Andersson, 2006; Remm et al., 2006; Aitken and Martin, 2007; Wesołowski, 2007; Boyle et al., 2008). Only a few comparable estimates are available for other vegetation zones. The average density of cavities was over 400 per hectare in old-growth tropical forests in Thailand (Pattanavibool and Edge, 1996), 110 per hectare in upland old-growth tropical forests in Costa Rica (Boyle et al., 2008), but only 17 per hectare in subtropical semi-deciduous Atlantic forests in Argentina (Cockle et al., 2008). In Australian jarrah (*Eucalyptus marginata*) forests with a Mediterranean climate, the average cavity density exceeded 100 cavities per hectare (Whitford, 2002). Even where cavities appear to be abundant, there may be a shortage of high-quality cavities (Lõhmus and Remm, 2005; Cockle et al., 2008).

4.1.3 Cavity-nesting birds

Birds comprise the most conspicuous and best-known animal group using tree cavities as nest sites; the terms 'cavity-nesting' and 'hole-nesting' have both been widely used in the literature. If only forest birds are considered, the proportion of cavity-nesting species is about 30% in northern Europe (Siitonen, 2001), 35% in central Europe (Wesołowski, 2007), 40% in North America (Scott et al., 1977), and 20–30% in tropical Central America (Gibbs et al., 1993). The high proportion of cavity nesters is a compelling indication of the importance of living mature and dead standing cavity trees in natural forest ecosystems. Considering whole-regional avifaunas, the proportion of obligate cavity nesters varies between 4% and 11% in different regions throughout the world, and the proportion of species known to use cavities is between 9% and 18% (Table 4.2). Since the total number of bird species is about 10 000, the number of obligate cavity-nesting birds is about 500–1000 species. Most obligate cavity nesters belong to the following orders: songbirds or passerines (Passeriformes), woodpeckers (Piciformes), parrots (Psittaciformes), near passerines, including rollers and hornbills (Coraciiformes), owls (Strigiformes), and waterfowl (Anseriformes) (Saunders et al., 1982; Newton, 2003).

Woodpeckers (Picidae) are the most important group of primary cavity nesters (Figure 4.1). In addition, some passerine birds such as nuthatches (*Sitta* spp.), tits and chickadees (*Parus* and *Poecile* species) are weak cavity excavators and can dig their own nest holes in decaying snags when the wood is soft enough (Martin and Eadie, 1999; Martin et al., 2004). In most cases, primary cavity nesters will use the hole only once. Cavity trees and cavity-using species are connected ecologically in nest webs analogous to food webs, where interspecific interactions (facilitation, competition etc.) are centred around nest sites (Martin and Eadie, 1999). Woodpeckers are keystone species in many forest ecosystems because they provide nest sites for secondary cavity nesters,

Table 4.2. *Numbers and percentages of bird species that use tree cavities for nesting in different parts of the world (based on Saunders et al., 1982; Newton, 2003; Monterrubio-Rico and Escalante-Pliego, 2006; Sandoval and Barrantes, 2009).*

	Total number of breeding species	Number (%) known to use cavities	Number (%) of obligate cavity nesters
Europe			
Passerines	169	28 (17)	9 (5)
Non-passerines	250	32 (13)	13 (5)
Total	419	60 (14)	22 (5)
North America			
Passerines	292	28 (10)	13 (4)
Non-passerines	192	30 (10)	11 (4)
Total	484	58 (10)	24 (4)
Mexico			
Total	657	112 (17)	81 (12)
Costa Rica			
Total	850[a]	94 (11)	–
Southern Africa			
Passerines	333	20 (6)	13 (4)
Non-passerines	310	35 (11)	23 (7)
Total	643	55 (9)	36 (6)
Australia			
Passerines	297	23 (8)	7 (2)
Non-passerines	234	71 (30)	50 (21)
Total	531	94 (18)	57 (11)

[a] Approximate number, calculated on the basis of the number of cavity users and their proportion of the total avifauna as given in Sandoval and Barrantes (2009).

and thereby influence the abundance and distribution of many other species (Johnsson et al., 1990; Daily et al., 1993; Martin and Eadie, 1999). However, the importance of woodpeckers varies depending on the region and forest type. In some forest types, woodpeckers can produce up to 80–90% of all available cavities (Remm et al., 2006; Aitken and Martin, 2007). Woodpecker excavation may play a particularly important role in conifer-dominated forests, where naturally occurring cavities are uncommon (Walter and Maguire, 2005; Blanc and Walters, 2008). In other forest types, non-excavated cavities may occur so abundantly that excavated holes are relatively unimportant for secondary cavity nesters (Carlson et al., 1998; Bai et al., 2003; Wesolowski, 2007;

Figure 4.1. Nest of a black woodpecker (*Dryocopus martius*) in European aspen. The black woodpecker is a keystone species because it produces large treeholes which can later be used by many secondary cavity nesters and other species. Aspen (*Populus tremula* in Eurasia and *P. tremuloides* in North America) is the most important tree species for cavity nesters across many forest types and extensive regions in the boreal and temperate zones (photo Alastair Rae).

Cornelius et al., 2008). Woodpeckers are absent in Australia and New Zealand, and in those places, fire damage, fungal decay, termites and branch breakage are the main agents producing cavities (Gibbons and Lindenmayer, 2002; Whitford, 2002; Blakely et al., 2008).

Different species of primary cavity nesters use different tree species, sizes and types of trees. Most woodpeckers use deciduous trees, but some species favour conifers. For example, the North American red-cockaded woodpecker (*Picoides borealis*) excavates its holes exclusively in living pines with heart rot caused by the polypore species *Phellinus pini* (Jackson, 1977; Conner and Locke, 1982; Blanc and Walters, 2008). Most species prefer to excavate nest cavities in trees with soft wood, i.e. with a relatively low wood density (Schepps et al., 1999). Furthermore, fungal infection is usually necessary for cavity excavation. Only a few woodpecker species can make cavities in healthy living trees, and most prefer snags or living trees with decaying heartwood (Conner et al., 1976; Hågvar et al., 1990; Jackson and Jackson, 2004; Losin et al., 2006). In the northern hemisphere, aspen (*Populus tremula* in Eurasia and *P. tremuloides* in North America) is a particularly important tree species for cavity excavators, because of its low wood density and its susceptibility to heart rot caused by the polypore *Phellinus tremulae* (Hågvar et al., 1990; Li and Martin, 1991; Jackson and Jackson, 2004; Martin et al., 2004; Losin et al., 2006). As well as excavating holes, it is possible that woodpeckers also contribute to the dispersal of wood-decaying fungi by carrying spores or mycelia from infested trees and inoculating them into uninfested trees while foraging (Farris et al., 2004; Jackson and Jackson, 2004).

Trees selected for nesting must be large enough to allow each species of bird to excavate a cavity of sufficient size. The smallest species, such as tits and chickadees (weighing about 10 g), can use snags as small as 10 cm in diameter, while the largest woodpecker species, such as the North American pileated woodpecker (*Dryocopus pileatus*, weighing about 300 g) prefer trees that are at least 50 cm in diameter (Scott et al., 1977; Martin et al., 2004; Bull et al., 2007). Almost all primary cavity nesters select trees that are larger than average trees (Bai et al., 2003; Martin et al., 2004).

Secondary cavity nesters are dependent on the existing supply of cavities. Most species show clear preferences for certain types of cavities in terms of entrance size, volume, height above the ground, orientation, and how well concealed the entrance is (Saunders et al., 1982; van Balen et al., 1982; Nilsson, 1984; Li and Martin, 1991; Carlson et al., 1998; Martin et al., 2004). Entrance size limits the range of species that can use a cavity. Smaller entrance holes are advantageous in deterring predators and in reducing the chance of eviction from the cavity by larger competitors. Cavity volume is also an important factor in explaining cavity use because it can affect reproductive success (Martin et al., 2004). Cavities with large internal size may allow for better thermoregulation and reduce competition for space among siblings. Therefore, the ideal cavity to maximize fecundity and minimize predation is a large-volume cavity with a small entrance (Martin et al., 2004).

Body size is generally the most important factor determining the niche overlap of secondary cavity users. Several bird species can use the same type of cavity, and competition for nest sites is frequent, particularly in managed forests. The number of dominant competitors can therefore affect the number and distribution of other species. A subordinate species may be either absent from areas where all the nest sites are occupied by dominant competitors, or the subordinate species will be forced to use cavities of suboptimal quality (Nilsson, 1984; Newton, 2003; Martin et al., 2004). Early-nesting resident species may occupy the best nest sites, while later-arriving migratory species may not find vacant sites of the right size. In unlogged old-growth forests, cavity availability does not appear to be a limiting factor for cavity nesters, and nest predation has a much greater influence than nest competition on the use of cavities (Brightsmith, 2005a; Wesołowski, 2007; Cornelius et al., 2008).

Some secondary cavity nesters have special requirements for their nest sites. Waterfowl such as goldeneyes (*Bucephala clangula, B. islandica*) occupy large cavities close to water; these cavities are often excavated by the black woodpecker (*Dryocopus martius*) or pileated woodpecker (*D. pileatus*) (Johnsson et al., 1990, 1993; Bonar, 2000; Brightsmith, 2005a; Wesołowski, 2007; Cornelius et al., 2008). Many tropical cavity nesters, such as macaws, toucans and hornbills, are large and therefore need big trees to sustain sufficiently spacious cavities. The largest among these species are the South American hyacinth macaw (*Anodorhynchus hyacinthinus*), which is about 100 cm long and weighs 2 kg, and the Southeast Asian great hornbill (*Buceros bicornis*), which can be 120 cm long and weigh up to 4 kg. In general, very little is known about the nest-site requirements of tropical cavity nesters (Cornelius et al., 2008). In tropical lowland forests in the Peruvian Amazon, two contrasting sites were found to be crucial for nesting parrots (an assemblage including 15 sympatric species): long-lasting emergent *Dipteryx micrantha* trees and short-lived dead *Mauritia* palms in palm swamps (Brightsmith, 2005b). The large macaw species preferentially used *Dipteryx* trees, which can live for over 1000 years and contain dozens of cavities usable for nesting.

Most cavity-nesting birds also rely on cavities for roosting. Roost cavities afford shelter from extreme climatic conditions, reduce heat and energy loss during cold nights, and provide protection from predators. The availability of suitable roost trees may be particularly important for resident birds inhabiting northern latitudes with harsh winters and short winter days, which restrict the time available for foraging. In such conditions, insulation and reduced exposure to wind and precipitation may become vital in nightly energy saving (Askins, 1981; Cooper, 1999; Paclík and Weidinger, 2007).

4.1.4 Mammals using tree cavities and logs

Mammal species from mice (McCay, 2000) to bears (Wong et al., 2004) use tree cavities and logs for shelter and nesting. Dependence on cavities is evidently as widespread among mammals as it is among birds. The proportion of cavity users varies widely between regions. There are relatively few mammal species relying on cavities in the boreal and temperate forests of the northern hemisphere. In North America to the north of Mexico, about 20 species were classified as cavity users (Aitken and Martin, 2004), which corresponds to only about 6% of the terrestrial mammal fauna. In contrast, the number of native mammal species in Australia recorded as using cavities is 83, amounting to 31% of the terrestrial mammal fauna (Gibbons and Lindenmayer, 2002). The latter figures may be exceptionally high, due to the evolutionary history of the Australian biota. However, in species-rich tropical forests with many mammals that are exclusively arboreal or scansorial (adapted to climbing), the proportion of species relying on tree cavities for shelter or nesting may be equally high. Because of the flexibility of mammals in the selection of den and nest sites, it is difficult to estimate the proportion of strictly obligate cavity users. Nevertheless, given that the total number of terrestrial mammal species is about 5000, the number of obligate cavity-using mammals is certainly several hundreds of species and may amount to up to 1000 species.

Cavity users are found in diverse mammalian taxa, the orders having the highest numbers of obligate cavity users being bats (Chiroptera), rodents (Rodentia) and several marsupial orders. More than half of the approximately 1100 species of bats rely on roosts in vegetation and, of these, the majority are associated with hollow trees (Kunz and Lumsden, 2003). In rodents, there are several arboreal or scansorial families with a high proportion of cavity users, such as squirrels (including flying squirrels) (Sciuridae), and several mainly tropical families such as New World porcupines (Erethizontidae) and scaly-tailed squirrels (Anomaluridae) (see, e.g., MacDonald, 2001). In marsupials, there are also several arboreal taxa including the Australian species of possums, gliders, antechinuses etc. (for a species list, see Gibbons and Lindenmayer, 2002), as well as New World opossums with about 90 Neotropical species. Lemurs and other prosimians also include many cavity-using species (Kappeler, 1998).

Most mammal species using tree cavities are nocturnal, and they use tree cavities both as diurnal dens and for rearing their young. Bats spend at least half of each day in a roost, and thus the availability of roost sites will strongly influence their survival and reproductive success. Because of their small size, bats have a high metabolic rate, and maintaining body temperature is energetically costly. The ability to use torpor as a means of conserving energy is a typical characteristic of many bats. In temperate forests, most cavity-using bat species prefer large-diameter trees that are taller than the surrounding canopy, and that are located in areas with a relatively open canopy and high snag density (Sedgeley and O'Donnell, 1999; Kunz and Lumsden, 2003; Kalcounis-Ruppel et al., 2005; Vonhof and Gwilliam, 2007). Large emergent trees and snags may provide larger cavities with better thermoregulation, and also provide easier relocation and access to the roost trees (Barclay and Kurta, 2007; Vonhof and Gwilliam, 2007). Some bat species roost mainly behind loose bark on snags (Kunz and Lumsden, 2003; Barclay and Kurta, 2007). Indiana bats (*Myotis sodalis*) tend to use loose bark on dead ash (*Fraxinus*), elm (*Ulmus*) and other broadleaved tree species (Foster and Kurta, 1999; Barclay and Kurta, 2007), whereas California bats (*Myotis californicus*) roost beneath loose bark on conifer snags, particularly Douglas fir (*Pseudotsuga menziesii*) (Vonhof and Gwilliam, 2007). The selection of cavities varies not only among bat species and according to the availability of roost sites, but during the breeding period males, reproductive females and non-reproductive females have different cavity requirements (Kunz and Lumsden, 2003; Barclay and Kurta, 2007). Most cavity-roosting bats form maternity colonies involving several to hundreds of individuals. Species that roost in colonies, or form maternity colonies, typically use large cavities in standing trees. They are able to accurately select cavities with favourable microclimatic conditions for saving energy (Sedgeley, 2001).

Studies in Australia have shown that the probability of cavity trees being occupied by vertebrates increases with the number of cavities per tree, diameter and crown senescence (Lindenmayer et al., 1990, 1991; Gibbons et al., 2002; Kunz and Lumsden, 2003; Barclay and Kurta, 2007; Koch et al., 2008b). At the level of individual cavities, the probability of a cavity being used increases with hollow size and depth (Gibbons et al., 2002; Koch et al., 2008b). Most cavities used by vertebrates have an entrance diameter of at least 10 cm and are at least 30 cm deep. However, different species of arboreal marsupials show clear preferences in their choice of cavities. Similar to cavity-nesting birds, the preferences may vary in terms of entrance diameter and cavity volume, height above the ground, position on the tree (main stem or branches) and number of cavities per tree (Lindenmayer et al., 1990, 1991, 1996; Trail and Lill, 1997; Gibbons and Lindenmayer, 2002). The range of cavity sizes used by mammals is even larger than in birds. Another difference from cavity-nesting birds is that den-swapping behaviour is more common in mammals, i.e. they frequently use more than one cavity for denning and regularly move between sites. The number of cavities used by an individual of a range of arboreal marsupials varied from two to over 20 (Gibbons and Lindenmayer, 2002). Reasons for den-swapping may include avoidance of predation, thermoregulation (different cavities have different microclimates), availability of food sources and dispersal.

4.1.5 Reptiles and amphibians using tree cavities and logs

Compared with birds and mammals, there is very little information on cavity use by amphibians and reptiles. Arboreal species are dominant in tropical forests, and it is reasonable to assume that many of these species use tree cavities more or less regularly on a daily basis. In tropical forests, many amphibian species breed in water-filled tree cavities (see Chapter 7), which represents a particular type of cavity use. In their review about phytotelm-breeding anurans, Lehtinen et al. (2004) list 20 species of frogs that have been found as tadpoles exclusively in tree cavities.

In other biogeographical regions (and perhaps also in tropics), it is more typical that amphibians and reptiles use cavities and logs seasonally, during periods of adverse climatic conditions, for hibernation or aestivation (inactive period during summer). Since they are poikilothermic (their internal temperature varies along with the ambient environmental temperature) or ectothermic (they control their body temperature through external means), the relatively stable temperature in tree cavities in comparison with the ambient temperature can be assumed to be important for many species. For example, the broad-headed snake (*Hoplocephalus bungaroides*) living in southwestern Australia appears to be restricted to rocky sandstone habitats. However, during the hottest part of the summer, when the temperature under rocks became too high, radio-tracked snakes were shown to leave the cliffs and seek out dead hollow trees hundreds of metres away, where they remained sequestered for long periods (Webb and Shine, 1997).

Logs are particularly important refuges for amphibians during periods of hot and dry weather because logs retain a lower temperature and higher relative humidity than ambient conditions. Sufficient moisture is vital to the lungless salamanders (Plethodontidae), which respire through their skin. Well-decayed logs are an important habitat component for many of these species, and some species, such as the clouded salamander (*Aneides ferreus*), which occurs in old-growth forests in Pacific North America, seem to be strictly dependent on dead wood (Aubry et al., 1988; Alkaslassy, 2005).

4.2 Invertebrates

There are many insect species that use dead wood only as a nesting site and not as a nutrition source. That is, they bring food to their nest from outside the decaying wood, or they utilize food brought into the wood by some other wood-nesting species. Both vertebrate and invertebrate wood nesters accommodate in their nests a range of associated species which can include parasites, parasitoids, kleptoparasites (species which steal the prey or other food stored by their host) and commensals.

4.2.1 Nesting in dead wood

All the wood-nesting invertebrate species belong to two insect orders, the first including wasps, bees and ants (Hymenoptera) and the second consisting of termites (Isoptera). Also, among the insects, there are primary hole excavators which dig their own nesting chambers in dead wood and secondary species which use existing holes produced by other species. However, insects have highly diverse methods of nest construction in wood (Figure 4.2). Three main nesting modes can be distinguished:

1. Solitary wasp and bee species dig their nest tunnels into wood, or use pre-existing holes made by other wood-boring insects.
2. Social wasps (Vespidae) and honey bees (Apidae) build their nests in large tree cavities using material that they bring from outside the tree.
3. Social ants (Formicidae) and some termites (Isoptera) excavate their nest systems consisting of tunnels and chambers into dead parts of living and dead trees.

Unlike hymenopterans, which use dead wood only for nesting, the termites also feed on dead wood, i.e. they belong to the saproxylic food web.

Several hymenopteran families have a large number of species nesting in wood. Foremost among these are sphecid wasps (Sphecidae), potter wasps (Eumenidae), carpenter bees (Xylocopidae) and mason bees or leafcutter bees (Megachilidae) (Krombein, 1967; O'Neill,

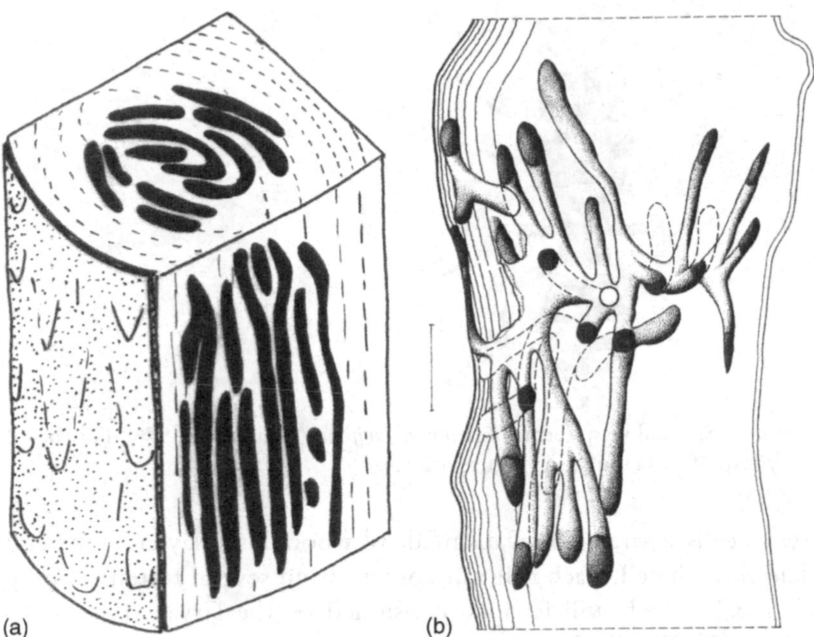

Figure 4.2. Different insect groups and species use dead wood for nesting, and excavate highly diverse and characteristic types of nest tunnels and chambers. (a) The nest of a carpenter ant species (*Camponotus* sp.) and (b) the nest of a sphecid wasp species (*Ectemnius cephalotes*). *Ectemnius cephalotes* excavates its nests in decaying wood, often as an enlargement of the nest system of the preceding year. Several females may use the same entrance hole, and the nest system branches in several planes. At the end of each short lateral tunnel, a larval chamber is constructed, and provisioned with medium-sized hoverflies (Syrphidae), of which 6–12 are placed in each cell. (a) Drawing by Juha Siitonen, (b) drawing by Ole Lomholdt (1975).

2001). Different species can either excavate their own tunnels into soft decaying wood, use pre-existing holes made by different saproxylic insects (particularly Anobiidae, Brenthidae, Cerambycidae, Scolytinae) emerging from the wood, or they can modify the existing holes to suit their own needs. Different-sized species select holes of different sizes, analogous to the hole-nesting birds but on a much smaller scale.

Sphecidae is the most species-rich group of wood-nesting insects (Figure 4.3). Sphecid wasps and potter wasps are predatory: they prey upon various invertebrates, e.g. aphids, hoverflies, muscid flies and spiders. Different species are specialized on different prey taxa. The prey are stored and are often tightly packed in the nests in wood, in most

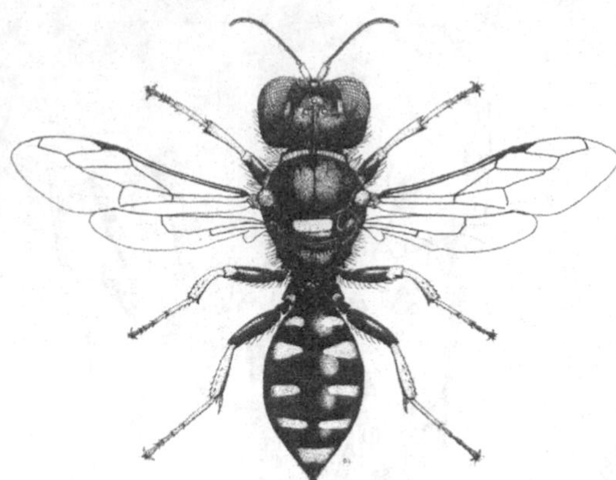

Figure 4.3. Sphecid wasp species *Ectemnius cavifrons* (Lomholdt, 1976). For the nest system of this family, see Figure 4.2 (b).

cases in cells separated by walls made of wood dust, clay etc. One egg is laid in each cell. Each cell can contain from several to tens of prey individuals which will later be consumed by the larvae (Lomholdt, 1975, 1976; O'Neill, 2001).

Carpenter bees and megachilid bees are vegetarians: they gather and store pollen and nectar for their larvae (Krombein, 1967). Similar to those of sphecid wasps, the nest tunnels are divided into separate cells, each containing one egg and sufficient provisions for the development of the larva. Carpenter bees have very strong mandibles and are able to dig their large nest tunnels even into hard undecayed wood. The nest generally consists of a short entrance tunnel and one or more horizontal tunnels running close to the wood surface. The same tunnel system can be used over several years, and new adults hibernate in the tunnels. Some megachilid bees dig their nest tunnels in soft, decayed wood, while other species use pre-existing holes. The walls separating the cells can be made of soil (hence the common name 'mason bees') or cut pieces of fresh leaves ('leafcutter bees'). Megachilidae are extremely important pollinators of various crop plants, and artificial nests (wooden blocks with holes of suitable size drilled in them) have successfully been used to increase populations or to introduce populations into areas where there is a shortage of pollinators.

All social wasps and paper wasps (Vespidae, subfamilies Vespinae and Polistinae) use dead wood as building material to construct their nests. Wasps use their mandibles to strip wood fibres from exposed wood surfaces, and chew the wood into a paper-like pulp. Some species, most notably the hornets (*Vespa* spp.), usually construct their nests inside hollow trees. Before domestication, the original nesting site of the common honey bee (*Apis mellifera*) and its close relatives was tree cavities.

A large number of ant species, e.g. carpenter ants (*Camponotus*), and many termite species, particularly in the families drywood termites (Kalotermitidae) and dampwood termites (Termopsitidae), dig their nests in dead wood. It is characteristic for carpenter ants and other wood-nesting ants to utilize the tree's annual growth rings in nest construction, so that they excavate the softer earlywood parts, and leave the denser latewood parts to form the walls that separate the chambers (see Figure 4.2).

4.2.2 Associates of insect nests

Wood-nesting insect species also have their own associated species. Unlike the nest-associates of vertebrate species, nest-associates of insects are often host-specific, at least at the level of particular host insect genus or family (Table 4.3). Kleptoparasitism is a particular type of species interaction between two species connected by nesting. Whereas each parasitoid larva consumes and kills one host larva, it is different with kleptoparasites, where the larvae mainly consume the stored prey items. In most cases even the kleptoparasite larvae first consume the egg or small larva of the host species before starting to use the stored prey insects.

Cuckoo wasps (Chrysidae) are kleptoparasites of solitary wasps and bees, and many species are specialized in nests in wood. These species, which have striking metallic blue, green and purple coloration, can often be seen running on the walls of old wooden buildings. Kleptoparasites are found in other insect orders too, including beetles (Coleoptera), e.g. *Zavaljus brunneus* (Languridae) (Palm, 1951; Lundberg, 1966) and flies (Diptera), e.g. *Eustalomyia* spp. (Anthomyiidae) and *Macronychia* spp. (Sarcophagidae) (Alexander, 2002).

The larvae of several skin beetles (Dermestidae), e.g. *Megatoma undata* and *Globicornis marginata*, occur mainly in old nests of solitary wasps, where they consume the dry remains of stored insects, but they can also live in other kinds of nests and under loose bark, where they feed on dead insects (Palm, 1951, 1959). Thus, these species are typical scavengers.

Table 4.3. *Examples of wood-nesting insect species, and insect species living in their nests in northern Europe. Hym = Hymenoptera, Col = Coleoptera, Dip = Diptera, Lep = Lepidoptera.*

Host insects		Associated insect species	
Taxonomic group	Species	Species (Order: Family)	Reference
Formicidae	*Camponotus herculeanum*	*Dermestes palmi* (Col: Dermestidae)	Ehnström (1983)
		Thiasophila wockii (Col: Staphylinidae)	Palm (1951)
		Niditinea truncicolella (Lep: Tineidae)	
	Lasius brunneus	*Scydmaenus perrisi* (Col: Scydmaenidae)	Palm (1959)
		Euryusa coarctata (Col: Staphylinidae)	Palm (1959)
	Lasius fuliginosus	*Thiasophila inquilina* (Col: Staphylinidae)	Palm (1959), Alexander (2002)
		Zyras funestus (Col: Staphylinidae)	Palm (1959)
		Milichia ludens (Dip: Milichidae)	Alexander (2002)
	Formica lehmani	*Eocatops lapponicus* (Col: Cholevinae)	Szymczakowski (1975)
Vespidae	*Vespa crabro*	*Velleius dilatatus* (Col: Staphylinidae)	zur Strassen (1957)
		Cryptophagus micaceus (Col: Cryptophagidae)	Alexander (2002)
Sphecidae	*Pemphredo*, *Passaloecus* spp.	*Omalus auratus* (Hym: Chrysidae)	Lomholdt (1975)
		O. puncticollis (Hym: Chrysidae)	Alexander (2002)
Megachilidae	*Osmia leaiana*, *Osmia* spp.	*Chrysura radians* (Hym: Chrysidae)	Alexander (2002)

4.2.3 Associates of vertebrate nests

Cavity-nesting vertebrates accommodate a great number of invertebrate visitors in their nests (Woodroffe, 1953; Hicks, 1971). Faeces, feathers, food remains and the carcasses of nestlings constitute a significant nitrogen input that can enrich the invertebrate fauna considerably. Most of the visitors are habitat generalists which are attracted by the rotting material inside the nests. However, some of the species are nest specialists, often termed nidicolous species.

Although most nidicolous species can utilize many kinds of nests, including exposed ones, certain species are more or less restricted to nests in tree cavities. A bird's nest that becomes saturated by water undergoes rapid fungal and bacterial decomposition and has a fauna similar to that of decaying plant material in a wide variety of situations (Woodroffe, 1953). Nests built in the open are subject to the flushing and wetting effects of rain after they have been abandoned by the birds. In contrast, nests in tree cavities are much better protected against weather and therefore constitute more stable and durable habitat patches.

Three wide categories of invertebrate species associated with cavity nests and roosts can be distinguished: ectoparasites of the nest-dwelling vertebrates; species consuming the decomposing nest materials (including detritivores, scavengers and fungivores); and predators and parasitoids of the first two groups of species. Ectoparasites include parasitic mites belonging to several families, chewing lice (Mallophaga) and louse flies (Hippoboscidae) on birds, fleas (Siphonaptera) on various mammals, and bat flies (Nycteribiidae) on bats. These feed on the blood, epidermis or feathers of their hosts. Most ectoparasitic species are host-specific.

Other nest-associated species form a complex food web which is based on the nest materials. They are generally not associated with a particular host species, but rather with particular nest types and nest materials. Moisture content is an important factor affecting species composition in a nest (Woodroffe, 1953). Wet nests are favoured by flies (Diptera) and predators of their larvae, while dry nests are favoured by skin beetles (Coleoptera: Dermestidae) and tineid moths (Lepidoptera: Tineidae). The larvae of both skin beetles and tineid moths feed on dry animal tissue including feathers and hairs, which are composed of a fibrous protein called keratin. Several indoor pest species that feed on stored products (foodstuffs and woollen fabrics), and that nowadays have a cosmopolitan distribution, were originally inhabitants of nests (Woodroffe, 1953). Beetle species typical of cavity nests are found among the small carrion beetles (Cholevinae), for example *Dreposcia umbrina*, *Nemadus colonoides*, rove beetles (Staphylinidae) such as *Haploglossa* spp., hister beetles (Histeridae) such as *Gnathoncus* spp., spider beetles (Ptinidae), and skin beetles such as *Anthrenus scrophulariae* (Palm, 1951, 1959). Rove beetles and hister beetles are predators of fly larvae occurring in the nests. Fly species inhabiting cavity nests are found in many families, including bird blow flies *Protocalliphora* spp. (Calliphoridae), Anthomyzidae, Chyromyidae, Fanniidae and Heleomyzidae, e.g. *Neossus nidicola* (Iwasa et al., 1995; Alexander, 2002).

There are many more invertebrates living in trunk cavities independently of the presence of cavity nesters. Most of these species are detritivores which feed on decaying wood and detritus accumulating at the bottom of cavities, or they are predators and parasitoids of the species mentioned above. The saproxylic food web in tree cavities is treated in detail in Chapter 7.

4.2.4 Invertebrates hibernating and aestivating in dead trees

Dead standing and downed trees are important sites for hibernation or aestivation for a large number of litter-dwelling arthropods. Logs in advanced stages of decay retain sufficient moisture even during dry periods and are used for cover, e.g. by molluscs, and for hibernation and aestivation by many ground beetles (Carabidae).

4.3 Epixylic species: life on the surface

A special group of saproxylic species are those which only use the surface of dead trees as their habitat. Epixylic species ('epi' meaning growing on) do not utilize the wood or other wood consumer as their primary energy source. Instead most are autotrophic and therefore get their energy through photosynthesis. Nevertheless, within this group we have species that are strictly dependent on dead trees, and most of these are either bryophytes or lichens.

4.3.1 Epixylic bryophytes

Most bryophytes belong to two phylogenetically distinct groups: the mosses and the liverworts. Within both groups, many species grow on dead trees. It is among the liverworts that we find most of the epixylic specialists. Liverworts are small plants with delicate leaves composed of a single cell layer. They lack the ability to conduct water and rely on whatever moisture is available in their immediate surroundings. Thus, many epixylic liverworts are confined to closed, moist and shaded forests, and seem to prefer logs in an advanced stage of decay with a high water content. The epixylic lifestyle is widespread among the liverworts and most families include at least some species living on dead wood. In some widespread genera, such as *Cephalozia*, *Lophozia* and *Scapania*, there are many species which are very common on logs.

The succession of bryophytes on logs has been relatively well studied (Muhle and LeBlanc, 1975; Söderström, 1988a; Kushnevskaya et al., 2007). Four different ecological groups of bryophytes growing on logs can be distinguished (Söderström, 1988b). The first group consists of facultative epiphytes that occur on the trunks of living trees, and after tree-fall they continue to grow on logs from early to mid-stages of decay. *Ptilidium pulcherrimum* is a typical example. Many lichen species also commonly occur as facultative epiphytes. The second group is composed of epixylic specialists that occur mainly on logs and stumps. Early epixylic species colonize fallen trees while they still have bark and grow on the bark of logs. Typical examples are *Anastrophyllum hellerianum* and several *Lophozia* species. As decay continues, and the logs lose their bark and begin to soften, late epixylic species become an important component of the community. Here we find species such as *Blepharostoma trichophyllum*, *Lepidozia reptans* and *Cephalozia* species. The third group consists of opportunistic generalists that occur on many kinds of substrate and can colonize logs at different stages of decay. Finally, there is a group of competitive ground flora species (epigeic species) that normally cover the forest floor, which gradually overgrow decomposed logs on the ground, thus displacing the epixylic species.

In a study of the succession on spruce in boreal old-growth forests, Kushnevskaya et al. (2007) measured the cover of the different groups of species. Their results confirmed the general patterns described by Söderström (1988a), but clearly showed that all the species groups are present during the full range of log decay, although their relative abundance varies (Figure 4.4).

A critical factor that influences the occurrence of epixylic liverworts is the size of the logs. When small trees fall, the ground flora species

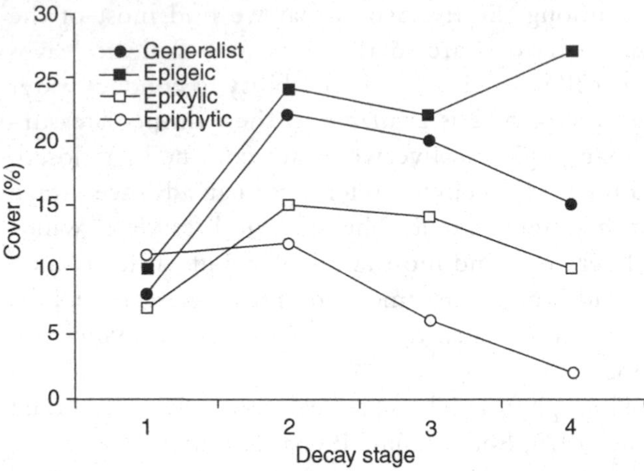

Figure 4.4. Percentage cover of the four main bryophyte groups during the decay succession of spruce (*Picea abies*) logs in old-growth forest in northwestern Russia (after Kushnevskaya et al., 2007). Note that all the four species groups are present throughout the full decay succession and that the epixylic species only constitute a relatively small proportion of the species growing on logs.

may rapidly overgrow the whole log and do not allow the establishment of the competitively inferior liverworts. Also, towards the end of the decay process, when the log is highly decayed and has a high water-holding capacity, the competition with ground flora species is a challenge for the true epixylics. They then often become confined to the vertical surfaces on the sides of logs. The ground-living species generally colonize the upper surfaces first, and only spread to the sides after the log has completely collapsed. The small 'windows' of wood surface without ground species on the sides of large old logs are often 'hotspots' for epixylic liverworts, and many species may co-occur at a very small spatial scale (Andersson and Hytteborn, 1991).

The relatively competition-free habitat that a log constitutes is potentially a key explanation for the occurrence of many liverworts on dead trees. Several of these species also occur on other exposed substrates such as boulders and disturbed ground. For many species that frequently occur on logs, it seems that a moist and competition-free substrate is the most important factor in their choice of habitat. In certain situations, as for instance on wet sandstone, species otherwise considered to be truly epixylic may also occur abundantly (Bengt Gunnar Jonsson, personal observation).

In boreal forests, and also in temperate zones, the flora of bryophytes growing on snags is very limited. The exposed position higher up on decorticated stems is usually too dry for the moisture-dependent bryophytes. The situation is different in tropical regions, where many epiphytic bryophytes may be abundant on dead standing trees (Pócs, 1982).

4.3.2 Epixylic lichens

Epixylic lichens, like all other lichens, consist of a fungus (mostly ascomycetes) and symbiotic algae or bacteria living within the fungal tissue. In contrast to the bryophytes, most epixylic lichens occur on standing dead trees and often on decorticated stems. Spribille et al. (2008) made a comprehensive review of epixylic lichens and compiled information from Fennoscandia and the Pacific Northwest of North America. They analysed the habitat affinities of 1271 epiphytic species (in its widest sense) and found that, of these, more than 40% occur on dead wood and about 10% are obligate epixylics. The obligate epixylic species are dominated by a large number of microfungus genera, while only one genus of macrolichens seems to include true epixylic species, namely *Cladonia*. The calicioid lichens stand out as the most specific of the epixylic lichens, and about a quarter of all species on dead trees are calicioid lichens (Figure 4.5).

For a few species, it has been suggested that the fungal component of the lichen may actually be saprotrophic, getting its energy from the decomposing wood. This is likely to be true, at least for species in the order Mycocaliciales (Tibell and Wedin, 2000). A partly saprotrophic lifestyle has been suggested also for the genera *Chaenotheca* (Tibell, 1997) and *Stictis* (Wedin et al., 2004). In their review, Spribille et al. (2008) compared the affinities of species occurring both in Fennoscandia and the Pacific Northwest. Interestingly, for several species there seems to be a niche shift, where some are obligate epixylic on one continent but not on the other. The reason for this pattern is not fully understood, but might relate to the extreme length of time that the continents have been separated, allowing evolutionary changes in habitat selection to take place.

The lichen flora of standing trees changes from when the tree is alive, through tree death and loss of bark, until finally the tree falls. All these stages have a partly different lichen flora. There are also distinct differences between tree species during some of the stages.

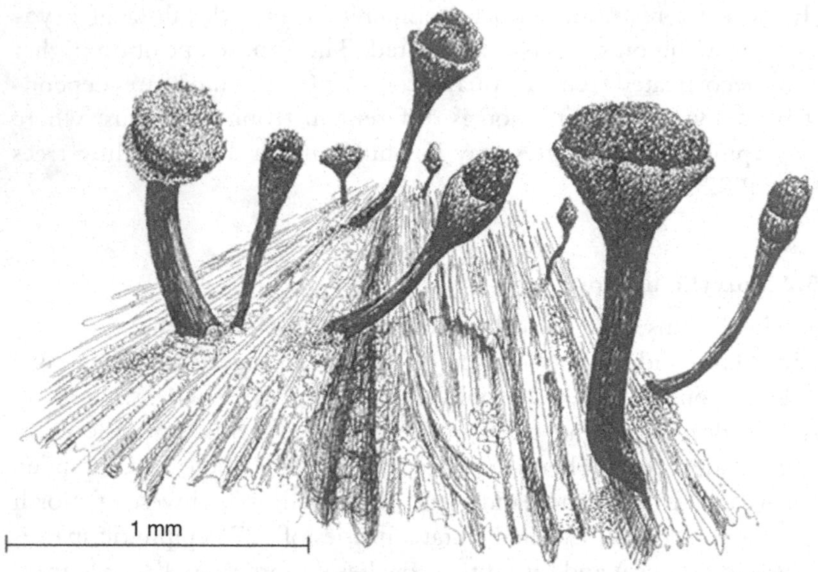

Figure 4.5. Calicioid lichens, sometimes called 'pin lichens', occur commonly on dead snags and constitute one of the more conspicuous groups of epixylic lichens. They typically bear their spore-producing organs (apothecia) on a short stalk (the 'pin'), while the lichen thallus spreads like a crust over the surface of the wood. The drawing shows the species *Chaenotheca ferruginea*, which is very widespread and is common on both bark and decorticated snags. It may also grow on burned wood and sometimes on old wooden buildings (drawing by Alexander Mikulin).

In a detailed analysis of epiphytic and epixylic lichens in Estonia, Lõhmus and Lõhmus (2001) compared four tree species (spruce, pine, birch and alder) and three stages of decay succession (living tree, snags with bark, and decorticated snags). Their study showed that the loss of bark represents a major shift in species composition, and many species are associated either with snags with bark or with decorticated snags. They also showed that the role of tree species was very pronounced for snags with bark, but this effect was almost absent for decorticated snags. Similarly, in a study focusing on calicioid lichens, Kruys and Jonsson (1997) showed that the stage in the decay succession is of central importance for species composition. However, they also showed that tree diameter is another crucial factor for the epixylic lichen flora.

Among the more conspicuous epixylic lichens are the so-called wolf lichen (*Letharia*) species. These bright yellow, fruticose species grow on decorticated conifer snags in open environments and are appreciated for their beautiful appearance. However, they also contain a hidden danger in the form of the poisonous substance, vulpinic acid. There is anecdotal evidence that the lichen used to be ground up, mixed with small pieces of glass, and placed inside a suitable carcass in order to kill the wolves that fed on it.

5 · *Host-tree associations*

Jogeir N. Stokland

Various textbooks have described the diversity of trees (Oldfield et al., 1998; Grandtner, 2005; Tudge, 2005) and another series of books have treated the internal anatomy, physiological functioning and defence mechanisms of trees (Blanchette and Biggs, 1992; Butin, 1995; Wagner et al., 2002; Schweingruber et al., 2006). In this chapter we describe trees from a different angle – how different tree properties have strong implications for the species composition of saproxylic species after the tree has died.

Like other topics in ecology, the host-tree associations of saproxylics must be understood in an evolutionary context. In Chapter 10 we examine the evolution of woody plants with an emphasis on structural innovations. Here we simply mention that the origin of coniferous trees dates from about 310 million years back in time, while different broadleaved trees first evolved 100–120 million years ago (mya). Thus, coniferous trees and broadleaved trees represent distinct plant groups which differ in many ways.

5.1 Conifers versus broadleaved trees

There is a striking lack of scientific review publications that provide quantitative information about host-tree associations among wood-inhabiting organisms. In a recent book on the ecology of wood-decaying basidiomycetes (Boddy et al., 2008), the topic of host-tree associations was only superficially treated. Only one chapter touched on this subject and quantified the proportion of fungi in Denmark that were specific, strongly selective, or weakly selective for different broadleaved tree species (Boddy and Heilmann-Clausen, 2008). The corresponding chapter on fungal communities in boreal, conifer-dominated forests did not mention host-tree associations at all. Similarly, a quite recent French book on forest insects (Dajoz, 2000), with a broad treatment of saproxylic insects, did not deal with host-tree association patterns explicitly. Dajoz was, of course, aware of such associations, since his chapter on community development during the decomposition process was subdivided into sections treating different tree species individually. Such treatments of community composition in wood from separate tree species are quite common (see Chapter 6). But it is only when information is brought together from many sources that we can get a broader overview of host-tree ranges and the specific preferences of wood-inhabiting species.

Even though there is an absence of scientific reviews on host-tree associations among saproxylic organisms, this does not mean that such knowledge is lacking. On the contrary, people who are interested in species diversity of wood-inhabiting fungi or insects rapidly learn that the species composition in wood from coniferous and broadleaved trees differs considerably. This knowledge has immediate practical applications. If you want to establish an extensive list of species from an area or collect a wide diversity of species for your collection, it is crucial to spread the sampling effort on both broadleaved and coniferous trees. In this way, species-oriented collectors have contributed much knowledge about host-tree associations among fungi and insects. Collectors typically make notes and publish their findings in native-language reports and journals of entomological, botanical or mycological societies. This has the cumulative effect that the information is widely scattered and difficult to summarize.

5.1.1 Host association patterns in northern Europe

Both the wide range of information sources that need to be consulted for uncommon and rare species and language difficulties probably explain the scarcity of comprehensive reviews of the host-tree association patterns of saproxylic organisms. Nevertheless, an undertaking to compile such information for a relatively large geographical area – the Scandinavian countries: Norway, Sweden, Finland and Denmark – has been initiated. A number of experts on various organism groups have collaborated to organize such information in a database and make the information available for various uses (Stokland and Meyke, 2008). In the following sections we present some analyses of this information that have previously been published only in rather obscure places (Dahlberg and Stokland, 2004; Stokland et al., 2004).

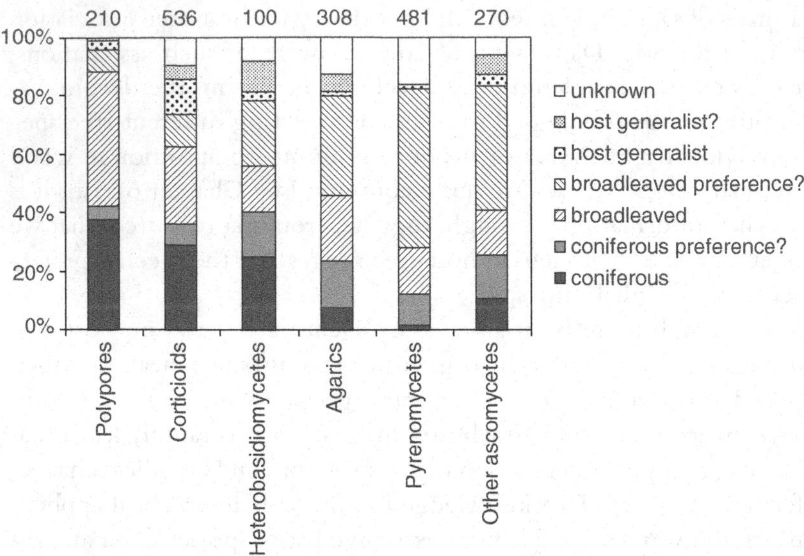

Figure 5.1. Proportion of species with different host-tree associations among different fungus groups in the Nordic countries (Norway, Sweden, Finland and Denmark). The number above the columns represents the number of species in each group. A question mark (?) following a host preference group indicates a weak empirical basis, with few recorded observations to document the host preference.

5.1.2 Wood-inhabiting fungi

In the Scandinavian database described above, Dahlberg and Stokland (2004) analysed 2021 fungal species, of which 1270 were basidiomycetes and 751 were ascomycetes. Figure 5.1 shows most of these, except for a few small groups of basidiomycetes. Figure 5.1 indicates that only 5% of all these species appear to be true generalists, occurring frequently on both coniferous and broadleaved trees. In addition, 4% of the species are potential generalists, but the empirical basis for classifying these species as generalists is too weak to be certain. The remaining 81% of the species are strictly confined to, or prefer, either coniferous or broadleaved trees. However, there are some interesting differences between fungal groups.

The polypores (210 species, see Figure 5.1) are well documented, and the host-tree associations are known for all species, at least when the question is restricted to a preference for either coniferous or broadleaved trees. Only 4% of these species are true or potential generalists. The remaining species are divided into species that are restricted to or strongly prefer either broadleaved trees (54%) or coniferous trees

(42%). These percentages should be considered as indicative rather than strictly accurate. In a similar classification of largely the same species, Junninen and Komonen (2011) found 45% specialists on broadleaved trees and 40% specialists on conifers when they used slightly different criteria for defining host-tree specialists.

The corticioids (536 species, see Figure 5.1) are also quite well known. In this group, the proportion of host-tree generalists is somewhat higher (16%) compared with the polypores. This may be related to the fact that some corticioids are mycorrhizal species that primarily use dead wood for positioning their fruiting bodies rather than as a nutrient source. Most of the corticioids are either associated with conifers (36%) or broadleaved trees (38%), and these species are mainly wood decomposers.

The wood-inhabiting agarics (308 species, see Figure 5.1) show a different pattern of host associations compared with those of the polypores and corticioids. In this group, the proportion of species with a preference for broadleaved trees is much higher (57%) compared with those that prefer coniferous wood (23%).

Among the ascomycetes, there is an even stronger skewness towards species that predominantly occur on broadleaved trees. This is especially the case for pyrenomycetes (481 species, see Figure 5.1), where 70% of the species seem to be associated with broadleaved trees and 12% with conifers. Just six species (1%) have been classified as potential generalists, but not a single one as a true generalist. These proportions of different host associations among the pyrenomycetes should be treated with caution, as these fungi are weakly documented and the species are difficult to identify. On the other hand, the numbers of species associated with coniferous and broadleaved trees are so different that they probably reflect a real pattern. The remaining ascomycetes, the so-called operculate ascomycetes (270 species, see Figure 5.1), are also strongly confined to broadleaved trees. Nearly 60% of them seem to be confined to the wood of broadleaved trees, whereas 25% seem to be associated with conifers. In this group there are some true generalists (perhaps up to 10%). But again it is important to treat these proportions with caution, due to a relatively weak knowledge base.

5.1.3 Wood-inhabiting invertebrates

Beetles are the best-known group of invertebrates that live in decaying wood. The majority of the species feed on woody material of different

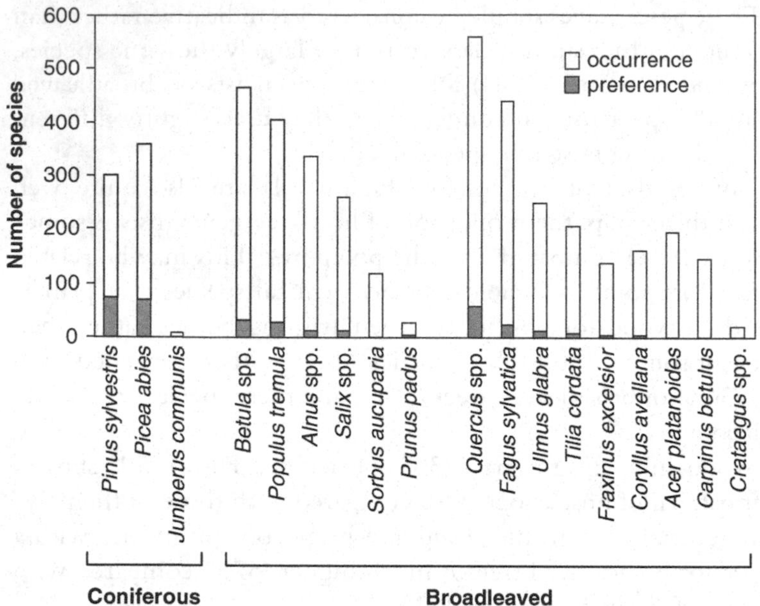

Figure 5.2. Number of beetle species that are associated with different host-tree species in the Nordic countries: Norway, Sweden, Finland and Denmark. Redrawn from Stokland et al. (2004).

kinds, such as inner bark, minimally decayed wood, or wood that is substantially decomposed by fungi. In addition there are species that feed on wood-decaying fungi, while still other species predate on the insects found in dead wood. Such feeding habits are described in detail in Chapter 3. Here we shall treat all these species collectively and explore their associations with different host trees.

Like the wood-decaying fungi, the beetles also show a sharp division between coniferous-associated species and broadleaved-associated species. In an analysis of 1257 saproxylic species, Dahlberg and Stokland (2004) found that 329 species (26%) showed a clear preference for individual tree species (Figure 5.2). But when grouping the trees into coniferous and broadleaved trees, at least 75% of the beetle species had a clear preference for one of these tree groups. At least 23% of the species preferred coniferous hosts and at least 52% preferred broadleaved hosts. Only 11% of all species were host-tree generalists occurring frequently on both types of trees. The remaining 14% were poorly known species, but it is likely that most of them are selective for either conifers or broadleaved trees. Thus, 75–90% of all saproxylic beetles in northern Europe separate trees into two different realms: coniferous trees and broadleaved trees. A similar split has also been found among saproxylic beetles in China (Wu et al., 2008) and it is most likely a universal pattern.

Under this broad categorization of host-tree associations, there are of course additional patterns, as the 329 species with more restricted host-tree preferences indicate. We will return to such refined host association patterns later in this chapter.

The difference of species assemblages in wood from coniferous and broadleaved trees is probably present in other invertebrate groups, but there are hardly any comprehensive overviews that quantify these patterns.

5.2 Diversity and phylogeny of trees

The estimated number of tree species on a global scale ranges from 60 000 to 100 000, of which the large majority are broadleaved trees (Table 5.1). In the following sections we present an overview of trees and their phylogenetic relationships. These relationships indicate which trees are closely or distantly related, and facilitate the interpretation of host-tree associations among saproxylic species.

5.2.1 Tree ferns

Tree ferns are remnants from the Carboniferous period (354–290 mya), when this plant group was very diverse. They have the fronds (leaves) on the top of a trunk and can grow up to 20 m high. But unlike coniferous and broadleaved trees, the tree ferns do not form new woody tissue in their trunk as they grow older. Instead the trunk is supported by a fibrous mass of roots that expands as the plant grows. Taxonomically, the tree ferns belong to the families Dicksoniaceae and Cyatheaceae in the order Cyatheales. They comprise about 500 extant species, mainly in humid forests in the tropics and in the southern hemisphere.

5.2.2 Ancient tree lineages

Cycads are another ancient group of trunk-forming plants from the Carboniferous period. They are often mistaken for ferns or palms but they are unrelated to both groups and belong in the division Cycadophyta, which now comprises about 300 species. They are at

Table 5.1. *Key plant orders where we find many or distinct tree species. For several orders, key families are listed (with examples of tree genera in parenthesis, + indicates that there are more genera in the family). The number of species is in the order or family at large, not necessarily the number of tree species. The table is not a complete overview of tree species. Growth forms are as indicated: trees – typical single-stem tree; woody – bush or liana; herbs – herbs and grasses.*

Taxonomic group	Number of species	Growth forms	Global distribution
Tree ferns	500	trees, woody	Tropical, southern hemisphere
Gymnosperms			
Cycadales (Cycads)	300	trees, woody	Tropical
Ginkgoales (*Ginkgo biloba*)	1	trees	China
Coniferales	630		
• Pinaceae (*Pinus*, *Picea*, *Abies*, *Larix*, *Tsuga*, +)	225	trees, woody	Nothern hemisphere
• Araucariaceae (*Araucaria*, *Agathis*, *Wollemia*)	41	trees	Southern hemisphere
• Cephalotaxaceae (*Cephalotaxus*)	11	trees	East Asia
• Podocarpaceae (*Podocarpus*, *Dacrycarpus*, *Prumnopitys*, +)	185	trees, woody	Mainly southern hemisphere
• Phyllocladaceae (*Phyllocladus*)	4	trees	SE Asia, Tasmania
• Sciadopityaceae (*Sciadopitys*)	1	trees	Japan
• Cupressaceae (*Cupressus*, *Taxodium*, *Sequoia*, *Juniperus*, +)	133	trees, woody	Worldwide distribution
• Taxaceae (*Taxus*, *Toreya*, +)	25	trees, woody	Mainly northern hemisphere
Angiosperms (flowering plants)			
Magnoliales	2 800	trees, woody	Mainly tropical, some temperate
Laurales	3 400	trees, woody	Mainly tropical and subtropical
Canellales	135	trees, woody	Southern hemisphere

Piperales	2 000	trees, woody, herbs	Tropical
Arecales (palms)	2 600	trees, woody	Tropical to warm-temperate
Poales			
• Bambusoidea (bamboo)	1 000	trees, woody	Nearly worldwide, not Europe
Proteales			
• Platanaceae (*Platanus*)	10	trees	Temperate, northern hemisphere
Trochodendrales	1	trees	Japan and Taiwan
Malpighiales			
• Euphorbiaceae (includes *Hevea*, rubber tree)	7 500	trees, woody, herbs	Trees mainly in the tropics
• Rhizophoraceae (red mangrove trees *Rhizophora*, +)	140	trees, woody	Tropical
• Salicaceae (*Salix*, *Populus*, +)	400	trees, woody	Northern hemisphere
Fabales			
Fabaceae (*Acacia*, *Dalbergia*, *Robinia*, +)	19 000	trees, woody, herbs	Worldwide distribution
Rosales			
• Rosaceae (*Malus*, *Prunus*, *Pyrus*, *Sorbus*, *Crataegus*, +)	3 000	trees, woody	Worldwide distribution
• Rhamnaceae	850	trees, woody	Mainly tropical, some temperate
• Moraceae (*Morus*, *Ficus*, +)	1 500	trees, woody, herbs	Mainly tropical, some temperate
• Ulmaceae (*Ulmus*, *Zelkova*, +)	40	trees	Temperate, northern hemisphere
Fagales			
• Fagaceae (*Fagus*, *Quercus*, *Castanea*, +)	900	trees, woody	Northern hemisphere
• Nothofagaceae (*Nothofagus*)	35	trees, woody	Southern hemisphere
• Betulaceae (*Betula*, *Alnus*, *Carpinus*, *Corylus*, +)	130	trees, woody	Temperate northern hemisphere
• Juglandaceae (*Juglands*, *Carya*)	50	trees	Tropical to temperate

Table 5.1. (*cont.*)

Taxonomic group	Number of species	Growth forms	Global distribution
Myrtales			
• Combretaceae (some mangrove tree genera)	600	trees, woody	Tropical, subtropical
• Myrtaceae (*Eucalyptus*, *Eugenia*, *Myrcia*, +)	5 600	trees, woody	Tropical to warm-temperate
Malvales			
• Malvaceae (*Hibiscus*, *Tilia*, *Bombax*, baobab, cocoa trees, +)	2 300	trees, woody, herbs	Tropical to temperate
• Dipterocarpaceae (*Dipterocarpus*, *Dryobalanops*, *Shorea*, +)	680	trees, woody	Mainly SE Asia, Africa,
Sapindales			
• Rutaceae (*Citrus*, +)	900	trees, woody	Tropical to warm-temperate
• Meliaceae (*Flindersia*, *Switenia*, *Azadirachta*, +)	550	trees, woody	Tropical to warm-temperate
• Anacardiaceae (*Anacardium*, *Pistacea*, *Rhus*, *Mangifera*, +)	600	trees, woody	Tropical to temperate
• Sapindaceae (*Acer*, *Aesculus*, *Sapindus*, *Pometia*, +)	2 100	trees, woody	Tropical to temperate
Cornales			
• Cornaceae (*Cornus*, *Davidia*, *Nyssa*, +)	110	trees, woody,	Mainly north temperate
Ericales			
• Ebenaceae (*Diospyrus*, +)	500	trees, woody	Nearly worldwide
• Sapotaceae (*Pouteria*, *Palaquium*, *Sideroxylon*, +)	1 100	trees, woody	Mainly tropical
• Theaceae (*Franklinia*, *Camellia*, +)	300	trees, woody	Tropical to warm-temperate

• Ericaceae (*Rhododendron*, *Arbutus*, *Calluna*, *Erica*, +)	2 700	trees, woody, herbs	Worldwide distribution
• Lecythidaceae (*Bertholetia* – Brazil nut, *Lecythis*, *Careya*, +)	400	trees, woody	Pantropical
Solanales			
• Boraginaceae (*Cordia*, +)	2 650	trees, woody, herbs	Worldwide distribution
Gentianales			
• Rubiaceae (*Coffea*, *Mitragyna*, *Nauclea*, +)	9 000	trees, woody, herbs	Mainly tropical and subtropical
• Apocynaceae (*Plumeria*, *Dyera*, +)	3 700	trees, woody, herbs	Mainly tropical and subtropical
• Gentianaceae (*Fragraea*)	970	trees, mainly herbs	Worldwide distribution
Lamiales			
• Myoporaceae (*Eremophila*, *Myoporum*, +)	150	trees, woody, herbs	Mainly Australia, South Pacific
• Oleaceae (*Fraxinus*, *Olea*, +)	600	trees, woody	Tropical to temperate
• Bignoniaceae (*Jacaranda*, *Paracetoma*, +)	800	trees, woody	Mainly South America
Aquifoliales			
• Aquifoliaceae (*Ilex*)	400	trees, woody	Mainly tropical mountains
Asterales (*Brachylaena*)	30 000	few trees, mainly herbs	Worldwide distribution
Dipsacales (*Sambucus*, *Viburnum*, *Caprifolium*, *Lonicera*, +)	1 100	trees, woody, herbs	Worldwide distribution

Source: Tudge (2005).

their most diverse in the tropics, especially in drier areas with relatively cool winters. Some cycads form secondary wood through multiple concentric rings of cambium tissue, whereas other cycads do not have secondary growth (Norstog and Nicholls, 1997; Schweingruber et al., 2006). The rings of secondary wood in the cycads do not correspond to the annual wood rings of conifer and broadleaved trees. Furthermore, the cycad stem has a relatively small proportion of secondary wood and a significant amount of central pith and outer cortex tissue. Thus, their anatomy is very different from that of other tree lineages.

The ginkgo tree, *Ginkgo biloba*, is a single species that represents a separate lineage of trees – Ginkgophyta. It is very special because it is the only gymnosperm tree that is broadleaved and deciduous (i.e. leaf-shedding). The species occurs naturally or is preserved in cultivation by Buddist monks in China. However, it is now commercially available worldwide as an ornamental tree. Ginkgo trees have the same trunk anatomy as conifers, i.e. with a layer of cambium tissue that produces wood tissue inwards and bark tissue outwards.

5.2.3 Conifers

The conifers represent yet another group of ancient trees that originated in the late Carboniferous, about 310 mya. The conifers are commonly considered equivalent to the gymnosperms, especially in areas with a temperate climate, where other gymnosperms – such as cycads and gingko – are typically absent.

Altogether, there are about 630 species of conifers (Table 5.1). Their most extensive distribution is in the northern hemisphere, especially in the boreal zone. But two families (Araucariaceae and Podcarpaceae) mainly occur on the southern hemisphere. All conifers are woody plants and most of them have a growth form with a single trunk. Most conifers are evergreen, but some exceptions exist, such as the larches (*Larix* spp.), which shed their foliage during winter. Many conifers are characterized by needle-shaped foliage, but some – such as most cypresses and some podocarps – have flat, triangular, scale-like leaves. Not all conifers have the typical woody cones composed of several scales protecting individual seeds. In the podocarps, the yews and one cypress genus (*Juniperus*), the scales are soft, fleshy and brightly coloured, and are eaten by birds. These morphological differences reflect the fact that conifers are composed of several lineages (see Figure 5.3). Later in this chapter we return to this topic, as these lineages also reflect different defence systems.

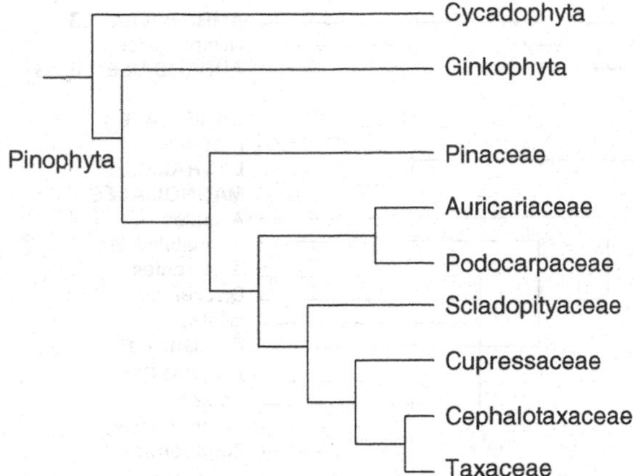

Figure 5.3. Phylogeny of the gymnosperms (cycads, ginkgo and conifers, here illustrated without the Gnetales lineage). Derived from Farjon (2003) and Quinn and Price (2003).

5.2.4 Broadleaved trees

Broadleaved trees belong to a large group called flowering plants (angiosperms). Flowering plants, as we know them today, first appeared about 130 mya (Magallón and Sanderson, 2005, see also Figure 10.1). They all have the same evolutionary origin and there is a fundamental phylogenetic split between the flowering plants and the gymnosperms, among which all the conifers belong (Crane et al., 2004; Magallón and Sanderson, 2005; Doyle, 2008). It is still quite mysterious from which ancient plant lineage the flowering plants originated (Taylor and Taylor, 2009). However, for the purpose of understanding host-tree associations among saproxylic organisms, it is sufficient to know that conifers and the ancestors of today's broadleaved trees have evolved separately for more than 300 million years.

Even though the flowering plants make up one phylogenetic lineage, the trees do not form a monophyletic group within this lineage. On the contrary, the tree growth form appears independently in widely separate orders and families (Figure 5.4, Table 5.1). Quite interestingly, the genetic basis for developing the cambium function and the tree growth form is not restricted to woody plants (Groover, 2005). Thus, woody growth has been gained and lost several times in plant evolution and it is readily enhanced or minimized in speciation among the flowering plants (Petit and Hampe, 2006). This has the consequence that there is no biological basis for defining trees as sharply distinct

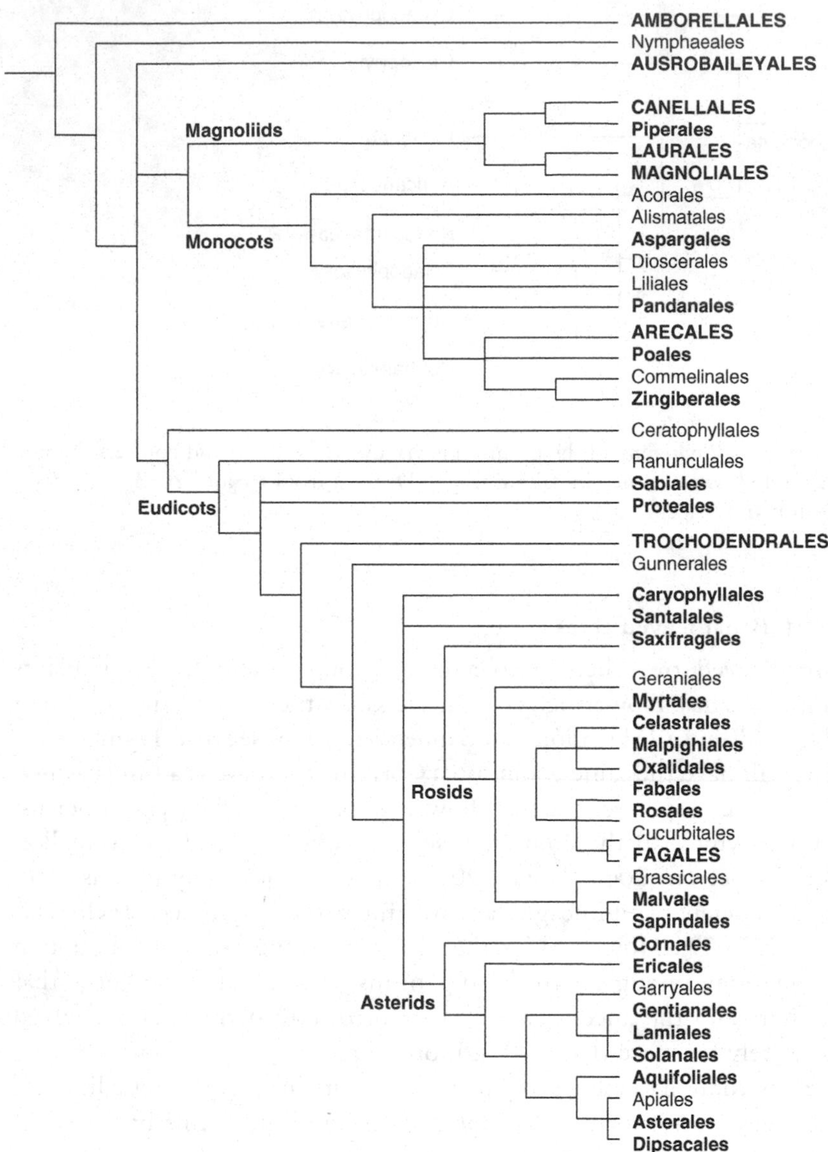

Figure 5.4. Phylogenetic tree of flowering plants (angiosperms). Orders that predominantly consist of woody plants are written with capital letters, orders with both woody and herbaceous plants are written in boldface, and orders essentially without woody plants are written in neutral typeface (derived from Judd et al., 2002; Tudge, 2005).

from other plants. In nature we find bush-like trees and vice versa. Some species even grow as trees or bushes under different environmental conditions.

The total number of broadleaved tree species is much larger and far more difficult to enumerate than that of the conifers. This is related to the practical difficulty of defining species as trees rather than other woody plants; the insufficient amount of taxonomic work carried out in the tropics, where tree diversity is highest; and the difficulty in sorting out species, subspecies and hybrids in several families. By using the definition of tree as 'a woody plant growing on a single stem usually to a height of over two metres', the Temperate Broadleaved Tree Specialist Group estimated that there are 21 000 tree species in plant genera that are predominantly woody and temperate in distribution (Hunt, 1996). On a global scale, estimates of tree species richness range from 60 000 (Grandtner, 2005) to 100 000 (Oldfield et al., 1998).

The total number of flowering species is roughly 300 000, which means that 20–30% of them are trees. The flowering plants are grouped into 48 orders (Judd et al., 2002), and one needs to have a great deal of botanical knowledge in order to get an overview of them, not to mention the more than 400 different families they comprise. There is, however, a very good book that focuses especially on the occurrence of trees across the whole plant kingdom (*The Secret Life of Trees* by Tudge, 2005). This book provides a broad and detailed account of trees in all the major plant lineages, often down to the family, genus and species level.

Most broadleaved trees have the same trunk anatomy and growth pattern as the conifers, with a single cambium zone and annual production of new wood causing a gradual girth increment as the tree grows older. But in one group of angiosperms we find trees with completely different growth forms – in the monocotyledons (usually shortened to 'monocots'). This is a monophyletic lineage of angiosperms where we find plants such as grasses, lilies and orchids, to mention just a few diverse families. Woody plants such as palms and bamboos also belong to the monocots. These plants do not have secondary thickening of the trunk, but instead grow primarily from the base (bamboos) or from the top (palms) while the stem remains uniform in thickness as the plant grows bigger. The lack of secondary stem thickening results in a very different wood structure in palms and bamboos compared with other trees.

5.3 Differences between the wood of conifers and broadleaved trees

In Chapter 2, we described in detail the differences between wood from conifers and broadleaved trees. Here we return to this topic, but now emphasize the consequences for host-tree associations among saproxylic organisms.

5.3.1 Lignin

The lignin in the wood of conifer and broadleaved trees differs both in quantity and qualitative composition. Conifers typically have a higher lignin content (25–33% of dry weight) than broadleaved trees (20–25%) (Sjöström and Westermark, 1998). Furthermore, conifer lignin is mainly composed of coniferyl subunits, whereas lignin in broadleaved trees has other subunit types. These differences make conifer lignin more resistant to microbial degradation than lignin from broadleaved trees.

The differences in lignin composition are probably important for host-tree preferences among brown-rot and white-rot fungi. The majority of brown-rot fungi are typically confined to coniferous trees (Ryvarden and Gilbertson, 1993) and brown-rot fungi are the principal wood decomposers of coniferous wood in boreal forests (Renvall, 1995). In a phylogenetic analysis of rot types, Hibbett and Donoghue (2001) showed that brown-rot fungi have at least six independent origins from separate groups of white-rot fungi. The analyses of Hibbett and Donoghue also suggested that conifer exclusivity has evolved five or six times, typically – but not necessarily – in combination with a switch to brown rot. This probably indicates that there has been a selective premium on decaying the cellulose fraction only from conifer wood.

If we turn to industrial studies, there are some interesting findings demonstrating the effect of lignin type on wood degradability. In a study designed to discover the optimal conditions for fungal degradation of lignin in pulp mass, Yang et al. (1980) used the white-rot fungus *Phanerochaete chrysosporium* and recorded the effect on various tree species while manipulating the amount of extra nitrogen added to the pulp mass. They already knew the temperature optimum for this fungus, and accordingly kept it at 39–40°C. After adding a small amount (0.12%) of nitrogen to alder (*Alnus*) pulp, lignin degradation increased from 5.2% to 29.8%. By adding the same amount of nitrogen to the pulp mass of hemlock (*Tsuga*), the lignin degradation only increased from 2.2% to 3.9%. In other words, this fungus was much more efficient in degrading lignin from broadleaved trees than from conifers. This enzymatic ability corresponds very well to the natural host-tree association of *P. chrysosporium*, which is strongly associated with broadleaved host trees (Karl-Henrik Larsson, personal communication).

5.3.2 Hemicellulose and cellulose

There are some differences in hemicellulose between conifers and broadleaved trees (see Chapter 2) but hardly any studies have suggested that these differences affect host-tree preferences among wood-inhabiting organisms. Cellulose is the common building material of cell walls in all plant groups. Thus, there is no difference in cellulose between conifers and broadleaved trees and no reason to assume that host specificity among saproxylics is due to cellulose.

5.4 Defence systems in trees

Longevity and large size are key features of trees. Thus, trees represent a massive potential resource for other organisms, and they are constantly challenged by different fungi and invertebrates. Over millions of years, the trees have developed various protection mechanisms that together form an impressive defence system. The ultimate function of these mechanisms is to prolong life and increase the reproductive output of the tree, but the defences also have implications for wood properties for many years after the tree has died.

Many investigations have documented the ability of trees to repair wounds that are superficial or that reach the vascular cambium and expose the naked wood (Biggs, 1992a, 1992b; Woodward, 1992). When the wound or damage extends deeper into the wood, the tree responds in different ways to restrict the damage and fungal spread (Pearce, 1996; Yamada, 2001). In Chapter 7, we deal with the topic of wounds in more detail. Here we just make the point that conifers, because of their resin defences, are able to seal wounds within a few hours and have very few wound-associated species living on them, whereas wounds in broadleaved trees can remain open for months or longer and host a rich diversity of saproxylic species. Thus, wound-repairing ability vary between different tree lineages and contribute to host-tree associations among saproxylic organisms.

Some saproxylic species can degrade physical or chemical tree defences, become immune, or otherwise break through the defence barriers. In the following sections we describe the most important tree defences against fungi and invertebrates, and give examples of species that have managed to overcome these defences.

5.4.1 Bark

In a living tree, the bark provides the first line of defence against intruding insects or fungi. It is common to use the term 'bark' loosely to refer to different tissues outside the vascular cambium. However, treating the bark as a single entity obscures important aspects of bark physiology, biochemistry and functionality. Thus, we describe the outer bark and inner bark separately. The outer bark is produced by the cork cambium (phellogen) and is mostly composed of dead cells. The inner bark is produced by the vascular cambium (normally referred to simply as 'cambium') and is located between the vascular cambium and the outer bark. It is composed of living tissue that transports photosynthetic products downwards from the canopy. But, as we shall see, the inner bark also has important defence functions.

Many publications have described the anatomy of bark (Borger, 1973; Biggs, 1992a) and bark defence components (Jensen et al., 1963; Biggs, 1992b; Woodward, 1992; Franceschi et al., 2005). In the following sections we draw upon these sources and provide a summary of the most important defence mechanisms.

5.4.2 Outer bark

The outer bark consists of cells with strongly lignified cell walls and high levels of suberin, which is a waxy substance that prevents water from penetrating into the tissue. Suberin is also highly resistant to fungal degradation. The outer bark frequently has layers of cells covered with calcium oxalate crystals that provide resistance to wood-boring insects. In addition, the cells typically contain large amounts of phenolic compounds and terpenes, which prevent fungal development. The combination of mechanical and chemical properties makes the outer bark highly resistant to attack from both insects and fungi. One can easily verify this by inspecting a fallen tree that has been dead for many years. Even if much of the wood is heavily decomposed, one can still find almost intact pieces of outer bark. Despite, or perhaps because of, the effective protection provided by the outer bark, there have been few detailed studies of the defence mechanisms used by the outer bark.

5.4.3 Inner bark

The inner bark also has strong defence systems. This is quite logical, since the concentration of nutrients is very high here. Thus, the inner bark is a true battlefield where invasive insects and fungi try to get at these resources and the tree fights back to protect itself.

The production of phenolic compounds is a basic defence component which is widely distributed across both conifer and broadleaved trees, and they can be found in nearly all types of plant tissue. These secondary metabolites are toxic to fungi, and different tree species produce their own types of phenolics. The chemistry of phenolics produced in conifer bark has not been extensively studied, but there is some evidence that the degree of resistance to invasive fungi is linked to the type of phenolics produced (Brignolas et al., 1995; Bonello and Blodgett, 2003). These compounds are produced and stored in special polyphenolic parenchyma (or PP) cells. The bark of all conifers has a great number of PP cells, which are produced as annual rings. The PP cells stay alive for many years; Franceschi et al. (2005) reported the presence of living PP cells that were more than 70 years old in a 100-year-old Norway spruce. Therefore, an insect boring into the inner bark will continuously destroy PP cells, and these will release the stored phenolics.

In addition to this standing defence, the PP cells increase their activity when the bark is wounded or invaded. They rapidly increase in size and accumulate more phenolics. But even more interesting is their ability to produce different and specific phenolics with greater toxicity to an invasive organism than the regularly produced phenolics (see Franceschi et al., 2005, and references therein). Thus, the PP cells represent a dynamic defence component of conifer bark. Since PP cells are the most abundant living cell type in the inner bark, Franceschi and colleagues argued that they represent the most important defence component in this tissue. It has also been reported that antifungal compounds are present in the bark of broadleaved trees, e.g. in *Populus tremuloides* (Flores and Hubbes, 1980) and *Morus alba* (Takasugi et al., 1979).

The resin defence system is another major protection mechanism in conifer bark. It consists of individual resin cells or small pockets of such cells (Franceschi et al., 2005). The pockets have a lining of cells that produce and secrete resin into the extracellular space, where it accumulates under pressure. When wounds or invading organisms damage the bark, the resin is released; this repels or flushes out the organisms, traps them in this sticky substance, or otherwise kills the invader due to its toxic nature. This defence system is at least 220 million years old, as evidenced by amber fossils from the Triassic period (Schönborn et al., 1999). Amber is crystallized resin that has been formed by physical forces in the environment (exposure to light, heat and pressure). There are countless examples of insects, other invertebrates, and even small vertebrates that have been trapped in resin before crystallization and amber formation.

It is interesting to note that the resin defence differs systematically between conifer families. Species in the Pinaceae family (e.g. *Pinus*, *Picea*, *Larix*, *Abies*) have permanent as well as inducible resin structures in both the inner bark and the sapwood, whereas other conifer families lack resin structures or only have inducible structures in either the sapwood or in the inner bark (Franceschi et al., 2005). The other conifer families have stronger mechanical defences based on parenchyma cells and calcium oxalate crystals, while the resin defence plays a subordinate role.

It seems paradoxical that the strong resin defence system that made the Pinaceae family so successful became a weakness when the bark beetles appeared on the scene around 100 mya. In Europe, the bark beetle *Ips typographus* is able to kill weakened, and even sound, *Picea abies* trees when bark beetle populations are high enough. In the same way, the southern pine beetle and the mountain pine beetle (*Dendroctonus frontalis* and *D. ponderosae*, respectively) are capable of killing several pine species (*Pinus* spp.) in different parts of North America. It is only the Pinaceae that have serious problems with aggressive bark beetles, and it is interesting that this conifer family has the most well-developed resin system. Conventional wisdom has been that resin is important in defending against bark beetle colonization by flushing out or entrapping and killing the insects. In comparison, conifers that produce no constitutive bark or wood resins, such as members of the Cupressaceae, have no aggressive bark beetle pests. Thus, some bark beetles have turned a defence into an opportunity. They have developed aggregation pheromones that gather thousands of bark beetles onto a tree and such mass attacks overcome the tree's defences. In addition they also have an intimate association with specific blue-stain fungi that cause lethal damage to the tree when they are inoculated in large quantities by the bark beetles. An interesting detail is that the aggregation pheromones seem to be derived from the resin molecules. This hypothesis is discussed further by Franceschi et al. (2005).

Structural defence systems include different patterns of sclerenchyma cells and calcium oxalate crystals that provide mechanical resistance, especially to wood-boring insects and browsing vertebrates. Sclerenchyma cells have strong lignified cell walls and can occur as massive, often irregularly shaped, stone cells clustered together, which is typical for many of the Pinaceae. Alternatively, they can occur as distinct concentric rows alternating with the PP cells throughout the inner bark. This arrangement is typically found in the non-Pinaceae conifers and appears to provide a stronger mechanical barrier than the stone cell arrangements (Franceschi et al., 2005). Calcium oxalate crystals occur as intracellular deposits in the Pinaceae and numerous extracellular wall deposits in the non-Pinaceae (Hudgins et al., 2003; Franceschi et al., 2005). Little is known about their formation and function. However, their tough physical nature and great abundance suggest a role against bark-boring insects, whereas their chemical inactivity makes them unlikely to affect fungi.

5.4.4 Sapwood

Healthy sapwood includes both living and non-living cells. The living cells are capable of metabolic activity and produce various antimicrobial substances that are soluble in water and therefore can be extracted from wood.

The amount of such extractives typically varies from 2% to 5% of the dry weight (Sjöström and Westermark, 1998); they comprise triglycerids, fatty acids, resin acids, steryl esters, and phenolic substances (Holmbom, 1998). It is generally believed that the extractives have antimicrobial effects, even though the exact effects of such extractives on fungi and insects are typically not documented in nature. On the other hand, there have been laboratory studies of fungi that document how various extractives slow down or inhibit fungal growth on artificial media (Shain and Hillis, 1971; Shortle et al., 1971; Shaw, 1985; Stenlid and Johansson, 1987; Witzell and Martín, 2008). Furthermore, higher amounts of extractives have typically been observed in reaction zones around wounds than in healthy sapwood (Shain and Hillis, 1971; Pearce, 1991) and also increased amounts around infections caused by wood-decaying fungi (Yamada, 2001). Although the circumstantial evidence presented above suggests that extractives have antimicrobial effects, there is still some uncertainty concerning their defensive role because the extractives are extremely diverse and also have other functions (see Witzell and Martín, 2008, and references therein).

Tannins are large complex molecules of phenolic compounds that occur in a wide range of vascular plants and tissue types (Hillis, 1987; Hernes and Hedges, 2004). Thus, we could equally well have mentioned these substances in other sections of this chapter. It is well documented that tannins are toxic to microorganisms (Scalbert, 1991), and rot resistance is a distinct property of tannin-rich materials. This has long been known by humans, who have selected certain timbers for their durability, e.g. oak and chestnut and some eucalypts – which are all rich in tannins (Scalbert, 1991). The tannins might also cause specific host associations among insects and fungi. Several studies have documented specific guilds of fungi and insects, both in temperate and tropical oak species (Lindblad, 2000; Ehnström and Axelsson, 2002; Heilmann-Clausen et al., 2005). There might even be intimate mutual relationships between fungi and insects such as the longhorn beetle *Anoplodera sexguttata*, which seems to develop specifically in oak sapwood decayed by the fungus *Hymenochaete rubiginosa* (Ehnström and Axelsson, 2002).

Resin also occurs in the sapwood of conifers, and here it operates in various defensive ways. The resin acids form a sticky substance that creates a physical barrier to fungal spread and insect boring. Small-molecule terpenes are the building blocks of resin, but they also occur in a gaseous phase and conifer wood is probably saturated with terpene vapours (Hintikka, 1970). These terpenes are toxic to fungi (Cobb et al., 1968; Shrimpton and Whimey, 1968; Flodin and Fries, 1978; Schuck, 1982) but their inhibitory effects vary substantially with the fungi or substances tested (Hintikka, 1970; Bridges, 1987). It is particularly interesting to note the results of Hintikka, who compared terpene effects on the fungi normally growing in the wood of coniferous and broadleaved trees. He found that most of the conifer-associated species were able to grow in air saturated with terpenes, whereas species from broadleaved trees were highly sensitive even to small amounts (Hintikka, 1970).

5.4.5 Heartwood

Many tree species have a dark-coloured zone of heartwood in the central part of their trunk. In addition to the difference in colour, the heartwood differs from the sapwood in having a higher concentration of extractives, the presence of dead cells, and a lower water content (especially in conifers). Many of the phenolic compounds that are found in sapwood are usually found in much higher concentrations in the heartwood of the respective species (Yamada, 2001). Heartwood results from profound changes in cell metabolism in the sapwood–heartwood transition zone. These metabolic changes involve a phasing up in the activity of several hundred genes (Yang et al., 2004). These changes result in the increased synthesis of many substances such as tannins, resins and several types of phenolic compounds (Hillis, 1987; Magel et al., 1994; Burtin et al., 1998). After this synthesis the cells undergo a form of programmed cell death (Magel, 2000) that finalizes the formation of heartwood.

The large quantities of extractives in the heartwood form a standing defence that is toxic or inhibitory to the great majority of wood-decaying fungi. But some fungi have managed to break through this defence – the so-called heart-rot fungi. These fungi can cope with the toxic environment in the heartwood and are typically selective for various host trees. Several species are specific for individual tree genera, such as *Piptoporus quercinus* and *Fistulina hepatica* on oak (*Quercus*) and *Rigidoporus ulmarius* on elm (*Ulmus*) (Phillips and Burdekin, 1982; Rayner and Boddy, 1988; Wald et al., 2004a). Especially in the polypore genus *Phellinus* we find many host-specific heart-rot fungi, such as *P. pomaceus* on *Prunus*, *P. tremulae* on *Populus*, and *P. pini* on *Pinus* (Niemelä, 2005). But there are other heart-rot fungi that clearly prefer tree species of one genus but occasionally occur on others, e.g. *Grifola frondosa*, *Inonotus dryadeus* and *Stereum gausapatum*, which primarily occur on *Quercus*, and *Phellinus chrysoloma*, which primarily occurs on *Abies* and *Picea* (Boddy, 1992).

The heart-rot fungi are not only able to tolerate an environment that is so hostile to most fungi, they even seem to utilize the extractives as a nutrition source. Thus, *Fistulina hepatica* use *Quercus* tannins as a carbon source (Cartwright, 1937).

5.4.6 Life-history strategies and defence systems

In this overview we have described where different defence mechanisms occur in the tree. But it is also a part of this story that it is resource-demanding to produce and maintain these defences, and trees seem to defend themselves to different degrees. In a very interesting publication on life-history strategies among North American tree species, Loehle (1988) found that species can be grouped according to two basic strategies. One group of trees typically correspond to pioneer species; these are fast-growing, early-maturing and relatively short-lived. Examples of such species are poplars (*Populus* spp.) and birches (*Betula* spp.). The other group consists of species that are slower-growing, later-maturing and long-lived. This group includes several oaks (*Quercus* spp.), hickories (*Carya* spp.) and several conifer genera (especially *Pinus* and *Sequoia*). Loehle attributed these differences to alternative strategies in defence investment. A substantial investment in structural and chemical defences in the early phase of life slows down the initial growth rate but pays off with extended longevity because the tree is more resistant to natural disturbances and pathogen attacks. In the other strategy, more resources are allocated to growth and photosynthetic tissue, which gives a competitive edge over other tree species at an early age. But the cost is that these species have a low resistance to disturbances and pathogens later on, when the senescence process starts.

These systematic differences in the amount of defences have clear implications for host-tree association patterns among saproxylic species. In trees with a relatively small defence investment, we would expect to find more generalist saproxylics compared with trees that invest more in defences. On the other hand, we should expect to find a higher proportion of host specialists on tree species that invest more in defences compared with less defensive trees. If we consider the Nordic beetles, we find indications that such patterns exist. Among the broadleaved trees, it is clearly *Quercus* that hosts the most specialists (Figure 5.2). Among the conifers, too, there is perhaps such a pattern. The Scots pine *Pinus sylvestris* is more long-lived than the Norway spruce *Picea abies*, and should be expected to invest more in defences. It is therefore interesting to find that spruce seems to host more generalist species than pine, while the number of specialists on pine is slightly higher than on spruce (Figure 5.2). We should perhaps add the proviso that the collecting effort was not equally distributed among all trees, and several species were not adequately documented concerning their host range and preferences. Thus, these patterns should be treated with caution.

There is also a very interesting dataset on host preferences among saproxylic species in the tropics. Over a three-year period, Tavakilian and colleagues (Tavakilian et al., 1997) reared insects from 690 felled trees and lianas representing more than 200 species in French Guiana, and found 348 longhorn beetle species. Both in the Brazil nut family (Lechytidaceae) and the Sapodilla family (Sapotaceae) they found a large number of beetle species that only occurred on trees from that family. At the same time, they hardly found any generalist beetles emerging from the wood of these trees. These findings triggered additional research on Brazil nut trees, and they found that the trees were rich in antimicrobial substances (Rovira et al., 1999) that might have affected the beetles either directly or perhaps through gut symbionts that seem to be involved in wood decomposition (Berkov et al., 2007). At the other extreme, the Tavakilian team found that trees of the order Malvales hosted rather few specialists, while the majority were generalist cerambycid species that utilized a wide range of host trees. Between these extremes they found that other tree families hosted a balanced mix of generalists and specialists.

5.5 Host-tree preferences and decay

Earlier in this chapter we highlighted the differences in species composition in wood from conifers and from broadleaved trees. But behind this broad grouping, there are several species with much narrower host ranges. This is especially the case for species that colonize dying or recently dead trees. They come to a table that is filled with food, especially the nutritious inner bark and cambium zone. But they also meet a challenge – the defence system of secondary compounds that is abundantly present in the cells that were alive a short time ago. Bark beetles are such a group of initial colonizers. A large proportion of the bark beetle species are specialized on a single tree species or on several species that belong to the same genus (Ehnström and Axelsson, 2002). It is likely that such narrow host associations are linked to the ability to cope with the defence system left over from the living tree.

Different patterns of host-tree preferences become evident when one combines information about host-tree preferences with information about decay preferences. This information is available for the Nordic beetles mentioned earlier in this chapter. Leaving out the facultative saproxylics (see Chapter 11), there are nearly 900 beetle species associated with woody material for which we know the host tree and decay preferences. Among these species it is evident that those that are strictly confined to a single host-tree genus occur on recently dead trees and wood at an early stage of decay (Figure 5.5). The beetles that prefer

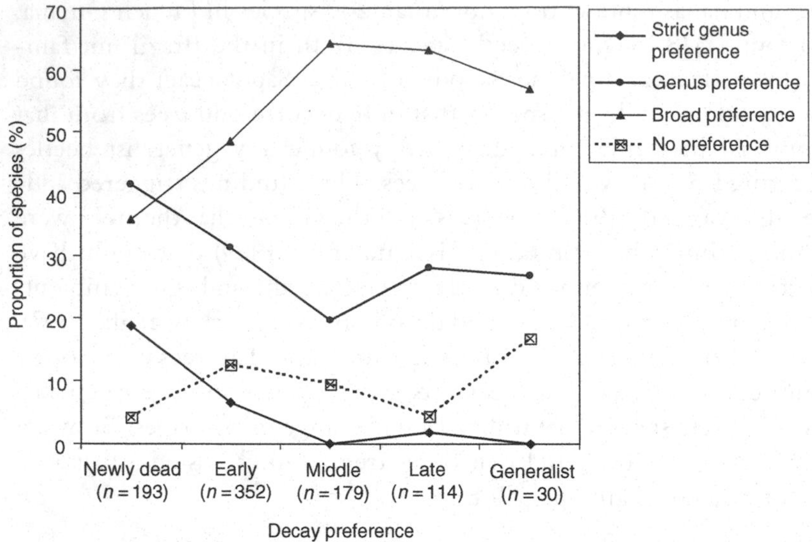

Figure 5.5. Host-tree associations of 868 obligate saproxylic beetle species in the Nordic countries, grouped according to their preference for decay status of the wood. *Strict genus preference* includes species that only occur on one genus of host trees; *Genus preference* includes species that predominantly occur on one genus of host trees, but also on other tree species; *Broad preference* includes species that prefer either coniferous trees or broadleaved trees; *No preference* includes species that occur on both coniferous and broadleaved trees. The Newly dead category includes species that can occur on living trees (healthy or weakened), but the species mainly occur on recently dead trees (0–1 years ago). *Early* decay comprises species that occur on trees that died more than 1–2 years ago. *Middle* and *Late* comprise species that occur on moderately and strongly decayed wood, respectively. *Generalists* are species with no clear preference for decay stages. In addition there are several obligate saproxylic beetle species in the region with unknown decay or host-tree preferences.

(but are not strictly confined to) a particular tree genus also show this pattern, but they also occur at later decomposition phases. Species preferring either conifers or broadleaved trees occur at all decay phases, but make up the largest proportion of those occurring in moderately and very decayed logs.

When these patterns are considered together, it is evident that specific associations with individual tree genera are most pronounced in recently dead trees but disappear as the wood becomes more decomposed. The difference in species composition between coniferous and broadleaved trees, on the other hand, is present throughout the decomposition process. It is likely that these differences are related to different durability of the chemical and physical properties of the wood. It has been shown, for example, that phenolic compounds have a shorter durability than resin in sound heartwood (Venäläinen et al., 2003). It is also well known that many white-rot fungi readily decompose various wood extractives (Gutiérrez et al., 1999; Dorado et al., 2000; Lekounougou et al., 2008). Structural wood components such as lignin, but also tannins and resins, are more enduring, and host associations caused by these substances are also evident in wood in advanced stages of decomposition.

5.6 Hypotheses about host-tree associations

We started this chapter by saying that there is a paucity of review papers on host-tree associations among saproxylic species. But the literature shows that much is known about trees (their phylogenetic relationships as well as their defence systems) and host associations among saproxylic species. It is therefore natural to end this chapter by addressing some challenges necessary to improve the knowledge base, and by presenting some hypotheses to be tested.

5.6.1 Empirical basis

There is ample scope to compile existing information and improve the documentation of host-tree associations among saproxylic species. Furthermore, the species should be classified into the following categories: true specialists (monophagous) using only a single tree species or tree genus; specialized (oligophagous), using a range of tree species but still being restricted to trees from one family or closely related families; and host generalists using trees from several plant families or orders. It is also important to document other aspects of their biology, especially their preference for different decay stages (and microhabitats) and their trophic associations with decaying wood. With such information at hand, we can learn much more about how saproxylic species are distributed across host trees.

5.6.2 Tree-based hypotheses

When host associations are considered from the host-tree side, there are at least two patterns that can be expected to occur. First, one should expect that closely related trees would host more similar guilds of saproxylic species than trees that are distantly related, due to similarities in wood structure and defence systems. It would be very interesting to see analyses of guild similarities across various tree species and compare these with the phylogenetic relationships between the same trees. Such analyses would be especially interesting if the saproxylic species are also grouped according to the attributes outlined under the species-based hypotheses listed below. The second pattern we hypothesize is that long-lived and 'defensive' trees should host more host specialists and relatively few generalist species compared with short-lived, pioneer trees that are potentially less 'defensive'. This pattern is primarily to be expected among species that are in close contact with the tree's defences, i.e. occurring in living or recently dead trees as well as at the lowest trophic level, directly consuming woody material (see species-based hypotheses).

5.6.3 Species-based hypotheses

When the host associations are viewed from the saproxylic side, we should also expect some distinct patterns. The most obvious pattern is that more host-tree specialists should be expected among species occurring in living and recently dead trees, where the defence systems are more or less intact. During the course of decomposition we should expect that the host-tree associations become more relaxed as the defences and wood structure are degraded. Another pattern we hypothesize is that host-tree associations are relaxed at higher trophic levels. As we move up the food chain, the various wood characteristics should not affect predators, fungivores or parasitoids to the same extent that they affect the detritivores. On the other hand, we know little about how species at higher trophic levels locate decaying wood, so it is still possible that they also show distinct host-tree associations.

5.6.4 Host diversity hypothesis

A final pattern that has not been discussed in this chapter is the relationship between host specificity and the diversity of alternative hosts. In forests with a low diversity of host trees, the dominant tree species represent large resource pools, whereas individual tree species in diverse forests represent scattered resources. In the first situation it might be rewarding to become a host-tree specialist, whereas the second situation hardly allows for host specialization. According to this line of reasoning, one should expect more host specialists to occur in forests with a low tree diversity compared with highly diverse forests.

At least among fungi, especially polypores, there is evidence for such relationships. In boreal forests where a few tree species dominate across thousands of square kilometres, we find many host-specific polypores (Niemelä, 2005), whereas polypores in tree-diverse tropical forests are weakly host-specific (Lindblad, 2000; Schmit, 2005). However, in tropical mangrove forests, with few and locally abundant tree species, we find many host-specific polypores (Gilbert and Sousa, 2002; Gilbert et al., 2008).

6 · *Mortality factors and decay succession*

Jogeir N. Stokland and Juha Siitonen

The way a tree dies has important effects on the species composition in decaying wood. It makes a great difference whether the tree dies suddenly, for instance because of a storm or a wildfire, or whether it dies gradually from competition, drought or old age. Different types of mortality produce dead trees with contrasting qualities, and therefore different species initiate the decomposition process. Later the decaying wood goes through major physical and chemical changes and the species composition changes completely several times until the wood is totally decomposed. The species themselves interact in many ways, and complete food webs build up and wane during the decomposition process. In combination, the different mortality factors and the decay succession have a great impact on the biodiversity in dead wood.

In addition to physically and chemically transforming the wood, the activity of wood decomposers also creates particular microhabitats such as sap exudations, insect galleries, space under loose bark, fungal fruiting bodies, rot holes and trunk cavities. Such microhabitats are important for a large number of species, and we describe them in detail in Chapter 7.

In this chapter we focus on the development of species communities during the decomposition of individual wood units with an emphasis on the qualitative differences of dead wood. In Chapter 12, we return to the topic of tree mortality and adopt stand and landscape perspectives to consider how different disturbances affect the amount of dead wood, with consequences for population dynamics and the long-term survival of saproxylic species.

6.1 Mortality factors and qualities of dead wood

Causes of tree death can be categorized in different ways, including dichotomies such as abiotic and biotic, or autogenic (driven by the internal development of a stand) and allogenic (driven by external natural disturbances). These classifications fail to account for the complex interactions between trees, their environment, and various mortality agents (Franklin et al., 1987). The most important factors affecting the saproxylic communities are the vitality of the tree at the time of death (sudden death of vigorous trees versus slow starving and weakening of trees), exposure of wood to initial fungus and insect colonization, and whether the wood dries up or retains its moisture.

Tree death is often a complex and gradual process with multiple contributing factors (Franklin et al., 1987; Waring, 1987; Manion, 1991). Pathogenic fungi and insects often become the proximate agent causing tree death (i.e. the final agent that kills the tree) when trees have already been weakened by other mortality factors, which are often called predisposing factors (Figure 6.1). Thus, in the continuous battle between trees and pathogenic species, the trees are usually the stronger

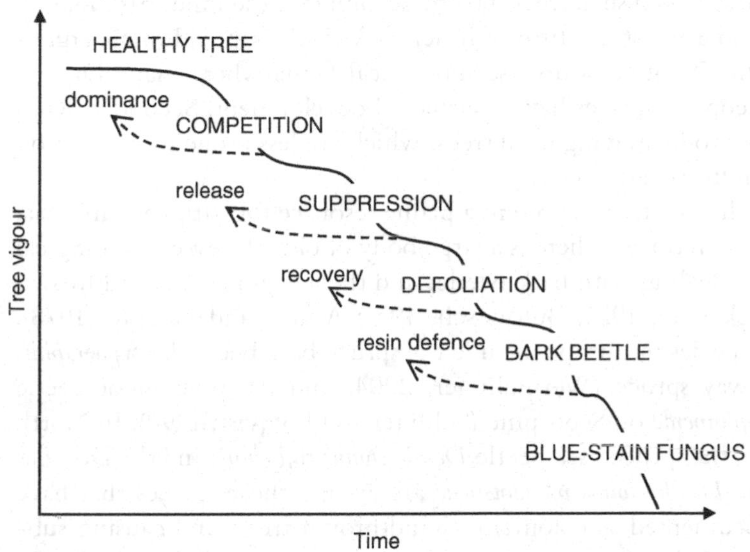

Figure 6.1. Mortality process of a coniferous tree, illustrating a potential series of events leading to its death (modified from Franklin et al., 1987). In this particular case, the growth of an initially healthy tree becomes suppressed by larger trees. If not released from competition, the tree can afford to allocate fewer resources to chemical defence and becomes predisposed to attack by defoliators. Once defoliated, the weakened tree becomes attractive to bark beetles. If the tree cannot block out the beetles by producing resin, the blue-stain fungi carried by the beetles will finally kill the tree. As the tree declines in vigour at each step, its opportunities to escape death become increasingly limited.

party. However, predisposing factors can reduce tree vigour and turn the battle in favour of the pathogens. The first fungi and insects invading the inner bark and sapwood of a weakened tree, thus killing it, will also initiate the decomposition process and be a part of the saproxylic community.

6.1.1 Wind

Strong winds often kill trees, either by uprooting them or by breaking the trunks. Sometimes, trees are predisposed to wind breakage by root-rot or heart-rot fungi, and these factors together create broken trees (Edman et al., 2007; Lännenpää et al., 2008). Two important features are characteristic of wind-induced mortality. Firstly, wind typically kills trees that have been actively growing until the moment of death. This means that the tree has a rich supply of energy and nutrients in the inner bark. When such trees start to decompose, the inner bark rapidly changes to a moist mixture of inner bark tissue, sap and microorganisms. This facilitates a diverse subcortical fauna where many Diptera and Coleoptera species have their larval development. Secondly, wind typically produces lying dead trees, which are less prone to desiccation than standing dead trees.

Windthrown trees represent a prime resource for bark and ambrosia beetles (Scolytinae). There is a large body of older forest entomological literature dealing with bark beetles and their enemies in windthrown trees (Lekander, 1955; Butovitsch, 1971; Annila and Petäistö, 1978). Typical species in Europe include the spruce bark beetle *Ips typographus* on Norway spruce (Wermelinger, 2004) and the pine shoot beetle *Tomicus piniperda* on Scots pine (Schlyter and Löfqvist, 1990). In North America, the spruce bark beetle *Dendroctonus rufipennis* and the Douglas fir beetle *Dendroctonus pseudotsugae* are among those species that have been documented as colonizing windthrown trees, and causing subsequent bark beetle outbreaks most frequently (Gandhi et al., 2007). Broader reviews of insect communities associated with wind-felled trees can be found in Bouget and Duelli (2004) and Gandhi et al. (2007). We are not aware of systematic fungal studies that relate species composition to causes of mortality, but – for example – the jelly fungi *Exidia saccharina* and *E. pithya* typically occur on recently wind-felled coniferous trees (Scots pine and Norway spruce, respectively) where the bark is still intact. In addition, members of the polypore genus *Trichaptum* are very common on wind-felled trees, where they decompose the inner bark and the outer sapwood. In summary, we can say that fungal and invertebrate communities in windthrown trees clearly differ from communities in trees that have died standing and have then dried up and lost their bark before falling.

6.1.2 Cutting

Cutting is another mortality factor that causes sudden tree death. In managed forests, most of the felled trees are healthy and growing, with a high nutritional value. In this respect the resulting dead wood has great similarity to wood from wind-felled trees. But there is, of course, a significant difference. While windthrow leaves a lot of dead trees in the forest, cutting is followed by the removal of the trunk and large branches for human use. Only the stumps and small-diameter logging residues are left behind in the forest. Nevertheless, such dead wood is utilized by fungi and insects, especially species that prefer, or regularly utilize, small-dimension wood (Kruys and Jonsson, 1999; Jonsell et al., 2007). Depending on the logging method (clear-cutting versus selective logging), the logging residues may be left behind in sun-exposed or shady conditions.

Although felled trees largely host the same species as windthrown trees, there are some differences in species composition. Species that favour small-diameter freshly dead wood benefit from logging; in bark beetles these include *Pityogenes chalcographus* on Norway spruce, and *P. quadridens* on Scots pine. Stumps may have a large diameter but they are often too short to accommodate galleries of the larger bark beetles or longhorn beetles, and not voluminous enough to sustain mycelia that are able to produce the fruiting bodies of some decay fungi. Because of the large cut surface, stumps vary more in moisture content than logs do. This may make stumps too stressful an environment for many decay fungi, and the species composition in cut stumps is clearly different from that in logs (Boddy and Heilmann-Clausen, 2008, and references therein). There are, however, many fungi that regularly utilize cut stumps, such as *Gloeophyllum odoratum* and *G. sepiarium*, *Heterobasidion* spp., *Trametes* spp. and *Phlebiopsis gigantea*.

6.1.3 Fire

Fire is a very distinct mortality agent that kills healthy trees. However, such trees differ markedly from those killed by wind or cutting. The obvious effects of fire include scorched bark and a charred wood surface. In addition, inside the trunk the wood may change through a process called pyrolysis, which is heating in the absence of oxygen. If the fire is intense enough and the temperature rises above 300°C, pyrolysis starts to change the chemistry of the wood, as large and complex molecule structures break down and new compounds are formed (Alén et al., 1996; Hosoya et al., 2009). Changes that may affect the decay of burned wood include the increased proportion of lignin, the breakdown or evaporation of decay-inhibiting substances such as resin, and the formation of new compounds that can either inhibit or stimulate the growth of saproxylic organisms. In addition to changes in the wood, the majority of fire-killed trees remain standing for several years. This makes the trees more sun-exposed – and this is further exacerbated by the absence of a shading canopy. Wikars (1992) pointed out that fire has three main effects on saproxylic species: it creates special substrate qualities, it creates competition-free substrates on existing dead wood by partly or completely destroying the pre-fire decomposer communities, and it changes the surrounding environmental conditions.

Forest fires have occurred ever since the first trees developed more than 350 million years ago and subsequently they have had a significant impact on forest ecosystems (Scott, 2000, 2009). Thus, it is no wonder that several species have special adaptations that enable them to locate and utilize burned wood. These include antennal receptors that detect smoke components, and organs that detect infrared radiation from fires (Evans, 1966; Schütz et al., 1999; Schmitz et al., 2000; Suckling et al., 2001).

Several studies have documented the colonization by insects of recently burned areas (Gardiner, 1957; Muona and Rutanen, 1994; Saint-Germain et al., 2004a). Cerambycid genera such as *Acmaeops* and *Arhopalus*, and the buprestid beetle *Melanophila acuminata*, are well-known examples of pyrophilous saproxylic taxa (Wikars, 1992; Saint-Germain et al., 2004b; Boulanger and Sirois, 2007). Such species are found at high densities in recently burned forests, but are uncommon or rare in unburned forests. In a recent paper, Saint-Germain et al. (2008) argued that pyrophilous insects also need unburned wood in situations where the fires are too far apart in space or time. Thus, Saint-Germain and colleagues suggested that pyrophilous insects are capable of using alternative wood qualities and do not completely depend on burned wood.

Many wood-inhabiting fungi are favoured by forest fires, as they occur frequently on burned wood. But most of them also occur regularly on unburned wood, especially in dry and open areas (Penttilä and Kotiranta, 1996; Penttilä, 2004). Some fungi, however, seem to be more strictly confined to burned wood, for example *Daldinia loculata*, *Gloeophyllum carbonarium* and *Phanerochaete raduloides* (Penttilä and Kotiranta, 1996; Johannesson et al., 2001). Similarly, among the lichens there are some species that only occur on burned wood, such as *Hypocenomyce anthracophila* and *H. castaneocinerea*.

6.1.4 Competition

Neighbouring trees compete increasingly for local resources as they grow bigger, and eventually the subordinate trees might die. Sometimes this causes substantial mortality in the form of self-thinning in dense, even-aged stands that develop after clear-cutting or stand-replacing disturbances. Trees dying from competition actually die from starvation due to lack of light, water or nutrients (Kozlowski et al., 1991). This is a slow process that may last for years or even decades. Trees dying from competition have some characteristics that make them a distinct saproxylic habitat. They are mainly small or medium-sized understorey trees in shaded conditions that remain standing after death. Because of starvation and slow growth, the annual rings are narrow and hence the wood density is high. The inner bark is thin and the nutritional value of the inner bark and sapwood is low. After death the inner bark dries up and becomes tightly attached to the wood surface.

The dry inner bark of suppressed, small trees is the favoured habitat of certain bark beetle and cerambycid species. In Norway spruce, for instance, these include *Xylechinus pilosus*, *Polygraphus subopacus*, *Phloeotribus spinulosus* (Scolytinae), *Molorchus minor* and *Obrium brunneum* (Cerambycidae). Most of these species can also utilize dead, self-pruned, lower branches of large living trees.

6.1.5 Drought

Water stress in trees develops when evaporation of water from leaves or needles exceeds water uptake by roots for a long period of time (Kozlowski et al., 1991). This occurs during extended periods of hot and dry weather, especially at locations with a thin soil layer or well-drained sandy soil. Two mechanisms have been proposed to cause mortality during drought (Bréda et al., 2006; McDowell et al., 2008). When reduced soil water availability is coupled with high transpiration, this may lead to air-filled sections and hydraulic failure in the water-transporting vessels in the roots or the trunk. This stops the transport of water and the sapwood will dry up and allow wood-decaying fungi to become active. The other potential mechanism causing mortality is carbon starvation caused by the closure of the pores in leaves or needles. This closure is a response that happens in order to prevent the hydraulic failure just described. However, at the same time, it reduces the uptake of carbon dioxide and therefore photosynthetic activity. The reduced production of sugar from photosynthesis leads to carbon starvation and reduced defensive capability against pathogenic fungi and insects. Carbon starvation is particularly likely if the drought is not intense enough to cause hydraulic failure, but lasts long enough to result in a depletion of plant carbon reserves. Suppressed trees are therefore more susceptible than dominant trees to drought. Reduced tree growth and increased mortality following a drought period can continue for several years, probably as a consequence of both water transport dysfunction and a deficit in carbon stores (Bréda et al., 2006).

In most cases, pathogenic fungi and insects are the final mortality causes of drought-stressed trees. Both hydraulic failure and carbon starvation can increase vulnerability to these biotic mortality agents, either as a result of reduced water pressure in the sapwood or reduced production of resin or other defensive compounds; there is also likely to be an increased emission of volatiles, such as ethanol, which attracts insects (Raffa et al., 2005; Desprez-Lousteau et al., 2006; Roualt et al., 2006).

Drought appears to favour canker-forming fungi in the ascomycete genera *Botryosphaeria*, *Sphaeropsis*, *Cytospora* and *Hypoxylon* (Desprez-Lousteau et al., 2006). These species are often present on or in the tree before drought stress without causing any symptoms. Reduced water pressure can trigger the development of these latent fungi. Canker fungi kill patches of bark, which often results in the death of branches – or the whole stem if it becomes entirely girdled. In broad-leaved trees, the lowered water potential in sapwood can also trigger the development of wood-decaying fungi that are present in the wood as latent invaders (Boddy, 1994, 2001).

In coniferous trees, certain bark beetles and their associated blue-stain fungi typically cause most of the mortality in drought-stressed trees. The degree of water stress appears to be crucial. Moderate drought does not necessarily reduce the resistance to bark beetles and their fungi (Christiansen, 1992; Dunn and Lorio, 1993; Reeve et al., 1995). However, conifers with resin defences gradually lose their capacity to flush out the attacking insects (Franceschi et al., 2005). Thus, severe drought stress can promote insect attacks and the growth of blue-stain fungi. These fungi cause internal damage in the sapwood and loss of water conductivity, resulting in tree mortality when they are brought in by the beetles in great numbers (Christiansen and Solheim, 1990; Croisé et al., 2001).

Alternative species may develop in drought-stressed trees under different situations. But it is typical that a single species, or a few interacting species, rapidly colonize large sections of the tree and consume the most valuable resources before other species arrive to utilize them. Bark beetles that colonize stressed but still living trees are often referred to as aggressive species. Examples of such species are *Ips typographus* in Europe, *Dendroctonus ponderosae* and *D. brevicomis* in North America, and *D. frontalis* in the southern USA and Central America. These beetles represent a small subset of all bark beetles, as most bark beetle species first colonize trees after they have died. Drought- and pathogen-killed trees typically lose their bark, remain standing, and soon dry up. Fungi with the ability to tolerate a low or fluctuating water potential will prevail in the subsequent decay of such trees. These include certain ascomycetes including *Hypoxylon* species, and certain jelly fungi such as *Dacrymyces stillatus* (Rayner and Boddy, 1988).

6.1.6 Senescence

When a tree grows old and approaches its full potential age, it shows increasing signs of senescence. Aging trees are characterized by a decreased metabolism, gradually reduced growth rate, loss of apical dominance, increased self-pruning, death of sections of cambium, reduced wound healing capacity, formation of trunk cavities, and increased susceptibility to injury from pathogens and unfavourable environmental conditions (Kramer and Kozlowski, 1979; Leopold, 1980). Finally, the trunk may only have a small sector of active inner bark, while most of it is dead. This gradual way of dying is typical for some broadleaved trees, e.g. oaks (*Quercus* spp.) (Alexander, 2008) and various eucalypts (Whitford, 2002), and may go on for more than a hundred years. Many coniferous trees also exhibit this type of gradual death when they get really old.

In the previous chapter we referred to Loehle (1988), who noted that the duration of the senescence phase varies systematically among tree species according to two basic strategies. Pioneer species such as poplars (*Populus* spp.) and birches (*Betula* spp.) are fast-growing, early-maturing and short-lived, with an abrupt senescence phase. Another group consists of species that are slower-growing, later-maturing, longer-lived and have a persistent old-growth phase. This group includes oaks (*Quercus* spp.), hickories (*Carya* spp.) and several conifer genera (especially *Pinus* and most Cupressaceae). Loehle attributed these differences to alternative strategies in defence investment. Investment in structural and chemical defences is energy-demanding and slows down the growth rate, but pays off with an extended longevity because the tree is more resistant to natural disturbances and pathogen attacks. Thus, it is long-lived tree species that typically have a long-lasting senescence phase.

Old, senescent trees are extremely important for saproxylic species, since they provide a much larger variety of microhabitats than young, vigorous trees (Speight, 1989; Winter and Möller, 2008) (see Chapter 7). The gradual death of trees also results in dead bark that is tightly attached to the wood, and there are some species that prefer the inner bark of trees that die in this way. In boreal forests, we find beetle species such as the cerambycid *Callidium coriaceum*, the anobiid (deathwatch beetle) *Ernobius explanatus*, and bark beetles in the genus *Carphoborus* (Ehnström and Axelsson, 2002). In nemoral forests of central Europe, the large longhorn beetle *Cerambyx cerdo* seems to attack the inner bark and outer sapwood of old, senescent *Quercus* trees (Buse et al., 2007).

The slowly dying trees also form long-lasting habitats after their death. For instance, old Scots pines (*Pinus sylvestris*) usually lose their vigour gradually when the tree reaches 300–500 years of age and die from senescence at an average age of 420–450 years in northern boreal Fennoscandia (Leikola, 1969; Niemelä et al., 2002), although the age of the oldest trees may exceed 800 years (Sirén, 1961). After death, it takes an additional 35–40 years before the trees lose their bark and become decorticated, silvery-grey snags that are called '*kelo*' trees in Finnish (Niemelä et al., 2002). The decay-resistant snags may remain standing for extended periods of time, with an average of *c.* 100 years and a maximum of over 250 years in the middle boreal zone (Rouvinen et al., 2002b), and over 700 years in the northern boreal zone (Niemelä et al., 2002). The extremely slow decay rate probably results from slow growth, producing narrow annual rings and dense wood, with scarring by fires increasing the resin content of wood, a high proportion of heartwood in relation to sapwood, and desiccation of the standing dead trees (Niemelä et al., 2002).

Standing kelo trees harbour very few fungi, but they host specialized epixylic pin lichens (Caliciales) such as *Calicium denigratum*, *Chaenothecopsis fennica* and *Cyphelium pinicola*. When kelo trees finally fall down and start to decay, the species composition of fungi is strikingly different from that of wind-felled trees (Niemelä et al., 2002). Specialist species include the polypores *Antrodia crassa*, *A. infirma*, *Gloeophyllum protractum* and *Postia lateritia*, and the corticioids *Odonticium romellii* and *Chaetoderma luna*.

6.1.7 Other mortality factors

As well as the mortality factors described above, many other factors can cause tree mortality and create dead-wood substrates. However, there are probably only a few species that would be specifically associated with these mortality factors. In the boreal zone, snow loads and ice formation can snap off tree tops and branches, and this creates lying dead wood similar to that from wind breakage or cutting. Avalanches and landslides also break down healthy trees and create similar substrates to those produced by wind damage.

Flooding is another mortality factor that causes complex physiological and biochemical changes in trees. The immediate consequence of flooding is poor soil aeration. Flooding tolerance, i.e. the capacity to survive anoxic conditions, varies widely among tree species, from several days to several months (Kozlowski, 1997; Glenz et al., 2006). Trees killed by flooding typically die in a standing position and in this respect resemble dead trees resulting from drought-induced mortality. Often the trees end up falling into the water, and they will decay partly or completely in a submerged position. Dead wood in water hosts distinct species assemblages that are described further in Chapter 9.

6.1.8 Partial mortality

The factors described above cause the death of the entire tree, but dead wood also occurs in live trees. Wounds, rot holes, trunk cavities

Figure 6.2. (a) An old and partly senescent *Quercus* tree. The left side shows several signs of senescence (partly decorticated trunk, substantial amount of self-pruning), whereas the right side of the tree appears more vital. (b) Close-up of two microhabitats that typically develop in senescent trees – a trunk cavity (tree hollow) with ground contact, and adjacent dying bark that will eventually fall off and expose dead sapwood, as in the background tree (photos Nikola Rahmé).

(hollow trees), and dry branches can represent a significant proportion of all the dead wood in a forest, and many saproxylic species prefer these dead wood qualities (see Chapter 7).

Wounds can appear in an otherwise healthy tree, either superficially in the bark or reaching deeper into the wood. Wounds are caused by falling nearby trees, breakage due to heavy snow loads, fire scars, or large animals that feed on or scratch the bark. In modern times, human activity also frequently produces wounds in trees. Local wounds do not usually kill the tree; the main reason is that the tree has several types of defence mechanisms in nearby living tissue, both in the form of high water pressure that prevents the development of decay fungi (Boddy and Rayner, 1983b) and the production of phenols and resins that are toxic to potential intruders (e.g. Shortle, 1990; Deflorio et al., 2007, see also section 5.4).

Dead branches represent a type of partial mortality that occurs regularly in most trees. When a tree grows bigger, some branches dry out and die attached to the tree (Figure 6.2a). This self-pruning is a natural process that results from the development of the tree itself. Internally in the trunk, the tree produces a barrier that separates the dead and living tissue (Shigo, 1985). Such dead branches may remain attached to the tree for several years, but they tend to decompose faster and more readily break off from broadleaved trees than from conifers. Attached branches in the canopy are exposed to sunshine and wind, and host distinct communities of both fungi and insects (see Chapters 7 and 9).

6.2 Decomposition pathways

When a tree dies, the community development or succession of saproxylic species does not follow a one-dimensional trajectory where species replace each other in a deterministic manner. Instead, the development can follow alternative decomposition pathways.

The mortality agent is probably the most important factor determining community development, as it opens up the wood for colonization in different ways and exposes it to different environmental conditions. This effect has often been commented upon in the literature and is best summarized by Boddy and Heilmann-Clausen (2008) for fungal development in broadleaved trees. They described how alternative fungal communities develop in wounds, dead attached branches, standing dead trees, uprooted trees and cut stumps. These mortality types also occur in coniferous trees and probably cause corresponding developments there.

In addition, the rot type resulting from wood-decaying fungi prepares the ground differently for later-arriving invertebrates and fungi. In the entomological literature there is scattered information about insect species that seem to prefer either brown-rotted or white-rotted wood; for example among click beetles (Martin, 1989), longhorn beetles and bark-gnawing beetles (Trogossitidae) (Ehnström and Axelsson, 2002), but there are hardly any comprehensive overviews. In addition, many intimate associations between specific fungi and fungivore insects are known (see Chapter 7). The species composition of yeasts is also very different in white- and brown-rotted wood (González et al., 1989).

Still another mechanism that contributes to decomposition pathways is the predecessor–successor association between distinct sets of wood-inhabiting fungi at the intermediate and late stages of wood decomposition (Niemelä et al., 1995). There is undoubtedly a regular pattern whereby certain species of fungi follow other specific fungi, and the interaction between these fungi most probably takes place between their mycelia inside the wood (Ovaskainen et al., 2010a). But whether the explanation is related to parasitism, saprotrophic use of dead tissue from the first species, or the fact that the successor species utilizes residual metabolites produced by the previous species is still unknown and hidden inside the wood.

Despite the lack of relevant quantitative data, we propose that alternative decomposition pathways cause systematic differences in species composition during the decomposition process (Figure 6.3). Initially, the systematic differences are largest due to the effects of alternative mortality factors. During the decay succession these differences gradually disappear due to the species turnover described later in this chapter, and the effects of mortality become negligible. On the other hand, different random aspects probably increase variation between individual logs within any of the particular mortality classes. We have not assigned any dissimilarity measure on the vertical axis in Figure 6.3, due to a lack of data; nor do we claim that the shape of the curves is correct. The main purpose is to illustrate that the effects of mortality factors are most pronounced in the early decay stages.

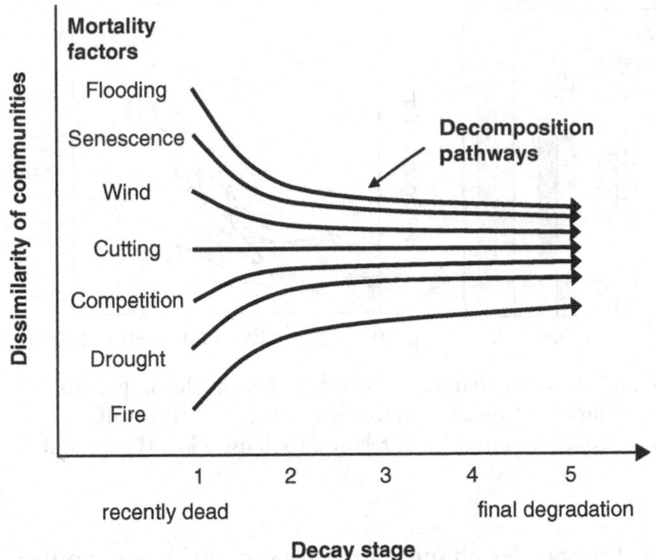

Figure 6.3. A conceptual illustration for the decomposition pathways of decaying wood resulting from different mortality factors. The vertical distance between the lines indicates differences in species composition. In the early decay stages, the species composition generally differs most. As the decay succession proceeds along the decomposition pathways, a continuous species turnover takes place. The communities then become more similar to each other until the differences between decomposition pathways may almost disappear in the final stages of wood degradation.

6.3 The decaying tree as a changing resource

During the decomposition process, a decaying tree undergoes many changes. These are of both a physical and a chemical nature, and develop in a predictable manner. The most important physical changes relate to bark cover and wood density. Typically the inner bark is rapidly consumed after the tree dies. The loss of bark can proceed quickly or slowly, depending on the mortality factor and whether the tree died standing or lying. Wood density is another important physical property: initially, the wood is hard, but it gradually becomes soft as a result of fungal decay (see Box 6.1).

Wood moisture increases steadily during the decomposition process (Dix, 1985; Sollins et al., 1987; Renvall, 1995; Figure 6.4a). This is partly a result of reduced desiccation, as the trunk sinks closer to the ground when losing its physical strength. More importantly, however, the decomposition process itself produces water as an end-product.

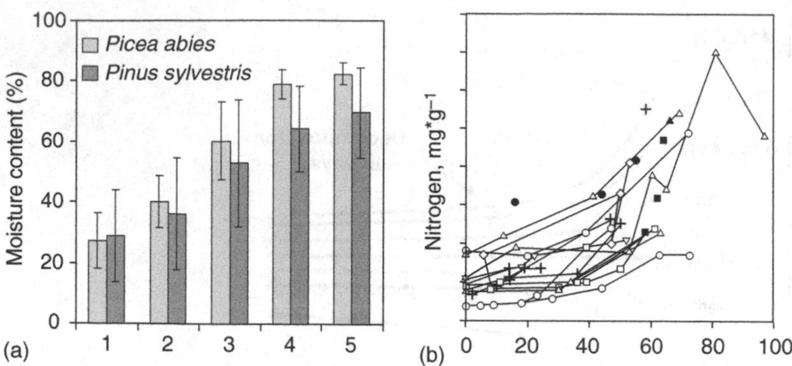

Figure 6.4. Physical and chemical changes in wood during the decomposition process: (a) moisture content (%) in relation to decay stages 1–5 (Renvall, 1995); (b) nitrogen concentration (mg/g) in relation to density loss (Laiho and Prescott, 2004).

Several chemical properties change substantially during the course of wood decomposition. As described in Chapter 5, a recently dead tree is full of chemical defence substances. These are decayed by the initial decomposers or they break down due to physical and chemical degradation. The structural cell wall components also undergo a systematic decomposition as a result of the enzymatic activity of wood-decaying fungi (see Chapter 2). Another chemical property that changes is the level of carbon dioxide. This is also an end-product of decomposition, and the level of gaseous carbon dioxide increases inside the wood – at least until fragmentation of the wood increases the ventilation. In addition, the concentrations of several mineral elements typically increase during decomposition, as the carbohydrates are decomposed (Harmon et al., 1986; Krankina et al., 1999; Laiho and Prescott, 2004).

Finally, the nutrient content of the wood changes during the decomposition process. Wood is a nitrogen-poor resource (except for the cambium zone). However, as cellulose and lignin are decomposed and the wood is filled with fungal mycelia, the nitrogen content actually increases (Figure 6.4b), because the concentration of nitrogen in a fungal mycelium is 7–10 times higher than in wood (Merrill and Cowling, 1966; Swift and Boddy, 1984). Nitrogen is mostly present in the form of chitin in the fungal cell walls. Nitrogen is a valuable element for wood-decaying fungi and they have the ability to break down and recycle nitrogen from their own cell walls as well as from the mycelia of earlier colonizers (Lindahl and Finlay, 2006). The amount of nitrogen can also increase in dead wood as a result of other processes. Nitrogen-fixing bacteria in wood can take up significant amounts of atmospheric nitrogen (Hendrickson, 1991; Brunner and Kimmins, 2003). Some wood-decaying fungi can transport nitrogen into wood from the surrounding soil using their mycelial networks (Tlalka et al., 2008). Furthermore, it has been shown that the larvae of certain wood-living beetles, such as Scarabaeidae and Lucanidae but also some bark beetles, have symbiotic gut bacteria that can fix atmospheric nitrogen (Jönsson et al., 2004; Kuranouchi et al., 2006; Morales-Jimenéz et al., 2009).

Box 6.1 Wood decomposition and decay stages

Wood decomposition

Wood decomposition is predictable in the sense that, as time passes, the dead tree loses mass, volume and density. The model most commonly used to describe this process is the single exponential model, which has the following form:

$$Y_t = Y_0 e^{-kt}$$

where Y_t is the density at time t, Y_0 is the original density and k is the decay coefficient. The decay coefficient is obtained by measuring the relative decay of a number of wood samples as follows:

$$k = -\ln(Y_t/Y_0)/t$$

The model is thus easy to parameterize and will give predictions of mass loss over time (Figure 6.5).

The decay rate constant provides a convenient tool to compare decay rates; for example between regions with different climate, localities with different microclimate, wood with different rot resistance, and different trunk diameters (e.g. Rock et al., 2008). However, although widely used, the model has some drawbacks. Primarily it assumes that the decay rate is constant during the whole decay process. Several studies have shown that this is not the case (Harmon et al., 2000; Mäkinen et al., 2006). This suggests that more elaborate mathematical models are needed to fully capture the decay process (see, e.g., Yin, 1999; Mackensen et al., 2003; Mäkinen et al., 2006, for more information).

Box 6.1 *(continued)*

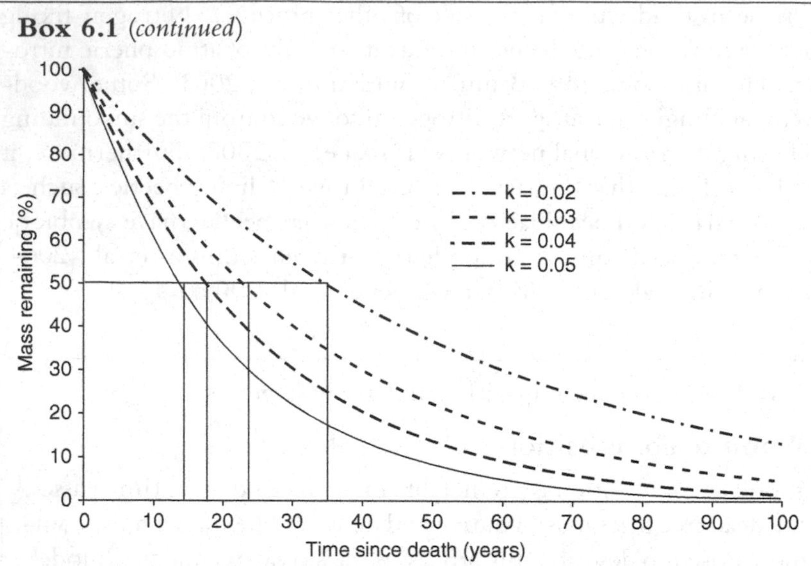

Figure 6.5. Mass loss over time, predicted according to a model of logarithmic decay, given four different decay constants (*k*). Time to 50% mass loss is indicated.

Decay stages

It is common to subdivide the decomposition of dead wood into five stages (see Table 6.1). Similar five-class systems of decay have been adopted in quite different research areas, such as wood decomposition studies (Sollins, 1982), research on species composition and succession (Renvall, 1995; Stokland, 2001; Heilmann-Clausen and Christensen, 2003), and national forest inventories (Waddell, 2002; Stokland et al., 2003). This does not mean that these classification systems are identical. There also exist classification systems with more decay stages, for example eight (Söderström, 1988a), or with fewer.

Most classification systems emphasize the visual appearance of each decay stage and the softness of the wood (Table 6.1, Figure 6.6). But it is also of interest to know the loss of biomass in each decay stage. Næsset (1999) estimated the remaining biomass across decay stages 1–4 in the system presented in Table 6.1, and Stokland (2001) extrapolated these results to cover the final stage. Alternative classification systems probably differ in the amount of biomass being lost during each decay stage.

Table 6.1. *Classification of live trees and dead wood at different stages of decay (from Stokland, 2001).*

Decay stage	Approximate % of live biomass	Characteristics
0 Weakened tree	100	Living weakened tree (by wounds, drought, senescence), green foliage
1 Recently dead	100–95	Recently dead tree, bark attached to stem; undecayed wood
2 Weakly decayed	95–75	Bark loose or fallen off, decay penetrating less than 3 cm into wood from surface, initial mycelium under bark
3 Medium decayed	75–50	Decay penetrating more than 3 cm into wood, core (or hollow tree surface) still hard
4 Very decayed	50–25	Stem rotten throughout, no (few) hard parts, ellipsoid cross-section, fragmented outline of stem
5 Almost decomposed	25–0	Stem outline heavily fragmented, completely decomposed in some places; wood disintegrates when lifted; wood mould in tree cavities

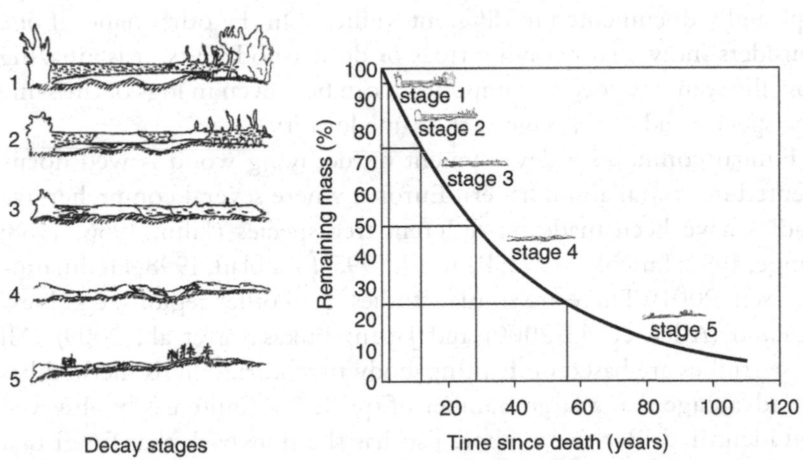

Figure 6.6. Visual appearance of the five decay stages from Table 6.1 (left). The diagram on the right shows the amount of remaining biomass and the relative duration of each decay stage.

Box 6.1 *(continued)*

The duration of each decay stage increases from the initial to the final stage (Figure 6.6) but the specific duration varies substantially as an effect of temperature, moisture and the specific organism(s) causing the decay. On a global scale, the total decomposition time of a fallen tree varies from less than 10 years in the tropics to several hundred years in boreal forests close to the timberline (see Chapter 14); however, microclimatic conditions around the log, the tree species, and trunk diameter also influence the decay rate.

6.4 Fungal succession

The development of fungal communities in decaying wood has sometimes been described as one of the most clear-cut cases of succession, where different species appear and replace each other in a predictable sequence (Park, 1968). A completely different view has been advocated by Boddy (1992, 2001), who considered the term 'succession' to be too simplistic and described the development of fungal communities in wood as a complex, multidimensional process which follows a diverse array of optional pathways.

Although these views seem to be conflicting, they both contain elements of truth. If one considers how different species prefer different decay stages (see Figure 6.7), there are clear regularities that have been repeatedly documented in different studies. On the other hand, if one considers individual decaying trees or dead wood units, it is amazing how different the species composition can be – even in logs of the same tree species and decay stage at the same locality.

Fungal community development on decaying wood is well documented in central and northern Europe, where several comprehensive studies have been made on different tree species (Jahn, 1966, 1968; Lange, 1992; Luschka, 1993; Renvall, 1995; Lindblad, 1998; Heilmann-Clausen, 2001). There are similar studies from other regions, e.g. New Zealand (Allen et al., 2000) and Japan (Fukasawa et al., 2009). All these studies are based on fruiting-body inventories. This method has the advantage that a large number of species are quite easily observed and identified. But the method also has the drawback that fungi that are present as a mycelium without fruiting bodies are overlooked. The actual number of species inside the wood is higher, and many species can be present for several years before they produce fruiting bodies. So

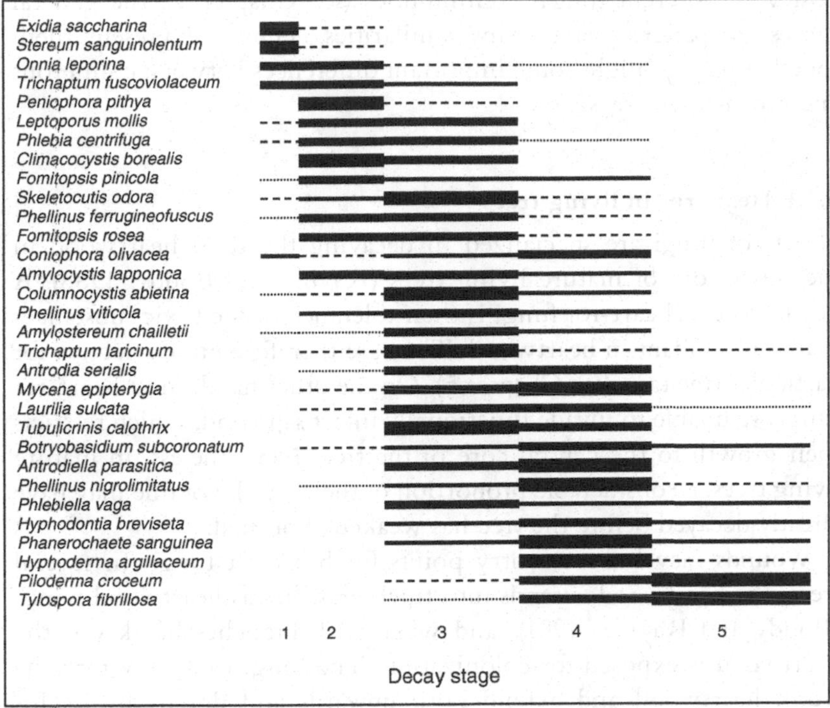

Figure 6.7. Species turnover of basidiomycetes from recently dead trees (decay stage 1) to strongly decayed wood (decay stage 5) on lying Norway spruce (*Picea abies*) trunks in northern Finland. The thickness of the lines corresponds to the proportion of fruiting-body observations that are made at different decay stages. Derived from Renvall (1995).

far, we have little knowledge about the duration of mycelium development before fruiting-body production, and how this varies between species.

Recent inventories of wood-inhabiting fungi have started to use DNA sequencing of mycelia inside the wood (Tedersoo et al., 2003; Lygis et al., 2004; Vasiliauskas et al., 2004; Allmér et al., 2006; Ovaskainen et al., 2010b). There are still many practical problems with this methodology, including the need for a complete library containing identified DNA sequences for all species (Stenlid et al., 2008). This methodology is nevertheless very promising for future research.

In the following outline, we build on the sources mentioned above and emphasize the regularities more than the complex aspects of community development. Even though coniferous and broadleaved trees host very different fungal communities (see Chapter 5), the general succession patterns have many similarities. Thus, we present them together but highlight some important differences between coniferous and broadleaved trees.

6.4.1 Heart rot in living trees

Heart-rot fungi are specialized in decaying the dead heartwood in the inner core of mature living trees (Rayner and Boddy, 1988). A key feature of heart-rot fungi is their tolerance of the toxic substances that are abundant in heartwood. They are therefore quite selective for particular tree taxa (see Chapter 5). On the other hand, most heart-rot fungi are unable to invade functionally intact sapwood, which restricts their growth to the central core of the tree. Nevertheless, in mature living trees, a considerable proportion of their trunk volume may have already decayed before the tree has weakened or died.

Wounds may become entry points for heart-rot fungi. In several trees, the heartwood extends into thick branches as heartwood wings (Boddy and Rayner, 1981), and when such branches break off, the heartwood is exposed for colonization. The fungi can grow into the trunk heartwood and extend both upwards and downwards. This phenomenon is common in broadleaved trees, for example in oaks (*Quercus* spp.), where *Laetiporus sulphureus* and *Fistulina hepatica* bring about heart-rot decay. Coniferous trees also have heart-rot fungi, such as *Phellinus pini* in pines (*Pinus* spp.) and *P. chrysoloma* in firs (*Abies* spp.) and spruces (*Picea* spp.). Other fungi, such as *Heterobasidion annosum*, enter through wounds in the root system and cause rot extending upwards from the base. The development of heart rot is a process that proceeds through all the decay phases described below while the tree is still alive, and the final result is a hollow trunk. Thus, one can often find the fruiting bodies of heart-rot fungi on living trees, but these fungi do not damage the functional sapwood.

6.4.2 Living weakened trees

Some fungi develop on living trees and seem to cause no harm to the tree; these include *Mycena corticola*, *Aleurodiscus disciformis*, *A. amorphus* and *Hymenochaete ulmi*. They are poorly known and largely overlooked because they are of no interest to forest pathologists, and most ecologists are only concerned about fungi growing on dead wood. One possibility is that these fungi decompose the dead outer bark, but there is no evidence to support this speculation. It seems that these fungi primarily develop on old and senescent trees and another more likely possibility is that they utilize dying inner bark that has poor defensive ability.

Live and healthy trees have several defence mechanisms against fungal invasion (see Chapter 5). But when a tree is wounded or otherwise weakened, the situation can easily switch in favour of the fungi. Open wounds and broken branches on broadleaved trees facilitate colonization by various sugar fungi (yeasts, sapstain fungi) and bacteria that are dispersed by air or by insects. Certain sap beetles (Nitidulidae) can bring pathogenic sapstain fungi (*Ceratocystis* spp.) into wounds, and these can cause lethal canker diseases in injured trees (Hayslett et al., 2008). Most conifer trees have an effective defence system – the resin ducts that occur throughout the inner bark and the sapwood. The sticky and toxic resin typically seals wounds within a few hours, and we find considerably fewer fungi in conifer wounds, if any at all.

Other fungi are transported by certain bark beetle species that can attack weakened living trees. Some of these fungi are pathogenic and cause wilt diseases such as Dutch elm disease (caused by *Ophiostoma ulmi* and *O. nova-ulmi*). These *Ophiostoma* species, as well as other sapstain fungi, consume sap in the inner bark, but may also follow the sap stream along ray parenchyma cells and cause stained wedges projecting into the sapwood (Christiansen and Solheim, 1990).

If a tree is weakened by drought or competition between trees, there is another suite of fungi that will kill the tree and initiate the decay succession. Strip cankers and other pathogenic fungi can become active and cause lethal damage to the tree. Also wood decomposers acting as latent invaders, such as *Fomes fomentarius* and *Piptoporus betulinus*, can rapidly grow in drought-stressed trees (Boddy, 2001).

6.4.3 Recently dead wood

This decomposition phase typically lasts 1–2 years and corresponds to decay stage 1 (see Box 6.1). As soon as a tree dies, it is rapidly colonized by fungi. Many of these fungi belong to various ascomycete genera including *Ceratocystis*, *Ophiostoma* and *Leptographium*, and they are transferred to recently dead trees by bark beetles (see Box 6.2). These fungi rapidly expand in the inner bark and sapwood, where they consume soluble carbohydrates and cell-content compounds. The fungal communities in bark beetle galleries can be surprisingly diverse. A total of 65 fungal taxa were obtained from the inner bark of Norway spruces colonized by *Ips typographus*, and an additional 36 taxa occurred in the sapwood under galleries in southern Poland (Jankowiak, 2005).

In addition, several basidiomycetes rapidly colonize recently dead wood. Many jelly fungi (*Exidia* spp.) are typical at this stage, and also some agarics (*Panellus* spp., *Crepidotus* spp.) and corticioids (*Peniophora* spp., *Phlebia rufa*, *P. radiata*, *Stereum* spp.). Unexploited and energy-rich newly dead trees allow these species to grow and reproduce rapidly. But they die quite soon, when the nutrients are depleted or when they become outcompeted by later-arriving fungi.

6.4.4 Initial to intermediate wood decay

This phase is dominated by fungi that effectively decompose cellulose and lignin in the wood. This decomposition phase starts in recently dead trees, but mostly takes place during decay stages 2 and 3 (see Box 6.1). The mycelium of these fungi expands throughout the sapwood, and after some years of decay it is easy to distinguish between brown- and white-rotted wood. It is typical that the decay extends into the heartwood in many tree genera (e.g. in *Populus* and *Betula*), but in other tree species one can find completely decomposed sapwood while the heartwood is almost intact. This is quite common in *Pinus* wood. Most standing dead trees fall down during this phase, since the rot reduces the structural integrity of the trunk.

This decomposition phase is the realm and battlefield of many polypores. Dead wood represents a huge resource base, and different species have developed alternative strategies to get their share of this cake (Rayner and Boddy, 1988). Some wood-decaying fungi, such as *Coriolus*, *Trametes* and *Stereum* species, have a strategy of primary resource capture. They rapidly colonize recently dead trees, where they start to decompose the inner bark and outer sapwood. These fungi have a typical *r*-selection strategy (see Chapter 14) and soon produce fruiting bodies in order to colonize new dead wood. Other species, such as *Lenzites betulinus*, *Sistotrema brinkmanni* and *Phanerochaete velutina*, have a strategy of selectively or non-selectively outcompeting initial wood colonizers, and take over their resource territory (Rayner and Boddy, 1988).

In a series of elegant laboratory studies, Holmer and co-workers (Holmer and Stenlid, 1997; Holmer et al., 1997) showed that there is a hierarchy of combative ability among different wood decomposers in spruce wood. In general, the initial colonizers are weak competitors, whereas later species are successively stronger competitors. Boddy (2001) described a similar hierarchy among wood decayers in broad-leaved trees. Within species there is also competition for resource capture. In an interesting study, Adams and Roth (1969) reported that several individuals (genets) of *Fomitopsis cajanderi* were present at the site of entrance in a broken Douglas fir stem. Further away, only a few individuals had expanded at the expense of the others. A similar finding has been reported for *Fomitopsis pinicola* in spruce trunks (Nordén, 1997). When individuals of the same species – or different species with similar combative strength – meet, they produce a confrontation zone that can be visible to the naked eye.

In addition to the polypores, which are the main decayers in this decomposition phase, there are many other structural wood decomposers; these include some corticioids (e.g. the genera *Phlebia, Phanerochaete, Hyphodontia, Tubulicrinis*), agarics (*Pholiota, Psathyrella*), jelly fungi (*Calocera, Dacrymyces*) and some ascomycetes (*Daldinia, Xylaria*).

6.4.5 Advanced wood decay

When most of the structural cell wall components have been decomposed, the succession proceeds to a new phase. This transition takes place when a log develops from decay stage 3 to decay stage 4 (see Box 6.1). Since cellulose and lignin have been broken down to a large extent, the wood has lost its structural strength. One can now pluck the wood apart into small pieces by hand.

In this phase the structural decomposers – including most polypore species – disappear, and another suite of fungi enter the scene. This is evidenced by the distinct shift from polypores to agarics during this phase (Figure 6.8). These agarics might include residual wood decomposers that utilize fragments of the cellulose, hemicellulose and lignin left behind by the structural wood decomposers. However, our empirical knowledge of the enzyme systems, metabolism – and hence correct classification – of these wood decomposers is very weak.

Some polypore species, such as *Trechispora hymenocystis* and *Phellinus nigrolimitatus*, still occur in strongly decomposed wood, but the fungal community is dominated by saprotrophic agarics such as *Pluteus, Galerina, Xerocomus, Resupinatus* and several *Mycena* species. Most of these might very well be residual wood decayers (see Section 3.1.3).

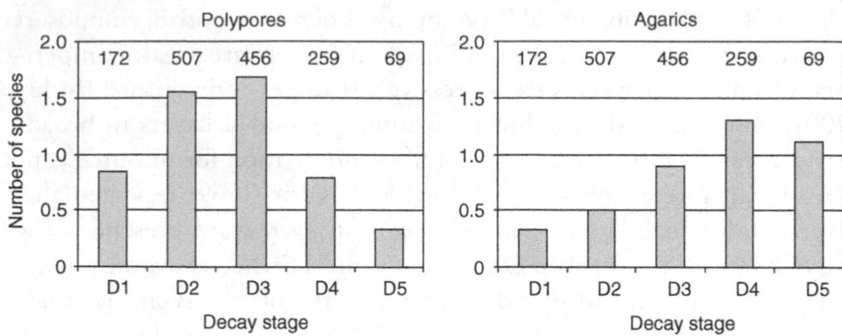

Figure 6.8. Average species numbers of polypores and agarics at different decay stages of fallen *Picea abies* logs in SE Norway. The numbers above the columns represent sample sizes (number of logs). Unpublished data from J. N. Stokland.

Several mycorrhizal fungi regularly appear as fruiting bodies on the surface of medium to strongly decayed wood (Renvall, 1995; Nordén et al., 1999; Tedersoo et al., 2003). One reason for their occurrence on a woody substrate is simply as a place to attach their fruiting bodies. This might be the case for many *Tomentella* and *Pseudotomentella* species (Basidiomycota, Telephorales) that have fruiting bodies loosely attached to the wood surface and that occur as mycorrhizal symbionts on the root tips of trees (Kõljalg et al., 2000). Another reason could be that they actually take part in wood decomposition at the later decay stages. Several mycorrhiza species have mycelia inside rotten wood, and at least some (*Piloderma fallax*, *Tomentellopsis submollis* and *Tomentella crinalis*) are able to grow saprotrophically on sterilized rotted wood at the medium decay stages (Tedersoo et al., 2003).

6.4.6 Final wood degradation

In the final decomposition phase, the stumps and lying wood completely disintegrate into small fragments and become pieces of organic matter in the soil or humus layer. This transition occurs during decay stage 5, which represents the final decomposition phase (see Box 6.1). In particular, brown-rot fragments rich in lignin can persist in the humus layer for a very long time. These wood fragments represent an important structural component of boreal forest soils.

The dominating species in this final decomposition phase are mycorrhizal fungi that started to occur regularly in strongly decayed wood.

Mycorrhizal fungi occurring on above-ground dead wood are the main root symbionts in boreal coniferous forests (Kõljalg et al., 2000). Wood fragments represent an important habitat for them, as they occur with well-developed mycelium inside such wood fragments buried in the soil (Harvey et al., 1979; Kropp, 1982a).

6.5 Invertebrate succession

The observation that invertebrate communities change with the advancement of wood decay has attracted the attention of researchers for a long time. Early works include those of Saalas (1917), Graham (1925) and Ingles (1933). These studies defined different successional phases and described the invertebrate communities typical of each. Saalas, who studied beetles inhabiting Norway spruce in Finland, divided the decomposition into three phases. Later, Schimitschek (1953, 1954) studied the succession of invertebrates on silver fir (*Abies alba*) in Austria and distinguished four phases.

The succession of invertebrates on European beech (*Fagus sylvatica*) in temperate deciduous forests was described in detail by Dajoz (Dajoz, 1966, 1977, 2000). He distinguished three phases in the decay process: the colonization phase, when the wood is invaded by primary saproxylics using intact wood; the decomposition phase, when the primary saproxylics are joined by secondary saproxylics; and finally the humification phase, when the saproxylics are progressively replaced by soil organisms.

Some excellent and very illustrative descriptions of invertebrate succession on Scots pine and Norway spruce have been provided by Bengt Ehnström (Ehnström and Waldén, 1986; Esseen et al., 1992, 1997). Ehnström separated four successional phases (Figure 6.9). These phases appear to be largely the same as those identified by Schimitschek, and as the three phases identified by Saalas, who did not separate the third and the fourth phase defined by Ehnström.

Most of the knowledge on invertebrate communities at different decay stages is built on qualitative observations. Up until now, only a few quantitative studies have covered the whole decay process and described the composition and turnover of invertebrate communities (Vanderwel et al., 2006; Saint-Germain et al., 2007). In the following sections we separate weakened and recently dead trees but otherwise we use the four succession phases defined by Ehnström and also include information from several other studies.

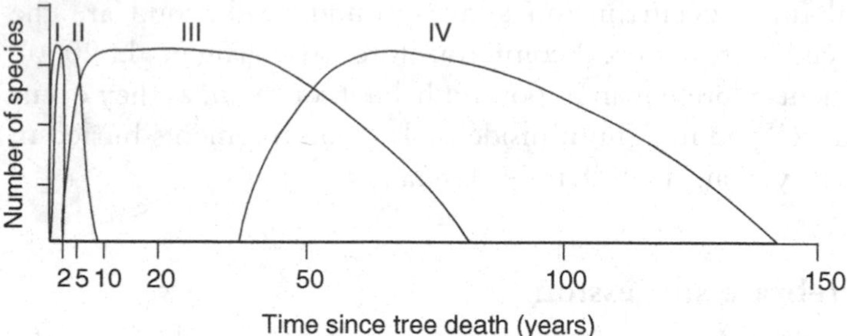

Figure 6.9. Succession phases of invertebrates in a wind-fallen Norway spruce (*Picea abies*) in northern Europe, as defined by Bengt Ehnström (redrawn from Esseen et al., 1992).

6.5.1 Live and weakened trees

Bark beetles (Scolytinae) capable of killing trees are often the first insect species that attack weakened trees. The beetles locate susceptible host trees by means of volatiles, such as monoterpenes, emitted by stressed trees (Byers, 2004). The first-arriving individuals release aggregation pheromones that attract additional individuals of the same species. This can initiate a mass attack where hundreds, or even thousands, of bark beetles bore into the tree and exhaust the resin defences (Raffa et al., 2008). The beetles also bring with them blue-stain fungi that contribute to the tree's death. Bark beetles that attack live trees are often called aggressive species. After the defence has been overwhelmed by a mass attack, the tree becomes available to colonization by other saproxylic invertebrates.

6.5.2 Recently dead trees

When the tree is already dead, there are other bark beetle species that colonize it. Such non-aggressive bark beetles are more diverse than the aggressive ones. Of the 84 bark beetle species that occur in Sweden, 14 species attack live, typically weakened trees (of which two – *Dendroctonus micans* and *Ips typographus* – also attack live, healthy trees), while 70 species utilize recently dead trees (Ehnström and Axelsson, 2002). Bark beetles establish their galleries within a couple of weeks, and newly hatched larvae start to excavate their own tunnels and consume the inner bark (Figure 6.10).

Bark beetles bring with them a diverse community of bacteria, fungi and invertebrates (see Box 6.2). These include blue-stain fungi

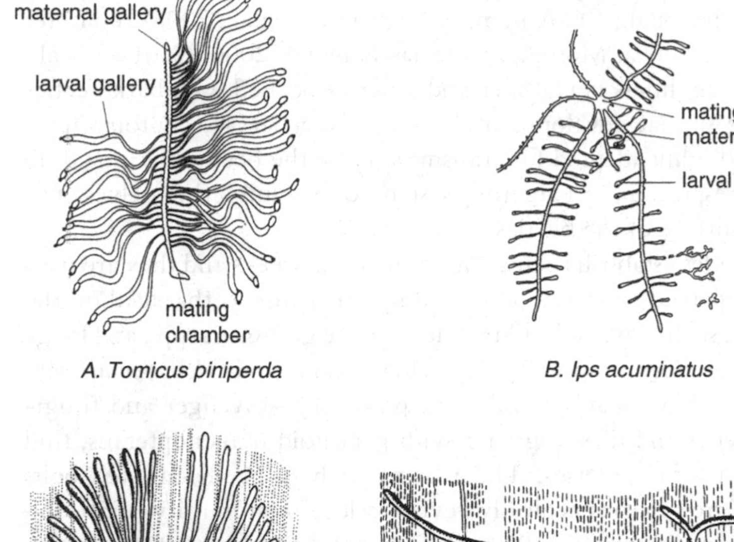

A. Tomicus piniperda

B. Ips acuminatus

C. Scolytus rugulosus

D. Tomicus minor

Figure 6.10. The shape of bark beetle galleries are distinct for most species and it is generally possible to identify all or most species from their galleries. (A) *Tomicus piniperda*, a monogamous species with one female, and (B) *Ips acuminatus*, a polygamous species with several (six are shown in the figure) females per gallery system. In monogamous species it is the female, and in polygamous species the male, that bores under the bark and establishes the first part of the gallery system, the mating chamber. Each female excavates its own maternal tunnel and lays eggs in small notches which it gnaws along the tunnel. The newly hatched larvae start to feed on the inner bark, each excavating its own larval tunnel which becomes wider as the larva grows. The length of the larval tunnels depends on whether the larvae feed mainly on inner bark or on blue-stain fungi. (C) In *Scolytus rugulosus*, the larvae appear to feed mainly on inner bark, judging by the long larval galleries. (D) *Tomicus minor* carries its own blue-stain fungus species (*Ophiostoma canum*), which causes strong staining of the wood (drawings by Juha Siitonen). The fungus appears to constitute the main part of the larval food. The small larvae make a short tunnel in the inner bark and later they bore a hole in the wood surface, where they stay for the rest of their development, feeding on the fungal tissue growing in the galleries.

and various other fungi, bacteria (Cardoza et al., 2006b), protozoa (Wegensteiner et al., 1996), nematodes (Cardoza et al., 2008) and mites (Moser et al., 1989a; Moser and Macias-Samano, 2000; Cardoza et al., 2008). The flightless nematodes and mites associated with bark beetles often have special developmental stages (dauer larvae, deutonymphs) which are highly adapted for transmission by the beetles by intruding into their respiratory or excretory system, or by gluing themselves onto the body surface of the beetles.

Bark beetles establish their galleries in a few weeks and they are usually accompanied or closely followed by other insects that feed on the inner bark, such as weevils (Curculionidae), e.g. *Pissodes* spp., and longhorn beetles (Cerambycidae), e.g. *Acanthocinus* and *Monochamus* spp. in conifers. After that a number of predatory, scavenger and fungivorous beetles and flies, together with parasitoid hymenopterans, find their way to the galleries. The rearing of bark-beetle-infested bolts of a given tree species typically reveals a local community of 50–150 insect species (Nuorteva, 1956; Dahlsten and Stephen, 1974; Langor, 1991; Weslien, 1992; Gara et al., 1995; Herard and Mercadier, 1996). About 40 mite species have been recorded as associated with the spruce bark beetle *Ips typographus* in Europe (Moser et al. 1989a), and almost 100 mite species with the southern pine beetle *Dendroctonus frontalis* (Scolytinae) and allied bark beetles in the southern USA (Moser and Roton, 1971).

Box 6.2 Interactions between bark beetles, fungi and invertebrates in recently dead trees

It has long been known that bark beetles carry blue-stain fungi with them (see, e.g., Craighead, 1928). Knowledge of the taxonomy, ecology and host specificity of these fungi increased markedly during the 1950s (e.g. Mathiesen-Käärik, 1953, 1960b). The role of blue-stain fungi in overcoming the host-tree resistance was intensively studied, particularly during the 1980s and 1990s (see Franceschi et al., 2005; Lieutier et al., 2009, for reviews). As a consequence of the unprecedented and devastating bark-beetle outbreaks during the 2000s in North America, affecting a total forest area of almost 50 million hectares (Raffa et al., 2008), research on bark beetles and their associated organisms has rapidly expanded, leading to many novel and surprising discoveries.

Ophiostomatoid (blue-stain) fungi represent a polyphyletic group of morphologically similar genera of ascomycetes, including *Ophiostoma, Grosmannia, Ceratocystiopsis, Ceratocystis, Gondwanamyces* and *Cornuvesica* (Spatafora and Blackwell, 1993; Zipfel et al., 2006). Several imperfect fungi, such as *Leptographium* and *Graphium* (Wingfield, 1993), have been associated with these genera.

Many blue-stain fungi are associated with bark beetles. Bark beetles carry spores, yeasts and other dispersal stages of fungi either in specialized organs called mycangia, in their gut, or on their body surface (Beaver, 1989; Berryman, 1989; Harrington, 1993; Paine et al., 1997). Mycangia are small cavities in the body, often supplied with glands secreting oily substances, which create optimal conditions for the survival of fungal spores. Those fungi which are symbiotically connected to the beetle are generally carried in mycangia, while the less specific associates are transmitted, often more incidentally, attached to the body surface (Six, 2003). Most of the primary bark beetles attacking weakened trees appear to have at least one or two specialized ophiostomatoid fungal partners (Mathiesen-Käärik, 1953; Solheim and Långström, 1991; Solheim and Saffranyik, 1997; Kirisits, 2004; Bleiker and Six, 2009; Masuya et al., 2009), though even one bark beetle species within a limited geographical region may carry between 10 and 20 ophiostomatoid associates (Kirisits, 2004; Jankowiak, 2005; Kim et al., 2005; Zhou et al., 2006; Alamouti et al., 2007; Yamaoka et al., 2009). Many ophiostomatoid and other scolytid-associated fungi are specific to a certain tree species but unspecific to the vector species; i.e. they can be carried by a range of bark-beetle species inhabiting the same host tree (Kirisits, 2004). Ophiostomatoid fungi produce asexual and sexual spores in slimy masses; the spores are well adapted to dispersal by sticking to insects (Harrington, 1993; Paine et al., 1997).

It is obvious that the fungi benefit from bark beetles as their vectors. But the beetles may also benefit from fungi because (1) the fungi may help them to overcome the host's resistance mechanisms; (2) the fungi produce an environment more favourable for the beetle brood, e.g. by lowering the moisture content in the inner bark; and (3) some of the fungal associates are an important nutrient source for the beetle larvae (Beaver, 1989; Paine et al., 1997; Six and Paine, 1998; Hsiau and Harrington, 2003; Franceschi et al., 2005; Adams and Six, 2007; Bleiker and Six, 2007).

Box 6.2 *(continued)*

Some blue-stain fungi are pathogenic and can help bark beetles to overcome the resistance of weakened trees and to kill the trees. This assumption is supported by the fact that tree death can be induced by artificial inoculation of these fungi. A roughly similar number of inoculations and beetle attacks is required to kill a tree (Franceschi et al., 2005). The mechanism by which the fungus kills the tree is probably hyphal growth into the sapwood, which blocks the water-conducting tracheas (Raffa and Berryman, 1983). Different strains of the same blue-stain fungus species may show highly variable pathogenicity (Krokene and Solheim, 1997; Plattner et al., 2008). Many blue-stain fungi extract the nutrients (particularly nitrogen) effectively from both inner bark and sapwood, and the fungal hyphae provide high-quality nourishment for bark-beetle larvae (Bleiker and Six, 2007). Some bark beetles feed mainly on fresh inner bark tissue, while others feed mainly on the fungal hyphae, with all intermediate cases being possible (Beaver, 1989). Newly hatched teneral adults may also feed on fungal spores before leaving the tree (Bleiker and Six, 2007).

Besides blue-stain fungi, bark beetles can carry several other fungal taxa such as Trichomycetes (Kirschner, 2001), Hypocreales (Kolarik et al., 2008), yeasts (Six, 2003) and other fungi (Beaver, 1989; Lim et al., 2005; Kolarik and Hulcr, 2009). They can also carry basidiomycetous fungi. Decay fungi isolated from bark beetles include, in particular, *Peniophora* and *Entomocorticium* species, but also some common decay fungi such as *Fomitopsis pinicola* and *Heterobasidion annosum* (Castello et al., 1976; Whitney et al., 1987; Hsiau and Harrington, 2003). Some of these may actually be more important for the nourishment of beetle larvae than the ophiostomatoid fungi (Whitney et al., 1987; Hsiau and Harrington, 2003). The common primary decomposer fungus, *Phlebiopsis gigantea*, produces windborne spores from its fruiting bodies growing on stump and log surfaces, but it also produces asexual spores (arthroconidia) in the pupal chambers of bark beetles. These spores provide food for the beetles and a mechanism of dispersal by insect vectors for the fungus (Hsiau and Harrington, 2003).

Fungi associated with bark beetles are not always useful; they may be pathogenic, opportunistic or antagonist as well. *Trichoderma* (Ascomycetes) and *Aspergillus* (mould) species occur in mountain pine beetle (*Dendroctonus ponderosae*) galleries and can reduce beetle survival and reproduction (Cardoza et al., 2006a). It was only recently discovered that the beetles have an antifungal weapon against the harmful fungi. The adult beetles exude oral secretions and smear them along their tunnels with their legs. This secretion contains several species of highly antifungal bacteria, such as the actinobacteria *Micrococcus luteus*, which efficiently inhibit the growth of harmful fungi (Cardoza et al., 2006a). The blue-stain fungi also change the composition of the volatile compounds released from bark-beetle-attacked trees; these volatiles may be used by parasitoids of the beetle to locate hosts (Adams and Six, 2008; Boone et al., 2008). Unlike predatory beetles, which use bark-beetle pheromones as kairomones to locate their prey, parasitoid hymenopterans and dipterans attack the life stages present only after the pheromone production of adult beetles has ceased (Boone et al., 2008).

Bark beetles also transmit many flightless invertebrate species into new trees. These species can have multiple roles and can belong to different trophic levels, including phloeophagous, mycophagous, predatory and parasitic species. Some of these invertebrates feed on the blue-stain fungi carried by the bark beetle, but some bring their own fungal associates with them. It was discovered in the 1980s that phoretic mites carried by bark beetles can transmit their own blue-stain fungi. Mites in the genus *Tarsonemus*, phoretic on the southern pine beetle (*D. frontalis*), were shown to carry the blue-stain fungus *Ophiostoma minus* (Bridges and Moser, 1986), and also to possess a special morphological adaptation called sporotheca for transporting fungal spores (Moser, 1985). It was assumed that the fungus would aid beetle development. However, later it became evident that *O. minus* actually reduces the larval survival of *D. frontalis* by outcompeting the two mutualistic blue-stain fungi carried by the beetle (Hoffstetter et al., 2006). It was recently shown that *D. frontalis* also uses actinobacteria against *O. minus*, a blue-stain fungi harmful to the bark-beetle brood (Scott et al., 2008). 'Hyperphoretic' spores of fungi have been found on many other mite species too (Moser et al., 1989b), but the relationships between these fungi, mites and their bark beetle vectors have not yet been studied in detail. In addition, the transfer of nematodes by bark beetles has been known for a long time, but it was only recently discovered that bark beetles may have special organs for carrying nematodes, consequently named nematangia (Cardoza et al., 2006b).

In their review, Kenis et al. (2004b) list a total of 115 insect species, mainly beetles and flies, recorded as potential predators of European bark beetles. Some predatory beetle species live exclusively in bark beetle galleries, among others several hister beetles (Histeridae), e.g. *Cylister, Platysoma* and *Plegaderus* spp., some root-eating beetles (Rhizophagidae) in the genus *Rhizophagus*, and bark-gnawing beetles (Trogossitidae) in the genus *Nemosoma*. Both adults and larvae of checkered beetles (Cleridae) in the genus *Thanasimus* are voracious predators; adults hunt bark beetles on the bark surface, while larvae prey on bark beetle larvae in their galleries. Many other beetles frequently found in bark-beetle galleries have traditionally been assumed to be predators of bark-beetle eggs or larvae. These include sap beetles (Nitidulidae), e.g. *Epuraea* spp., many rove beetles (Staphylinidae), e.g. *Placusa*, and *Phloeonomus* spp., darkling beetles (Tenebrionidae) in the genus *Corticeus*. However, many of these species are probably fungivorous and may act only facultatively as predators or scavengers (see Sections 3.4 and 3.5.2). For instance, larvae of *Placusa* have been shown to be fungivores based on both the structure of their mouthparts and analysis of their gut contents (Ashe, 1990). Dipteran larvae often occur abundantly in bark beetle galleries, e.g. long-legged flies (Dolichopodidae) in the genus *Medetera*, and lance flies (Lonchaeidae) in the genus *Lonchaea*.

Kenis et al. (2004b) list a total of 175 species of hymenopteran parasitoids recorded from different European bark beetle species. Coniferous and deciduous trees host almost completely separate parasitoid assemblages. The host specificity of different parasitoid species varies greatly. Some of them are polyphagous and attack various bark beetles in different tree genera, whereas others are linked to a particular bark beetle host species. The most important parasitoid families of bark beetles are Braconidae, Pteromalidae and Eurytomidae. The larvae of longhorn beetles (Cerambycidae) and weevils (Curculionidae), which live in fresh inner bark, are generally larger than bark beetle larvae and therefore have a completely different set of parasitoids (Kenis and Hilszczanski, 2004; Kenis et al., 2004a).

6.5.3 The subcortical space and initial wood decay

In the boreal zone, a new successional phase usually starts during the second year after tree-fall, and lasts up to a few years, depending on environmental conditions. This is phase II in Figure 6.9 and corresponds to decay stage 2 in Box 6.1. It is characterized by secondary phloem-feeders that consume the remaining inner bark, and species associated with fungi growing between the bark and the wood surface. These species usually make no recognizable galleries or tunnels and move freely under the loose bark. The phloem-feeders and fungus-feeders have their own predators, parasitoids and other associates that differ from those associated with the bark beetles.

Many beetle species in this successional phase have a flattened body form, both as larvae and adults, which is an adaptation to the subcortical habitat. Typical secondary phloem-feeders include the larvae of fire-coloured beetles in the genera *Pyrochroa*, *Schizotus* and *Dendroides* (Pyrochroidae) as well as larvae of *Pytho* spp. (Pythidae). The larvae of predatory beetles may look morphologically similar, but they move much faster and can be separated from the phloem-feeders (with blunt mandibles) by their sharp, pointed mandible apices (Smith and Sears, 1982). Many flat bark beetles (Cucujidae), such as *Cucujus*, *Dendrophagus* and *Laemophloeus* (Cucujidae), are typical for the subcortical space. A wide variety of dipteran species belonging to many different families can be found in these circumstances, including lance flies (Lonchaeidae), long-legged flies (Dolichopodidae), wood-soldierflies (Xylomyidae) in the genus *Solva*, and awl-flies (Xylophagidae) in the genus *Xylophagus*.

6.5.4 Intermediate wood decay

The third successional phase starts when the inner bark has been consumed, and the subcortical habitat disappears because the bark falls off. This phase corresponds closely to the fungal phase 'initial to intermediate wood decay', and is characterized by fungivore insects feeding on fungal fruiting bodies or mycelium-ridden wood. Even larvae that bore tunnels through the wood may actually feed mainly on fungal mycelia (Tanhashi et al., 2009). When the first polypore species start to produce fruiting bodies, the number of fungivorous invertebrates increases rapidly (see Chapter 7, especially Box 7.2). The main decomposer fungus, usually a polypore, may determine which invertebrate species inhabit the trunk. For example, birch snags decayed by different polypores (*Fomes fomentarius*, *Phellinus igniarius* or *Fomitopsis pinicola*) were found to host distinct beetle assemblages (Kaila et al., 1994).

The larvae of longhorn beetles (Cerambycidae), stag beetles (Lucanidae), and bark-gnawing beetles (Trogossitidae), e.g. *Peltis grossa*, are efficient decomposers advancing the physical breakdown of wood at this stage. In drier wood, the larvae of wood-boring Buprestidae and Anobiidae may be dominant. In Diptera, the larvae of certain hoverflies (*Temnostoma* spp.) and crane flies (Tipulidae) bore tunnels in the decaying wood of broadleaved trees. It seems that the proportion of fungivore species increases in relation to wood-boring detritivore species as the decay advances (Vanderwel et al., 2006).

6.5.5 Advanced wood decay

The fourth successional phase starts when much of the sapwood has been consumed and the heartwood starts to decay. Relatively few wood-feeding insects and their predators are present at this stage. But some of the stag beetles (Lucanidae) that colonize wood in the intermediate decay phase can stay in the wood for several generations until much of the wood is completely consumed. Click beetles (Elateridae) is another group that regularly occurs in strongly decayed wood. Wood-consuming species are slowly replaced by litter-dwelling invertebrates that use the trunk for cover (e.g. molluscs), hibernation and aestivation (many carabids), preying (centipedes) or nesting (ants).

6.6 Succession of mosses and lichens

Bryologists have often used more fine-tuned decay stage classifications, with up to eight decay stages (Söderström, 1988a). These decay stages have more emphasis on bark cover and the surface structure of fallen logs. In the succession of epixylic bryophytes, four different species groups have been distinguished (see, e.g., Söderström, 1988a).

1. *Facultative epiphytes* grow on living trees, and continue to occur on logs until the intermediate stages of decay.
2. *Epixylic specialists* have the highest number of species at the mid-stages of decay.
3. *Opportunistic generalists* occur on many kinds of substrates, and can colonize logs in different decay stages.
4. *Competitive epigeics* cover the forest floor, and gradually overgrow decomposed logs and displace epixylic species.

The succession of epixylic mosses and lichens is described in more detail in Chapter 4.

6.7 Overview of the decay succession

6.7.1 Fungal and invertebrate interactions

The successions of fungi and insects in decaying wood have typically been studied independently of each other. But these organism groups interact in several ways. Here we comment on these interactions and examine whether the succession phases of fungi and invertebrates are synchronous and whether the same factors bring about the transition from one phase to another. We also point out that the enzyme systems of fungi and the dietary specialization of insects closely match the resources that become available during the decomposition process (see section 3.2). These differences represent the keys to understanding the main patterns in species turnover during the succession stages.

Phase 0–I. From a nutritional point of view, the fungi and invertebrates occurring in living, wounded or recently dead trees are very similar. Both groups consume inner bark tissue, the nutrients being transported in the sapwood, or the cell contents of living or recently dead sapwood cells (see Chapter 3 for a detailed discussion). These nutrient types are composed of relatively small organic molecules that are easy to decompose or digest.

In this phase we find intimate associations between insects and fungi. We have already mentioned that different sapstain (including blue-stain) fungi are transported by sap beetles and bark beetles to new wounds and trees. Furthermore, we find a gradient from loosely associated fungi and insects occurring in the same tree to intimate symbiotic dependence between ambrosia beetles and ambrosia fungi (see Chapter 3).

Most of the inner bark is consumed very rapidly, typically during the first summer in the boreal and temperate zones. In addition, the ambrosia beetles boring into the sapwood of dying or dead trees and the ambrosia fungi rapidly deplete their resources and stay for just one generation in a single tree. Thus, the species in phase one have a very rapid turnover and this phase is quite similar in fungi and insects.

Phase II. In the invertebrate succession, this phase is mainly characterized by the species living under detached bark. This phase does not correspond to a similar phase of fungal succession, although there might be distinct fungi (different from those occurring in phase I) living under detached bark representing a community that has been neglected by mycologists. Typical fungi are moulds and mycelia of wood-decaying fungi that often occur abundantly under loose bark. Neither moulds nor sterile mycelia of basidiomycetes in decaying wood have received much attention from mycologists, so we know little about these fungi.

Inside the wood, we find that polypores and corticioid fungi have started to decompose structural wood components (cellulose, hemicellulose and lignin) during the period corresponding to insect phase II. Their activity is hardly affected by the loss of bark that terminates insect phase II. Instead, their activity continues for several years or decades, paralleling insect phase III.

Phase III. This phase in the insect succession coincides with the peak activity of fungi breaking down the structural wood components (see Figure 6.6). The fungi are the main wood decomposers (see Chapter 2), and the fruiting bodies, hyphae and mycelium-ridden wood constitute the main nutrition source of many invertebrates. However, the insects that bore into weakly to moderately decayed wood also contribute to the physical decomposition of the wood (Ausmus, 1977; Hanula, 1996), and provide routes of access and food for yet other organisms.

It is possible that invertebrates contribute significantly to fungal dispersal by transmitting fungal spores and conidia (asexual dispersal units) to new dead trees. Until now, there is mainly circumstantial evidence suggesting this role of the invertebrates. Entomologists are well aware that large numbers of adult insects are present and feed on sporulating fruiting bodies during the period when they are visiting dead trees for egg-laying. Many of these insects have hairs and bristles, and may be almost completely covered with spores attached to them. But to what extent different fungi are dispersed by wind or by insects is still largely unknown.

While the species turnover is rather abrupt from living to recently dead trees, and from recently dead to weakly decayed trees, the transitions become less clear-cut between the subsequent succession phases. This is illustrated in Figures 6.7 and 6.9 by a larger overlap in species composition between intermediate and late succession phases.

Phase IV. This final phase in the insect succession corresponds more or less to the last two phases of fungal succession described above. We have much less knowledge about the enzymatic activity and feeding modes for insect species occurring towards the end of the decomposition process.

The breakdown of the structural integrity of the wood is probably an important factor that determines the transition from invertebrate phase III to phase IV. During most of phase III the wood was still quite hard and was inhabited by insects with sclerotic head capsules, forceful mandibles, and the ability to bore tunnels or excavate chambers inside the wood. But when the wood disintegrates and cracks up into smaller pieces it becomes accessible to soft-bodied invertebrates such as gnats (Diptera), springtails (Collembola) and earthworms (Annelidae). It might also be that the transition of invertebrate species is related to the microorganisms decaying the wood.

6.7.2 Species richness patterns

In addition to the distinct species turnover with advancing wood decay, there are also distinct species richness patterns during the decomposition process. In the basidiomycetes we find a distinct hump-shaped pattern, with the greatest diversity in the intermediate decay stages. This pattern is typical for number of species per individual log and total number of species associated with each decomposition stage (Figure 6.11). This pattern has been demonstrated in several studies of coniferous trees (Renvall, 1995; Lindblad, 1998) and broadleaved trees (Heilmann-Clausen and Christensen, 2003). A similar pattern has also been found for yeasts (González et al., 1989).

The explanation for the hump-shaped pattern in basidiomycetes seems to have two components. Firstly, there are more species that are confined to intermediate decay stages rather than to early and late stages. This is evident in the study of Renvall (1995), who scored each

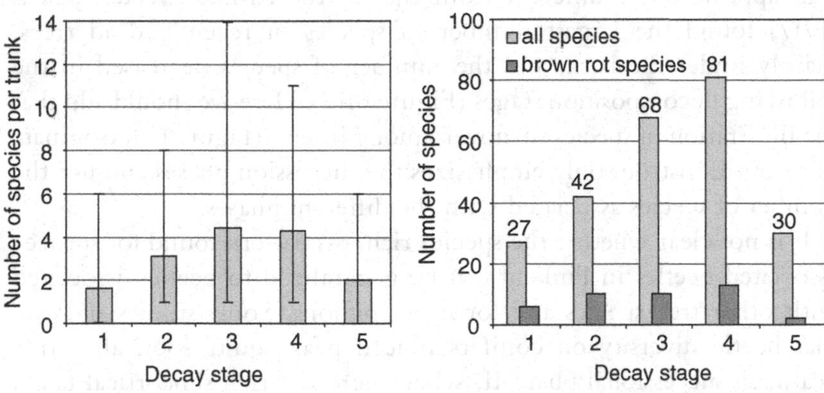

Figure 6.11. Number of basidiomycete species with fruiting bodies on decaying pine logs in Finland (Renvall, 1995).

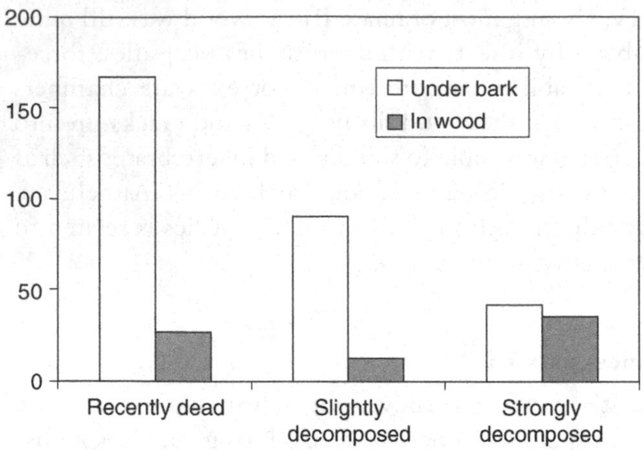

Figure 6.12. Numbers of beetle species found in the first three decomposition stages of Norway spruce in Finland (Saalas, 1917). The study was based on an extensive survey (hundreds of trees examined in different parts of Finland), with 295 beetle species assigned to the preferred decomposition stage. Note that the total in the figure (based on a table given in Saalas 1917) is 374, so there is some overlap in species composition between the stages.

species according to decay preference and found a higher number of species preferring intermediate decay stages than early and late decay stages. Secondly, in addition to those species preferring intermediate decay stages, both early and late decay species can be found (with lower frequency) on logs in the intermediate decay phase (Figure 6.7).

In invertebrates, or at least among beetles, the species richness pattern appears to be different from that of the basidiomycetes. Saalas (1917) found the largest number of species in recently dead trees, mainly under the bark, and the number of species decreased in the following decomposition stages (Figure 6.12). Here we should add that the illustration of species turnover among insects (Figure 6.9) originating from Ehnström only emphasizes the succession phases and not the number of species associated with the different phases.

It is not clear whether the species richness pattern found for spruce-associated beetles in Finland can be generalized to beetles associated with other tree species and for other regions. Some studies indicate that beetle diversity on conifers indeed peaks quite soon after tree death, in successional phase II, when there is a rich subcortical fauna (Howden and Vogt, 1951; Ulyshen and Hanula, 2010). However, other studies indicate that, in broadleaved trees, species richness may increase with decay, at least until the mid-decay stages (Hammond et al., 2004; Lindhe and Lindelöw, 2004; Saint-Germain et al., 2007). Successional patterns have been far less studied in Diptera than in Coleoptera, but since most saproxylic dipterans are fungivorous, it is probable that their diversity is at its highest in mid- or late stages of decay. The few studies available indeed show that species richness of Diptera, particularly fungus gnats (Mycetophilidae etc.) and Sciaridae, increases with decay and is highest in the late stages of decomposition (Irmler et al., 1996; Hövenmeyer and Schauermann, 2003).

7 · *Microhabitats*

Juha Siitonen

Both living and dead trees provide a number of distinct microhabitats for saproxylic species. By 'microhabitats' we mean discrete parts of a tree that host different species assemblages. Some of the microhabitats are only present in living trees, typically in mature or old individuals (Figure 7.1), and additional ones become available after the trees die. Wounds, rot holes and cavities; attached dead branches and roots; phloem (inner bark), sapwood and heartwood; fruiting bodies and mycelia of decomposer fungi etc. each host very different species assemblages. In this chapter, we describe in more detail the various dead-wood microhabitats, and the many ways in which saproxylic species can be dependent on their particular microenvironments. We present the microhabitats in the order in which they appear in a tree, starting from wounds and sap exudations that can occur even in young trees, followed by cavities and other microhabitats that develop as a result of decay in mature living trees, and ending with those microhabitats, such as subcortical space, that become available only after a tree has died. Some microhabitats, especially sap exudations and cavities, are far more common in broadleaved than in coniferous trees.

7.1 Wounds and sap exudations in living trees

Many factors can cause injuries to trees. Mechanical damage includes branch and stem breakages by wind, wounds made by falling trees hitting adjacent trees, lightning strikes, fire scars, frost cracks and snow breakage. Pathogenic fungi and bacteria can create necrotic patches, termed cankers, in the phloem layer. Most of the fungi that induce cankers kill the phloem and do not bring about sap flows. 'Bleeding cankers' are generally caused either by bacteria (Schmidt et al., 2008) or by *Phytophthora* species (Oomycetes) (Brown and Brasier, 2007) and produce only small amounts of exudates. Some vertebrate species, such as woodpeckers, and many invertebrate species can injure the bark,

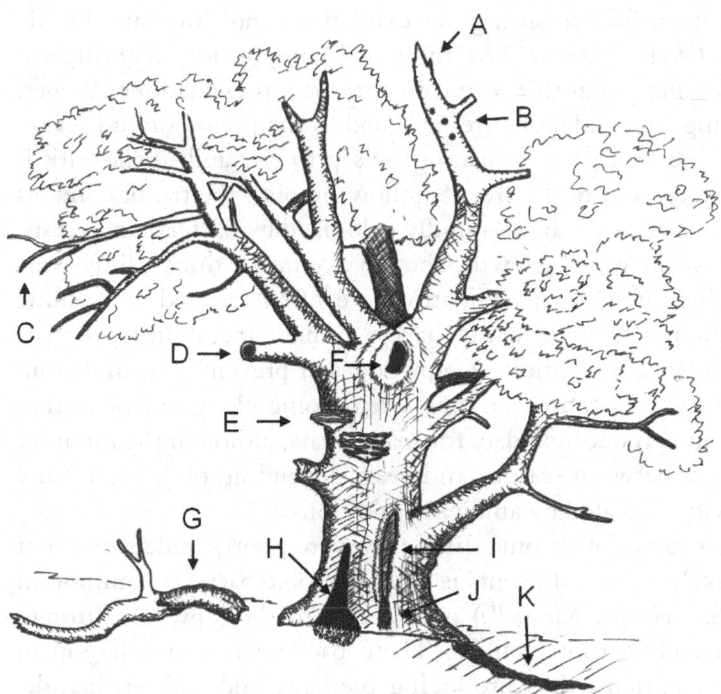

Figure 7.1. 'Arboreal megalopolis' (*sensu* Speight, 1989). As the tree ages, parts of it become damaged or dysfunctional, thus providing microhabitats which saproxylic species can colonize. Microhabitats within a living, senescent tree are marked with arrows and letters in the figure. **A** = dead sun-exposed limb, **B** = woodpecker holes, **C** = dead attached branches in the canopy, **D** = branch cavity, **E** = polypore fruiting bodies, **F** = trunk cavity, **G** = fallen branch on the ground, **H** = basal cavity, **I** = open wound surrounded by callus tissue, **J** = sap exudation, **K** = dead root in the soil (drawing by Juha Siitonen).

which may result in sap exudation and provide an opportunity for microbial colonization. For instance, cicadas (Hemiptera: Cicadidae) are known to cause sap exudation by piercing the bark with their stylet-like mouthparts to feed on sap (Yamazaki, 2007), and carpenterworm (Lepidoptera: Cossidae) larvae can initiate and maintain sap exudation by gnawing the bark (Yoshimoto and Nishida, 2007).

7.1.1 Sap exudations

Two kinds of sap exudation ('sap exudation' and 'slime flux' have been used as synonymous terms) can be distinguished on broadleaved trees:

vigorous but short-lived spring sap exudations and slow but chronic exudations (Weber, 2006). The microbial composition in spring sap exudations differs from that found in chronic sap exudations (Weber, 2006). Spring sap exudations are confined to the period of bud break, when the high root pressure causes the sap to rise and, if the bark is injured, to bleed out of the tree. Sap flowing onto the trunk contains sugars and amino acids, and is rapidly colonized by bacteria and yeasts. As a result of microbial growth, the sap assumes a thick, slimy consistency, which is why sap exudations are often referred to as slime fluxes. Although trees generally have a remarkable ability to repair injuries, bacterial infestation of sapwood can prevent a wound from healing. The sap exudation can therefore become a long-lasting feature of the tree and produce exudate for many years, although the intensity of flow varies between seasons and years depending on bacterial and insect activity mediated by ambient conditions.

The exact causes of chronic slime fluxes are poorly understood, but the most likely course of events is that anaerobic bacteria common in soil (e.g. *Enterobacter, Klebsiella*) gain access to the sapwood through wounds, including root wounds. These bacteria ferment the sap in anaerobic conditions, thus producing methane and carbon dioxide. The gases cause pressure in the wood, which forces the exudate out. In some cases the bleeding has been associated with the so-called 'wetwood syndrome' of trees (Schink et al., 1981; Murdoch and Campana, 1983; Murdoch et al., 1983). Many of the microbial species occurring in chronic slime fluxes seem to be specialized to this microhabitat, as they have never been found in other substrates (Phaff and Knapp, 1956; Bowles and Lachance, 1983; Pore, 1986, Kerrigan et al., 2004; Weber et al., 2006). Furthermore, there is a high degree of host-tree specificity, which may be enforced by selection due to the nutrient composition of the sap or the presence of host defence compounds (Lachance et al., 1982; Weber, 2006). Yeast assemblages also differ between individual sap exudations of the same tree species, depending upon the environmental conditions (Bowles and Lachance, 1983) and which insect species are the first to colonize the exudation and inoculate different fungi into it (Ganter et al., 1986).

Sap exudations rapidly attract a large number of insect species (Wilson and Hort, 1926; Sokoloff, 1964; Ratcliffe, 1970; Yoshimoto et al., 2005). This assemblage mainly consists of opportunistic non-saproxylic species using the resource only as adults for feeding. For instance, over 100 insect species were recorded visiting sap exudations of oaks (*Quercus acutissima*) in Japan; the most abundant groups being fruit flies (Drosophilidae), ants (Formicidae), sap beetles (Nitidulidae) and rove beetles (Staphylinidae) (Yoshimoto et al., 2005). Other typical adult visitors include nymphalid butterflies (Nymphalidae), moths (Noctuidae etc.), horntails and other wasps (Vespidae), but also certain saproxylic species such as stag beetles (Lucanidae), flower beetles (Cetoniinae), longhorn beetles (Cerambycinae), and hoverflies (Syrphidae). Most adult guests use other sources of food too – such as nectar produced by flowers, or honeydew secreted by aphids – and they feed on sugars or ethanol and other fermentation products that the sap contains. The availability of sap exudations may be important for the adult survival and reproductive success of some saproxylic insects such as flower beetles and longhorn beetles.

Some insect groups are entirely specialized to live in the slime fluxes that accumulate in bark crevices and under the loose bark around wounds. Dipteran species which live as larvae in slime fluxes are found in many families, such as hoverflies (Syrphidae), e.g. *Brachyopa* spp., long-legged flies (Dolichopodidae), fruit flies (Drosophilidae), and some gnats (Mycetobiidae) (Rotheray and Gilbert, 1999; Rotheray et al., 2001; Alexander, 2002). Beetle species with similar ecology include sap beetles (Nitidulidae), e.g. *Cryptarcha*, *Soronia*, *Carpophilus* and *Amphicrossus* spp., and tree-wound beetles (Nosodendridae). The sap-living beetle larvae have morphological adaptations such as the spiracles (openings leading to the respiratory system) placed at the ends of tubular projections, enabling the larvae to live immersed in the liquid substrate (Lawrence, 1989). Moreover, structures such as comb-hairs in the larval mouthparts indicate adaptation to filter-feeding (Lawrence, 1989). The main food of larvae living in sap exudations is not the sap itself but the bacteria and yeasts growing in it.

Because of the presence of fly and beetle larvae, sap exudations attract predatory and parasitoid insects. Most of these are generalists that also occur in other media, such as decomposing fungi, carcasses etc. Typical species include hister beetles (Histeridae) and many rove beetles (Staphylinidae). A small, histerid-like beetle family (Sphaeritidae) with predatory larvae appear to be true specialists of sap exudations (Crowson, 1981).

7.1.2 Wounds

Small wounds usually heal completely and are overgrown by callus tissue, i.e. secondary cambium that grows over the wound and forms secondary sapwood inwards and secondary bark outwards. However, sometimes larger wounds do not heal completely but remain open. Some saproxylic species breed mainly in barkless wounds of living trees and the secondary wood formed around them, e.g. the beetle species *Anobium nitidum* (Anobiidae) and *Leptura revestita* (Cerambycidae), and the clearwing moths (Sesiidae) *Sesia melanocephala* and *Synanthedon myopaeformis* (Ehnström and Axelsson, 2002). Other species frequently found in barkless lesions of living trees more often colonize dead standing, barkless trees, or sometimes the hard inner walls of trunk cavities. However, barkless lesions on living trees can offer important microhabitats in environments where dead standing trees are absent, such as in urban parks. Typical wood-boring invertebrates colonizing exposed areas of hard dead wood include many deathwatch beetles (Anobiidae) and powderpost beetles (Bostrichidae, Lyctidae), as well as certain jewel beetles (Buprestidae) and false click beetles (Eucnemidae).

7.2 Cavities and hollow trees

Rot holes and trunk cavities represent another type of microhabitat – or, rather, a range of microhabitats – present in living trees. Trunk cavities often start to develop as a consequence of branch breakage providing free access for heart-rot fungi. Cavities develop through the combined action of heart-rot fungi, invertebrates and physical breakdown of the decaying wood (see also Chapter 4). Depending on the size and exposure of the opening, cavities can be dry, moist or wet. This, in turn, affects the species composition of both fungi and insects colonizing the treehole and its further development. Each cavity carries a fauna consisting of organisms that continue to enlarge it, others that arrive to predate them, and yet others that use microhabitats created within the treehole by the accumulating debris and remains of the diversifying saproxylic community (Speight, 1989).

Hollow trees with wood mould (see below) at the bottom of the cavity are exceptionally long-lasting and stable dead-wood microhabitats. Some temperate tree species with high longevity, such as oaks and limes, can reach a maximum age of well over 500 years. These tree species start to become hollow and contain wood mould at the age of *c.* 150–200 years (Ranius et al., 2009a). This means that some cavities in living trees may be continuously available for several hundred years (see Figure 16.3). Saproxylic species adapted to hollow living trees appear to have low dispersal propensity as compared with species associated with snags and logs (McLean and Speight, 1993; Nilsson and Baranowski, 1997; Hedin et al., 2008).

7.2.1 Cavity development

Heart-rot fungi are the key organisms in the development of cavities and hollow trees. They are generally divided into butt rots which originate in the roots and spread upwards, and top rots which originate in the crown and spread downwards. Thus, top rots form cavities that are not connected to the ground, whereas butt rots cause the trunk to become hollow at the base, and the cavity is connected to the ground. Microclimatic and other physical conditions, and thereby species composition, differ clearly between top-rot trunk cavities and butt-rot basal cavities. Top rots can colonize the tree by spores through wounds that are large enough to expose dead heartwood. Large enough limbs and branches already have a core of dead heartwood which is connected to the heartwood of the main trunk, and branch breakage thus provides an avenue of entry.

Although it is generally well known which fungus species cause heart rot in different tree species, and whether they cause butt or top rot, detailed knowledge on the relative importance of different fungus species in causing different types of cavities is largely lacking. In the European temperate deciduous forests, important fungus species forming cavities include *Laetiporus sulphureus* mainly in oaks (*Quercus*); *Polyporus squamosus* in elms (*Ulmus*), European ash (*Fraxinus excelsior*) and sycamore (*Acer platanoides*); and *Ganoderma* species causing butt rot on oaks, limes (*Tilia*), European beech (*Fagus sylvatica*) and many other broadleaved tree species (Rayner and Boddy, 1988; Schwarze et al., 2000b). In most cases there are several fungus species decaying the hollow trees at the same time or in succession. For instance, *Ganoderma lipsiense* was the principal fungus decaying urban limes (*Tilia*) and making the trees hollow in southern Finland. In hollow lime trees, *Pholiota* spp. and *Hypholoma* spp. were frequently isolated from the hollow walls (Rayner and Boddy, 1988; Schwarze et al., 2000b; Terho et al., 2007; Terho and Hallaksela, 2008). Besides the main decayer species, tree hollows can host large numbers of more inconspicuous fungal species. No less than 186 morphospecies of wood-decaying fungi (none of which could be matched with the reference collection of 130 known wood-decaying basidiomycete species producing sporocarps) were found in decay columns of living, hollow *Eucalyptus obliqua* trees in Tasmania (Hopkins et al., 2005).

Tree cavities typically develop from recently established small rot holes into large cavities. However, the possible trajectories and rates of cavity development are poorly known. They may vary a lot depending on tree species, fungus species causing the decay, invertebrate species inhabiting the cavity, and environmental conditions. Park et al. (1950) were among the first to draw attention to the successional development of treeholes. They also presented an outline of the invertebrate food web occurring in treehole microhabitats. Kelner-Pillault (1974) and Jansson (1998) presented a schematic illustration for the development of cavities and differentiated four main development stages (Figure 7.2).

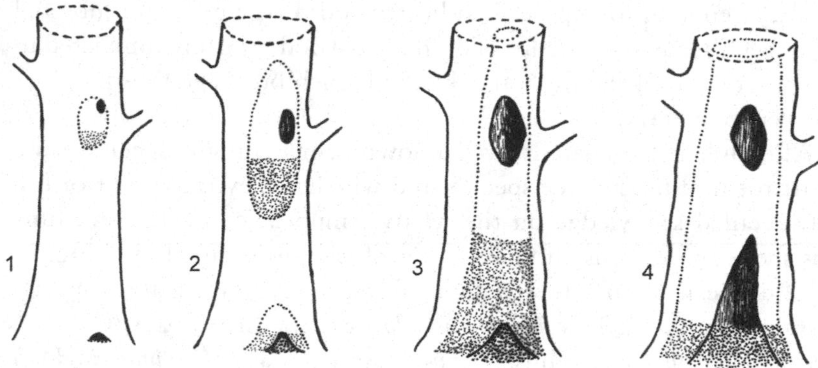

Figure 7.2. Four different development stages (1–4) of hollows in oak. (**1**) The development of a cavity starts, in most cases, as consequence of branch breakage. In the first stage, the size of the opening is small (diameter *c.* 5 cm) and the amount of wood mould small. (**2**) The decay has extended downwards and upwards in the heartwood, but the hollow is not connected to the ground. The opening is medium-sized (diameter *c.* 15 cm) and the amount of wood mould is usually large. The upper layer often consists of nest material of cavity-nesting birds, and the mould is moist because water is efficiently retained in the hollow. Another cavity may start to develop at the base of the tree. (**3**) The whole basal part of the trunk becomes hollow and connected to the ground. The opening is large (diameter 30 cm or larger) and the amount of wood mould is at its greatest. However, the wood mould (as well as water) may begin to leak out of the hollow through openings formed near the ground. (**4**) Part of the wall has decayed and the hollow is open. Wood mould lays over the soil and is easily flushed out of the hollow. Modified from Jansson (1998) and Hultengren and Nitare (1999).

7.2.2 Wood mould and other microhabitats in hollow trees

As a cavity develops and becomes larger, it becomes structurally more complex, and species diversity in it generally increases. Heartwood decayed by fungi is subsequently colonized by invertebrates boring into the wood, thus converting the wood into frass (borings and excrement produced by insect larvae) which starts to accumulate at the bottom of the expanding cavity. In addition, tree cavities that are open to the outside will annually receive an input of litterfall, including dead leaves, twigs and seeds. Birds and mammals can bring in a varied assortment of plant and animal debris. Furthermore, animal faeces and carcasses enrich the cavities with a considerable nitrogen input. The decomposing layer of loose wood remains and other debris formed inside a hollow tree has been referred to as treehole mould (Park et al., 1950), tree humus (Speight, 1989) or wood mould (Ranius and Nilsson, 1997; Dajoz, 2000).

Wood mould is the principal substrate in hollow trees. The quality of wood mould varies depending on whether the main decomposer fungi cause brown rot or white rot, how much litter from outside enters the cavity, and what kind of microclimatic conditions, particularly moisture and temperature, prevail in it. The internal cavity environment is much more stable with respect to temperature and relative humidity than the exterior (Park and Auerbach, 1954; Kelner-Pillault, 1974; Sedgeley, 2001). If the water content of wood mould is sufficiently high, a cavity is cooler during the day because of evaporation, and warmer during the night than the exterior. In about 50 cavities studied by Park and Auerbach (1954), the water content of wood mould varied from only a few percent to 90% but was between 40% and 80% in most cases. The nitrogen content of wood mould is about 1%, which is three or four times higher than in undecayed wood, and two or three times higher than in decaying wood (Kelner-Pillault, 1974; Jönsson et al., 2004).

The wood mould fauna is taxonomically very diverse. It consists of both generalist species that can dwell on the forest floor (these are more dominant in cavities that are connected to the ground), and strict specialist species confined only to wood mould. The average density of arthropods is in the order of 2500 individuals per kilogram of wood mould (Park and Auerbach, 1954), with mites and springtails (Collembola) being the numerically dominant groups. Beetles (Coleoptera), dipterans (Diptera) and parasitoid wasps (Hymenoptera) are found in practically all trunk cavities, and beetle larvae often dominate the community in terms of biomass. Most beetles appear to favour the reddish-brown, crumb-like or powdery mould typical of oaks decayed by *Laetiporus sulphureus*. On the other hand, the moist, dark wood mould typical of several other broadleaved tree genera, e.g. elms, ashes and horse chestnuts, seems to be favoured by dipterans (Andersson, 1999; Alexander, 2002). Among other taxa, nematodes (Nematoda), isopods (Isopoda), centipedes (Chilopoda), spiders (Aranea), harvestmen (Opiliones) and pseudoscorpionids (Pseudoscorpionida) (Park and Auerbach, 1954) are frequently found in trunk cavities.

The larvae of certain flower beetles (Scarabaeidae, Cetoniinae) are among the most characteristic and functionally important inhabitants of hollow trees. In Europe, the principal species are the hermit beetle (*Osmoderma eremita* complex) (see Box 7.1), *Gnorimus* spp. and *Liocola* spp. The larvae live deep in the wood mould, in the border zone between soft and hard wood, and consume the decaying walls of the cavity. By doing so they expand the cavity efficiently and produce large amounts of frass. Other typical beetle families inhabiting hollows with wood mould include comb-clawed beetles (Alleculidae) – e.g. *Allecula*, *Mycetochara*, *Prionychus* and *Pseudocistela* spp., click beetles (Elateridae) – e.g. *Ampedus*, *Elater*, *Ischnodes* and *Limoniscus* spp. and darkling beetles (Tenebrionidae) – e.g. *Neatus picipes*, *Tenebrio obscurus* and *Uloma culinaris* (Kelner-Pillault, 1974; Martin, 1989; Ranius and Jansson, 2000; Ranius, 2002a). The larvae of all these species are morphologically convergent, being cylindrical and strongly sclerotized, of the wireworm type. This form is evidently an adaptation enabling the larvae to penetrate rapidly through their friable substrate in search of food. The larvae of some click beetles are predatory, while others are probably omnivorous or mainly detritivorous. For instance the large larvae of *Elater ferrugineus* prey upon the larvae of *O. eremita* and other cetonids (Martin, 1989; Svensson et al., 2004). The alleculid and tenebrionid larvae are detritivorous and consume the frass produced by other beetle larvae. In the beetle families Pselaphidae (Park et al., 1950) and Scydmaenidae, as well as in pseudoscorpions (Arachnida: Pseudoscorpionida) (Ranius and Wilander, 2000), there are many small species which are strict specialists of treeholes and which prey on mites, springtails and other small arthropods dwelling in the wood mould.

Box 7.1 The hermit beetle (*Osmoderma eremita*) – a flagship species of hollow trees

Osmoderma eremita is one of the most impressive inhabitants of hollow trees in Europe. Adult beetles are 2.5–3.5 cm long, and full-grown larvae can reach 6 cm (Schaffrath, 2003). The males emit a distinctive odour resembling that of apricots or plums. The odour is caused by a compound (decalactone) that functions as a pheromone attracting females (Larsson et al., 2003). The hermit beetle inhabits hollows of mature oaks, limes, beeches and many other deciduous tree species. The favoured tree species varies between regions, depending on the availability of suitable host trees and local conditions (Ranius et al., 2005; Oleksa et al., 2007). The larvae live in wood mould and feed on decaying wood in the walls of the cavity. The development time is usually 3 years, and tens of successive generations can develop within the same tree.

In Sweden, hermit beetles live mainly in oaks (Antonsson et al., 2003), and their occurrence probability is higher in sun-exposed sites, in cavities with openings directed towards the sun, and in hollows with large amounts of wood mould (Ranius and Nilsson, 1997). The average number of new adults produced per inhabited tree per year is 10–20 (Ranius, 2000, 2001). However, variation between trees is large, with some of the best trees producing most of the adults (up to 100 individuals) in the local population each year (Ranius, 2000, 2001; Ranius et al., 2009b). Population size in each tree depends on the volume of wood mould, which sets the carrying capacity of the host tree (Ranius, 2007). Oaks with the largest *O. eremita* populations are generally 300–400 years old (Ranius et al., 2009b).

Based on both mark–release–recapture studies and radio-tracking of adult beetles, it seems that about 85% of new adults remain in their natal trees and only about 15% disperse to nearby hollow trees (Ranius and Hedin, 2001; Hedin et al., 2008). This suggests that each hollow tree sustains a local population, with only a limited exchange of individuals between neighbouring trees. Furthermore, all the observed dispersals were within a range of 30–190 m and occurred within the same stand (Ranius and Hedin, 2001; Hedin et al., 2008). However, recent studies based

Box 7.1 *(continued)*

on mark–release–recapture using pheromone-baited traps (Larsson and Svensson, 2009), radio-tracking (Dubois and Vignon, 2008), and flight experiments in laboratory conditions (Dubois et al., 2010) indicate that at least half of the population may perform flights from their natal trees, and the dispersal capacity can be over 1000 m.

The proportion of occupied trees was higher in larger stands with many hollow trees than in smaller ones, but there was no correlation between occupancy and isolation of stands (Ranius, 2000). This pattern suggests that dispersal is important within stands but not between stands. *O. eremita* was found in most of the larger stands with hollow oaks, but was systematically absent from single trees and very small stands, probably because of local extinctions and poor long-distance dispersal (Ranius, 2000). The population structure of *O. eremita* can be described as a metapopulation (see Chapter 14), where individual trees each host a local population, and trees within the same stand constitute the metapopulation. However, instead of a classic metapopulation, in which local extinctions and recolonization of empty patches may take place repeatedly, the population structure more closely resembles a habitat-tracking metapopulation, in which local extinctions are deterministic when the habitat patches (host trees) become unsuitable (Ranius, 2007). *O. eremita* populations also have features of a mainland–island metapopulation, in which local extinctions due to demographic or environmental stochasticity are very rare in trees with large amounts of wood mould.

The presence of *O. eremita* indicates a high species richness of other beetle species specialized to hollow trees (Ranius, 2002b; Jansson et al., 2009). By conserving *O. eremita*, many other species (including those that are small, elusive and poorly known) associated with the same microhabitat will also be conserved. Thus, *O. eremita* can serve both as an indicator and an umbrella species (Ranius, 2002b).

In many small and isolated stands where *O. eremita* still occurs, the extinction risk is high. This is because habitat loss, i.e. the loss of large hollow trees at the landscape scale, has taken place relatively recently as compared with the long persistence time of local populations (Ranius, 2000). Extinction risk is low in stands with at least 10 trees (Ranius, 2007), and a stand with at least 20 suitable trees is sufficiently large to secure a viable metapopulation of *O. eremita* even in the long term (Ranius and Hedin, 2004). The limited conservation resources should be targeted to sites with a high potential for restoration, to increase their quality (in terms of number of hollow trees and openness) and connectivity between high-quality sites. Cessation of grazing and other traditional uses of woodlands is not only harmful to the species, but also threatens to shorten the lifespan of ancient trees because of regrowth and subsequent competition by young trees (Ranius et al., 2005).

The hermit beetle originally inhabited most of the temperate forests in Europe, from Sicily in the south to southwestern Finland in the north. The present distribution of the species is very fragmented and reflects the remnant occurrence of semi-natural woodlands and cultural habitats such as wooded meadows, pollarded hedgerows, old parks and alleys created by traditional forms of land use. Ranius et al. (2005) compiled a comprehensive account (including over 200 references) of the known records and observations on the biology of the hermit beetle in Europe. Of the total of over 2000 known localities, the beetle has been found in less than half since 1990. Although many populations obviously remain to be found, particularly in southwestern Europe, the majority of the known localities are small and isolated, and the species seems to have declined throughout its distribution area. Consequently, *O. eremita* is included in Annex IV of the Habitats Directive of the European Union, i.e. the species and its habitats are strictly protected.

Recently, Audisio et al. (2007, 2009) showed, based on molecular genetic studies, that the species known as the hermit beetle is actually a complex containing four distinct species. *O. eremita* occurs in western Europe, *O. barnabita* is distributed from eastern Europe to central Russia, *O. cristinae* is confined to Sicily, and *O. lassallei* to Greece and the European parts of Turkey. This speciation may have occurred in refugial areas of temperate forest in the Italian and Balkan peninsulas and Sicily before and during the Pleistocene. Similar patterns of genetic diversity (though not necessarily speciation) might be found in other saproxylic species with similar ecology. This emphasizes the need to preserve species through their whole distribution area in order to maintain genetic diversity.

Among the Diptera, species specialized in wood mould are found in many different families, including some crane flies (Tipulidae) – e.g. *Ctenophora ornata*, the wood-soldier fly *Xylomya maculata* (Xylomyidae), hoverflies (Syrphidae), long-legged flies (Dolichopodidae) – e.g. *Systenus* spp., and stiletto flies (Therevidae) – e.g. *Pandivirilia melaleuca* and *Thereva nobilitata* (Andersson, 1999; Alexander, 2002). The larvae of crane flies, soldier flies and hoverflies are detritivorous, whereas the narrow and cylindrical larvae of stiletto flies are predatory.

As well as the wood mould, other subsections of the cavity – the ceiling, walls and floor – are each used by a particular set of species (Figure 7.3). Wood in the ceiling of a hollow is soft because of the heart-rot decay. Partly decomposed lumps of wood frequently fall from the ceiling and become buried in the wood mould. After the heartwood has completely decayed, the walls of the hollow consist of sapwood which often remains undecayed and hard. The floor consists of wood (or, in the case of basal cavities, soil under the tree), but is usually covered by a layer of wood mould that can be several metres thick.

7.2.3 Water-filled treeholes

Rainwater trickling down the trunk (stemflow) will enter most treeholes. If the cavity is not connected to the ground, water will stay in until it completely evaporates. In deep cavities, water can persist throughout all or most of the year and form pools. Such semi-permanent pools of water, mixed with tree exudates, woody debris and leaf litter, inside living trees is a special kind of saproxylic microhabitat (Figure 7.4). The term phytotelm (from the Greek: *phyto* = plant, *telm* = pond; plural *phytotelmata*) has been collectively used to refer to water bodies held by plants, of which water-filled treeholes are a special case (Kitching, 1971, 2000). The invertebrate fauna of these has been relatively intensively studied, compared with other kinds of treeholes, starting from the beginning of the 1900s, for two reasons. Firstly, the larvae of several mosquito species carrying human diseases live in phytotelmata, making them interesting from the point of view of medical entomology (Jenkins and Carpenter, 1946; Barrera, 1996). Secondly, phytotelmata comprise a clearly defined microecosystem allowing experimental manipulations, with relatively simple macroinvertebrate communities, which makes them a suitable object for ecological studies on community structure and functioning (Jenkins et al., 1992; Kitching, 2001; Srivastava, 2005; Ellis et al., 2006).

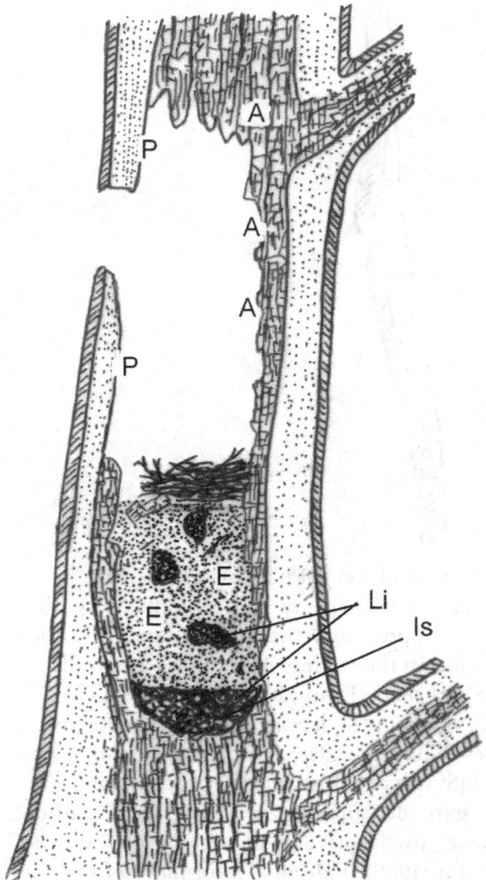

Figure 7.3. Section of a tree with a large trunk cavity. Click-beetle (Elateridae) species living in different parts of the cavity are indicated with letters. *Ampedus cardinalis* larvae (**A**) live in dry brown-rotted parts of the ceiling and walls, often high up in the trunk, where they prey on the larvae of wood-boring deathwatch beetles (*Dorcatoma* spp., *Anitys rubens*). *Procraerus tibialis* larvae (**P**) prefer hard, white-rotted parts of walls, where they prey on the larvae of wood-boring weevils (*Rhyncolus*, *Phloeophagus* and *Cossonus* spp.) which tunnel the cavity walls. *Elater ferrugineus* larvae (**E**) live in the loose wood mould and prey on the larvae of *Osmoderma eremita* and other cetoniids. The larvae of *Limoniscus violaceus* (**Li**) and *Ischnodes sanguinicollis* (**Is**) live in moist, dark and humified wood mould in the bottom of cavities, often below ground level. They possibly prey on larvae of Diptera. Modified from Iablokoff (1943); additional information from Martin (1989).

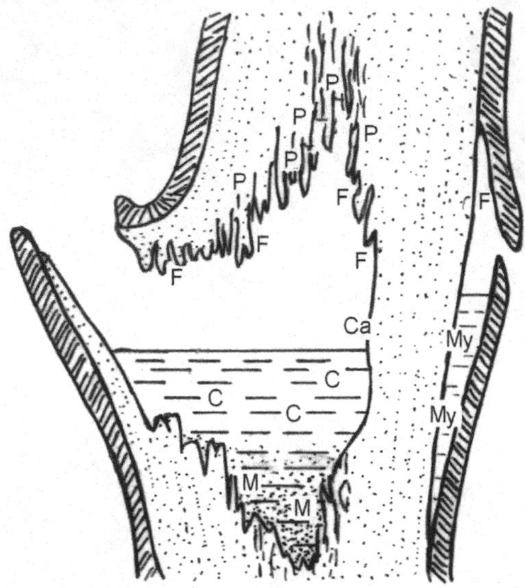

Figure 7.4. A water-filled cavity, and a small wet rot hole under the bark. Dipteran species occurring in different parts of the cavities are indicated with letters. *Phaonia* (Muscidae) larvae (**P**) are predatory and can be found in various parts of the cavity, including wet wood in the ceiling. The ceiling and walls above the water level are inhabited by *Fannia* (Fanniidae) larvae, which are scavengers. *Mallota* and *Callicera* (**Ca**) (Syrphidae) larvae feed on wet detritus. The larvae of ceratopogonid midges (Chironomidae) (**C**) live freely in the water and filter microbes. *Myathropa florea* (Syrphidae) larvae (**M**) are detritivores capable of feeding under water; the extended anal segment works as a breathing tube. *Myolepta* larvae (**My**) (Syrphidae) are detritivores and live in small water-filled rot holes. Modified from Speight (1989); additional information from Rotheray and Gilbert (1999) and Alexander (2002).

Leaf litter is the energy base for food webs in most water-filled tree-holes (Kitching, 1971; Paradise, 2004). The submerged litter is a substrate for surprisingly diverse communities of microfungi: a total of 45 aquatic hyphomycete species were found in just 13 water-filled treeholes studied in Hungary (Gönczöl and Revay, 2003). Some treehole macroinvertebrates directly consume decaying litter (Paradise and Dunson, 1997) but most feed by grazing on decomposer microbes from litter surfaces or filtering microbes in the water (Carpenter, 1983; Kitching, 2000). The dominating taxa in phytotelmata include larvae of various families, such as scirtid beetles (Scirtidae), chironomid (Chironomidae) and ceratopogonid midges (Ceratopogonidae), mosquitoes (Culicidae) and hoverflies (Syrphidae) (Kitching, 2000; Schmidl et al., 2008). In all these cases, the treehole specialist species have evolved from aquatic or semi-aquatic ancestors living in small water pools on the ground. In European temperate forests, there are only a handful of invertebrate species (and in boreal forests probably none) that are dependent on water-filled rot holes. The assemblage consists of larvae of the scirtid beetle *Prionocyphon serricorne*, the chironomid *Metriacnemus cavicola*, several ceratopogonids belonging to the family *Dasyhelea*, the mosquito *Aedes geniculatus* and the hoverfly *Myathropa florea* with, less commonly, microcrustaceans (Kitching, 1971; Schmidl et al., 2008). In warmer regions, both the species richness and food-web complexity increase, with additional detritivores, predators and even top predators such as dragonfly (Odonata) larvae (Kitching, 2000). In tropical forests, water-filled rot holes are an important microhabitat (Yanoviak, 2001). In addition to a diverse insect fauna, they are typically inhabited by amphibians.

7.2.4 Nests of vertebrates and insects in wood

Cavities are used by many vertebrate species for nesting, roosting and hibernating (Chapter 4). Woodpeckers excavate their own nest holes in trees with heart rot, and these holes are subsequently used by secondary hole-nesting birds. Vertebrates can carry large amounts of twigs, grasses and other nest materials into the cavity. Nitrogen input in the form of faeces, food remains and carcasses of nestlings is an even more important factor that can enrich the invertebrate fauna considerably.

A number of solitary wasps and bees also construct their nests in wood, but they use exit holes produced by the adults of wood-boring beetle species (Anobiidae, Cerambycidae etc.) for this purpose. Large nests of social insects (wasps, ants, termites) in cavities represent yet another type of microhabitat. Different insect hosts each have their own associated species, which can be predators, parasitoids, commensals or kleptoparasites (see Chapter 4).

7.3 Dead branches and roots

When a tree dies, clearly different species communities will colonize its branches, trunk and roots. However, the division can be much more refined, so that the thinnest twigs, thin branches, thick branches, upper, middle and basal parts of the trunk, root necks and underground roots each host species specialized to that particular part. There are two main reasons for the differentiation of species assemblages in the vertical dimension of a tree: stem diameter and exposure. The effects of diameter on species composition are treated in more detail in Chapter 8. Exposure is another important factor. Dead branches in the upper canopy are sun-exposed, while underground roots represent the other end of the exposure continuum. The effects of surrounding environment on species composition are treated in Chapter 9. In this part of the chapter we highlight three types of microhabitat typical for living trees: dead attached branches in the canopy, branches that have fallen to the ground, and dead roots.

7.3.1 Dead attached branches

Dead attached branches are a naturally occurring, essential part of nearly every tree crown. Self-pruning and fungal pathogens cause mortality of branches all the time, thus providing an evenly distributed and reliable resource. Compared with dead wood on the ground, dead wood in canopies is subject to frequent desiccation and to large variations in temperature and humidity. There are distinct microclimatic gradients from the forest floor to the upper canopy. The upper canopy contains mainly young and thin twigs exposed to solar radiation and wind. The inner and lower canopy layers contain thicker branches in shaded and more stable conditions.

Many fungus species are specialized on dead attached branches. For instance, over 100 species of fungi were recorded on dead branches within a small plot of temperate broadleaved forest in central Europe (Unterseher et al., 2005). Corticioid fungi (Corticiaceae, Stereaceae, Hymenochaetaceae, etc.) are the most species-rich group and occur mainly in the lower canopy, whereas ascomycetous fungi belonging to the Pyrenomycetes (Sordariales, Xylariales, Diaporthales, etc.) dominate the fungal assemblage on thin, exposed twigs in the upper canopy (Unterseher and Tal, 2006). Fungus species that grow on newly dead branches are usually specific to a particular host-tree species (Boddy and Rayner, 1983a; Boddy et al., 1987; Chapela and Boddy, 1988; Chapela, 1989; Griffith and Boddy, 1990; Unterseher et al., 2005). Branch fungi exhibit two contrasting strategies to protect themselves against drought: some species have minute and short-lived fruiting bodies which are produced under humid conditions, while other species grow sterile, tough and leathery sporocarps that start to produce spores only during longer periods of humid weather (Nuñez, 1996;

Unterseher et al., 2005). Furthermore, many branch fungi grow vegetatively as endophytes in the bark and wood of living branches. These latent species await the death of branches, or expand into the wood and kill branches suffering from water stress (Chapela and Boddy, 1988; Chapela, 1989; Griffith and Boddy, 1990).

There are many canopy specialists among saproxylic invertebrates too. The average density of arthropods in dead branch wood is of the order of 500 individuals per litre (Paviour-Smith and Elbourn, 1993), with mites and springtails (Collembola) being the numerically dominant groups. Ubiquitous detritivores make up a large proportion of the individuals, but both groups also contain saproxylic species, some of which are probably branch specialists. Thrips (Thysanoptera) and gall midges (Cecidomyiidae) are strikingly abundant in the canopy (Paviour-Smith and Elbourn, 1993). Both groups contain fungivorous saproxylic species, and these species obviously feed on the fungi growing on dead branches.

Beetle species attacking weakened and freshly dead branches are found among bark beetles (Scolytinae) – e.g. *Pityophthorus* and *Trypophloeus* spp., jewel beetles (Buprestidae) – e.g. *Agrilus* spp., and longhorn beetles (Cerambycidae) – e.g. *Grammopterus*, *Poecilium* and *Ropalopus* spp. Branches already infested with ascomycetous fungi are favoured by longhorn beetles (Cerambycidae) in the subfamily Lamiinae, e.g. *Exocentrus*, *Leiopus* and *Pogonochaerus* spp., and fungus weevils (Anthribidae). These branch-inhabiting species are typically small and have a remarkable camouflage consisting of brownish, greyish and whitish hair tufts, which efficiently conceals the adults amongst the epiphytic lichens growing on branches. Other beetle species typical to decaying branches are found among the false flower beetles (Anaspidae), soft-winged flower beetles (Melyridae) and narrow-waisted bark beetles (Salpingidae) (Paviour-Smith and Elbourn, 1993; Stork et al., 2001; Schmidt et al., 2007).

7.3.2 Branches on the ground

Dead branches falling to the ground have usually already been colonized by fungi and insects while still in the canopy. Because of the change in microclimatic conditions, the most specialized drought- and heat-tolerant canopy species will soon disappear. Other species are able to persist for longer and continue their decomposition work. It is probable that few, if any, species are strictly specialized on branches lying on the ground. The significance of fallen branches is due to the continuous input of both small- and coarse-diameter dead wood onto the forest floor.

7.3.3 Dead roots

Roots of living trees are the main microhabitat of several fungi, e.g. *Heterobasidion*, *Ganoderma* and *Armillaria* species. These species are known to colonize new living trees mainly by infecting and killing the roots, thereby gaining access to the main trunk (Rayner and Boddy, 1988; Schwarze et al., 2000b). The role of dead roots of living trees for saproxylic invertebrate species is poorly known. Insect species living in roots have been studied mainly in connection with dead standing trees or cut stumps; in both cases all the roots have died at the same time. However, similar to dead branches, living trees produce dead roots all the time, and these may actually be the main habitat of some underground saproxylic species that are treated in more detail in Chapter 9, which deals with the surrounding environment.

7.4 Bark, sapwood and heartwood

A tree trunk is composed of four different layers of tissues that perform specific functions. Because of the very dissimilar physical structure and chemical composition of outer bark, inner bark (including phloem and cambium layers), sapwood and heartwood, these layers constitute clearly different substrates and host different species assemblages. After the tree dies, new microhabitats will become available or will be formed by the saproxylic community. The development of these microhabitats – subcortical space, sapwood and heartwood – is closely connected to the decay succession of trees, which is described in more detail in Chapter 6.

7.5 Fruiting bodies of fungi

7.5.1 Fungal fruiting bodies as a microhabitat

A multitude of invertebrate species are associated with the fruiting bodies (also known as fruit bodies, sporocarps, sporophores or carpophores) and mycelia of various wood-inhabiting fungi. Fungal tissue constitutes a highly nutritional resource with much higher concentrations of important nutrients than the wood they grow on (Merrill and

Cowling, 1966; Martin, 1979; Boddy and Jones, 2008). For instance, the nitrogen content of sporocarps varies from about 0.7% to 4% in wood-decomposing fungi (Merrill and Cowling, 1966; Vogt and Edmonds, 1980; Gebauer and Taylor, 1999) which is about 2–10 times more than in undecayed wood.

All wood-decaying fungi provide three distinct food sources, each used by a separate assemblage of species. Spore feeders dwell on the hymenial layer (between the gills or on pore surfaces) of sporocarps; fruiting-body feeders live inside sporocarps; and mycelium feeders live in the mats of mycelia growing under bark or in sapwood crevices (Table 7.1) (see also Chapter 3). As well as fungivores, decomposing fungi are often inhabited by detritivores which feed on a variety of decaying organic matter, including fungal fruiting bodies. Both fungivores and detritivores have their own, more or less specialized, predators and parasitoids.

The fruiting bodies of gilled mushrooms versus polypores provide fundamentally different types of microhabitats and host quite dissimilar species assemblages. The soft fruiting bodies of most gilled mushrooms (Agaricales *sensu lato*) as well as sac fungi (Ascomycota) are short-lived, often lasting only a few days. On the other hand, the more or less tough fruiting bodies (brackets, conks) of polypores (poroid Aphyllophorales) are durable and can last from weeks in annual species to several years in perennial species. Furthermore, the fruiting bodies of perennial polypore species differ from most other fungi in that their year-to-year occurrence is more predictable, and some of the most common species can be very abundant in natural forests: e.g. 3000 fruiting bodies of *Fomitopsis pinicola* per hectare were recorded in a southern Norwegian spruce forest (Økland and Hågvar, 1994).

The differences between agarics and polypores are reflected in their species assemblages: wood-inhabiting agarics are usually inhabited by polyphagous dipteran larvae, whereas oligophagous beetle larvae dominate in polypores. However, many different factors affect the host selection patterns and host specificity of fungivores. These include the hyphal structure (Paviour-Smith, 1960; Lawrence, 1973), size (Midtgaard et al., 1998), toughness and temporal durability (Paviour-Smith, 1960; Schigel et al., 2006), chemical composition (Guevara et al., 2000a, 2000c), successional stage (Thunes et al., 2000; Jonsell et al., 2001), moisture of sporocarps (Paviour-Smith, 1960; Midtgaard et al., 1998; Jonsell et al., 2001), and environmental conditions (Komonen and Kouki, 2005).

Table 7.1. *Examples of polypore species and their associated beetle species, feeding either on spores, fruiting bodies or wood decayed by the polypore, in northern Europe.*

Polypore species	Associated beetle species	Food source	References
Antrodia sinuosa, A. xantha	*Calitys scabra* (Trogossitidae)	Fungous wood	Ahnlund and Lindhe (1992)
Diplomitoporus lindbladi	*Phryganophilus ruficollis* (Melandryidae)	Fungous wood	Lundberg (1993)
Fomes fomentarius	*Dorcatoma robusta* (Anobiidae)	Fruit bodies	Süda and Nagirnyi (2002), Nikitsky and Schigel (2004)
	Cis jaquemarti (Ciidae)	Fruit bodies	Reibnitz (1999), Jonsell and Nordlander (2004)
	Ropalodontus strandi (Ciidae)	Fruit bodies	Nikitsky and Schigel (2004)
	Bolitophagus reticulatus (Tenebrionidae)	Fruit bodies	Nilsson (1997), Midtgaard et al. (1998)
	Melandrya dubia (Melandryidae)	Fungous wood	Nikitsky and Schigel (2004)
Fomitopsis pinicola	*Gyrophaena boleti* (Staphylinidae)	Spore feeder	Økland and Hågvar (1994)
	Dorcatoma punctulata (Anobiidae)	Fruit bodies	Süda and Nagirnyi (2002), Nikitsky and Schigel (2004)
	Peltis grossa (Trogossitidae)	Fungous wood	Nikitsky and Schigel (2004)
	Pteryngium crenatum (Cryptophagidae)	Spore feeder	Nikitsky and Schigel (2004)

Funalia trogii	*Sulcacis bidentulus* (Ciidae)	Fruit bodies	Reibnitz (1999)
Gloeophyllum sepiarium	*Curtimorda maculosa* (Mordellidae)	Fruit bodies	Nikitsky and Schigel (2004)
Inonotus radiatus	*Abdera flexuosa* (Melandryidae)	Fruit bodies	Nikitsky and Schigel (2004)
Laetiporus sulphureus	*Eledona agaricola* (Tenebrionidae)	Fruit bodies	Nikitsky and Schigel (2004)
	Pentaphyllus testaceus (Tenebrionidae)	Fungous wood	Nikitsky and Schigel (2004)
Phellinus conchatus	*Baranowskiella ehnstromi* (Ptiliidae)	Spore feeder	Sörensson (1997)
Piptoporus betulinus	*Diaperis boleti* (Tenebrionidae)	Fruit bodies	Nikitsky and Schigel (2004)
	Tetratoma fungorum (Tetratomidae)	Fruit bodies	Paviour-Smith (1964)
Trametes spp.	*Cis boleti*, *C. micans*, *Octotemnus glabriculus*, *Wagaicis wagae* (Ciidae)	Fruit bodies	Reibnitz (1999), Nikitsky and Schigel (2004)
Trichaptum abietinum, *T. fuscoviolaceum*	*Zilora ferruginea* (Melandryidae)	Fungous wood	Saalas (1923), Nikitsky and Schigel (2004)
	Xylita livida (Melandryidae)	Fungous wood	Saalas (1923)
	Abdera triguttata (Melandryidae)	Fruit bodies	Saalas (1923), Nikitsky and Schigel (2004)
	Cis punctulatus (Ciidae)	Fruit bodies	Saalas (1923), Reibnitz (1999), Nikitsky and Schigel (2004)

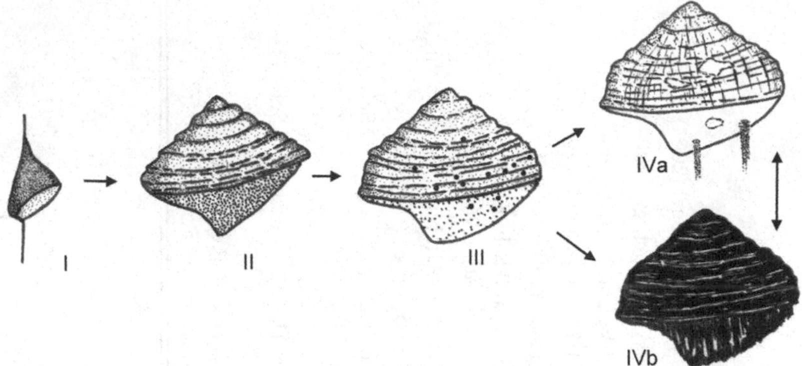

Figure 7.5. Development stages of a polypore sporocarp. (**I**) Immature, undeveloped hymenium, does not produce spores; (**II**) mature, living, capable of producing spores; (**III**) recently dead (small round exit holes are made by the deathwatch beetle *Dorcatoma* sp.); (IV) decomposing; (**IVa**) dry, desiccated (the hanging silk-frass tubes underneath the sporocarp are made by tineid moth larvae); (**IVb**) wet, rotting stage. Stage V includes rotten sporocarps after they have fallen to the ground. Modified from Graves (1960).

Individual fruiting bodies of fungi undergo a successional development through aging, insect attack and decomposition, which all bring about physical and chemical changes in the fungal tissue. Consequently, the invertebrate community inhabiting fruiting bodies also changes over the course of time. The succession is very rapid in agarics and particularly slow in perennial polypores. Successional phases in the maturation and decomposition of mushrooms, and subsequent changes in the beetle fauna, were first described and discussed by Scheerpelz and Höfler (1948) and Benick (1952) in their pioneering works. The successional development of polypore sporocarps was described by Graves (1960), who described five successional stages (Figure 7.5).

7.5.2 Fruiting bodies of perennial polypores

The first development stage covers the period from the appearance of the fruiting body up to the development of the hymenial layer and the beginning of spore production. Perennial sporocarps at this stage are not generally attacked by fungivores (e.g. Økland and Hågvar, 1994).

Sporocarps enter the second development stage when they reach maturity and start to produce spores. Perennial polypores develop a new layer of pores each year, and can produce huge amounts of spores during a period of many years. Mature fruiting bodies host some very specialized spore feeders. The world's smallest beetles belong to the family feather-winged beetles (Ptiliidae), subfamily Nanosellinae. These species are only 0.3–0.6 mm long, narrow and elongated, and most are exclusively known, both as adults and larvae, from the hymenial pore tubes of different polypore species (Dybas, 1956, 1976; Newton, 1984; Hall, 1999). The subfamily, with several genera (e.g. *Nanosella, Cylindrosella, Porophila*) was only known from the New World until an undescribed species, later named *Baranowskiella ehnstromi* (Sörensson, 1997), was unexpectedly found in southern Sweden, in the tiny pores of the polypore *Phellinus conchatus*, which only grows on goat willow (*Salix caprea*).

In mites (Acari) there are many minute species belonging to different families (e.g. Nanacridae, Cheyletidae) which inhabit pore tubes, and which are either fungivorous or predatory (Graves, 1960; Matthewman and Pielou, 1971). Even some fungivorous dipteran larvae are small enough to fit into pores. These include the larvae of gall midges (Cecidomyiidae) in the subfamilies Porricondylinae (Økland and Hågvar, 1994; Økland, 1995) and possibly also Lestremiinae. Rearing of these tiny larvae in laboratory conditions generally fails, which has prevented the identification of species and reliable assessment of their abundance. The hidden arthropod community living in the pore tubes of perennial polypores is probably much more diverse and abundant than currently known. According to Hågvar (1999), some sporocarps of *F. pinicola* contained mites and dipteran larvae in more than half of the pores. This may explain why certain predatory beetles often occur in large numbers on living, seemingly empty fruiting bodies; e.g. the brightly coloured *Lordithon* species (Staphylinidae) are known to prey on dipteran larvae (Newton, 1984).

Other spore feeders do not enter the pores but dwell on the surface of the hymenium layer. Rove beetles (Staphylinidae) in the genus *Gyrophaena* are highly specialized spore feeders which consume spores both as adults and as larvae. The apices of their maxillae are truncate and have a dense mat of short hairs that serve as spore brushes (Ashe, 1984). Most *Gyrophaena* species live between the gills of various agarics, but some species live on perennial sporocarps, e.g. *Gyrophaena boleti* on *F. pinicola* (Økland and Hågvar, 1994). Larvae of the fungus gnat genera *Keroplatus* (Keroplatidae) and *Sciophila* (Mycetophilidae) spin sticky webs on the hymenial layers of polypores and corticioid fungi. It has been assumed that these webs would catch other insects and provide shelter, but it is more likely that the webs function as spore traps, allowing the larvae to catch, store and feed on spores.

Large numbers of various saproxylic beetles visit perennial sporocarps when they are sporulating (Kaila, 1993; Økland and Hågvar, 1994; Hågvar and Økland, 1997; Hågvar, 1999). A few of the visitors belong to the specialized spore feeders described above, while others breed in living or dead fruiting bodies of the same polypore species. However, the great majority of frequent visitors live as larvae in quite different dead-wood microhabitats, such as in soft fruiting bodies of wood-decaying agarics or annual polypores (e.g. Erotylidae, Endomychidae), under bark in sappy phloem or bark-beetle galleries (e.g. Nitidulidae, Rhizophagidae), or in fruiting bodies of myxomycetes (Leiodidae: *Anisotoma* and *Agathidium* spp.). Most guests probably consume spores, but some (Lathridiidae, Cryptophagidae) may also feed on moulds growing on the surface of sporocarps. The high species richness and abundance of beetles visiting sporocarps indicate that they are important in terms of behavioural ecology and act as attraction centres (Økland and Hågvar, 1994; Hågvar and Økland, 1997). Of the approximately 60 beetle species that visited sprorulating sporocarps of *Fomes fomentarius* and *Fomitopsis pinicola*, over 30 species were observed to continuously arrive and leave the sporocarps around midnight (Hågvar, 1999). It is possible that saproxylic beetles seek out living sporocarps not only for feeding, but also for mating, to find dead wood for breeding, or to locate potential prey.

Mature living fruiting bodies can occasionally be attacked by fruiting-body feeders. However, most species colonize perennial sporocarps only after they have died. The beginning of tissue breakdown inaugurates the third stage, during which primary fungivores colonize the sporocarp. The functionally most important group is ciid beetles (Ciidae), which live both as adults and as larvae inside the fruiting bodies; several partly overlapping generations can complete their development in the same sporocarp (Box 7.2). Other functionally important species tunnelling in the recently dead fruiting bodies include some darkling beetles (Tenebrionidae), e.g. *Bolitophagus reticulatus* in Eurasia and the forked fungus beetle (*Bolitotherus cornutus*) in North America, which both live in the tinder fungus (*F. fomentarius*); the anobiid genus *Dorcatoma* (Box 7.2); and some individual species in other beetle families. In addition to beetles, primary sporocarp feeders are also found in Lepidoptera, mainly in the family tineid moths (Tineidae) (Lawrence and Powell, 1969; Rawlins, 1984). For instance, the tineid moth *Agnathosia mendicella* was the main fungivore on the perennial polypore species, *Fomitopsis rosea* in old-growth forests in eastern Finland (Komonen, 2001).

Box 7.2 Host-use patterns and life strategies of polypore-dwelling ciid (Ciidae) and deathwatch (Anobiidae: *Dorcatoma*) beetles

Ciidae is a cosmopolitan beetle family (tree fungus beetles) with over 500 described species. The whole family has specialized in living in polypore fruiting bodies. Some of the species live in other wood-decaying fungi or in fungous wood. It has long been known by coleopterists that certain ciid species are invariably found on particular polypore species (Weiss, 1920; Weiss and West, 1920; Saalas, 1923; Donisthorpe, 1935). However, Paviour-Smith (1960) was the first to draw attention to the more general host-use patterns of ciids. She noticed that the beetle species and the host fungi divided into two mutually exclusive breeding groups, named the *Polyporus betulinus–Cis bidentatus* group and the *Polystictus* (= *Trametes*) *versicolor–Octotemnus glabriculus* group. Each group consisted of a set of host polypore species, and a group of ciids breeding in them. Ciids in one group would only occasionally occur in fungi of the other group. Paviour-Smith (1960) suggested that the cause of the host affinity was the hyphal structure of sporocarps, with one ciid group breeding in monomitic and dimitic (softer) sporocarps, and the other group breeding in trimitic (harder) ones. She also assumed that the choice of host would not be based on chemical attractants, because fruiting bodies were often colonized long after their death, after periods of wetting and drying.

The division of breeding groups by Paviour-Smith (1960) was based on only 10 ciid species and a restricted number of potential host species occurring in Wytham Wood, near Oxford, England. The host records of North American ciids (74 species in 117 species of basidiomycete fungi) were summarized by Lawrence (1973). Ciids appeared to be host-group-specific, preferring to breed in one or two hosts but occurring on several others, which were often phylogenetically related. The majority of host fungi could be placed into one of four host preference groups, on the basis of shared ciid inhabitants. Most ciid species, except the most polyphagous ones, were associated with one of these host preference groups only. Host preferences of ciids were later discussed by other authors (Thunes, 1994; Fossli and Andersen, 1998; Jonsell and Nordlander, 2004).

Recently, Orledge and Reynolds (2005) have compiled and analysed comprehensive data on British, German, North American and Japanese ciids. Cluster analysis, and subsequent cross-dataset

Box 7.2 *(continued)*

comparisons, demonstrated the existence of host-use patterns of wide geographical occurrence. Six Holarctic host-use groups and two subgroups were identified. There was a strong tendency for closely related taxa (both in fungi and beetles) to belong to the same host-use group. The authors suggested that host preferences are defined ultimately by host chemistry, so that the ciids belonging to a particular host-use group recognize, and respond positively to, volatiles that are common to all the fungi belonging to the same host group. Modern phylogenies of fungi may explain the host selection patterns of ciids and other fungivores better than older classifications, since fungivores use hosts that are closely related to each other (Jonsell and Nordlander, 2004). Polyphagous species generally colonize fruiting bodies only after they have reached a certain stage of decay, thus escaping their chemical defences (Jonsell and Nordlander, 2004).

The hypothesis that dispersing ciids locate suitable hosts by olfactory stimuli (Lawrence, 1973; Orledge and Reynolds, 2005) has gained support from experimental studies. Flying individuals of ciids breeding in *Fomitopsis pinicola* were strongly attracted to chopped, living fruiting bodies of their host (Jonsell and Nordlander, 1995). Furthermore, both sexes were equally attracted to the odour of the host, whereas the prior presence of males or females did not increase the attraction (Jonsson et al., 1997). This implies that host volatiles are used for locating suitable sporocarps, and that there is no evidence for pheromone attraction. The volatile composition of different polypore species, e.g. *F. pinicola* and *Fomes fomentarius*, has been shown to differ (Fäldt et al., 1999). Furthermore, the volatile composition released by fruiting bodies changes when they are sporulating and when they are recently dead. These volatiles may act as signals for species living in different developmental stages of the sporocarp (Fäldt et al., 1999). In laboratory experiments, the ciid species *Octotemnus glabriculus*, *Cis boleti* and *Cis nitidus*, with narrow host selection, were specifically attracted to the odour compounds of their preferred host polypore, in contrast with the generalist *Cis bilamellatus*, which was attracted to several polypore species (Guevara et al., 2000c).

Two or three ciid species often occupy a single sporocarp at the same time. Competitive displacement in sympatric species preferring the same host fungus can take place (for examples, see Lawrence, 1973; Thunes, 1994). However, several species preferring the same host species can coexist, in which case different types of resource partitioning may occur. *O. glabriculus* and *C. boleti* were found to coexist within *Trametes versicolor* because they temporally partition the host fungi: *O. glabriculus* breeds in young and expanding fruiting bodies in early summer, and *C. boleti* colonizes fully developed fruiting bodies later in summer (Guevara et al., 2000a). The species were shown to have differential behavioural responses to odour compounds from young and mature fruiting bodies of their host (Guevara et al., 2000a). In addition to the temporal resource partitioning, different species may show niche differentiation and prefer fruiting bodies at different development stages, with different moisture or height above ground etc. (Jonsell et al., 2001), or they may show different environmental preferences (Komonen and Kouki, 2005). *O. glabriculus* and *C. boleti* favoured clusters of *Trametes* growing in closed forest, while *Sulcacis affinis* and *Cis hispidus* occurred more frequently and were more abundant in clearcuts (Komonen and Kouki, 2005).

Ciids exhibit a range of conspicuous adaptations which allow them to efficiently use polypore sporocarps. The egg-laying period can be several months long, and several partly overlapping generations, often occurring in large colonies, can complete their development in the same sporocarp (Lawrence, 1973). Some species can withstand complete drying out of their substrate, and these can continue breeding for years, even in herbaria specimens. Males of many ciid species have spectacular teeth or spikes on their clypeus or pronotum, or their mandibles have developed into curved horns (*Octotemnus* spp.). These armatures are most probably used to dislodge rival males from the sporocarp (see Miller and Wheeler, 2005). All the above features enable ciids to monopolize those sporocarps which they have managed to colonize. Ciid species are the dominant fungivores in different polypore species, and the most common species can occupy 30–90% of the available sporocarps of the preferred host (Jonsell and Nordlander, 1995; Guevara et al., 2000b; Jonsell and Nordlander, 2004). In some cases, ciids can be so abundant that they reduce the reproductive output of their host sporocarps (Guevara et al., 2000b).

Deathwatch beetles (anobiid beetles) of the genus *Dorcatoma* are as strict specialists of polypore sporocarps as ciids, yet they have quite different life strategies. Most species appear to have a 2-year development time (at least in the boreal zone), and only one generation develops in the same fruiting body. *Dorcatoma* species

Box 7.2 *(continued)*

are generally much less frequent than ciids, occupying less than 10% of the available sporocarps of the preferred host (Jonsell and Nordlander, 1995, 2004). Unlike ciids, which do not appear to have long-range aggregation pheromones, *Dorcatoma* species evidently do. In *D. robusta*, only females were attracted to the odour of the host polypore (*F. fomentarius*), whereas males were strongly attracted to conspecific females (Jonsson et al., 1997). It has been suggested that *Dorcatoma* species are inferior competitors, but superior colonizers, of scattered resources, compared with ciids (Jonsell et al., 1999). In general, the pheromone strategy may be more efficient for species occurring at low densities or adapted to colonize isolated patches (Jonsson et al., 2003).

Interestingly, the differences in host-use strategies, and the consequent differences in frequency, abundance and adult detectability are reflected in the degree of taxonomic and faunistic knowledge of these groups. While most north European ciids had already been described before 1850, half of the *Dorcatoma* species were first described after 1900, and new species are still being discovered over the past 30 years (Baranowski, 1985; Zahradnik, 1993; Büche and Lundberg, 2002). Taking into account that the north European beetle fauna is exceptionally well known, one can assume that many undescribed species inhabiting polypores could be found in other poorly studied groups (e.g. mites, gall midges) and in other parts of the world.

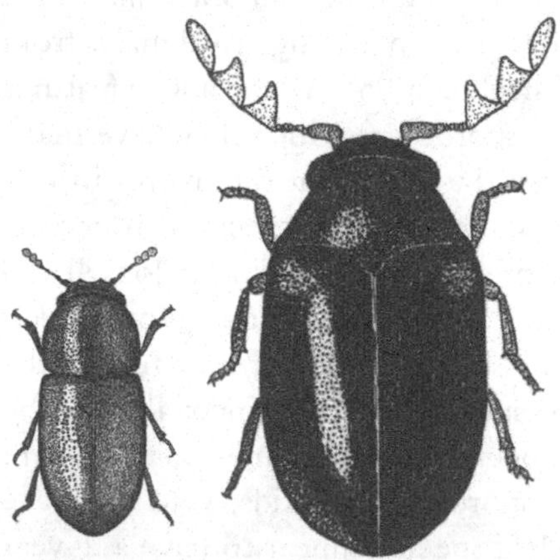

Cis jaquemarti (left) and *Dorcatoma robusta* (right).

The primary fungivorous species excavate the inner parts of fruiting bodies, thus opening the way for other, secondary, fungivores, predators, scavengers and parasitoids. The most frequent and numerically dominant groups are mites (Acari), beetles (Coleoptera), and springtails (Collembola) (Graves, 1960; Matthewman and Pielou, 1971). A species-rich and taxonomically highly variable assemblage of mites, including fungivores, predators and detritivores, occurs in all the development stages of sporocarps. Most of the mite species are also found in other dead-wood habitats and in litter, but some species are strict specialists on sporocarps (Gwiazdowicz and Łakomy, 2002; Makarova, 2004; Mašán and Walther, 2004).

There is no clear-cut transition between the third and fourth development stages, and the changes in the fruiting body and its fauna take place gradually. After the primary fungivores have consumed most of the sporocarp, the inner parts are riddled with insect burrows or completely hollowed out. In dry weather, the fruiting body becomes desiccated. On the other hand, in moist conditions it may absorb water and pass into the rotting stage, during which it rapidly decays due to the action of bacteria, moulds and feeding by generalist fungivorous and detritivorous arthropods. An abundance of barklice (Psocoptera) is characteristic for the desiccated fruiting bodies, whereas springtails (Collembola) and mites dominate in the wet stage (Graves, 1960). Colonies of ciid beetles will persist in the fruiting bodies (even in those that have fallen on the ground) as long as there are any patches of unused fungal tissue.

7.5.3 Fruiting bodies of annual polypores

The successional development of annual polypores is generally similar to that of perennial species but faster. The life strategies of annual polypores vary, and three different strategies (sporocarp types) can be distinguished: ephemeral, annual sturdy, and annual hibernating (Schigel et al., 2006). Ephemeral species (e.g. *Amylocystis lapponica*, *Oligoporus* and *Postia* species) follow the time- and resource-saving strategy similar to agarics, i.e. they emerge quickly and produce small, short-lasting fruiting bodies which may appear only during suitable years. Annual sturdy species (e.g. *Laetiporus sulphureus*, *Piptoporus betulinus*, *Inonotus* and *Polyporus* species) are usually more voluminous and robust, and both their growth and spore production period last longer. After their sporulation and death in the autumn, the fruiting bodies remain on the substrate for at least the following summer. Annual hibernating (e.g. *Bjerkandera*, *Trametes* and *Trichaptum* species) complete their growth in the autumn, stay alive during the winter, and sporulate in spring, after which they die.

The durability, softness, size, and most likely also the chemical composition of the fruiting bodies of different species affect the insect communities. Some annual species producing large fruiting bodies, such as *Piptoporus betulinus*, *Polyporus squamosus* and *Laetiporus sulphureus*, host very species-rich insect communities (Pielou and Verma, 1968; Klimaszewski and Peck, 1987; Nikitsky and Schigel, 2004). But some smaller annual species also have rich faunas and several specialized species. For instance, *Bjerkandera adusta* and *Trametes* spp. host monophagous (at the genus level) species of flat-footed flies (Platypezidae) (Chandler 2001), while *Trametes* and *Trichaptum* spp. each have several monophagous beetle species (Table 7.1).

7.5.4 Fruiting bodies of agarics

Fruiting bodies of wood-decomposing agarics represent the saproxylic microhabitat with possibly the largest numbers of facultative saproxylic and non-saproxylic species. Many generalist fungivores which mainly live in agarics growing on the ground can also utilize wood-inhabiting agarics. Decomposing fruiting bodies are utilized by detritivorous species, particularly by dipteran larvae, which can develop in any kind of putrefying material. Other species use dead fruiting bodies as a hunting ground. Nevertheless, many widespread and common wood-decomposing agarics (e.g. *Armillaria*, *Pleurotus*) have their own associated species, which are therefore obligate saproxylics (Table 7.2).

Living fruiting bodies of wood-decomposing agarics (e.g. *Armillaria*, *Pleurotus*, *Megacollybia*, *Pluteus*), especially mature sporulating ones, are visited by many beetle species, most of which belong to the rove beetles (Staphylinidae) (Schigel, 2007). Most agarics decompose so fast that there is not enough time for the beetle larvae to finish their development. However, fruiting bodies of *Armillaria* and *Pleurotus* often grow in large clusters which may become dry and persist for weeks. *Pleurotus* hosts a particularly rich beetle fauna, including many monophagous (at genus level) species, e.g. *Triplax* and *Eutriplax* species (Erotylidae) and *Cyllodes ater* (Nitidulidae) (Cline and Leschen, 2005; Schigel, 2007).

Table 7.2. *Examples of wood-inhabiting fungus species and their associated insect species in northern Europe. Col = Coleoptera, Dip = Diptera, Lep = Lepidoptera, Hem = Hemiptera, Hom = Homoptera.*

Host fungi		Associated insect species	
Taxonomic group	Species	Species (Order: Family)	References
Aphyllophorales	*Ganoderma applanatum*	*Agathomyia wankowiczii* (Dip: Platypezidae)	Chandler (2001)
	Polyporus squamosus	*Bolopus furcatus* (Dip: Platypezidae)	Chandler (2001)
	Gloeophyllum odoratum	*Aradus erosus* (Hem: Aradidae)	Ahnlund and Lindhe (1992)
	Antrodia sinuosa, *A. xantha*	*Cixidia confinis* (Hom: Achilidae)	Ahnlund and Lindhe (1992)
Agaricales	*Kuehneromyces mutabilis*	*Atomaria umbrina* (Col: Cryptophagidae)	Benick (1952), Palm (1959)
	Pleurotus spp.	*Triplax rufipes* (Col: Erotylidae)	Schigel (2007)
		T. aenea (Col: Erotylidae)	Schigel (2007)
		Cyllodes ater (Col: Nitidulidae)	
		Lordithon trimaculatus (Col: Staphylinidae)	Palm (1959)
		Hirtodrosophila trivittata (Dip: Drosophilidae)	Jakovlev (1994), Ševčík (2006)
	Tricholomopsis spp.	*Mycetophila finlandica* (Dip: Mycetophilidae)	Jakovlev (1994), Ševčík (2006)
Ascomycetes	*Daldinia loculata*	*Platyrhinus resinosus* (Col: Anthribidae)	Wikars (1997b)
		Cryptophagus corticinus (Col: Cryptophagidae)	Wikars (1997b)
		Paranopleta inhabilis (Col: Staphylinidae)	Wikars (1997b)
		Apomyelois bistriatella (Lep: Tortricidae)	Wikars (1997b)
Tremellales	*Auricularia auricula-judae*	*Platydema violacea* (Col: Tenebrionidae)	Palm (1959)
		Hirtodrosophila lundstroemi (Dip: Drosophilidae)	Jakovlev (1994), Ševčík (2006)
		Camptodiplosis auriculariae (Dip: Cecidomyiidae)	Jakovlev (1994), Ševčík (2006)
Myxomycetes	*Trichia decipiens*	*Agathidium pulchellum* (Col: Leiodidae)	Laaksonen et al. (2010)

7.5.5 Fruiting bodies of ascomycetes and myxomycetes

Fruiting bodies of ascomycetes are generally small and short-lived. Saproxylic beetles and dipterans associated with ascomycetes usually have larvae that live under the bark of fungus-infested trees and feed mainly on hyphae or conidial stages (Crowson, 1984). Beetles of this type are generally small, flat, and often have obscure colours and bristles; they include representatives of many families such as Cucujidae, Cryptophagidae, Biphyllidae, Colydiidae, Salpingidae and Anthribidae (Crowson, 1984). Compared with the insect fauna living on basidiomycete fungi, the insects living on ascomycete fungi have been very little studied. The coal fungi *Daldinia concentrica* and *D. loculata* are among the largest species; the black ping-pong ball-like fruiting bodies host quite a large number of specialized beetle, fly, gnat and moth species (Hingley, 1971; Wikars, 1997b; see Table 7.2).

Finally, the ephemeral fruiting stages of myxomycetes (which do not belong to fungi) generally host relatively few species, but most of these are strictly specialized to feed on myxomycetes. The round fungus beetle (Leiodidae) genera *Agathidium* and *Anisotoma* are the most species-rich group of saproxylic insects strictly associated with slime moulds (Wheeler, 1984; Wheeler and Miller, 2005). Other 'myxomycetophagous' beetle species can be found in Lathridiidae, Sphindidae and Eucinetidae (Blackwell, 1984; Stephenson et al., 1994). Some fungus gnat species are also known to be specialists on myxomycetes (Jakovlev, 1994; Ševčík, 2006).

7.6 Wood surface

After the bark has fallen off, the decorticated surface of logs and snags constitutes a new microhabitat. Logs in contact with the ground retain their moisture better and can host a rich flora of epixylic bryophytes, while snags usually become dry and host mainly epixylic lichens. These epixylic species do not get their nourishment from dead wood but use the snags and logs as their growing substrate only. Epixylic species are described in Chapter 4.

8 · *Tree size*

Juha Siitonen and Jogeir N. Stokland

Many saproxylic species are only able to use dead wood of a particular size or diameter. Some species prefer large trunks, while others favour small trees or thin branches. Some can use dead wood of many sizes while others are specialized within a narrow diameter range. In an individual dead tree, trunk sections with different diameters tend to be used by different species.

In this chapter, we describe the factors that contribute to the niche separation of species according to tree size. In general, the basal diameter of a tree is closely correlated with other dimensions such as height, surface area and volume, and each of these correlated factors can be important for individual species. For simplicity, the terms 'diameter' or 'size' are used in the text to refer to all the diameter-related effects, and the other factors (height, surface area, volume) are mentioned only when their effects are specifically considered. The preferences of individual species are reflected in species richness and species composition patterns that can be observed in dead-wood units belonging to different diameter classes. These patterns are reviewed in this chapter.

8.1 Factors causing diameter effects on species preferences

8.1.1 From seedling to senescent tree

The properties of both wood and bark change considerably in living trees as the trees develop from saplings into mature trees and then into senescent trees. After a tree dies, most of these properties are retained and affect the quality of the tree as a substrate for saproxylics. Tree age and diameter are generally closely correlated, and it is difficult to distinguish the individual effects of these two factors when preferences of species are related to the diameter of trees.

One of the most visible characteristics that changes as a tree grows older is the thickness of the outer bark. Young trees generally have thinner and smoother bark than old trees, whose bark is thick and rough. Bark thickness also changes from the base to the top of the trunk. Since the outer bark acts as a barrier against colonization by saproxylic fungi and invertebrates (see Chapter 5), bark thickness – and its variation between trees and along the trunk – is likely to be a significant substrate factor for many species during the first decay stage. Once these species have managed to penetrate through the bark, a thick layer of bark provides better insulation against changes in temperature and moisture, and may protect subcortical species from predators and parasitoids better than thin bark does.

Both the physical and chemical properties of wood change as trees grow older. Wood density, latewood content and heartwood content all increase with tree age and diameter (Wilhelmsson et al., 2002). The most notable change that occurs during tree development is the formation of heartwood. Heartwood production usually starts at an age of about 10–20 years, with a large variation between different tree species, and the proportion of heartwood increases steadily as the tree grows older (Gjerdrum, 2003). Heartwood contains substances that inhibit the growth of most wood-decomposing fungi (see Chapter 5), although the heart-rot fungi manage to cope with this environment. These fungi create trunk cavities and hollow trees that are important microhabitats for a large number of species (see Chapter 7). In short, the age of the tree and the proportion of heartwood in the trunk will have a substantial effect on which species are able to colonize the tree when it dies. Logs with a high proportion of heartwood originate from large old trees.

The growth rate of a tree affects both the rate of diameter increase and wood composition. Trees whose growth is suppressed or that are situated in poor sites grow more slowly, have higher wood density, and can invest fewer resources in defensive substances than vigorously growing trees. Furthermore, there is a complex interaction between tree diameter, tree age, growth rate and mortality agents. Trees of different sizes often die from different causes, and this will influence species composition during the decay succession (see Chapter 6).

8.1.2 Physical properties connected with trunk diameter

Surface–volume ratio changes with diameter, so that large trees have less surface area per volume unit than small trees. The surface–volume ratio also influences the moisture and temperature regimes inside the trunk. Small logs are susceptible to desiccation and large variations in moisture content and temperature. Large logs can retain moisture even during dry periods, and temperature within the log remains relatively stable.

Yet another important factor related to diameter is the persistence time of logs and other dead-wood units of varying size. Small logs decompose quickly and will soon be overgrown by ground vegetation. Large logs decompose more slowly and provide a more persistent habitat for the saproxylic species that colonize and inhabit the trunk during different stages of decomposition.

8.2 Diameter preferences of individual species

8.2.1 Fungi

The diameter preferences of saproxylic fungi have not received as much attention as their preferences for different host trees and decay stages. Nevertheless, it is well known that many species preferentially occur on logs with large dimensions. Renvall (1995) reported that several basidiomycete species favour large logs in the boreal zone, including *Climacocystis borealis*, *Laurilia sulcata*, *Phlebia centrifuga*, *Skeletocutis odora* and *Antrodia crassa*. Large logs, in this context, typically refer to logs that are more than 30–50 cm in diameter. There are also some studies that have quantitatively assessed the diameter preferences of individual wood-inhabiting fungus species. Stokland and Kauserud (2004) made a thorough analysis of *Phellinus nigrolimitatus*, a white-rot polypore fungus that is strictly confined to coniferous trees, especially spruce. This species rarely occurs on spruce logs smaller than 20 cm in diameter, but is quite common on larger logs, especially in unmanaged forests. They suggested that variations in moisture content and energy limitations were possible explanations for the absence of this species on small-diameter logs. They concluded that little can be learnt from observational data alone, and called for controlled experiments to elucidate the causal relationships between log size and diameter preferences of different fungi.

Fomitopsis rosea is another boreal polypore species that seems to prefer large-diameter spruce logs. In the southern and middle boreal zones it is strongly associated with trees larger than 30 cm in diameter, but in the northern boreal zone close to the timber line it also occurs frequently on spruce logs with a diameter smaller than 20 cm (J. N. Stokland, unpublished data taken from more than 2000 spruce logs).

This corresponds well with a laboratory study by Edman et al. (2006; see also Section 9.5), showing that *F. rosea* is more efficient than *F. pinicola* in decaying dense wood (i.e. from slow-growing trees). Thus, the species is not confined to large trees *per se*, but seems to have a competitive advantage on dense wood. Such wood develops in the trunks of old trees when growth slows down. On marginal sites, such as near the timber line, slow growth is characteristic of all ages and diameter classes.

In temperate forests one has found that some species need particularly large-diameter trees. Heilmann-Clausen and Christensen (2004) have reported that *Ischnoderma resinosum*, a heart-rot fungus, occurs on beech logs with a minimum diameter of 70 cm, and they made the point that heart-rot fungi, in general, are confined to large trees.

It is also worth mentioning that a preference for large-diameter trees is a typical feature of many red-listed fungi. In Sweden, Kruys et al. (1999) found that red-listed fungi had a strong preference for spruce logs with a diameter greater than 30 cm. A similar trend has also been observed in Denmark, where Heilmann-Clausen and Christensen (2004) compared beech logs varying from 20 to 139 cm in diameter. They found that the majority of 27 red-listed fungi were only recorded on logs larger than 70 cm in diameter.

From the examples presented above, one could gain the impression that large diameters are necessary for most saproxylic fungi. However, all the species mentioned above are basidiomycetes and most of them are polypores. Later in this chapter we shall see that many polypores have diameter preferences that are different from those of other fungal groups.

8.2.2 Invertebrates

Some of the best-known examples of diameter preferences and niche separation according to trunk diameter come from the field of forest entomology. The diameter preferences of several bark beetle species have been studied in detail (e.g. Paine et al., 1981; Schlyter and Anderbrant, 1993; Amezaga and Rodríguez, 1998; Ayres et al., 2001; Kolb et al., 2006; Foit, 2010). Results of these studies show that different species colonizing the same tree generally occupy different sections of the trunk according to diameter and bark thickness, which allows the coexistence of competing species. In closely related species, body size is often correlated with the diameter of the trunk sections that they use (Hespenheide, 1976; Table 8.1). Even when competitors are absent,

Table 8.1. *Examples of bark beetle (Scolytinae) species breeding in fresh inner bark of different vertical parts of Norway spruce* (Picea abies) *and Scots pine* (Pinus sylvestris) *trees in northern Europe. Mainly based on Ehnström and Axelsson (2002).*

Tree species and beetle species	Beetle size (mm)	Tree part	Preferred diameter (cm)
Norway spruce (*Picea abies*)			
Pityophthorus tragardhi	1.5–1.8	Thin branches	0.3–0.8
Pityophthorus micrographus	1.0–1.5	Branches, top	1–10
Pityogenes chalcographus	2.0–2.3	Thick branches and upper trunk	5–15
Ips duplicatus	3.5–4.0	Upper trunk	10–20
Ips typographus	4.2–5.5	Lower trunk, under thick bark	>20
Hylastes cunicularius	3.5–4.5	Roots	1–10
Scots pine (*Pinus sylvestris*)			
Pityophthorus lichtensteinii	1.8–2.0	Thin branches	1–4
Pityogenes quadridens	1.7–2.2	Branches	2–10
Ips acuminatus	2.5–3.7	Thick branches and upper trunk, under thin scaly bark	5–20
Tomicus minor	3.2–4.8	Upper trunk, under thin scaly bark	5–20
Tomicus piniperda	3.2–5.2	Lower trunk, under thick shield bark	>15
Ips sexdentatus	6.2–7.8	Lower trunk, under thick shield bark	>25
Hylastes brunneus	3.5–4.5	Roots	1–10

species do not colonize entire trees but show clear niche boundaries. For instance, minimum thresholds for trunk diameter and bark thickness, and a maximum threshold for height above the ground were detected for the emerald ash borer (*Agrilus planipennis*) attacking young ash trees in Canada (Timms et al., 2006).

Similar to basidiomycetous fungi, many invertebrate species inhabiting logs in the later stages of decay favour large trunks. For instance, the large longhorn beetle species *Tragosoma depsarium* mainly inhabits large pine logs that are at least 25 cm in diameter

(Wikars, 2004). The favoured trunks are more than 200 years old, with a large proportion of heartwood and slow rate of decay. It seems that the properties of dead wood that only occur in old and large-diameter trees are more important for the occurrence of the species than the diameter itself.

8.3 Species richness and composition patterns in relation to diameter

8.3.1 Species richness patterns

There are several studies documenting the effects of trunk diameter on the species richness of wood-inhabiting fungi and epixylic bryophytes. Most of these studies have shown that the number of species on a log increases with log size (Bader et al., 1995; Renvall, 1995; Lindblad, 1998; Kruys and Jonsson, 1999; Schmit, 2005; Ódor et al., 2006; Stokland and Larsson, 2011). As with the species–area relationship, there are two main explanations for the species–diameter relationship. Firstly, a large log can offer more microhabitats than a small log, including different diameters and decay stages along the trunk, dry and moist conditions on the surface and deep in the wood, and so on (the 'microhabitat diversity' hypothesis). Secondly, larger volumes of wood provide more space and resources for coexisting species (the 'volume *per se*' hypothesis). A third factor that may affect species richness is the longer persistence time of large logs, allowing for a longer colonization period that may be important, particularly to species that disperse passively through airborne spores.

Schmit (2005) explicitly tested the relationship between the number of species per log and log volume, surface area and initial wood density. Volume and wood density together give an estimate of the amount of energy available in each log. In agreement with this 'species–energy' hypothesis, when logs were grouped by tree species, the total wood volume and density of live wood had a significant positive effect on total species richness and abundance.

But which patterns emerge if we consider equal amounts of wood belonging to different diameter classes? Using this approach (known as 'rarefaction'), cumulative species richness curves are constructed to standardize the sampling effort, i.e. identical wood volumes, wood surface areas or numbers of individuals are compared. To our knowledge, the first study of this kind was carried out by Kruys and Jonsson (1999), who compared the species richness of fungi in boreal coniferous forest on spruce logs that were smaller or larger than 10 cm in maximum diameter. They found that although the average species richness per log was higher on the large logs, the small and large logs did not differ when equal surface areas were compared. However, when they compared similar volumes made up of small or large logs, the species richness was significantly higher on the small logs. Heilmann-Clausen and Christensen (2004) found that the number of fungal species was higher on small logs than on large logs when equal volumes were compared in temperate broadleaved forests. They also assessed the species richness for similar numbers of fungus individuals sampled (classical individual-based rarefaction), and found that the number of species was almost identical on different diameter classes.

Thus, it seems to be a general rule that the number of species per unit volume decreases with increasing diameter of the logs. At least two processes may explain these patterns. Firstly, small-diameter logs have a larger surface-to-volume ratio than large-diameter logs, and hence can trap more spores and can support more fungal sporocarps per volume unit. Heilmann-Clausen and Christensen (2004) called this effect 'the surface-area factor'. Secondly, when equal volumes are compared, small-diameter logs include more individual wood pieces than larger logs. This allows the logs to represent a wider range of substrate types. They called this effect 'the number-of-items factor'. Another aspect of this numerical effect is that more logs represent more colonization units. This increases the stochasticity in colonization and allows ecologically similar species to coexist in separate pieces of wood (see also Section 11.3.2).

Among invertebrates, it has also been shown that small pieces of wood together have a higher species richness than an equal volume of larger trunks. In a Swiss study of beech (*Fagus sylvatica*) wood, Schiegg (2001) found more species of both beetles and gnats (Diptera) in branches than in large logs with the same volume. When branches and trunks were compared based on similar numbers of insect individuals, the species richness was almost identical for beetles, whereas the branches had more dipteran species than the trunks. A similar result was obtained by Jonsell et al. (2007), who compared the number of beetle species reared from logging residue branches of three diameter classes. They found that the species number was very similar in each diameter class when they compared equal numbers of individuals. According to Lindhe et al. (2005), diameter had no general positive effect on the population densities of saproxylic beetles breeding in artificially created high stumps of different tree species.

When we consider these studies together, it may superficially appear as if the diameter of dead wood has no particular importance for the diversity of saproxylic organisms. But one essential aspect is hidden when one compares such gross species numbers – namely the identity of the species. When the species richness is more or less similar on small and large pieces of dead wood, the species assemblages can consist of largely the same or completely different species in each diameter class. When this aspect was considered in the above-mentioned studies, it was invariably revealed that the species composition was distinctly different between the size classes being compared.

Furthermore, there is a positive relationship between the diameter of logs and the total number or volume of fruiting bodies growing on the log (Urcelay and Robledo, 2009). Although this relationship might seem trivial, it probably has significant ecological consequences. The number and size of fruiting bodies is connected to the amount of spore production which, in turn, affects the dispersal (emigration rate), colonization and dikaryotization opportunities Thus, fruiting-body production is closely connected to the population dynamics of polypore species. It is possible that large trunks produce most of the individuals in a local population. This may be equally true for many saproxylic invertebrates.

8.3.2 Species composition patterns

Although the diameter preferences of individual species have been studied, very few studies have explicitly focused on the effect of tree dimension on entire species assemblages. A very informative study is that by Nordén et al. (2004b), who collected fungi on branches and logs in temperate broadleaved forests and compared the species composition on wood units smaller or larger than 10 cm in diameter. There was a striking difference in the occurrence pattern between ascomycetes and basidiomycetes. Among the ascomycetes (102 species) the large majority (75%) were exclusively found on small-diameter substrate units and just two species (2%) were only found on large-diameter units (Figure 8.1). The basidiomycetes (309 species), on the other hand, were more or less equally distributed on small- and large-diameter units: 30% were only found on small-diameter units, 26%

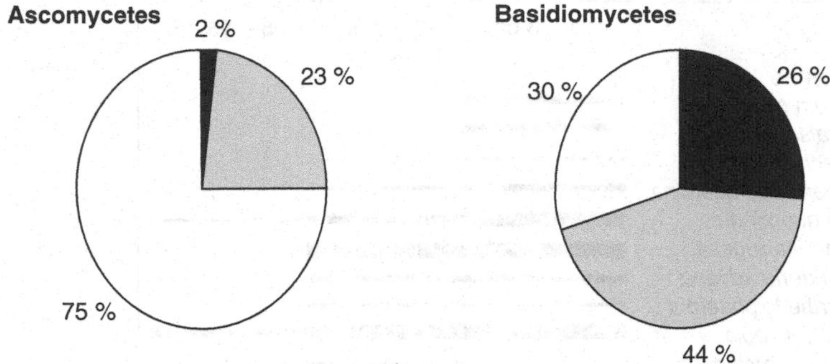

Figure 8.1. The occurrence of ascomycete and basidiomycete species on dead wood with different dimensions (smaller or larger than 10 cm in diameter), showing the proportion of species found exclusively on small-diameter substrate units (white), exclusively on large-diameter units (black) or on both (grey). Reprinted from Nordén et al. (2004b) with permission from Elsevier.

were only found on large-diameter units, and 44% were found on both size classes (Figure 8.1).

Many species that are specialized to use small-diameter dead wood are found living in dead but still attached branches in the canopies of living trees (see Chapter 7). These species have adapted to the harsh environmental conditions in the canopies, including frequent desiccation, wide temperature variations and strong solar radiation (see Chapter 9). It is difficult to determine whether diameter or environmental conditions play the major role in shaping these branch communities. Nevertheless, the species composition of the fungal community changes within the canopy according to branch diameter. Ascomycetes dominate in thin twigs, and the proportion of basidiomycetes increases in thicker branches (Figure 8.2).

These results clearly show that thin, small-diameter branches are very important for a large number of species. They also show that ascomycetes and basidiomycetes are highly segregated with regard to their preferences for different diameters of dead wood. This is important to bear in mind, as nearly all the studies concerning species richness of fungi on dead wood have been restricted to basidiomycetes. Similar systematic differences in diameter preferences can also be found among saproxylic invertebrates, mainly at the level of genera or families but possibly even at higher taxonomic ranks, such as is the case for the ascomycetes and basidiomycetes described above.

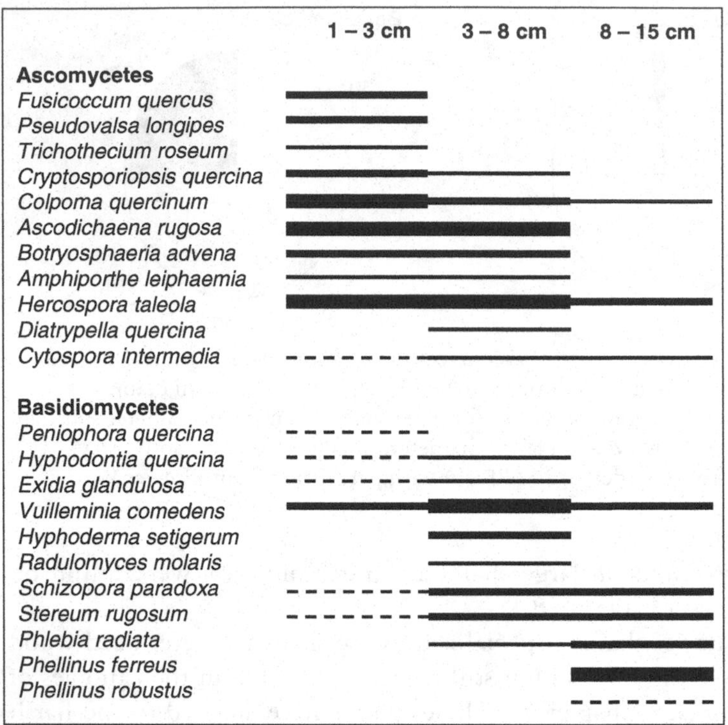

Figure 8.2. Species occurring on dead attached *Quercus robur* branches with different diameters (derived from Table 3 of Butin and Kowalski, 1983b).

8.4 Importance of large trunks for species diversity

The studies based on species accumulation curves reviewed above suggest that large-diameter and small-diameter dead trees host about equal numbers of species when equal surface areas or volumes are compared. But all these studies are based on sampling dead-wood units from rather few forest stands. Such rarefaction curves primarily represent local species pools or, to put it another way, a relatively small proportion of all the species occurring in the region. Thus, they cannot answer the question of whether large dead trees support a larger regional species pool than small trees. But there are at least two general arguments which suggest that large trees should host a higher overall species richness than small ones.

8.4.1 Wider diameter variation along large trunks

Large trees have both thick and thin parts, and they can therefore host species requiring both large and small diameters. On the other hand, small trees have only thin parts and cannot host species requiring large diameters. Many species that mainly live in the trunks of small dead trees can also colonize the top parts and branches of large trees. For instance, Renvall (1995) studied polypore assemblages inhabiting large conifer trunks and showed that some of the species were restricted to the thick basal parts, while other species were confined to the thin crown area. The middle parts were occupied by the largest variety of species, because species with preferences for both thick and thin parts could be found there, as well as many other species with no clear diameter preferences.

8.4.2 Large trunks make up most of the dead-wood volume

In natural forests, large trees always make up most of the dead-wood volume. For instance, in boreal spruce-dominated old-growth forests, large-diameter (≥30 cm) dead trees accounted for an average of 42–54%, and small-diameter (<10 cm) dead trees only 1.7–2.7% of the total CWD (coarse woody debris) volume (Siitonen et al., 2001). Also in temperate and tropical forests, large trees make up most of the dead wood volume. So from an evolutionary viewpoint it makes sense that there should be more species which have adapted to use large-diameter dead wood than species which have adapted to use small-diameter dead wood.

9 · *The surrounding environment*

Bengt Gunnar Jonsson and Jogeir N. Stokland

The surrounding environment strongly influences the conditions inside the wood and is fundamental to determining whether a saproxylic species is able to utilize a certain piece of dead wood. Many species show a clear preference for wood in sun-exposed and dry habitats, while others prefer shady and moist conditions. The tree's position, whether it is standing or lying, also determines the degree of sun exposure, temperature and moisture in the wood. In addition, the species composition varies according to the surrounding medium. In terrestrial habitats, the vast majority of species utilize the above-ground wood, although some species are specialized to use dead roots buried in the soil. Other species only utilize submerged wood from trees that have fallen into rivers or lakes, and yet others occur on wood in marine waters. In addition, man-made wooden constructions create opportunities for saproxylic species. When these species occur inside houses, they can attack and severely damage the wooden construction materials (see Box 9.1).

In addition to the direct effects, the surrounding environment also has an indirect effect on dead wood through the conditions experienced by the living tree. The local conditions determine the annual growth increment and wood density, and events such as physical injury and insect attacks affect the chemical characteristics of the wood. These wood properties may strongly influence the saproxylic species that later utilize the dead tree. Some of these aspects have partly been addressed in Chapter 6, but deserve some additional attention in this chapter.

9.1 The abiotic environment

The abiotic environment strongly influences the conditions for the species living inside decaying wood, and several textbooks provide a thorough treatment of the factors that determine the microenvironment inside wood (Rayner and Boddy, 1988; Schmidt, 2006). The most important factors are temperature, moisture and the gaseous regime (particularly oxygen pressure). Here we give a brief overview of these factors.

9.1.1 Temperature

As with all organisms, saproxylic species have limits to the range of temperatures they can tolerate. In nature the range extends from deeply frozen wood to strongly sun-exposed wood and, in extreme cases, high temperatures during forest fires.

At low temperatures, body fluids and cell contents might freeze. At high latitudes and altitudes, this poses a particular challenge to the growth, development and even survival of saproxylic species. The physiological adaptations are not unique to saproxylic species, but below we give some specific examples from wood-living insects and fungi.

There are two broad strategies that allow insects to cope with low temperatures, either by avoiding being frozen or by allowing themselves to freeze. In the former case, their physiology prevents the body fluids from freezing and keeps the insect in a supercooled condition at low temperatures (down to −20°C). In the latter case, the insects actually freeze, but are able to revive when the temperature increases again (Sinclair, 1999; Zachariassen et al., 2008). Both strategies are present among saproxylic insects.

The ability to withstand low temperatures varies both between different development stages and over the course of the year. In an experimental study, Vernon and Vannier (2001) explored the freezing tolerance of beetles living in wood mould, including the hermit beetle *Osmoderma eremita* (Figure 9.1). Both the supercooling temperature and lethal temperature showed distinct variation relating to life stage and time of the year. The first-instar larvae were frost-tolerant down to –15 °C during late winter, but were unable to survive in frozen conditions during the autumn.

Freezing tolerance is associated with three different biochemical components: so-called ice-nucleating agents (INAs), different sugars, and antifreeze proteins. The INAs are proteins which are primarily produced during autumn at temperatures around −2°C to −10°C, and then circulated in the haemolymph. These help the individual to survive the formation of extracellular ice in the body fluids. The alternative to supercooling is to allow the body to freeze. This ability is less common than supercooling, but still occurs regularly among wood-inhabiting species. Freezing-tolerant species can survive temperatures

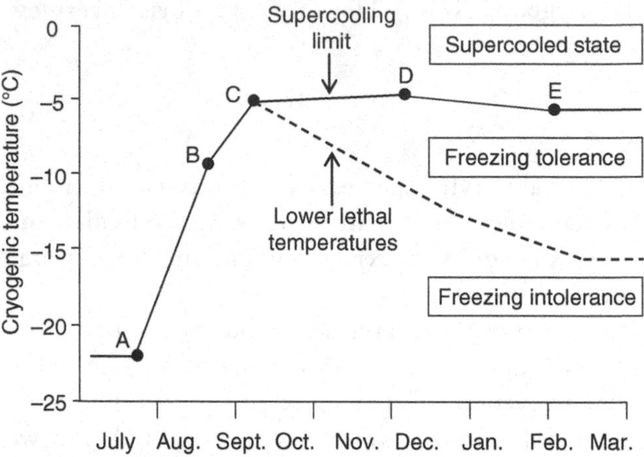

Figure 9.1. Supercooling ability of different life stages of the beetle *Osmoderma eremita* from July to March. Newly laid eggs (**A**), newly hatched larvae with empty gut (**B**), first-instar larvae with food-filled gut (**C**, **D**, **E**) (after Vernon and Vannier, 2001).

much lower than their supercooling point (Block, 1991; Vernon and Vannier, 2001).

Mycelia of wood fungi are very resistant to low temperatures; in laboratory environments they are regularly stored for long periods in a frozen state, even in liquid nitrogen at −196°C (Schmidt, 2006). Thus, survival in natural conditions is rarely a problem. However, mycelial growth normally ceases at temperatures below zero. In some species the production of antifreeze agents, such as trehalose and glycerol, occurs and allows metabolic activity at temperatures below 0°C. This has been observed for some blue-stain and mould fungi that have a lower limit for mycelial growth of −7°C to −8°C (Riess, 1997, cited in Schmidt, 2006). Relative growth at low temperatures varies between species, and the ability for sustained growth at low temperatures may give a competitive advantage in cold climates.

The optimal temperature for growth also varies. Among the fungi, tropical species seem to be better adapted to grow at higher temperatures than temperate species (Magan, 2008). In the tropics, some species, such as *Trametes cervina*, *T. cingulata* and *T. socotrana*, have growth optima as high as 37–40°C and are able to grow at temperatures up to 55°C (Mswaka and Magan, 1999). This contrasts with temperate species, which mostly have growth optima around 20–30°C and are usually unable to grow at temperatures above 35–40°C (Boddy, 1983; Schmidt, 2006).

When the temperature rises well above the optimum temperature, it will finally become lethal. Our knowledge about temperatures lethal to saproxylic insects mostly stems from experimental studies on controlling the spread of unwanted pest species. FAO (2002) provides regulations on treating wooden packing material and demands treatment at a minimum temperature of 56°C for at least 30 minutes to ensure that all larvae and eggs are killed. In a specific study of the emerald ash borer (*Agrilus planipennis*), Nzokou et al. (2008) found that 100% mortality was not obtained below 65°C. This temperature level is congruent with the fact that most proteins start to denaturize between 40°C and 50°C, given that the wood itself may confer some protection against the heat.

In wood-inhabiting fungi, current knowledge of lethal temperatures is based on laboratory experiments on mycelia grown on agar or wooden discs. Both the temperature itself and the exposure time seem to matter. Schmidt (2006) compiled a number of studies, which suggested that mycelia are more sensitive when grown on agar than when they grow on pieces of wood. Agar-grown colonies of most species did not survive temperatures above 60°C for more than 1 hour, while colonies of several species grown on wooden discs survived temperatures above 80°C or 90°C up to 4 hours.

It appears that resistance to high temperature varies greatly between species. In a recent study, Carlsson et al. (in press) exposed wooden discs with mycelia to varying temperatures (100–220°C) for 5–25 minutes. The experiment aimed to emulate conditions during forest fires. The results showed that many species were surprisingly resistant. Several species (e.g. *Dichomitus squalens*, *Gloeophyllum protractum* and *Antrodia infirma*) even survived a temperature of 220°C for 5 minutes. An interesting observation was that fire-associated species were much more resistant to high temperatures than generalist species (Figure 9.2). This suggests that adaptation to forest fires may include the ability of mycelia to survive high temperatures.

9.1.2 Moisture

The moisture content of the wood is regulated primarily by the local environment but also by the wood decay process itself. A minimum level of water is needed for decay to occur, since wood-degrading enzymes are dissolved in water and because water is needed for internal nutrient transport and cell metabolism. Above a certain level, water

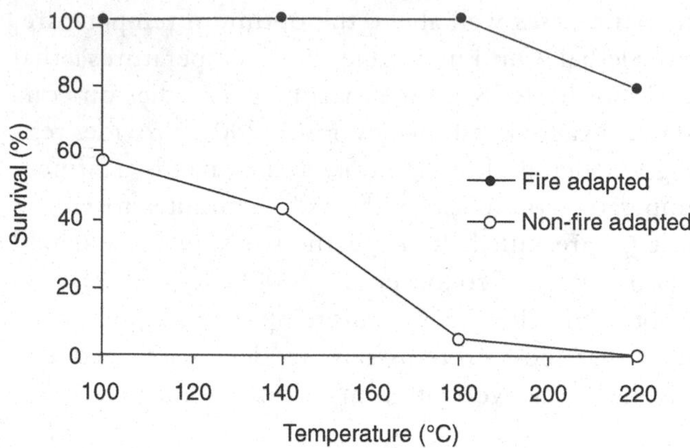

Figure 9.2. Percentage survival of mycelia grown on wooden discs in nine fire-adapted species (black circles) and seven non-fire-adapted species (white circles) after 5 minutes of heat treatment (after Carlsson et al., in press).

content may be too high and will limit access to oxygen (see below). The relative wood moisture is usually measured as the percentage difference in weight between moist wood and dry wood; i.e. wood moisture = 100 × (moist weight − dry weight)/dry weight. The critical moisture level is set by the so-called 'fibre saturation point' below which free water is no longer available in the cell lumen. This saturation point is, in most cases, around 30% wood moisture (Zabel and Morrell, 1992; Schmidt, 2006). At moisture levels below this point, fungal decay will cease unless the mycelium is able to obtain water from alternative sources. The latter is rare but does occur in some species, such as the dry-rot fungus (*Serpula lacrymans*; see Box 9.1). Other species may survive dry conditions by forming resting asexual spores.

Several studies of wood in service (i.e. wood modified for human usage such as buildings, fence poles, wooden boats) have measured the moisture in wood that has been soaked in water, buried in soil, or exposed to different degrees of solar radiation, in order to investigate the effects on wood decay (Levy, 1982; De Belie et al., 2000; Kim and Singh, 2000). These studies show that wood moisture has a profound effect on fungal decomposition and which fungi bring about the decay. The methodology and knowledge from these studies has direct relevance for studies of dead wood under natural conditions, and can explain how the environment affects the condition of the wood (see Figure 9.3).

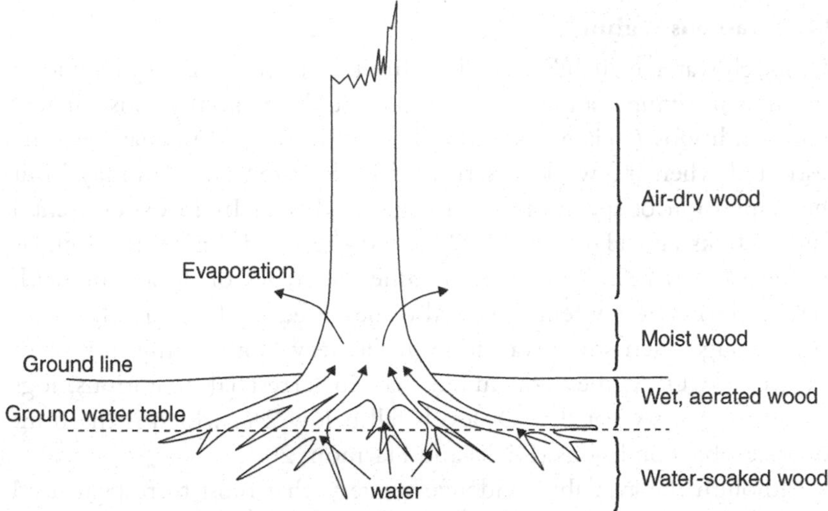

Figure 9.3. Water movement inside a standing dead tree. Water is absorbed by the roots until equilibrium is reached between the wood and the surrounding soil. From here, water moves upwards by capillary action and evaporates above ground. The result is a strong gradient of wetness inside the wood (derived from Levy, 1982).

In general, dead trees buffer the microclimatic variation and, compared with the surroundings, dead wood tends to provide more stable and moist conditions. There are groups of species, such as many epixylic liverworts and mosses, which depend on moist logs for their survival. Also among the invertebrates, some species depend on moist conditions in logs. For instance, the soft-bodied velvet worm *Euperipatoides rowelli* (Onychophora) is unable to control its respiration due to its permeable skin and permanently open spiracles in the tracheal system. This species lives in the forests of southeastern Australia and is confined within the sheltered habitat inside decaying logs; it only moves outside into open conditions during the night or during rainfall (Woodman et al., 2007).

Besides being a prerequisite for the occurrence of some species, moisture content may also contribute to niche segregation among species occurring in the same general habitat. A clear example comes from two longhorn beetles (*Callidiellum rufipenne* and *Semanotus bifasciatus*) which occur together in conifer logs in eastern Asia. Larvae of the species differed in their utilization of logs, where the larvae of *C. rufipenne* occurred in the dry upper part of the logs while *S. bifasciatus* larvae inhabited the moist lower areas (Iwata et al., 2007).

9.1.3 Gaseous regime

Although water availability can be a limiting factor, at the other end of the moisture range, an excess of water could potentially cause anaerobic conditions (lack of oxygen). The diffusion of O_2, may become restricted when the wood is saturated with water, and this may limit the respiration of species living in the dead tree. In an experimental study, Hicks and Harmon (2002) tested the O_2 diffusion in Douglas fir (*Pseudotsuga menziesii*) wood at different stages of decay and with varying moisture content. They also measured O_2 levels in the field. Even though their study was done in the very moist conifer forests of the western USA, they concluded that, in terrestrial conditions, logs rarely become so wet that O_2 levels fall below 2%, which is the limit for anaerobic conditions (Hicks and Harmon, 2002).

Although the available evidence suggests that most terrestrial dead wood does not experience anaerobic conditions, a high moisture content may still restrict decay rates. Several studies have shown that dry wood normally decays faster than wet (Schmidt, 2006; Barker, 2008). The optimal water content for many species is actually close to the fibre saturation point and, especially in brown-rot fungi, the wood decay rate is known to decrease in wet conditions (Käärik, 1974). That is why water spraying is regularly used to limit decay when storing industrial wood (Bjurman and Viitanen, 1996).

Of the major groups of wood-decaying species, it is mainly soft-rot fungi and certain bacteria (see Chapter 2) that are able to decay wood in water-saturated situations with low levels of available oxygen. These groups replace wood-decaying basidiomycetes in moist conditions.

9.2 Above-ground environments

In nature, the average temperature, determined by climate, is probably a major determinant for the global distribution patterns of saproxylic species. But the occurrence of species also varies greatly on a local scale. Here the sun exposure is a key factor influencing the temperature conditions inside the wood. Sun exposure varies substantially both within and between forests. Sun-exposed wood occurs in upper parts of the tree canopy, sun-exposed hillsides, after stand-replacing disturbances, in naturally open forests (savannas, low-productive forests, along lakes and mire edges), and in different types of cultural habitats (see Chapter 16).

An experienced collector knows very well that different saproxylic species prefer wood situated in different positions. Many species occur in wood lying on the ground; others prefer the trunks of standing dead trees; and still others are confined to a restricted zone close to the trunk base. Higher in the tree we find another set of species in dead branches, and in the opposite direction there are completely different species in roots and wood fragments buried in the soil. These differences in species composition are equally valid for fungi and invertebrates, and the major underlying factor is presumably wood moisture.

9.2.1 Above-ground invertebrates

In a review of habitat preferences of 542 red-listed saproxylic invertebrates in Sweden, Jonsell et al. (1998) analysed their preference for shaded or sun-exposed conditions (Figure 9.4). A significant proportion (30–40%) of species were indifferent to sun exposure, i.e. they were able to use suitable host trees in both open and shaded conditions. A much higher percentage of species preferred open (about 25%) than shaded (about 10%) conditions. The proportion of species preferring sun-exposed sites was higher in recently dead than decayed trees and the proportion of species preferring shade increased with wood decay.

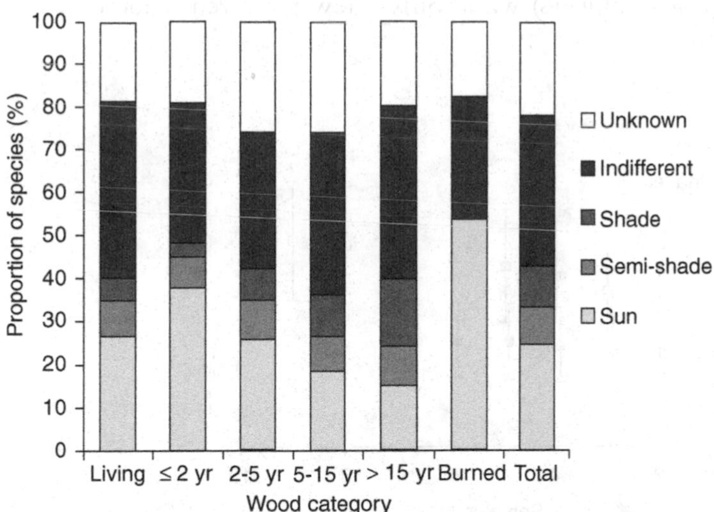

Figure 9.4. The preferences for sun-exposure of Swedish red-listed invertebrates. Species are categorized according to their substrate requirements (time since tree death and association with burned trees). Of the species showing a preference, most preferred open sunny conditions. Data from Jonsell et al. (1998).

Metabolically, invertebrates are often more restricted by temperature than by moisture. It is therefore reasonable to assume that many saproxylic insects prefer sun-exposed wood at higher latitudes, but do not have this preference at latitudes closer to the equator. Sun-exposed snags and dead trees in open areas – such as clearcuts – host a different and often more species-rich fauna than similar habitats in closed forests (Kaila et al., 1997; Lindhe et al., 2005). For some tree species, e.g. oaks, which often occur in open and semi-open forests, it is also likely that selection pressures have favoured species preferring sunny habitats (Gärdenfors and Baranowski, 1992).

To explore the role of sun exposure, Vodka et al. (2009) designed baits made out of freshly cut oak wood (bundles including pieces of stem, branches and twigs) and placed these in different settings. They showed that the diversity of jewel beetles (Buprestidae) and longhorn beetles (Cerambycidae) was highest in the understorey in open sunny positions, while shaded baits attracted fewer species (Figure 9.5). Their results also showed that samples placed in the canopy had fewer species than samples placed in the understorey. Species composition also differed between open and shaded baits, and between understorey and canopy. Most species strongly favoured sunny conditions (e.g. *Agrilus angustulus*, *A. obscuricollis*, *Cerambyx scopolii*, *Clytus arietis*, *Plagionotus arcuatus*) while only a few preferred shade or were

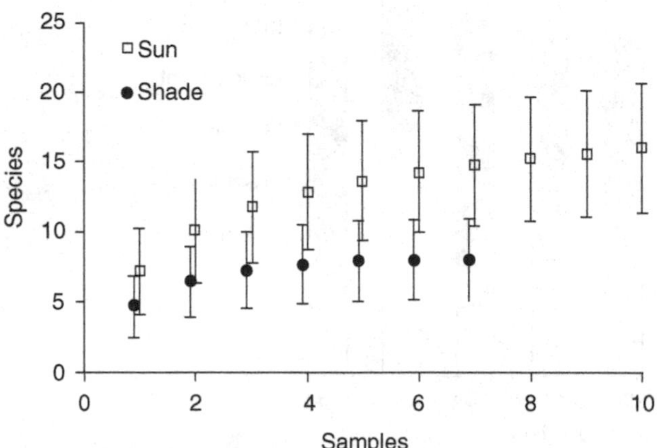

Figure 9.5. Number of xylophagous beetles (Buprestidae, Cerambycidae) reared from oak-wood baits in a temperate woodland (from Vodka et al., 2009). Baits in sunny conditions had significantly higher species richness than baits placed in the shade.

indifferent (e.g. *Leiopus nebulosus*). This neatly exemplifies that similar substrates serve as habitats for different groups of species when located in contrasting environments.

Investigations of invertebrates also demonstrate that the species developing in wood on the forest floor and in the canopy are quite different (Vodka et al., 2009; Bouget et al., 2011). Both studies demonstrated that species richness was highest in wood on the forest floor. But the canopy fauna was not a simple subset of that found in the understorey, as some species clearly favoured canopy wood. Bouget and colleagues, who also reviewed additional studies, attributed these species differences partly to distinct types of decaying wood being present in the canopy and the understorey, and partly to microclimatic differences.

In another interesting study, Foit (2010) showed how different saproxylic beetles developing in standing, recently dead Scots pine trees have distinct vertical distribution patterns (Figure 9.6). The pine trees (18–25 m high) were felled and divided into 20 trunk sections of equal length, while the branches were sorted into diameter classes. One group of species mainly developed in the lower part of the trunk, including *Arhopalus rusticus*, which was strictly bound to the basal part. Another group was associated with the lower middle part of the trunk. A third group was confined to the upper part of the trunk and branches, and within this group there was a subset that developed mainly or exclusively in thin branches (*Pityogenes bidentatus*, *Magdalis* spp., *Ernobius nigrinus*). Foit found that tree section was the single factor that best explained differences in species composition. However, he noticed that other factors such as trunk (branch) diameter, bark thickness and height above ground were highly correlated with tree section and so it was difficult to assign the observed preferences to a specific factor.

9.2.2 Above-ground fungi

It appears that many wood fungi thrive in open areas as long as sufficient dead wood is present (Junninen et al., 2006). Obvious examples are found in areas after forest fires, where many species, including red-listed species, may be abundant (Penttilä, 2004). Many species can also find suitable conditions on clearcuts. It seems that fruiting-body production is relatively indifferent to microclimate (Heilmann-Clausen et al., 2005; Junninen et al., 2007), and that dead wood at early decay stages after forest fire or clear-cutting is suitable for many wood fungi.

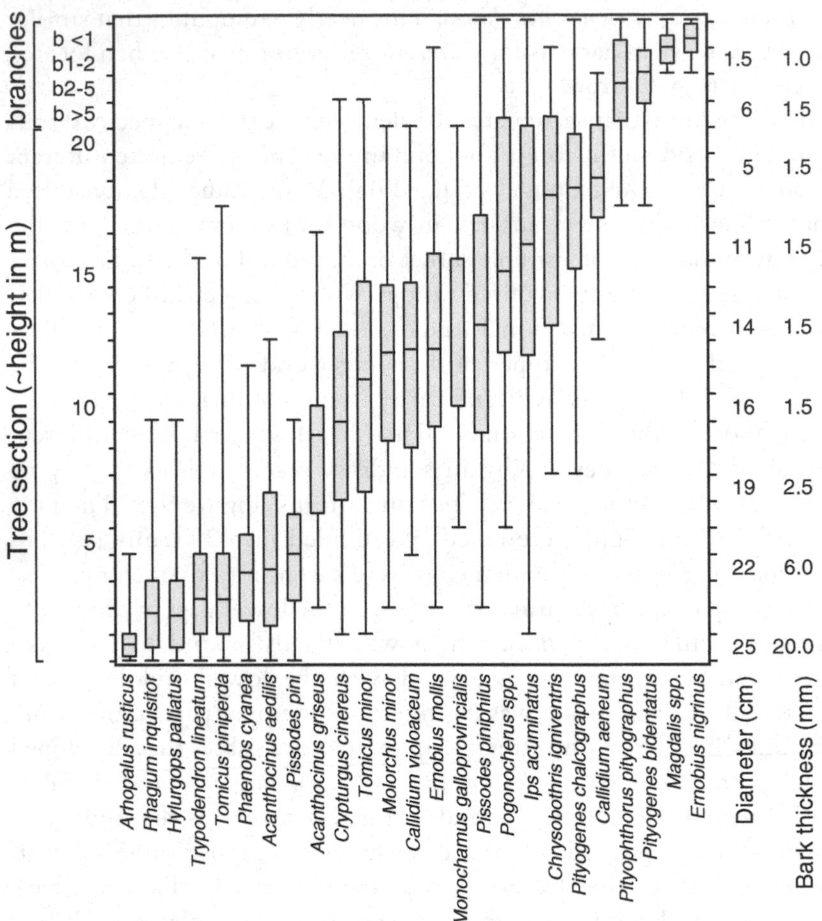

Figure 9.6. Associations of different beetle species with different height sections of standing dead Scots pines (18–25 m high). The box and whisker plots indicate minimum, lower quartile, median, higher quartile and maximum (from Foit, 2010).

Some species even show a strong preference for sun-exposed dry wood, such as *Gloeophyllum sepiarium* (mostly on *Pinus sylvestris* and *Picea abies* logs). The brightly coloured *Pycnoporus cinnabarinus* is another boreal fungus that strongly favours dead broadleaved trees in open areas.

Nearly all studies of wood-inhabiting fungi are restricted to downed wood and standing dead wood up to 2–3 m above the ground. However, a few studies have explored the lower canopy by using a ladder, and the upper canopy by means of a balloon (Ryvarden and Nuñez, 1992) or a crane (Unterseher et al., 2005; Unterseher and Tal, 2006). So, which fungi utilize dead wood in the canopy? In their study of fungi in canopies of broadleaved trees in Germany, Unterseher and colleagues revealed a diverse and distinct species assemblage. They found that the species composition changed from upper, thin twigs to thicker branches in the middle and lower canopy. Typical species on twigs were *Nectria cinabarina*, *Steganosporium acerinum*, *Eutypa maura*, *Nitschkia cupularis* and *Episphaeria fraxinicola*. These species are all ascomycetes. Further down in the canopy, different corticioid fungi such as *Stereum rameale*, *Peniophora quercina* and *Vulleminia comedens* (all belonging to the basidiomycetes) became frequent. Another finding was a conspicuous absence of polypores in the canopy. Only 7 out of 118 taxa were polypores and none of them were common in the canopy. Only *Phellinus contiguus* was recorded several times (5 records in total) while the other polypores occurred accidentally with only 1–2 records in 703 dead wood samples.

In a comprehensive German–Polish study of fungi on dead branches in the lower canopy, Butin and Kowalski (1983a, 1983b, 1986, 1989, 1990) found that different ascomycete species by far outnumbered basidiomycete species. This pattern was observed repeatedly on different tree species. Also in this study there was a conspicuous lack of polypores. The only place where they observed polypores was on thick *Quercus* branches.

A distinct feature of dead branches is their strong tendency towards desiccation, due to a large surface-to-interior ratio, strong sun exposure, and aeration from winds (Parker, 1995). The ability to withstand desiccation is probably the main reason for the distinctness of the species composition in the canopies. In a paper entitled 'Hanging in the air: a tough skin for a tough life', Maria Nuñez (1996) succinctly described how fungi on dead branches overcome desiccation. One strategy is to develop long-lasting fruiting bodies with a waterproof surface and to produce spores under conditions of high humidity. Another strategy is to produce small, short-lived fruiting bodies only when water is available. A third, intermediate, strategy is to tolerate the situation where the fruiting bodies dry up and collapse, but later absorb water and reactivate spore production, as in some groups of Auriculariales and Tremmellales (Sherwood, 1981).

9.3 Wood buried in the soil

Nearly all studies of wood-inhabiting species deal with their occurrence above ground. But a large amount of dead wood occurs on the forest floor and buried in the soil, such as tree roots and wood fragments that have become embedded in the litter. In such buried wood, we find a distinct community of saproxylic species. Information about these species is typically scattered across papers dealing with one or a few species, and there seem to be no review papers on the species diversity in buried wood.

9.3.1 Subterranean fungi

Among the best-known and most-studied fungus species in tree roots are the honey fungi belonging to the genus *Armillaria.* They have a worldwide distribution and comprise about 40 species that are ecologically significant in many forest ecosystems. These fungi are best known as pathogens causing root and butt rot in forests, orchards and gardens (Schmidt, 2006). However, they also comprise important saprotrophs that are significant decomposers of buried wood (Shaw and Kile, 1991). Similar to other pathogenic fungi, they primarily attack weakened or wounded trees. In a very interesting study, Prospero and colleagues (2003) demonstrated how two *Armillaria* species rapidly colonized recently dead wood of Norway spruce. They produced a dense network of rhizomorphs enveloping the roots of live trees. When the trees were cut, the fungi rapidly colonized roots and stumps along the cambial zone, and within 3–4 years after tree felling most of the spruce stumps were colonized. Following colonization of roots and stumps, *Armillaria* species can produce new rhizomorphs, spreading through the soil until the food base is completely consumed (Stanosz and Patton, 1991). Individual *Armillaria* clones can obtain impressive sizes of several hectares; the world record is a clone in Oregon that colonized an area of about 9 km^2 (Ferguson et al., 2003).

Another well-known fungus occurring in tree roots is the root-rot fungus *Heterobasidion annosum.* This is a complex composed of several inter-sterile groups that differ in distribution, host-tree associations and fruiting-body morphology (Schmidt, 2006). These fungi enter old roots through wounds and get into young uninjured roots through the thin bark. Once inside the root, the mycelium spreads to the trunk and upwards in the heartwood, where it causes substantial heart-rot decay in live trees. In coniferous forests throughout the northern hemisphere it is likely to be an important decayer of dead roots (Woodward et al., 1998).

Very few studies have documented the broader diversity of fungal species occurring in dead roots. In one study, von Sydow (1993) isolated and identified fungal mycelia from recently dead (within the last 1–2 years) conifer roots at three points: (1) near the base of the stump, (2) roots with 3–10 cm diameter, and (3) roots thinner than 3 cm. She found that wood-decaying basidiomycetes were common below ground close to the stump base but were virtually absent deeper in the soil, suggesting colonization through the cut surface. The only exception was *Armillaria*, which also occurred frequently in thin roots. The most frequent fungi in the thin roots were different sapstain fungi (e.g. *Leptographium* spp., *Ceratocystis* spp.) and various types of hyaline mycelium representing other ascomycete species. In another study of the above-ground parts of cut stumps, Vasiliauskas et al. (2002) described the community of wood-decaying basidiomycetes based on almost 4000 spruce stumps (aged 1–5 years old) in Lithuanian forests. They found a total of 25 species that are otherwise typical of above-ground logs.

All the species mentioned above occur in wounded or recently dead wood. But there are additional fungi decomposing buried wood that may also utilize soil resources to some extent. An interesting feature of these fungi is their ability to develop mycelial cords, which are linear organs formed by hyphal aggregation that interconnect old and new wood resources (Boddy, 1999). Some of these species, such as *Phanerochaete velutina*, *Phallus impudicus* and *Hypholoma fasciculare*, have been investigated in detail concerning their growth dynamics and nutrient translocation, especially *P. velutina* as a model species. A typical growth pattern of cord-forming fungi is to develop fan-shaped mycelium stretching in all directions from a piece of wood. When encountering a new piece of wood, the growth pattern changes remarkably; cords are formed between the resource units, while the non-connected mycelium stops growing. Multiple studies of such cord systems have demonstrated that carbon components and mineral nutrients are reallocated between different parts of the mycelium–cord network, typically from woody material to the mycelium and from an old to a newly colonized piece of wood (Boddy, 1999; Cairney, 2005).

During the search for new wood units, some cord-forming fungi also utilize soil resources (Boddy, 1993) and their performance varies between different soil types. In a laboratory experiment, Donnelly and Boddy (1998) showed how the mycelium networks of two species (*Phanerochaete velutina* and *Stropharia caerulea*) were differently influenced by enrichment with phosphorus and nitrogen of the soil surrounding the wood bait (Figure 9.7). *P. velutina*, which is a strict wood decomposer, was almost unaffected by different soil enrichment levels. *S. caerulea*, on the other hand, received a strong benefit from increased nutrition levels, as shown by increased mycelium growth and increased decay rate of the wood bait. This species has a wider resource base than *P. velutina*, as it also decomposes soil organic material other than wood.

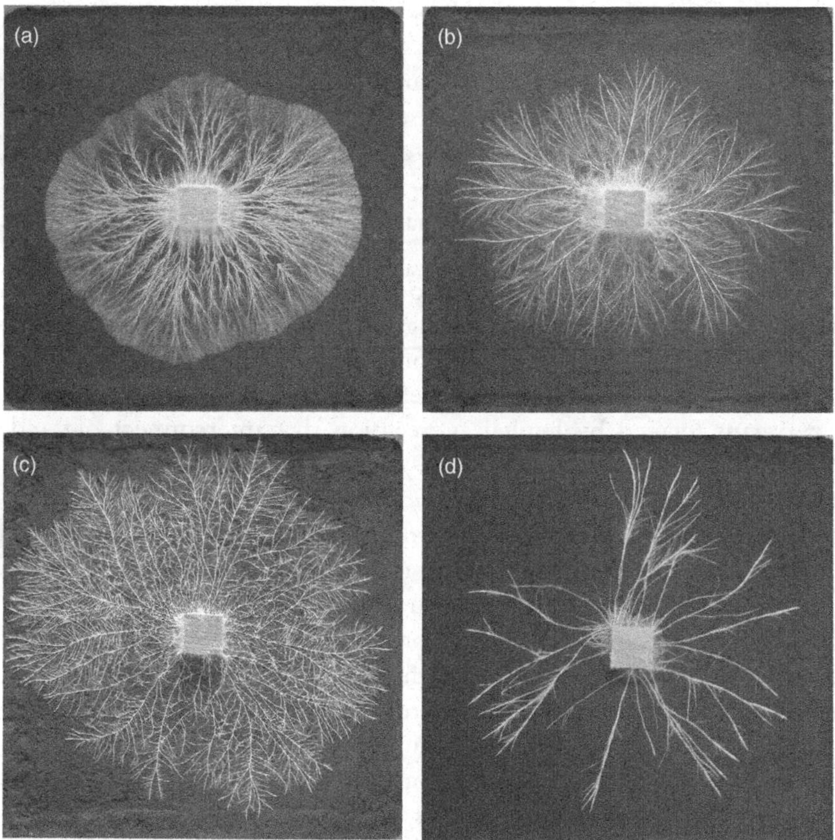

Figure 9.7. Effect of increased soil phosphorus on mycelial cord systems of *Stropharia caerulea* (a, c) and *Phanerochaete velutina* (b, d) after 22 days of growth on: control soil (a, b) or soil enriched with phosphorus (c, d) (from Donnelly and Boddy, 1998).

In the field, the effects of changed soil conditions on wood-inhabiting fungi have also been observed. A number of studies have followed the effect of liming on forest stands (Fielder and Hunger, 1963; Garbaye et al., 1979; Veerkamp et al., 1997). These studies suggest that species richness on conifers may decrease due to liming, while species confined to broadleaved trees may increase. In a liming experiment in the Netherlands, species richness generally increased after liming (Veerkamp et al., 1997). However, the frequency of several conifer-associated species decreased (e.g. *Botryobasidium subcoronatum*), while species associated with broadleaved trees showed a strong positive response to liming. The positive response to liming was considered most likely to have been an effect of increased nitrogen dynamics in the soil.

In summary, the ecological functioning of below-ground saproxylic fungi is quite well documented for selected model species, especially for the economically important *Armillaria* and *Heterobasidion* species. But the diversity of subterranean saproxylic fungi is poorly known. There are probably many saprotrophic species in forest soils with different degrees of wood association. For example, in the Nordic countries, more than 200 agaric species have been found on wood fragments as well as litter and other decaying organic matter on the forest floor (Knudsen and Vesterholt, 2008).

9.3.2 Subterranean invertebrates

Some invertebrates seem to be strictly confined to buried wood but, except for beetles, these invertebrates are even less well explored than subterranean fungi. Among the best known are several bark beetles in the genus *Hylastes* and weevils in the genus *Hylobius*. These species develop in the bark at or below the soil surface (down to 0.5 m depth in sandy soils) of recently dead coniferous trees (Ehnström and Axelsson, 2002). A similar breeding site is used by the longhorn beetles *Pachyta lamed* and *Lamia textor*. Also, in more decomposed roots, we find beetle species such as the stag beetle *Lucanus cervus* and the longhorn beetle *Prionus coriarius*, which develop in *Quercus* roots. In a thorough account of Swedish beetles that make distinct galleries in bark and wood, Ehnström and Axelsson (2002) reported that 15 out of 482 species had a strong preference for woody substrates below the soil surface.

In addition, some species in the Diptera family hoverflies (Syrphidae) primarily develop in buried wood. Larvae of *Criorhina berberina* have been found in decaying tree roots, and the females have been observed to lay eggs repeatedly into the soil exactly above decaying roots (Rotheray, 1994). Decaying roots are probably also a major breeding site for *Xylota sylvarum*, as numerous larvae have been encountered underground in the wet decaying roots of *Fagus* stumps (Rotheray, 1990). Additional *Criorhina* and *Xylota* species have been found as larva in decaying roots and other wet sites (Rotheray, 1991). Knowing that the hoverflies only constitute a small proportion of saproxylic Diptera, it is reasonable to assume that several other species also develop in this environment.

9.4 Submerged wood

Dead wood is an important structural component in aquatic environments. Branches and whole trees regularly fall into rivers and lakes, and the rivers transport a lot of wood to marine waters (see Chapter 12). In addition, the mangrove forests represent an ecosystem that is particularly rich in submerged wood.

Since many species have colonized the wood before it falls into the water, it can be difficult to determine whether a particular species is aquatic or not (Shearer et al., 2007). Nevertheless, the aquatic environment is very different from terrestrial habitats and many species have adaptations to this environment, showing that they are truly aquatic. In many organism groups, there is little or no overlap in the species composition between the two environments. Furthermore, there are major differences on higher taxonomic levels. In fungi, the aquatic medium has selected strongly against almost all types of basidiomycetes. In a recent review Shearer et al. (2007) found that of 3047 aquatic fungal species, 2313 were ascomycetes, while only 11 were basidiomycetes from freshwater and 10 were basidiomycetes from brackish or marine water. Many of these fungi occurred on other substrates than woody material but, as we will see below, there are hundreds of ascomycete species that bring about decay in submerged wood.

Among animals, there are also saproxylic species that strictly occur only in aquatic environments. Many of these belong to taxonomic groups that are totally aquatic, such as marine bivalves and crustaceans, or predominantly aquatic such as the freshwater chironomid mosquitoes. In these groups, which are generally non-saproxylics, some species have evolved to utilize wood. In other cases, there are species-rich groups of terrestrial saproxylic species in which a small number of species have evolved to utilize wood in an aquatic environment.

9.4.1 Freshwater fungi

There is a rich diversity of fungi acting as wood decomposers in freshwater. It was the British mycologist C. T. Ingold who established that freshwater ascomycetes are a distinct ecological group with special adaptations to aquatic environments (Ingold, 1954). These adaptations include floating spores with special appendages to hook on to their substrates, and soft rot, which is an effective decomposition mode for submerged wood. Since Ingold's pioneering work during the 1950s–1970s, the number of known freshwater fungi has risen to 530 (Shearer et al., 2007). About 90% of these species have been recorded predominantly or exclusively on woody substrates. New species of freshwater ascomycetes are continuously being described and recently a new order (Jahnulales) was established for a monophyletic group of freshwater wood-inhabiting fungi (Pang et al., 2002).

9.4.2 Freshwater invertebrates

Among invertebrates there are several freshwater species that are closely associated with decaying wood. The intimacy of these associations ranges from opportunistic to obligate association (Dudley and Anderson, 1982). In a thorough review of central European freshwater invertebrates, Hoffmann and Hering (2000) classified the associations into obligate xylophagous, facultative xylophagous and non-xylophagous (species that are closely associated with wood without feeding on it). The obligate xylophagous species are species that bore into and feed solely on wood, and comprise mostly Diptera species in the families Chironomidae and Limoniidae, but also a few beetles and caddisflies (Trichoptera). The facultative xylophagous species exploit the wood by chewing or gouging. These species typically feed on leaf litter in autumn, but they expand their diet to include wood from early spring and throughout the summer, when leaves are in short supply. They probably rely on fungi and bacteria to obtain their nutrients from wood, as they do not have the ability to digest cellulose themselves. These facultative species mostly include caddisflies, chironomids, but also some snails (Gastropoda) and beetles. The group of non-xylophagous species consists partly of surface-feeding species grazing on the biofilm of bacteria, cyanobacteria, algae and fungi on the wood surface, and partly of predatory species using the wood as a hunting ground.

Altogether, Hoffmann and Hering listed slightly more than 100 species from central Europe, of which only 15 were classified as obligate xylophagous. These numbers correspond well with the findings of Dudley and Anderson (1982), who reported 185 wood-associated species from the northwestern USA. Only a few of these species were feeding obligately on wood.

9.4.3 Marine fungi

A lot of wood-inhabiting fungi can be found in marine waters, especially in mangrove forests, which represent a biodiversity hotspot for marine fungi. In a comprehensive review, Schmit and Shearer (2003) listed 625 species from mangrove environments and the great majority of these were ascomycetes. This number included species from terrestrial environments and sediments, but about 50% of the species only occurred on permanently or periodically submerged substrates (wood and leaves). Among the ascomycete and asexual (anamorph or mitosporic) species, 181 were only found on woody material, 24 on wood and leaves, and 115 on leaves only. The species that occurred on wood were predominantly species representing the sexual stage, while the species that occurred on leaves and in sediments were mostly asexual species. It is worth noting that the sexual spores of marine fungi have appendages or gelatinous sheets that help them to attach to wood (Rees and Jones, 1984; Jones, 1985). The asexual spores largely lack such appendages, which may explain why asexual fungi are more common in sediments. Schmidt and Shearer also showed that most of the ascomycetes on woody substrates had a wide geographical distribution. More of these species have been found in multiple ocean basins (two, or all three, of the Atlantic, Pacific and Indian basins) than in just one ocean basin. Several of the species from mangrove forests have also been found on driftwood far outside the tropical zone. Examples are *Marinosphaera mangrovei* and *Savoryella lignicola* reported from driftwood in Denmark (Koch and Petersen, 1996).

9.4.4 Marine invertebrates

Several groups of invertebrates occur in marine wood, but very few of these have close relatives among freshwater or terrestrial invertebrates, which demonstrates that they have evolved their wood associations completely independently. The shipworms (Bivalvia, Teredinidae) constitute the most important group both ecologically and in terms of species diversity. These organisms got their name from their ability to bore into the submerged part of wooden boats. The shipworms colonize the wood as soon as the bark falls off. During only 1–2 years, they can completely consume 4 cm thick prop roots of mangrove trees (Kohlmeyer et al., 1995). Furthermore, they do not discriminate between wood from different tree species and bore into any kind of submerged wood from mangrove forest to driftwood across all the oceans. The shipworms belong to a mollusc family where 14 genera are known. There are also examples of marine species among crustaceans and beetles. This includes some isopod species in the genus *Limnoria* that bore into and digest wood (King et al., 2010) and the cosmopolitan weevil *Pselaphys spadix*, which can cause severe damage to wooden constructions in harbours (Oevering and Pitman, 2002).

The wood borers described above are common in surface waters. But the oceans hide more saproxylic species in deep waters. At depths of several thousand metres, different species of *Xylophaga* (Bivalvia, Pholadidae) bore into and feed on wood in a similar manner to the shipworms (Turner, 1973; Distel and Roberts, 1997). There are also crustaceans that have an obligate association with sunken wood. In an experiment by Maddocks and Steineck (1987), they positioned twelve 85-cm wooden cubes at depths of 1800–4000 m in five different locations in the Pacific and Atlantic basins in Central America. On retrieving the wood after a year, they found 14 species of *Ostracoda*, which are tiny 0.5–1 mm sized crustaceans. Twelve of these were new to science and four of them belonged to a new genus. Interestingly, the different locations had several species in common, suggesting good dispersal ability.

9.5 Tree growth rate, wood density and secondary substances

The morphological and chemical characteristics of the wood are strongly shaped by the external environment during the life of the tree. This is a 'hidden' quality aspect that has evaded most research on saproxylic species up to the present.

The variation in growth rate between individual trees of the same species may be very large. When growing on productive sites, annual rings are wider and volume increment fast. This results in many structural and chemical effects on the wood such as, for instance, lower density (Figure 9.8), thinner cell walls, less heartwood and changes in lignin content (Mäkinen et al., 2002; Sarén et al., 2004). Furthermore, growth rate seems to influence the allocation of resources to secondary

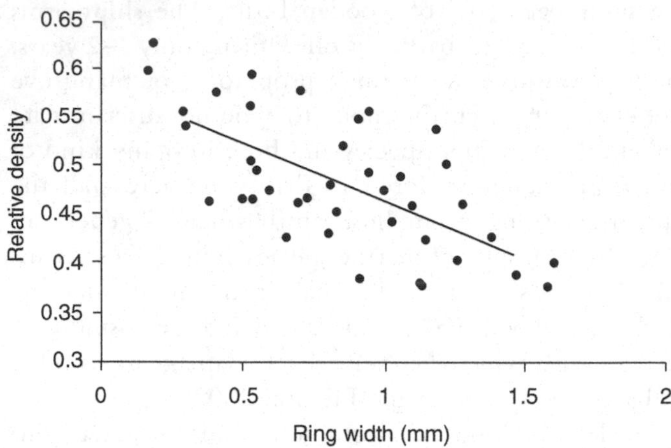

Figure 9.8. Changes in relative density of annual rings of Norway spruce (*Picea abies*) as a function of ring width. The regression line is highly significant ($R^2 = 0.45$, $p < 0.001$). Data from Swedish old-growth forests (Nic Kruys and Bengt Gunnar Jonsson, unpublished data).

substances such as different types of phenols and sterols, which play a central role in herbivore defence. The major point for saproxylic species is clear: different wood densities and levels of secondary substances translate into varying levels of decay resistance (see, e.g., Venäläinen et al., 2003), and these differences influence important habitat characteristics for the species colonizing the tree when it is dead. Some of these effects have already been described, but we still do not fully understand the importance of these factors in shaping saproxylic communities.

In an experiment with three fungal species, Edman et al. (2006) showed that all species decayed fast-growing wood better than slow-growing wood. However, the relative decay rate in slowly growing wood differed substantially between species (Figure 9.9). Thus, *Fomitopsis rosea*, considered as a species associated with old-growth forests, was significantly better at decaying slow-grown spruce trees than its close relative *F. pinicola*. This suggests that growth rate may be an important habitat factor and that the relative difference between species may be large enough to explain their occurrence patterns. One hypothesis is that the serious decline of *F. rosea* in managed forests is an effect of the loss of a substrate type (old and slowly grown spruce wood), where it is a stronger competitor.

There are clear relations between growth, site conditions and secondary compounds (Wainhouse et al., 1998; Lombardero et al., 2000),

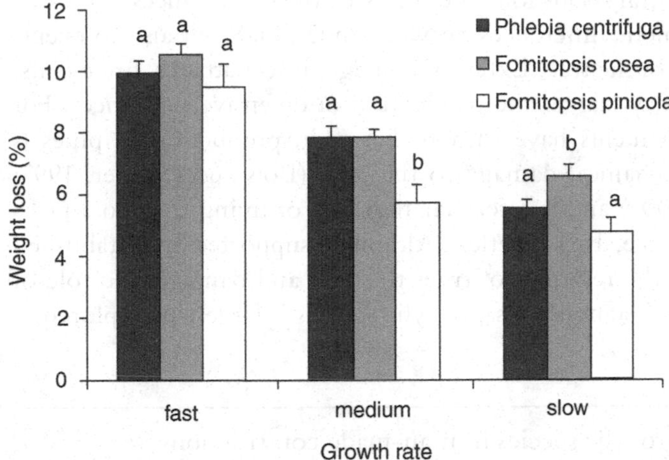

Figure 9.9. Decay rate in the form of percentage weight loss of inoculated wooden discs after 5 months. The wooden discs are taken from trees with different growth rates, ranging from 1.6 to 12 annual rings per centimetre. Two of the species are confined to old-growth forests (*Phlebia centrifuga* and *Fomitopsis rosea*) while one species (*Fomitopsis pinicola*) is a generalist species also common in managed forests (after Edman et al., 2006).

and a general pattern seems to be that slowly growing trees and tree species have a greater resistance to decay. However, the patterns are complex and may differ between groups of species. For instance, in a study of white spruce (*Picea glauca*), it was shown that increased wood density increases resistance to decay by white-rot species, while brown-rot species – by contrast – were better at decaying wood of high density (Yu et al., 2003). In this study it was also shown that the heritability of decay resistance in white spruce was present in relation to some species (*Gloeophyllum trabeum* and *Trametes versicolor*) but not for others (*Fomitopsis pinicola*). This suggests that decay resistance is complex and is not only related to properties of the tree species and environmental conditions during growth but also to interactions with the actual wood fungi responsible for the decay.

Decay resistance is further strongly linked to the types and amounts of so-called secondary substances in trees (Harju et al., 2003; Gierlinger et al., 2004). There has been extensive research in this field, because it is of considerable economic importance. Decay-resistant timber is highly valued as a building material, and to choose plant material and forest management methods that produce such timber is of silvicultural interest. In natural conditions the levels of these substances vary, due not only to genetic differences between individuals but also to events during the life of the tree. External damage, insect attacks, fire events, drought etc., may trigger the production of defensive substances. For example, experiments have shown that resin production in pines is induced by mechanical damage to the bark (Bois and Lieutier, 1997; Harju et al., 2009) and is a regular response of living trees to attacks from, for instance, bark beetles. Although supported by established knowledge on the response of trees to stress and damage, the role of these induced substances for saproxylic species is largely unexplored.

Box 9.1 Saproxylic species in man-made constructions

This book is devoted to the richness and value of saproxylic biodiversity in natural environments. But there are also some species with a bad reputation. Below we have collected a few 'villains' that may be found in your neighbourhood.

The dry-rot fungus, *Serpula lacrymans,* is a wood-decaying fungus that rarely occurs in natural environments. Instead it is a widespread, aggressive indoor fungus being a major problem where it is present. It has the particular ability to transport water through mycelial cords from one part of the mycelium to another and thereby to increase the water content in completely dry wood and facilitate colonization in unfavourable conditions. The decomposition subsequently creates additional water as a by-product. In order to get rid of the fungus, one needs to remove absolutely all the wood that has been attacked by the fungus, which is normally very costly as the fungus attacks load-bearing constructions. The fungus has probably been a problem in wooden houses for several thousand years; the correct treatment of houses with 'leprosy' was even described in the Bible (Leviticus).

Old-house borer *Hylotrupes bajulus* is a longhorn beetle (Cerambycidae) that has spread worldwide. It occurs indoors and its larvae, which eat and tunnel through wood, take several years to develop. It prefers conifer wood and requires a relatively high temperature to develop (>20°C). The larval excavates galleries within the wood and it is detected either by the emergence holes of adults (5–10 mm in diameter) or actually by the scraping sound the larvae make while feeding. It may cause extensive damage in bearing constructions and is very difficult to exterminate. The worldwide expansion of the species has most likely happened through packing material and wooden crates. Currently it is viewed as a major and expanding threat in warmer climates, such as in Australia.

Termites are common in tropical and subtropical regions, where they are important wood decomposers. Termites also occur in several temperate regions. Their ability to decompose wood and other cellulose-rich material includes the ability to feed on houses and other man-made constructions. Dampwood termites are mostly subterranean, and they build nests below the ground. They will search for cellulose sources in the vicinity of their nests, including wood in houses. To process the cellulose, they live in symbiosis with protozoans in their guts, and with this lifestyle they are efficient in decomposing wood. Over a short period of time, they may cause severe damage on the load-bearing parts of a house. When established in buildings they do not only attack the construction wood but may also feed on other available cellulose-containing materials such as paper, cloth and furniture.

10 · *Evolution of saproxylic organisms*

Jogeir N. Stokland

The evolution of dead wood inhabitants has taken place on our planet ever since woody plants appeared at least 385 million years ago (mya). This evolution has brought about numerous adaptations to enable these organisms to utilize the resources in decaying wood. Furthermore, several fascinating co-evolved systems have developed between the species that live in this microcosmos.

The basic approach in this chapter is to assess the phylogenetic trees of various organism groups, their sequence of branching, and the evidence indicating when these lineages originated. Next, we superimpose different aspects of saproxylic lifestyles upon these phylogenies to identify roughly when different saproxylic life forms originated. But first we need to summarize some key events in the evolution of woody plants.

10.1 Evolution of woody plants

Fossil records show that plants colonized the land from freshwater some 450–480 mya (Kenrick and Crane, 1997) (Table 10.1). During their early evolution, the land plants developed a secondary cell wall strengthened with lignin. This was probably a response to the new challenge of growing on land – the loss of support from water. Lignin is an extraordinarily strong structural component that enables plants to grow upright; it is present in all vascular land plants but absent in aquatic algae. Furthermore, land plants increase their lignin production in response to gravity (Chen et al., 1980), and large trees have a higher proportion of lignin deposits than bushes and herbaceous plants. This has particular significance for wood-decaying fungi, as there is a selective advantage in becoming an effective lignin decomposer.

The first woody plants appeared during the Devonian period. They were arboreal lycopsids, a group of plants akin to the extant club mosses (Lycopodiaceae). Towards the end of the Devonian some representatives reached heights of 10–15 m, and during the Carboniferous some

Table 10.1. *Important aspects of woody plant evolution.*

Million years ago (mya)	Period	Plant forms and elements of palaeoclimate
144–65	Cretaceous	First occurrence, rapid radiation and sharply increased abundance of angiosperms, including broadleaved trees.
206–144	Jurassic	
248–206	Triassic	Initiated by a massive collapse of terrestrial and marine ecosystems during the first 10 million years. Delayed recovery of forests. Expansion of conifers.
290–248	Permian	Expansion of gymnosperms, forming extensive forests on dry land, while arboreal lycopsids and glossopterid seed ferns formed swamp forests.
354–290	Carboniferous	Huge areas of swamps and swamp forests dominated by lycopsids and seed ferns. Initial development of gymnosperms in drier areas. The first conifers.
417–354	Devonian	Several-metre-high lycopsids and arboreal ferns. Progymnosperms (*Archaeopteris* = *Callixylon*) were the first true trees with cambium and secondary diameter growth.
443–417	Silurian	Earliest fossils of vascular plants: mosses and horsetails. Lignin as additional cell wall component.
510–443	Ordovician	Aquatic algae with cellulose and hemicellulose as cell wall building blocks. The plants colonized land late in this period.

lycopsids grew even higher. The famous 'earliest forest' dates back to 385 mya and belonged to the plant group cladoxylopsids (Stein et al. 2007).

In the late Devonian, at least 370 mya, the first true trees evolved – the *Archaeopteris* (Meyer-Berthaud et al., 1999). While the lycopsids grew only in a forking manner at the tip of their main axes, the *Archaeopteris* had multiple branching points along the trunk. Furthermore, they had secondary lateral growth of the trunk brought about by a cambium layer. *Archaeopteris* had trunk diameters larger than 1 m and grew up to 30 m high, with voluminous canopies. On the other hand, they had fern-like leaves and reproductive organs like those of the ferns. For quite some time the complete morphology of these trees was not understood because leaves and trunks were found as separate fossils. *Archaeopteris* was used to name the leaves and *Callixylon* was used for fossil trunks until it became clear that these parts originated from the same plant (Beck, 1960). Beck then proposed a new taxonomic group, the progymnosperms, to accommodate such plants with fern-like reproduction and conifer-like wood. *Archaeopteris* expanded rapidly to different continents, and towards the end of the Devonian there were extensive floodplain and coastal forests dominated by *Archaeopteris* (Meyer-Berthaud et al., 1999).

Quite mysteriously, *Archaeopteris* went extinct at the end of the Devonian, while woody lycopsids evolved and expanded during the Carboniferous period. The name 'Carboniferous' comes from England and refers to the rich deposits of coal that occur there. We find such coal deposits throughout central Europe, Asia, and mid-western and eastern North America. Coal is a carbon-rich rock packed with fossil leaves, branches and tree trunks. Such formations develop in wet areas only, so the Carboniferous deposits indicate the presence of widespread peatlands and swamp forests.

During the Carboniferous period a new plant group evolved: the gymnosperms or the true seed plants (Figure 10.1). The progymnosperms are widely regarded as the ancestors of the seed plants, but it is uncertain whether early seed plants evolved once or several times from the progymnosperms. The development of seeds with a store of nutrients gave a head start to the germinating plants in new locations. Furthermore, the pollen grains were wind-dispersed and not dependent on moisture like the gametes of mosses and ferns. This facilitated their expansion into drier areas, while lycopsids and arboreal seed ferns (glossopterids) were restricted to swampy areas. The conifers are an important group of gymnosperms that evolved in this period. They have been found as fossils dated back to 310 mya, and by the end of the Carboniferous the conifers were rather diverse (Miller, 1999).

The massive coal formations from the Carboniferous were mainly tropical and subtropical. During the Permian, coal-forming swamp forests expanded into the temperate climate belts of higher latitudes, where arboreal seed ferns and other plants formed distinct forests in the south and north (Retallack et al., 1996). The conifers expanded

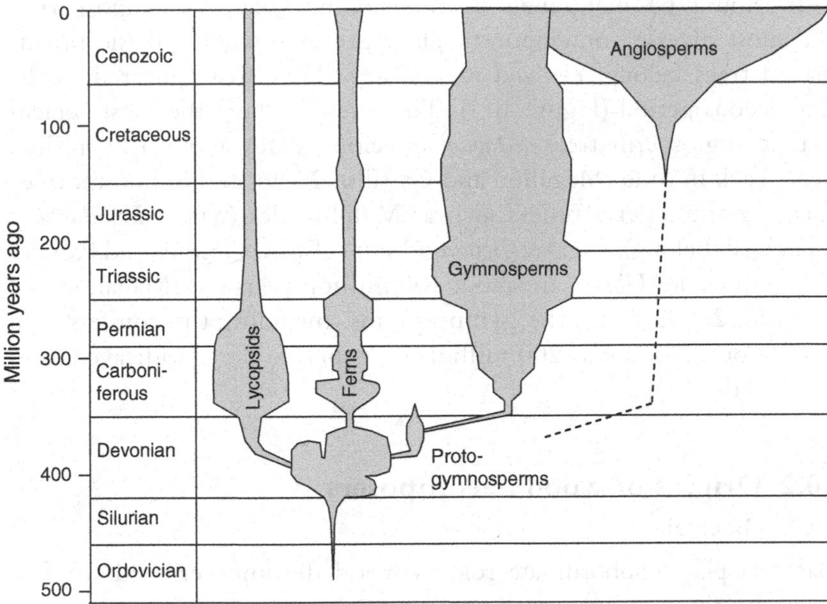

Figure 10.1. Relative species diversity in major terrestrial plant groups since the beginning of the Ordovician. Redrawn and modified from Schweingruber et al. (2006).

rapidly into drier areas and towards the continental interiors during the Permian.

At the Permian–Triassic boundary 250 mya, our planet experienced the worst extinction episode ever. Coral reefs disappeared totally and more than 90% of all marine species went extinct. Land plants and animals also went through a catastrophic period. Peat areas and swamp forests collapsed totally and 10 million years passed with a complete lack of coal formation (the 'coal gap') followed by another 10 million years with very thin coal formations (Retallack et al. 1996). There was also a massive dieback of equatorial coniferous forests during the Permian–Triassic transition, followed by 4–5 million years of terrestrial ecosystem degradation in Europe (Looy et al., 1999). After this episode, the conifers expanded rapidly and they have been successful ever since.

The Permian–Triassic transition must have been a major setback in the evolution of saproxylic organisms. Just as new plant forms replaced the extinct Permian flora, there was a series of new saproxylic lineages that appeared among the insects during the Triassic and the Jurassic.

It should be mentioned in this context that the angiosperms, the most diverse contemporary plant group – where all the broadleaved trees belong, evolved much later. They first appeared in the Cretaceous period (Figure 10.1). The lineage where the most ancient extant angiosperm trees (*Magnolia*) belong dates about 120 million years back in time (Magallón and Castillo, 2009), and important tree-forming angiosperm orders such as Malphigiales (where *Populus* and *Salix* trees belong), Fagales (*Betula, Corylus, Fagus, Quercus*) and Rosales (*Prunus, Sorbus, Ulmus*) are about 100 million years old (Magallón and Castillo, 2009). Thus, the gymnosperms, including the conifers, had been around for about 200 million years before the broadleaved trees entered the scene.

10.2 Origins of wood decomposers

10.2.1 Bacteria

Bacteria play a subordinate role as wood decomposers and we find them mostly as decomposers in aquatic and subterranean environments (see Chapter 2). Bacteria originated about 3 billion years ago; some became photosynthetic cyanobacteria or sulphur bacteria, while others developed a parasitic or saprobic nutritional mode. These have been able to utilize woody substrates to some extent, but never as effectively as the wood-decaying fungi.

Bacteria have evolved distinct lineages that decompose woody material in various ways, some even with the ability to degrade lignin fractions. But these erosion and tunnelling bacteria have mostly been identified from their traces in degraded wood and not from their genetic composition. So even if the phylogeny of bacteria is rather well developed, the identity of wood-decaying bacteria is not sufficiently known to discuss their evolutionary origins.

10.2.2 Fossil fungi

Reports of fossil fungi are very sparse, but this does not mean that fungi are uncommon as fossils. They have simply received much less attention than other groups of organisms. One reason is that fossil fungi are difficult or impossible to identify. They tend to be microscopic hyphae, and very few fossil fruiting bodies have been found.

Well-preserved terrestrial fungi are known from *c.* 400 million-year-old Devonian fossils from the Rhynie chert in Scotland (Taylor T. N.

et al., 2004), which was a time when woody plants had not yet evolved. These preservations document the existence of different taxonomic groups, e.g. Chytridiomycota, Glomeromycota and Ascomycota, but not Basidiomycota. Taylor and colleagues attributed little significance to the absence of the basidiomycetes from the Rhynie chert, since the first fossil ascomycete has only recently been reported from these deposits. A likely explanation for the absence of basidiomycetes is that they have not yet been identified (Taylor T. N. et al., 2004).

A particularly interesting discovery from the Rhynie chert was that of ascomycete fruiting bodies in the vascular plant *Asteroxylon* (Taylor et al., 1999, 2005). The fungus occurred in the stem cortex just beneath the epidermis. The presence of nearby necrotic tissue led Taylor and colleagues to suggest that the fungus was perhaps a pathogen. The fruiting-body appearance led Taylor and colleagues to place the fungus close to the pyrenomycetes. This is the earliest certain ascomycete and it is used as an important calibration point for fungal molecular clocks. The classification of this fungus as a pyrenomycete has puzzled various researchers. Several have suggested that the fungus was misclassified, since a pyrenomycete at this time makes other time estimates very unlikely. Some years later, Taylor and Berbee (2006) opted for the possibility that the fungus might belong to an extinct lineage with a more basal position in the ascomycete phylogeny.

The oldest evidence for wood-decaying fungi dates back to the late Devonian, 375–360 mya (Stubblefield et al., 1985). This fossil material consists of superbly preserved wood from the progymnosperm tree *Callixylon newberryi*. Fungal hyphae occur abundantly in tracheid and ray cells that show extensive decay in the cell walls. The excellent preservation also allowed comparison with microscopic aspects of extant rot types. Stubblefield and colleagues concluded that the wood had a kind of white rot brought about by an ascomycete or a basidiomycete fungus. In the original paper, they did not place the fungus in a particular taxonomic group, but later one of the co-authors and others have referred to the fungus as a basidiomycete (Taylor and Osborn, 1996; Hibbett et al., 1997b). The oldest definitive wood-decaying basidiomycete is from the second half of the Carboniferous, about 300 mya (Dennis, 1970).

10.2.3 Phylogeny of wood-decaying fungi

Phylogenetic analyses and molecular clocks represent another source of information to date the possible origins of wood-inhabiting fungi. The most comprehensive high-level phylogenetic analysis of fungi is that of James et al. (2006), which is based on six different genes sampled from nearly 200 species scattered across a wide range of fungus groups. They suggested that diversification in major fungal lineages (phyla) probably occurred in terrestrial environments.

James and colleagues did not assign any time scale to their phylogeny, but this can be retrieved from other sources. In the early 1990s, Berbee and Taylor (1993) published a groundbreaking work that estimated the divergence between Ascomycota and Basdiomycota to have occurred about 390 mya. Later Taylor and Berbee (2006) used the phylogenetic tree from James et al. (2006) and introduced different calibration points. When they assigned the 400 million-year-old ascomycete from the Rhynie chert to the Ascomycota as such and not to the pyrenomycetes, they estimated that the Ascomycota–Basidiomycota divergence occurred about 450 mya. Alternative calibration points produce quite different age estimates, so the age of these important fungal groups remains quite uncertain.

10.2.4 Ascomycota

Within Ascomycota we find the true yeasts (Saccharomycotina) as a basal branch (Figure 10.2). Today, many of the yeasts live as saproxylic fungi, consuming sap from tree wounds and sugar compounds from decaying trees, while others have developed as gut symbionts of saproxylic insects (Suh et al., 2005, 2006). Yeast colonization of wounded and decaying trees must be very old. We know this from a distinct genome duplication that is estimated to be about 100 million years old (see Suh et al., 2006). Species possessing this duplication belong to a relatively young clade and there are at least five older clades of saproxylic yeasts in the Saccharomycotina phylogeny. Further phylogenetic analyses are needed to resolve when different yeast clades specialized to utilize wounded and decaying trees.

The Saccharomycotina is followed by Orbiliomycetes and Pezizomycetes as subsequent saprotrophic lineages near the base of the ascomycete phylogeny. Also in the Orbiliomycetes we find interesting links to wood decomposition. Species in the genus *Orbilia* are common on decayed wood in temperate and tropical regions. In the Orbiliomycetes we find several types of nematode-trapping fungi. This has been interpreted as a mechanism for wood decomposers to expand their nutrition base, since they live on a nitrogen-poor diet (see Section 3.6 – Predatory fungi).

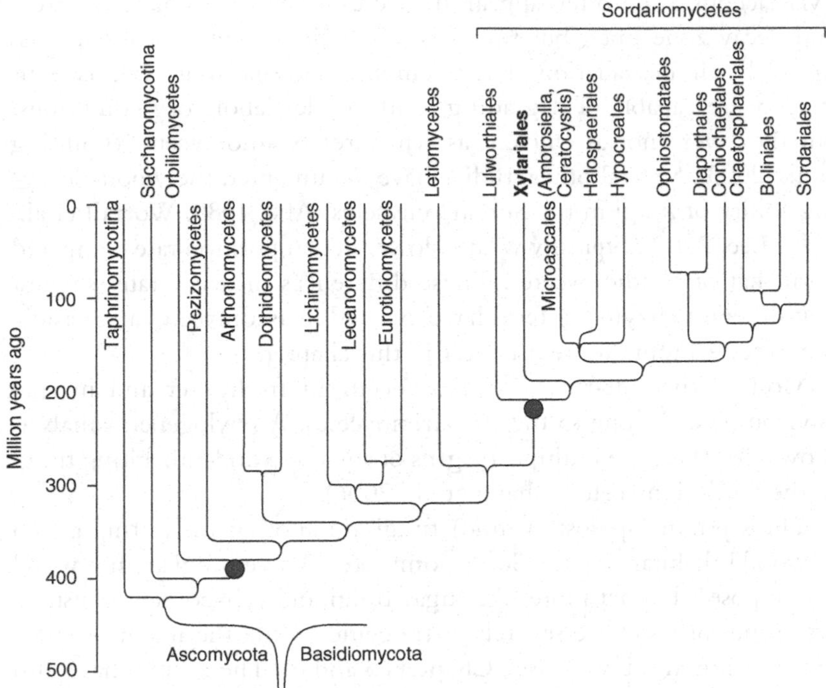

Figure 10.2. Coarse phylogenetic tree of Ascomycota based on James et al. (2006) and Zhang et al. (2006). Taxa on top of branches have many saproxylic species (printed in boldface if most species are wood decomposers). Taxa placed along branches are predominantly non-saproxylic. The time-scale positions of the branching points are highly uncertain, as no branch age is supported by fossil evidence. The dark circles indicate divergence dating based on estimates from Taylor and Berbee (2006).

The majority of wood-decaying ascomycetes belong to the class Sordariomycetes. This is a large monophyletic lineage in Ascomycota with more than 600 genera and 3000 known species. The nutritional mode of these fungi is very diverse, comprising litter and wood decomposers, endophytes, plant and animal parasites, as well as mycoparasites. Unfortunately we do not know the age of this lineage. In the analysis of James et al. (2006), the Sordariomycetes appear to be a rather derived group.

Near the base of the Sordariomycetes we find the Xylariales, containing the most effective ascomycete wood decayers, such as the genera *Hypoxylon* and *Xylaria*. Several of these fungi have been classified as white-rot fungi, but this is mainly based on field observations. Pointing and co-workers (2003) showed that wood colonized by xylariaceous fungi often appeared 'white-rotted', with bleached areas and decay zone lines, but this was not indicative of substantial mass loss or lignin degradation. They demonstrated that some xylariaceous fungi were capable of degrading lignin under laboratory conditions, but they were not as efficient as white-rot basidiomycetes (Pointing et al., 2003). Several other studies have documented the wood-decaying ability of fungi in the Sordariomycetes (Abe, 1989; Worrall et al., 1997; Lee, 2000), typically with a slower decomposition rate compared with that of basidiomycetes. These differences in decay rates suggest that different enzyme systems have evolved in ascomycetes and basidiomycetes – a topic we revisit later in this chapter.

Most of the soft-rot fungi that occur in freshwater and marine environments belong to the Sordariomycetes. A phylogenetic analysis shows that there are multiple origins of aquatic wood-inhabiting fungi in the Sordariomycetes (Zhang et al., 2006).

The sapstain (ophiostomatoid) fungi are another interesting group of wood inhabitants in the Sordariomycetes. These fungi are not wood decomposers but act more like sugar fungi and cell-content consumers. Some of them are severely pathogenic, while the majority colonize recently dead wood (see Chapters 3 and 6). These fungi have also developed interesting co-evolved systems with bark beetles and wood-boring ambrosia beetles. Like the soft-rot fungi, the sapstain fungi have multiple origins that have evolved at different points in time (Zhang et al., 2006). The split between the lineages leading to *Ceratocystis* and *Ophiostoma* occurred more than 170 mya, while *Ophiostoma*, as such, might be more than 85 million years old (Farrell et al., 2001).

It is tempting to speculate that the whole class of Sordariomycetes originated, and to a large extent has diversified, in decaying wood. Evidence supporting this possibility is the basal position of Xylariales in the phylogeny. Furthermore, there are many saproxylic lineages in this order. We find many wood decomposers in terrestrial environments, there have been repeated colonizations of wood decomposers into freshwater and marine environments, the sapstain fungi have probably evolved as specialists on dying or recently dead inner bark and sapwood, and some of the sapstain fungi have developed intimate associations with saproxylic insects. A proper test of whether decaying wood has been the environment for Sordariomycetes evolution would be to sort out the position of saproxylic lineages and to thoroughly investigate the development of wood-degrading enzyme systems in these lineages.

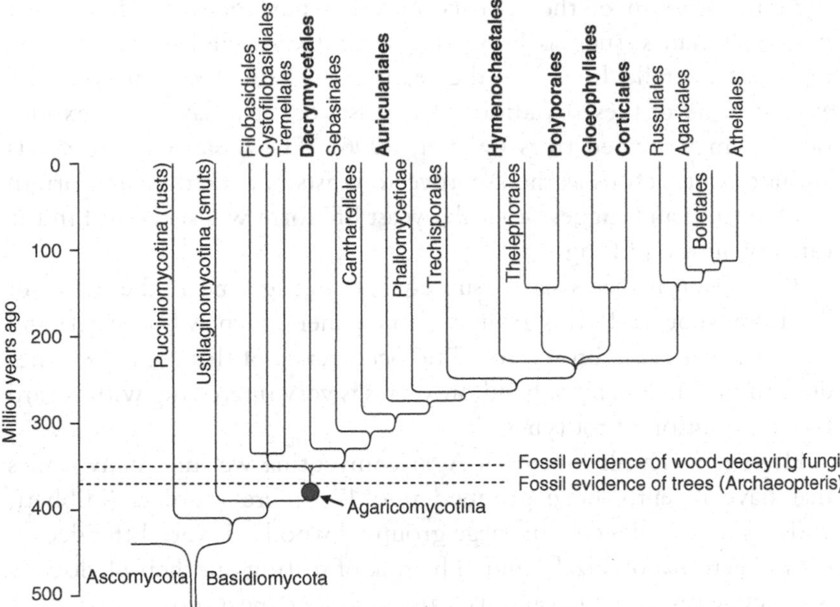

Figure 10.3. Coarse phylogenetic tree of Basidiomycota based on James et al. (2006). Taxa on top of branches include many saproxylic species (printed in boldface if most species are wood decomposers). Taxa placed along branches are predominantly non-saproxylic. The time-scale positions of the branching points are highly uncertain, as no branch age is supported by fossil evidence. The dark circle indicates divergence dating based on estimates from Taylor and Berbee (2006).

10.2.5 Basidiomycota

Within Basidiomycota (Figure 10.3), the rusts (Pucciniomycotina) and smuts (Ustilaginomycotina) come out as basal lineages of obligate plant pathogens. The next major clade is Agaricomycotina, where all the macroscopic basidiomycetes belong. The basal positions of rusts and smuts led James et al. (2006) to suggest that some plant pathogen was the ancestor of Basidiomycota. The possibility exists, however, that there is a direct line from a saprobic Basidiomycota origin to Agaricomycotina, while the rusts and smuts are early side-branches. A closer inspection of their life cycle and phylogeny provides some evidence for this. In their haploid phase the smuts are saprobic yeast-like fungi, while the diploid phase is pathogenic. The majority of the rusts are pathogenic both in the haploid and diploid phase, but close to the base of their phylogeny we find some saprobic species such as *Pachnocybe ferruginea* and many species in the Erythrobasidiales (Aime et al., 2006).

At the bottom of the Agaricomycotina phylogeny we find some yeast-like clades, such as Filobasidiales and Cystofilobasidiales. They represent a parallel lineage to the yeasts at the base of Ascomycota, and at least some of these basidiomycete yeasts occur today in sap exudations from wounded trees (Weber, 2006). The basidiomycete yeasts are not as numerous as the ascomycete yeasts but their ancient origin is interesting and suggests that the yeast life form was successful in the early evolution of fungi.

The Dacrymycetes is a subsequent lineage near the base of Agaricomycotina. In this group we find genera, such as *Dacrymyces* and *Calocera*, that cause brown rot. The occurrence of this rot type rather deep in the phylogeny of basidiomycetes is very interesting with regard to the evolution of rot types.

Higher in the phylogeny of Agaricomycotina we find many clades that have recently been grouped into 17 different orders (Hibbett, 2006). These clades contain large groups of wood decayers, litter decayers and ectomycorrhizal fungi. The task of sorting out their phylogeny is an active research field and we can certainly expect future rearrangements (see Hibbett, 2006, for an overview). Even if the basidiomycete phylogeny is not settled, it is obvious that the wood decayers do not form a monophyletic group. Instead they dominate, or occur as distinct subsets, in different orders, such as Auriculariales, Hymenochaetales, Polyporales, Russulales and Agaricales.

The unsettled phylogeny and few fossils make it difficult to know when different wood decomposers in Agaricomycotina evolved. Furthermore, the few fossils that do exist, including the much cited *Phellinites digiustoi*, are surrounded by misidentifications (Hibbett et al., 1997a). Molecular clock studies indicate, however, that the origin of Agaricomycotina could be 380 million years old (Taylor J. W. et al., 2004). This corresponds quite well with the origin of woody plants and the earliest fossil evidence of wood-decaying fungi described above.

Even if the dating of wood-decaying fungi is uncertain, there are some very interesting molecular studies that suggest how these fungi originated. In Chapter 2 we described different heme peroxidase enzymes that are crucial for lignin degradation. Peroxidase enzymes, as such, are distributed across bacteria, plants, animals and fungi and probably evolved more than 2 billion years ago as a response to oxidative stress when atmospheric oxygen levels increased. Today, they form three distinct enzyme classes: class I peroxidases are found in bacteria, archaeans and fungi; class II peroxidases are found exclusively in fungi; and class III peroxidases are found in plants. The basic reaction mechanism is identical in all peroxidases. However, they depend on different organic substrates to complete the catalytic cycle, and this may have facilitated the development of new functions (Morgenstern et al., 2008).

By analysing the phylogeny of genes coding for the peroxidase enzymes, Morgenstern and co-workers (2008) found that the class II peroxidases represent a monophyletic clade that probably arose within the basidiomycetes after the split from ascomycetes. The class II peroxidases further include different subtypes: lignin peroxidases (LiP), manganese peroxidases (MnP) and versatile peroxidases (VP). A more detailed analysis by Morgenstern and colleagues revealed that the MnPs occur in at least four groups of Agaricomycotina (Hymenochaetales, Polyporales, Agaricales, Corticiales – and probably also in Russulales). The VPs have so far been found in Agaricales and Polyporales and seem to have evolved independently in these groups. The LiPs, on the other hand, are monophyletic and occur only among members of Polyporales. This pattern suggests that MnPs are phylogenetically older than LiPs, and originated before the major lineages of Agaricomycotina split up.

Morgenstern and co-workers hypothesized that ancient fungal peroxidase enzymes used various types of organic substrates to complete the catalytic cycle. Following gene duplication and mutation events, these enzymes might then have been refined by natural selection to become efficient lignin degraders. As plants gradually became enriched in lignin and formed woody plants, natural selection would then favour fungi with the ability to utilize this increasing resource base. The new function of peroxidase enzyme varieties might explain why different lignin-degrading enzymes have evolved independently in different clades of wood decomposers. Furthermore, the varieties of peroxidases appear to have subfunctions characterized by using narrower substrate ranges; for example, lignin with different degrees of decay (see Chapter 2). This subfunctionalization might also be advantageous from an evolutionary perspective.

10.3 Ancient and derived saproxylic invertebrates

10.3.1 Mites

The oldest well-documented signs of saproxylic invertebrates date back to the Carboniferous, from which many fossil wood borings

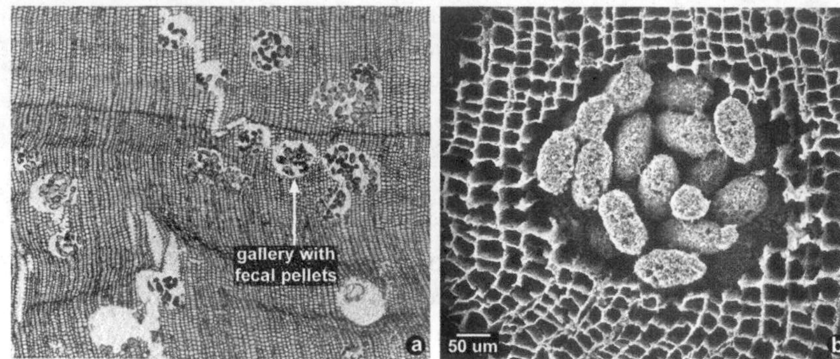

Figure 10.4. Scanning electron micrographs of galleries in fossilized *Premnoxylon* (gymnosperm) wood from the Carboniferous. The galleries were probably formed by mites, and they are filled with fossilized fecal pellets (frass). Photos T. & E. Taylor, University of Kansas.

exist (Labandeira et al., 1997). These borings consist of 0.1–0.4 mm wide holes in various woody plants, often filled with fossil fecal pellets Figure 10.4). The trace fossils contain no body parts of the animals that produced the holes, but the hole diameters and fecal pellet sizes led Labandeira to attribute the borings to oribatid mites. Mite body fossils, as such, are known from the middle Devonian, about 380 mya (Labandeira et al., 1997).

Mites continued to be frequent, even predominant, wood borers until the middle Jurassic (Kellogg and Taylor, 2004). In these fossils it is evident that the mites preferred decayed woody tissue, probably decomposed by fungi. Kellogg and Taylor further suggested that mites were principal wood borers in the Jurassic swamp forests, because they found no insect borings in more than 200 plant specimens. They suggested that insect wood borers perhaps occurred primarily in drier environments. After the Jurassic, the fossil records provided hardly any cases of mite borings, whereas insect borings took over as the dominant and diversifying trace fossils (Labandeira et al., 1997, 2001).

Today, we find a rich diversity of saproxylic mites, and they have developed several functional roles including wood borers, detritivores, predators and parasites. The task of sorting out mite phylogeny has received relatively little attention in terms of phylogenetic analyses based on comprehensive character matrices and molecular data sampled across a wide range of taxonomic groups. Even less effort has been devoted to systematically documenting the various associations of mite taxa to woody material. Thus, substantial work needs to be done before we can present a broader picture of how saproxylic mites have evolved and radiated through evolutionary time.

10.3.2 Early radiation of insects

The oldest insect fossils are from the early Devonian, *c.* 400 mya (Grimaldi and Engel, 2005). Most of these insects were flightless and some extant representatives, such as bristletails (Archaeognatha) and silverfish (Zygentoma), belong to the same basal insect group. Bristletails occur worldwide in diverse habitats although mainly under loose bark and stones. Their diet is typically composed of algae and lichens. Silverfish, on the other hand, have a diet of cellulose from different types of plant material, and they might even possess an enzyme system for cellulose degradation (see Chapter 2).

There are hardly any fossil records of potential saproxylic insects from the Carboniferous. Instead, the Carboniferous insect fauna was dominated by various herbivores (Labandeira, 1998). The diversity and abundance of fossil insects increased substantially during the Permian, when ancestors or representatives of extant orders with saproxylic species occurred, such as termites (represented by primitive roachoids), Thysanoptera, Hemiptera and Protocoleoptera (Grimaldi and Engel, 2005).

In this context it is relevant to highlight a somewhat overlooked paper, entitled 'Evolution and diversity under bark' by the great evolutionist William D. Hamilton (1978). Hamilton is certainly more famous for his theories of inclusive fitness and the evolution of insect eusociality. In this 1978 paper he suggested that several major insect groups have originated and diversified specifically in dead wood. This hypothesis was substantiated by the observation that disproportionally many 'primitive' representatives of various insect groups live in dead trunks and branches. Furthermore, he pointed out that evolutionary novelties seem to be more common in dead-tree insects than in soil and litter insects, such as high rates of wing polymorphism, male haploidy, sex dimorphism and at least two origins of advanced social life (termites and ants). Building on the work of Hinton (1948), he even suggested that holometaboly (the metamorphosis of a larva to an adult insect through a pupal stage) evolved in decaying wood. Hamilton also suggested a mechanism to explain why different insect groups have evolved in dead wood. His main point was that dead trees offer a protected space where insect groups can live for several generations isolated from other related groups. In this situation, evolutionary rates can differ markedly between groups and such a breeding structure is favourable for rapid evolution.

Hamilton's hypothesis was received with scepticism, as most insect taxonomists thought instead that the habitat complexes of soils and plant detritus in general were the most likely environments for insect evolution. But subsequent studies on insect evolution, especially from extensive phylogenetic analyses, have provided support for Hamilton's view, as we shall see. The present knowledge base about insect evolution is outstandingly summarized in a comprehensive and well-referenced book by Grimaldi and Engel (2005). In the rest of this section, we mostly build on this source, but include some other interesting references as well.

10.3.3 Beetles

Beetles (Coleoptera) are the most species-rich insect group in the world today. They include several saproxylic families and a high proportion of saproxylic species in total. The phylogeny of the beetles is not completely settled, but the group is undoubtedly monophyletic and four suborders have been recognized for a long time (Figure 10.5). It is widely accepted that a small suborder, the Archostemata, is the most basal among the living groups in the phylogenetic tree. The extant Archostemata are obligate saproxylic species, boring in dead wood as larvae. Furthermore, in the family Cupedidae, with approximately 30 species distributed almost worldwide, individuals have a flattened body and distinctively sculptured forewings. This ornamentation has facilitated the identification of numerous cupedoid fossils back to the Triassic and allowed interpretation of the early phylogeny of beetles. The oldest group of beetles, the Protocoleoptera, which lived in the early Permian about 280 mya, had an aspect similar to the later cupedoids (Figure 10.6). Based on this link and the hypothetical life mode of the initial oligoneopterans (a subset of ancient winged insects), Ponomarenko (2003) suggested that the first beetles most probably lived in decaying wood.

By the end of Permian, some Coleoptera had evolved into aquatic forms that became abundant in the Triassic and Jurassic. Today they are represented by descendants in the suborder Adephaga. The ancestors of the hyperdiverse suborder Polyphaga probably evolved in the late

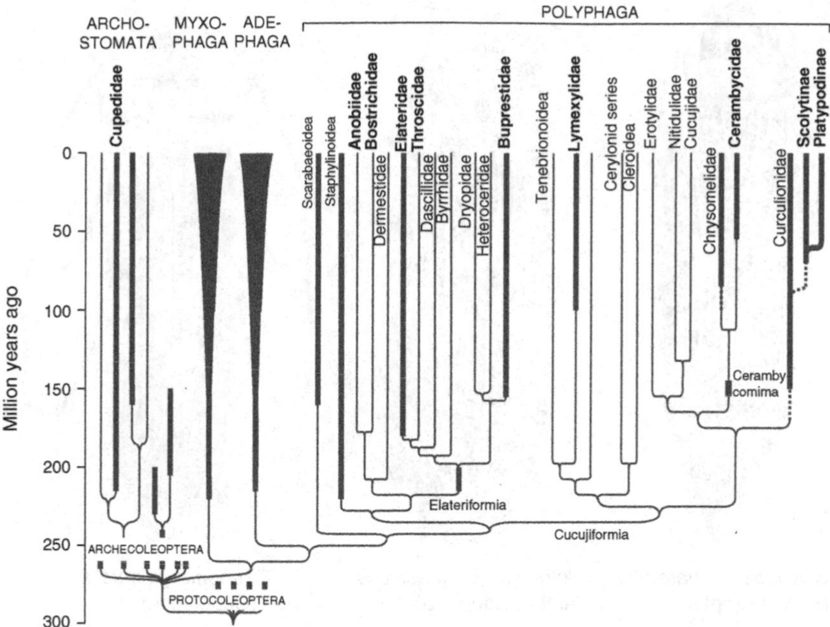

Figure 10.5. Phylogenetic tree of Coleoptera based on Hunt et al. (2007) and Grimaldi and Engel (2005). Taxa on top of branches include many saproxylic species (printed in boldface if most species are saproxylic). Taxa placed along branches are predominantly non-saproxylic. Thick lines indicate known occurrence in the fossil record. Dotted thick lines indicate fossils probably belonging to the lineage.

Permian or early Triassic, since carnivorous Staphylinidae, primitive elateriforms, together with herbivorous Polyphaga groups appeared in the late Triassic (Ponomarenko, 2003; Grimaldi and Engel, 2005). It might be that the basal polyphagan group(s) developed in aquatic or semi-aquatic environments. Subsequently, saproxylic forms started to appear among the polyphagans, such as jewel beetles (Buprestidae) and click beetles (Elateridae) in the late Jurassic, about 150–160 mya (Grimaldi and Engel, 2005). It has also been suggested that the first scarab beetles (Scarabaeidae) from the late Jurassic were most probably saproxylic (Nikolajev, 1992). As of today, no fossils have been found to link the Jurassic saproxylic forms to the ancient Permian saproxylic beetles (Ponomarenko, 2003).

If we consider the fossil wood borings, the earliest *possible* beetle borings are from Permian trees in Antarctica, which at that time had a seasonal, temperate climate. The earliest *definitive* beetle borings are

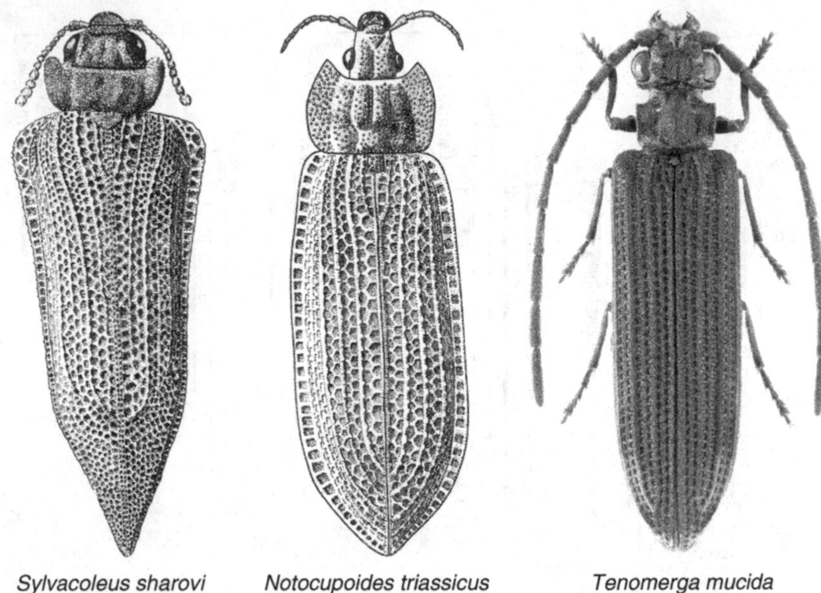

Figure 10.6. Fossil and extant representatives of Protocoleoptera and basal Coleoptera: (a) *Sylvacoleus sharovi* (Protocoleoptera from Permian); (b) *Notocupoides triassicus* (Archecoleoptera from Triassic); (c) *Tenomerga mucida*, an extant representative of Cupedidae (photo Kirill Makarov). Illustrations (a) and (b) are reproduced from Ponomarenko (2003), with permission from *Acta Zoologica Cracoviensia*.

from the Triassic of Europe and Arizona about 200 mya (Grimaldi and Engel, 2005). In one of these, Ash and Savidge (2004) even found a fossil larva superficially resembling a deathwatch beetle (Anobiidae) larva. After the Triassic, the beetle borings became more diverse with the radiation of various beetle families (Labandeira et al., 2001).

10.3.4 Diptera

The order Diptera, comprising the gnats and flies, dates back to the early Triassic, about 240 mya. By the late Triassic most of the major dipteran infraorders had evolved. The phylogeny of the basal lineages is not fully settled (for an updated phylogeny, see Bertone et al., 2008), but it is likely that much of the early evolution took place in aquatic or semi-aquatic environments. This is based on the observation that most of the early lineages have diverged into extant groups where the larvae predominantly live in such wet habitats (Grimaldi and Engel, 2005).

In one of the ancient dipteran infraorders, the terrestrial Bibionomorpha, we find several families with a large proportion of saproxylic species, such as fungus gnats (Mycetophilidae), gall midges (Cecidomyiidae), and wood gnats (Anisopodidae). Again, as for the Diptera as a whole, the relationships between the deep evolutionary lines are not fully resolved and different phylogenetic trees exist as alternative hypotheses. Thus, the knowledge base is not sufficiently developed to infer whether the base of Bibionomorpha was associated with dead wood or not. However, in one of the oldest lineages we find clear links to dead wood – the one leading to Brachycera, where all the higher flies belong. Brachycera is a well-established monophyletic group with fossil records back to the early Jurassic. All basal families in Brachycera have larvae living as obligate saproxylics in decaying wood, such as Pantophthalmidae, Stratiomyidae, Xylomyidae and Xylophagidae. Thus, it appears that this major Diptera lineage originated in decaying wood about 200 mya. Later, the flies radiated substantially and colonized almost every conceivable environment for larval development, including secondary establishments back to decaying wood.

A key feature of the Brachycera is the development of the larval mouthparts. In these larvae the mouthparts are held parallel to each other and move up, out and down, rather than with a scissoring action as in more primitive Diptera. This feature probably led to predation as a major innovation in Diptera larvae (Grimaldi and Engel, 2005). The larvae of most basal Brachycera are predators feeding on insect larvae and soft-bodied invertebrates (Krivosheina and Zaitzev, 2008).

10.3.5 Hymenoptera

Hymenoptera (e.g. ants, wasps, bees) is another large insect order where major lineages have evolved in dead wood. The origin of this order dates back to the late Triassic, about 220 mya, from which the earliest definitive fossil Hymenoptera occurs. Unlike the situation in beetles, there are relatively few fossil records to document the sequence of new lineage formations. On the other hand, the phylogenetic tree of Hymenoptera is well developed and there seems to be broad agreement about the sequence of branching through evolutionary time (Grimaldi and Engel, 2005; Sharkey, 2007).

The base of the Hymenoptera phylogeny consists of herbivorous insects – the sawflies that have traditionally been grouped as Symphyta. But nearly 200 million years ago a particular group of Hymenoptera evolved – the woodwasps (Siricidae). The woodwasps have a fascinating biology, as the larvae feed on a mixed diet of fungal mycelia and wood decomposed by various basidiomycetes.

From the woodwasps there is an evolutionary transition to the families Xiphydriidae and Orussoidea (Figure 10.7), where all species live in decaying wood. The biology of the xiphydriids is very similar to

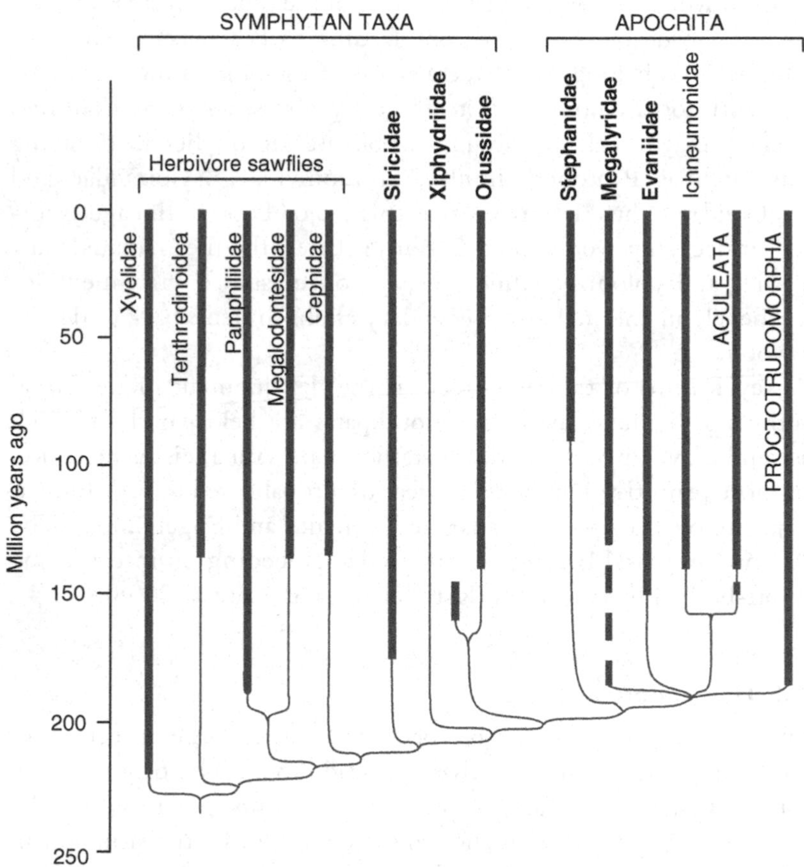

Figure 10.7. Phylogenetic tree of Hymenoptera based on Grimaldi and Engel (2005). Taxa on top of branches include many saproxylic species (printed in boldface if most species are saproxylic). Taxa placed along branches are predominantly non-saproxylic. Thick lines indicate known occurrence in the fossil record. Dotted thick lines indicate fossils probably belonging to the lineage.

that of the woodwasps, as the larvae depend upon symbiotic fungi in their tunnels. The Orussidae is especially important as it constitutes a direct transition to the most basal ectoparasitic parts of Apocrita. Here we should mention that Apocrita is the large, hyperdiverse section of Hymenoptera where all the parasitic wasps belong. The orussids themselves share many primitive morphological structures with the woodwasps and sawflies – most notably the broad attachment of the thorax to the abdomen. But, unlike these, the orussids have a long, slender ovipositor and are external parasitoids attached to the larvae of wood-boring beetles and woodwasps (Grimaldi and Engel, 2005). If we follow the Hymenoptera phylogeny further, the basal parts of major Apocritan branches are parasitic on other saproxylic insects such as the families Stephanidae, Megalyridae and Evaniidae. Higher up in the phylogeny we see a radiation to parasitism on insects (as well as spiders and other invertebrates) living in quite different macrohabitats.

The development of parasitism was undoubtedly a major feature in the evolution of Hymenoptera. All the evidence suggests that this development took place in decaying wood. Furthermore, the formation of the 'wasp waist' – the slim junction between the first and second abdominal segment – is coupled with the environment of decaying wood. The thin waist allows for greater flexibility in controlling a long ovipositor or the egg-laying tube. This is essential for accurate egg positioning in the larvae of other insects boring inside the wood (see Figure 10.8).

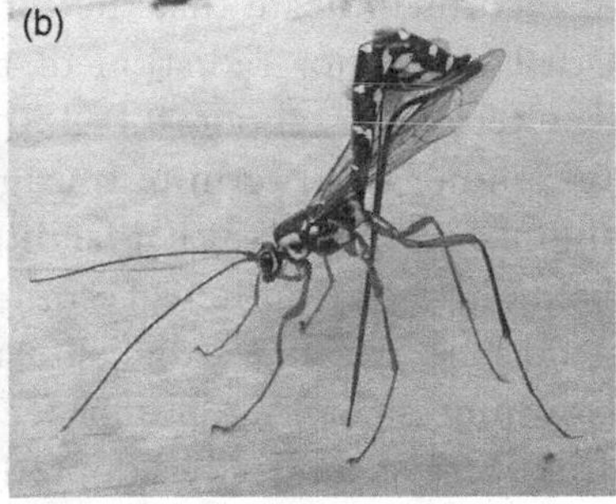

Figure 10.8. Two saproxylic hymenopterans showing different degrees of flexibility in controlling the ovipositor when inserting eggs into dead wood: (a) the woodwasp *Urocerus gigas*, which bends the abdomen to direct the ovipositor into wood (photo Jean-Marie Mouveroux); (b) the ichneumomid wasp *Rhyssa persuasoria*, which bends the abdomen but also flexes the abdomen at an angle to the thorax in order to better direct the ovipositor into the wood where it is parasitizing the larvae of *Urocerus gigas* (photo Brian Hansen).

10.3.6 Termites and wood roaches

Dictyoptera is a reappraised taxon including cockroaches, termites and mantises as well as ancient, now extinct, groups resembling the cockroaches. A popular – but nevertheless erroneous – belief that cockroaches date back to the Carboniferous is understandable because of cockroach look-alikes, termed 'roachoids', from this period.

Modern taxa of Dictyoptera first appeared in the Cretaceous fossils (145–65 mya) and probably derived from one of the ancient groups in the Jurassic. The termites comprise a major branch that dates back to the early Cretaceous. Quite interestingly, the termites have a sister group, the relict wood roaches (*Cryptocerus*). These are wingless, long-lived roaches that inhabit and consume soft rotten logs. Like the most ancient (lower) termites, the wood roaches digest lignocellulose with the help of mutualistic hindgut protists. The wood roaches live in family groups where the nymphs feed on liquids exuded from the anus of an adult for approximately one year, thereby acquiring the mutualistic protists of the hindgut. In return, the young nymphs groom the older nymphs and adults, a unique social trait among the cockroaches.

The phylogeny of termites has been sorted out in several morphological and genetic studies that constitute a stable framework for evolutionary interpretation (Grimaldi and Engel, 2005; Engel et al., 2009). Furthermore, there is a strong correspondence between the termite phylogeny and their chronology of fossils, with the most basal families appearing first in geological time. The phylogeny and the fossil records indicate that the termites appeared in the late Jurassic and that the major radiation of termite families took place in the early Cretaceous, about 100–150 mya. The ancestral feeding type found among basal termite lineages is wood digestion aided by gut symbionts (protozoans) that facilitate cellulose digestion. In the most derived clade of termites (Termitidae), the protozoan symbionts are lost and here we find a diversification into numerous feeding types such as fungus farming, humus feeding in organic-rich soils, true soil feeding, and secondarily adopted feeding on decaying wood, epiphytes or leaf litter. Judging from the phylogenetic position of wood-feeders among the Termitidae, it seems that these have developed in several places among the ground-dwelling humus feeders (Inward et al., 2007).

10.3.7 Snakeflies

The order snakeflies (Raphidioptera) consists predominantly of saproxylic predators (Aspöck, 2002). Snakeflies have an almost worldwide distribution, mostly occurring in temperate regions and in mountainous areas closer to the equator. The oldest definitive fossil dates back to the early Jurassic, about 200 mya, whereas older reported fossils have been attributed to other related insect groups. Grimaldi and Engel (2005) suggested that the order may perhaps date back to the Triassic, based on the known ages of related insect groups, and they anticipated that future fossil records will probably refine the estimated age of the group.

The earliest snakeflies from the Jurassic already had typical features of the order, and it is reasonable to assume that they evolved as a saproxylic group based on their extant ecology. Today, the snakeflies constitute a small order with about 220 living species, but they were more common during the Jurassic and Cretaceous, when at least five different families existed compared with the two of today.

10.3.8 Zorapterans

Finally, in this review of ancient saproxylics, we shall mention a very small insect order called Zoraptera. Zorapterans are minute insects, *c.* 3 mm long, and live in small colonies under the bark or in crevices of moist decaying logs in tropical regions of America, Africa and Asia. Here they feed principally on fungal hyphae, but they can also be scavengers or predators on nematodes, mites and other tiny invertebrates (Grimaldi and Engel, 2005). Only 32 species are known worldwide. We mention the order since it supports a point made below, and also because of the curiosity that the term 'saproxylic' was introduced in the description of this order (Silvestri, 1913). The oldest known fossils are from amber and date back to the mid Cretaceous, more than 100 mya. It is probably older than this (perhaps more than 200 million years old), since such tiny insects can hardly be identified from other material than amber. At present, insect-bearing ambers are only known back to the early Cretaceous (less than 150 mya). The taxonomic position of zorapterans is quite uncertain (Grimaldi and Engel, 2005) but since all extant species live in decaying wood it is reasonable to assume that they evolved in this medium.

To conclude this section on ancient saproxylic insects, we would like to give additional credit to William Hamilton. Today there is strong evidence that he was right in his claim that whole insect branches have evolved specifically in dead wood. We find support for this at the very base of Coleoptera, in major innovative steps leading to parasitism in Hymenoptera, in the predatory Raphidioptera and the Brachycera in Diptera, in the order Zoraptera, and in the termites – which were among Hamilton's favourite organisms.

10.3.9 Derived saproxylic lineages

Saproxylic invertebrates are not all of ancient origin. In some cases we know that diverse groups such as longhorn beetles and bark beetles have evolved and diversified after the Cretaceous, i.e. less than 65 mya. Although the longhorn beetles can be traced back to more than 150 million-year-old ancestors, there are essentially no cerambycid fossils from the Cretaceous. Longhorn beetles first appear as a diverse group in about 50 million-year-old Baltic amber (Grimaldi and Engel, 2005). Bark beetles originated about 100 mya (Cognato and Grimaldi, 2009). Later in this chapter we will emphasize that these beetle groups originated from herbivore ancestors. Here we make the point that they evolved after the origin of the angiosperms – the plant group to which the broadleaved trees belong. The bark beetles most probably originated as conifer feeders, but there have been multiple shifts to broadleaved trees and back again. Each shift to broadleaved hosts has been accompanied by a substantial diversification in host selection and an accompanying increase in species richness (Farrell et al., 2001). The diversification of longhorn beetles has also mostly occurred in association with broadleaved trees.

A parallel to the herbivore-derived saproxylic beetles occurs among the moths (Lepidoptera). The moths, as such, are almost completely herbivorous, where the larvae feed on live plants (Grimaldi and Engel, 2005). In some lineages we find saproxylic species, such as wood-boring species in the families Cossidae and Sesiidae. We also find saproxylic species in the Tineidae; these are fungivorous species living in the fruiting bodies of wood-decaying fungi. These families merely represent exceptions to the herbivorous diet of most Lepidoptera.

Just as a few groups of saproxylic species occur in the Lepidoptera, we find many examples of saproxylic species groups embedded within large organism groups that mainly live in other habitats. Examples are predatory and some fungivorous staphylinids. The staphylinids, as such, form a hyperdiverse beetle family, where the species are generally ground-dwelling in decaying litter. Another example is fungivorous fruit flies (*Drosophila*) that feed on short-lived fruiting bodies of wood-decaying fungi (Lacy, 1984), whereas the majority of the species feed on a wide range of decaying fruits and other decaying plant material.

It is interesting to note that nearly all the species in ancient saproxylic lineages are obligate saproxylics. In more derived lineages, it is common to find a mixture of obligate saproxylics, facultative saproxylics and non-saproxylic species (for definitions of obligate and facultative saproxylics, see Section 11.1.1). This indicates that species in these derived lineages have not developed such strong adaptations to utilize dead wood as a habitat.

10.4 Evolution of functional roles

10.4.1 Fungal rot types

In Chapter 2 we briefly mentioned that the major fungal rot types, i.e. brown, white and soft rot, represent an artificial classification. This becomes evident if one wants to understand how these rot types have evolved. It is obvious that a large number of enzymes are involved in wood decomposition. A deeper understanding of how the wood-decaying ability has evolved among fungi can only be achieved by sorting out which enzymes are involved and how they are phylogenetically related.

Earlier in this chapter we described how lignin-degrading basidiomycetes probably evolved through multiple gene modifications of ancient peroxidase enzymes. Here we shall assess the subsequent development of different rot types. One might perhaps assume that brown rot is the ancestral decay mode, since it involves fewer enzymes and predominantly occurs in conifer (i.e. gymnosperm) wood. Phylogenetic studies indicate, however, that the ancestor of most wood-decaying basidiomycetes was a white-rot fungus (Hibbett and Donoghue, 2001). Furthermore, Hibbett and Donoghue showed that the ability to decompose conifer wood is the ancestral state, while conifer exclusivity is a specialized, derived condition. This is not surprising, considering that the ability to decompose lignin arose before the conifers evolved (see Stubblefield et al., 1985).

Hibbett and Donoghue (2001) further suggested that the brown-rot mode has at least six independent origins within different white-rot clades. A possible mechanism is that the lignin-decaying enzymes have simply become inactive. The analyses of Hibbett and Donoghue also suggest that conifer exclusivity has evolved five or six times, typically (but not necessarily) in combination with a switch to brown rot. This may reflect the more decay-resistant nature of conifer lignin compared with angiosperm lignin, and that there has been a selective premium on decaying the cellulose fraction only.

Unfortunately, neither Hibbett and Donoghue (2001) nor Morgenstern et al. (2008) included representatives of Dacryomycetes in their phylogenetic analyses. These are brown-rot fungi with a more basal phylogenetic position than the basidiomycetes that they included in their studies. It would certainly be interesting to know whether genes for lignin-degrading peroxidases are present (but not expressed) in the Dacrymycetes, or if these peroxidases might have evolved at a later stage.

The wood-decaying ascomycetes have not been as thoroughly studied as the wood-decaying basidiomycetes. We find the majority of wood-decaying ascomycetes in the Sordariomycetes. These include the Xylariales, which are effective wood decomposers in terrestrial environments. Many xylariaceous fungi have been classified as white-rot fungi, and several of them can indeed decompose lignin (Abe, 1989; Pointing et al., 2003). On the other hand, they are not as effective wood decayers as the basidiomycete white-rot fungi. So which enzymes do these ascomycetes use? Several species evidently possess laccase enzymes (Lee, 2000; Pointing et al., 2003). The presence of laccase enzymes is interesting, as some of them can degrade certain lignin elements (see Chapter 2). Lee also found indications of peroxidase activity in some ascomycetes, but this was based on indirect evidence (a colour reaction when adding hydrogen peroxidase). Thus, it remains to be conclusively demonstrated that ascomycetes possess peroxidases to decompose lignin like the white-rot basidiomycetes do.

Laccases are also commonly present in a wide range of ascomycete soft-rot fungi in aquatic environments (Rohrmann and Molitoris, 1992; Luo et al., 2005), while peroxidases seem to be absent (Luo et al., 2005). This explains the poor lignin-degrading ability of these soft-rot fungi. We also find different wood-decaying ascomycetes classified as soft-rot fungi in terrestrial environments (Råberg et al., 2007). Just like the xylariaceous fungi described above, these ascomycetes are less efficient wood decomposers than the basidiomycetes.

When we consider the full picture, it seems to be as Morgenstern and co-workers (2008) suggested; namely that effective lignin-degrading peroxidase enzymes have never evolved among the ascomycetes. However, we find numerous wood-decaying ascomycete lineages in terrestrial and aquatic environments. These show different enzyme activities as well as structural variations in their rot types (cf. soft rot type 1 and type 2). Thus, we can expect that future research will reveal various interesting enzyme types and subtypes among the soft-rot fungi.

10.4.2 Wood-consuming detritivores

Just as a large number of fungi have become specialized decomposers of woody material, there are thousands of insect species that utilize the same resource. But an intriguing part of this evolution is that no wood-consuming insect seems to produce the enzymes necessary to decompose cellulose, let alone lignin. Instead they depend largely on symbiotic microbes residing in their digestive channel. It is therefore relevant to ask what circumstances have favoured the evolution of symbiont-mediated cellulose digestion.

Before we ask this question we shall reiterate the nutritional differences between bark and wood. Even though fossil bore-holes in wood represent the first evidence of wood-inhabiting insects, it is not necessarily true that wood itself was the first tissue type to be utilized by saproxylic insects. It is more likely that the first insects colonized the subcortical space of recently dead gymnosperm trees and consumed the decaying inner bark, as many insects do today. The flattened body shape of the Protocoleoptera and the derived cupedoid beetles suggests this microhabitat association. The key point here, however, is that decaying inner bark contains a lot of easily digestible sugars and more nitrogen than firm wood, which is almost completely composed of cellulose and lignin. Thus, the subcortical space represents an entry point to decaying wood with a nutritional composition similar to that of decaying ground vegetation and canopy litter.

So, let us return to the question about symbiont-mediated cellulose digestion. There is a very interesting discussion of this topic by Martin (1991). He outlined a scenario starting with a general scavenger or detritivore with a nutritional dependence on hindgut bacteria without cellulose-degrading capability. It should be added to this explanation that the occurrence of intestinal microbes taking part in food degradation is very common across insects and other animal groups. So, when this general detritivore consumed decaying plant matter (not necessarily wood) its intestine would easily become colonized by microbes capable of cellulose degradation, creating a cellulose-digesting scavenger or detritivore. Such insects could then expand their feeding niche to include items high in cellulose content. Some of these cellulolytic scavengers or detritivores with expanded feeding options might subsequently specialize on a narrower food range such as wood. This specialization could, for example, be to chew or scrape off pieces from a wood surface softened by wood-decaying fungi. The specialization could develop further into wood-boring or other specific adaptations that result in an obligatory wood-ingesting specialist, now with restricted feeding options. This kind of development could easily have brought about the many obligatory detritivores existing today.

The evolution of obligatory wood-ingesting detritivores must have occurred independently over and over again at various times throughout history. This is evidenced by specialized wood consumers in widely separate branches of the insect phylogeny, such as some crane flies (Diptera: Tipulidae), all *Temnostoma* flies (Diptera: Syrphidae), some aquatic chironomid gnats (Diptera: Chironomidae), species in the beetle suborder Archostemata (the family Cupedidae), all stag beetles and some scarab beetles (Lucanidae, Scarabaeidae), all bess beetles (Passalidae), nearly all deathwatch beetles (Anobiidae), all woodwasps (Hymenoptera: Siricidae) and basal termites (Isoptera: various families). Additional evidence documenting the independent origins of these specialized detritivores comes from the gut symbionts that also have widely separate taxonomic origins. These include protozoans in wood roaches and termites, various types of bacteria in crane flies and scarab beetles, different yeasts in the families of bess beetles and deathwatch beetles, while several longhorn beetles and woodwasps depend upon enzymes from distinct basidiomycetes (see Chapter 2 for details and references).

10.4.3 Herbivore-derived detritivores

The detritivores just described typically have a diet of cellulose-rich wood from weakly to strongly decayed trees. While those detritivores most probably developed from less specialized detritivores, there are other bark- or wood borers with a quite different origin. The species-rich beetle family weevils (Curculionidae) diversified from a herbivore lineage that originated in the middle Jurassic, about 160–175 mya (Oberprieler et al., 2007). At that time the weevil ancestors had developed a distinct snout (rostrum) with a pair of mandibles at the tip. The females probably used this snout to make an opening deep into green cones and other plant organs, where they placed their eggs. This invention probably gave the weevils a competitive advantage over other herbivore beetles and contributed to their later success, as evidenced by their species diversity (Oberprieler et al., 2007).

In the mid Cretaceous we find the first fossil bark beetles (Cognato and Grimaldi, 2009). Thus, about 100 mya, some particular weevils specialized on a diet of inner bark of coniferous trees and formed the origin of today's bark beetles. A distinct feature of the bark beetles is that they lack cellulose-degrading gut symbionts and they only occur in living or recently dead trees. Thus, they act as herbivore cell-content consumers that feed on live or recently dead inner bark, containing sap from the photosynthetic canopy and nutritious cell contents. In other words, they are cytoplasm feeders *sensu* Abe and Higashi (1991; see also Section 3.2.1).

We find a similar origin of wood borers in a very different herbivore insect group – the moths (Lepidoptera). In Chapter 3 we described how the wood-boring goat moth (*Cossus cossus*), and probably also the wood-boring clearwing moths (Sesiidae), have digestive enzymes typical of consumers of living plant cells. These wood-boring moths are confined to living (often wounded) trees, and essentially have the same diet as herbivores – the nutritious cytoplasm or cell contents, while the cellulose-rich cell wall material passes undigested through the alimentary tract.

The wood-boring longhorn beetles (Cerambycidae) also most probably originated as herbivores. They seem to share a common ancestor with the herbivorous leaf beetles (Chrysomelidae). This is evidenced by a 152 million-year-old fossil of *Cerambycomima* that has typical features of both families (Grimaldi and Engel, 2005). Recent phylogenetic analyses even suggest that the leaf beetles might be embedded within the longhorn beetles (Hunt et al., 2007). Thus, one should not be puzzled by the fact that several longhorn beetles are herbivores feeding on green plants, and that the family is not completely saproxylic like several other wood-boring families. Unlike the bark beetles, however, some longhorn beetles have developed the ability to utilize fungal enzymes for cellulose digestion (see Chapters 2 and 3). Such species can digest weakly, or even strongly, decayed wood.

10.4.4 Co-evolution between insects and fungi

There is a particular type of insect–fungus association that has fascinated biologists for a long time: the cultivation of certain fungi by the so-called ambrosia beetles (Farrell et al., 2001; Mueller et al., 2005). The ambrosia beetle habit has evolved among the bark beetles and has also formed the closely related subfamily Platypodinae. A few species have also developed this lifestyle in the completely separate family ship-timber beetles (Lymexylidae). All these beetles bore into the wood of living or recently dead trees. Here the larvae feed on a carpet of fungus mycelium that grows on the inner wall of their galleries. This mycelial carpet is attended by the female, who daily inspects her gallery and probably removes other fungi that compete with the ambrosia fungus. When the new generation of adults leave the gallery they take some of this mycelium to a new tree, where it is introduced into the wood. The beetles have developed special pockets (mycangia) on their body that are particularly suited to carrying the fungus. In addition, the fungi show adaptations to match the life cycle of the beetle. The most prominent adaptation is the formation of the mycelial carpet. However, at the time the insects leave their galleries the fungus develops elongated threads that produce asexual conidiospores. These threads protrude into the gallery and brush the surface of the departing insects.

The ambrosia beetle habit has evolved repeatedly since about 60 mya, and today we have some 3400 species in 10 different lineages, mostly in the Platypodinae and the bark beetle tribes Xyleborini and Corthylini (Farrell et al., 2001). This habit evolved some time after the bark beetles themselves occurred. The ambrosia fungi, on the other hand, belong in the Sordariomycetes and are closely related to the sapstain *Ophiostoma* fungi, which appeared more than 85 mya (see Farrell et al., 2001). Thus, the fungus was present as a saproxylic group when the first bark beetles evolved. It is worth commenting on the biochemical capability of the sapstain fungi, which lack the ability to decompose cellulose and therefore mainly degrade sugars and other easily degradable compounds (see Chapter 3). The inability to decompose cellulose is retained among the ambrosia fungi and explains why the ambrosia beetle needs to shift host tree every generation – it is because the energy base of the fungus rapidly becomes depleted.

10.5 Prospects

We can now look back on 30 years of phylogenetic research based on advanced computer methods and detailed genetic studies. Major changes have taken place in our understanding of evolutionary pathways during this period, and we can expect further progress, as many organism groups have not yet been thoroughly investigated. The raw material is available in the form of species waiting to be scored for morphological characters and to be screened for genetic composition. There are also large amounts of fossils, both in palaeontological museums and in nature, that hold clues about the evolutionary history of insects and fungi. New and exciting findings of ancient evidence will almost certainly appear, and these will improve the age estimates of various phylogenetic lineages.

However, in order to learn more about the specific evolution of saproxylic organisms we cannot just wait for the answers from taxonomists and palaeontologists. There is also a need to determine which species are saproxylic and to find out how wood-degrading enzyme systems, feeding modes and habitat specialization are distributed across organism groups. A lot of existing evidence needs to be compiled and analysed, and species with unknown biology need to be documented. There is indeed a great deal of work that must be carried out so as to reconstruct the evolution of the rich diversity of saproxylics that we see today.

11 · *Species diversity of saproxylic organisms*

Jogeir N. Stokland and Juha Siitonen

It is evident that planet Earth hosts several million species. The actual number of species is unknown but a much cited study has calculated a global figure of 12.5 million species (Hammond, 1992), while a recent detailed revision came close to 11 million species (Chapman, 2009). In addition to these calculations, there are also estimates that have arrived at global figures as high as 30–100 million species (Erwin, 1982). Nobody has tried to estimate the number of saproxylic species on a global scale. This is quite understandable, since many groups are poorly investigated and large areas are minimally explored for wood-inhabiting species. But there is one region where most saproxylic species are well documented – in the Nordic countries of Europe. We therefore highlight some of this knowledge and present an overview of the diversity in various groups. Despite major knowledge gaps on a global scale, we also make an attempt to calculate the relevant numbers indicating the global diversity of saproxylic species.

11.1 Saproxylic diversity in northern Europe

There is a long and strong tradition of documenting species diversity in the Nordic countries of Sweden, Finland, Denmark and Norway. This tradition is rooted in the work of Carl von Linné, who made Sweden the European centre for alpha-taxonomy (i.e. the description of new species) in the 1700s. The Linnean school also had a great local impact in Sweden and the neighbouring countries. During the 1700s and 1800s, the majority of terrestrial species in the Nordic region were described and identification keys were made for large groups of insects and fungi.

All this taxonomic work had the result that by the beginning of the 1900s most species had been described in this region. Several scholars and collectors then turned their interest towards the ecology of species, and some specialized in documenting the substrate preferences of different saproxylic groups. Subsequently, there has been a tradition of collecting specimens and documenting the biology of saproxylic species in different organism groups. This accumulated knowledge has been transferred to the next generation of collectors and researchers through joint excursions and written publications until the present. Now we find ourselves in a situation where it is very time-consuming to obtain an overview of all the species-specific information from different publications and collections. This is the reason why some researchers started to compile information about Nordic saproxylic species in a database, to make the knowledge more available for different uses (Stokland and Meyke, 2008). Although this database is expanding in content, it already contains sufficient information to present quite a reliable picture of the species richness in most saproxylic groups.

11.1.1 Obligate and facultative saproxylics

In Chapter 1 we defined a saproxylic species as 'any species that depends, during some part of its life cycle, upon wounded or decaying woody material from living, weakened or dead trees'. In Chapters 3 and 4 we defined different functional roles and explained how species depend, directly or indirectly, upon wounded or decaying wood. Now we shall elaborate on another aspect of this dependency, namely to what degree a species is associated with decaying wood.

Most saproxylic species are strictly confined to woody material, either directly as a wood consumer or as a consumer of other saproxylic species, or by strictly using woody material for breeding or some other essential activity to complete its life cycle. Such species are *obligate* saproxylic species. But there are additional species utilizing wood (or other saproxylic species) that also use alternative resources as well. Many fungivorous insects with larval development in wood-decaying fungi can instead breed in non-saproxylic mycorrhizal fungi. Especially among beetles, flies and gnats it is common to find such species. In the same manner, many predatory insects that hunt larvae in decaying wood can also search for prey in other decaying material such as litter on the forest floor. Such species that regularly utilize, but do not completely depend upon, the presence of decaying wood are *facultative* saproxylic species. Species that visit and utilize dead wood but predominantly use alternative resources should not be classified as saproxylic species. The logic behind these definitions is that obligate saproxylics would not be able to exist if all wounded and decaying wood were to be removed from a forest; the facultative saproxylic would suffer a significant population decline (but not go locally extinct); while non-saproxylic species (dead wood visitors) would be only negligibly affected by the removal of dead wood.

At this point, a reader who prefers strict definitions would probably not be satisfied with the general wordings above, such as 'regularly utilize', 'do not completely depend upon' and 'predominantly use' in the preceding paragraph. But a problem with exact definitions in this context is that they require much information about each species to classify them as obligate, facultative or non-saproxylic. The practice we have adopted when classifying Nordic species into these categories is that a species is classified as obligate if all records of that species (in a particular part of the life cycle) are from woody material. If not all, but still a significant proportion (at least 30% of the records) are from woody material, the species is classified as facultative. Species that have predominantly (typically more than 70% of the records) been found in non-woody material are classified as non-saproxylics because they are clearly more frequent in other habitats. During the species-by-species deliberations behind the figures in Table 11.1, we have classified hundreds of species reported from decaying wood as non-saproxylic. But for many species we were unable to classify them as either obligate or facultative. This was typically the case for species with just a few records from woody material and no other records. Even though two out of two is 100%, a third record from non-woody material would represent more than 30% of all three records. So, for many species, one must await additional information in order to classify it as obligate or facultative. This is the main reason why the sums for obligate and facultative species in Table 11.1 do not add up to the total numbers. Other reasons are that a species list has been established but the classification into obligate and facultative has not been carried out (Hymenoptera), or a species list has not yet been established (myxomycetes).

With these definitions and practical rules of thumb in place, we can present rather accurate numbers of saproxylic species, and for many groups also the number of obligate and facultative saproxylics in the Nordic countries. Altogether there are about 7500 known saproxylic species in these countries (Table 11.1). The numbers of saproxylic Ascomycota, Nematoda, Hymenoptera and perhaps also Diptera in the table are probably well below the actual numbers. Thus, the total of 7500 species represents a conservative value. To put this number in

Table 11.1. *Number of individual saproxylic species from the Nordic countries currently recorded in the Nordic Saproxylic Database (see Stokland and Meyke, 2008). In most groups, the real number is probably higher, see text for details.*

	Total	Obligate	Facultative
Fungi			
Ascomycetes	893	614	3
Basidiomycetes	1461	1252	209
Lichens	281	112	169
Plants			
Mosses	98	19	79
Myxomycetes[a]	200		
Animals			
Acari[b]	545	199	79
Pseudoscorpiones	12	4	8
Coleoptera	1447	1087	360
Diptera	1550	675	184
Hymenoptera[c]	803		
Lepidoptera	66	50	5
Hemiptera	26	24	
Thysanoptera	23		
Collembola	27	12	15
Raphidioptera	4	4	
Nematoda[a]	100		
Teredinidae	7	7	
Limnoriidae	1	1	
Vertebrates	45		
Total	7589	4060	1111

[a] This number is an estimate, not an enumeration from the database.
[b] Includes species known from Poland.
[c] Not well documented, the actual number of species is higher.

perspective, saproxylic species make up 20–25% of all forest species in this region (Siitonen, 2001).

11.1.2 Basidiomycetes

The basidiomycetes include the wood-inhabiting fungi with the largest fruiting bodies, such as many polypores and agarics, where several species have fruiting bodies larger than 10 cm in diameter. During the last two decades, different DNA-based methods have provided a new taxonomy that replaces the traditional way of grouping fungi. According to this system, the most diverse group of saproxylic basidiomycetes is the order Polyporales, where we find many of the traditional polypores such as *Antrodia*, *Fomitopsis*, *Polyporus*, *Oligoporus* and *Trametes*, but also a large number of corticioids such as *Corticium*, *Phlebia*, *Sistotrema* and *Tubulicrinis*. Other orders with many saproxylic species are Agaricales (including genera such as *Armillaria*, *Pholiota*, *Pleurotus*), Hymenochaetales (e.g. *Hymenochaete*, *Inonotus*, *Phellinus*), Russulales (with many corticioid fungi such as *Aleurodiscus*, *Peniophora*, *Stereum*), Thelephorales (e.g. *Amaurodon*, *Pseudotomentella*, *Tomentella*) and the jelly fungi, which are now divided into the orders Dacrymycetales, Tremellales and Tulasnellales.

The great majority of basidiomycetes have a saprobic nutrition mode, and in this group we find the most effective white- and brown-rot fungi. But not all basidiomycetes are wood decomposers. In *Tremella* and *Antrodiella* we find several species that are parasitic on the mycelium or fruiting body of other wood-inhabiting fungi. In Telephorales we find many mycorrhizal species, and many species in this group have fruiting bodies on the surface of lying dead wood in close ground contact. The majority of the basidiomycetes are obligate saproxylics. The facultative saproxylics are mostly agaric species that occur on moderately to very decayed wood and also regularly on other decaying plant material.

The basidiomycetes are well known in the Nordic countries and we consider that the total number of 1461 saproxylic species deviates by less than 10% from the actual number of species.

11.1.3 Ascomycetes

The ascomycetes are less conspicuous and have received far less attention from collectors and researchers than the basidiomycetes. Nevertheless, close to 900 saproxylic species are known from the Nordic countries. The great majority of these species belong in the Sordariomycetes, where we find important orders such as Xylariales (with genera such as *Daldinia*, *Hypoxylon*, *Xylaria*), Ophiostomatales (*Ophiostoma*, *Ceratocystis*), Hypocreales (*Hypocrea*, *Nectria*) and Pleosporales. But Sordariomycetes also includes several small orders and quite a number of species that have not yet been formally positioned in specific orders. Other groups of ascomycetes with saproxylic species are Letiomycetes, especially in the diverse order Helotiales (e.g. *Ascocoryne*, *Bisporella*, *Lachnum*, *Mollisia*), and the class Orbiliomycetes, with only two genera (*Orbilia*, *Hyalorbilia*).

Many of the ascomycetes are poorly understood concerning their relationships with decaying wood. But, at least among the Xylariales, many species are wood decomposers, even if the exact rot type is not fully known (see Chapters 3 and 10). In the Ophiostomatales, the species are definitely not wood decomposers. These species primarily occur in living, weakened or recently dead trees, where some species cause wilt diseases and tree mortality (e.g. Dutch elm disease and blue-stain diseases in conifers; see Chapter 3). However, most saproxylic ascomycetes have attracted little attention concerning their nutritional mode and we do not know much about that.

Just as our taxonomic and ecological knowledge is incomplete for the ascomycetes, so also is our knowledge about their species diversity. The current number of nearly 900 saproxylic species from the Nordic countries could easily be well below the actual diversity, and we would not be surprised if their number increased by 20–30% or more.

11.1.4 Lichens and mosses

Although lichens and mosses are completely unrelated, we treat them under the same heading as they are ecologically similar, growing on the wood surface without taking nutrition from the wood itself. Instead they obtain nutrition from the air or the water on the wood surface (with the possible exception of some pin lichens, which might bring about some wood degradation). In Chapter 4 we treated these organisms in some detail. Here we shall only make the point that there are relatively more facultative saproxylics in these groups compared with most other wood-inhabiting organisms. This is quite understandable, as these organisms are not nutritionally dependent on wood and several can grow on the bark of live trees, on rocks or on bare soil.

The lichens and mosses are quite well known and we consider the number of saproxylic species in Table 11.1 to be quite reliable. However, the choice of which species to include or omit as facultative saproxylics has been rather arbitrary. Thus, a change in these numbers would result from updated judgements rather than new species being detected as wood-associated.

11.1.5 Slime moulds

The slime moulds (myxomycetes) are a remarkable group of organisms that have traditionally been treated as fungi because their sporocarps (fruiting bodies) resemble those of small fungi. It is now apparent that it is a polyphyletic group, but none of the taxa are included in the fungal kingdom. The life cycle of a typical myxomycete consists of a uninucleate amoebal stage, a multinucleate plasmodium stage that moves freely across or within the substrate, and a sessile sporocarp stage that releases uninucleate spores which develop into uninucleate amoebae (Figure 11.1). Both the uninucleate amoebae and the plasmodia feed upon bacteria, but the plasmodium stage of some species can also feed upon algae (including cyanobacteria), yeasts, and fungal spores and hyphae (Ing, 1994; Stephenson and Stempen, 1994; Keller and Braun, 1999).

Slime moulds are typically observed in nature when their plasmodia form sporocarps. The substrate association of the sporocarps reflects where the plasmodia have been feeding, and different species show quite distinct associations with different types of substrates, including various types of decaying wood. Some species occur on moist rock surfaces, others on the bark of living trees, still others on different types of ground litter, but the majority of species are more or less strictly associated with decaying wood (Stephenson, 1988; Schnittler and Novozhilov, 1996). In extensive myxomycete surveys, it has been reported that wood-associated species make up 68–80% of all encountered species from different parts of the world (Stephenson, 1988; Schnittler and Novozhilov, 1996; Härkönen et al., 2004).

A recent checklist of Nordic myxomycetes shows that there are about 300 species in this region. These species have not yet been classified as obligate and facultative saproxylics, but considering that Schnittler and Novozhilov (1996) found 93 species in this region and 73% of them occurred on decaying wood, it seems reasonable to assume that there are about 200 saproxylic myxomycetes in the Nordic countries. A careful examination of the individual species will certainly change this number, but it will probably stay within a range of 170–230 species.

11.1.6 Mites, spiders and pseudoscorpions

Mites (Acari) are a hyperdiverse class of arthropods. They are very diverse in decaying wood, and different species have all kinds of functional roles (detritivores, fungivores, predators, parasites). Saproxylic

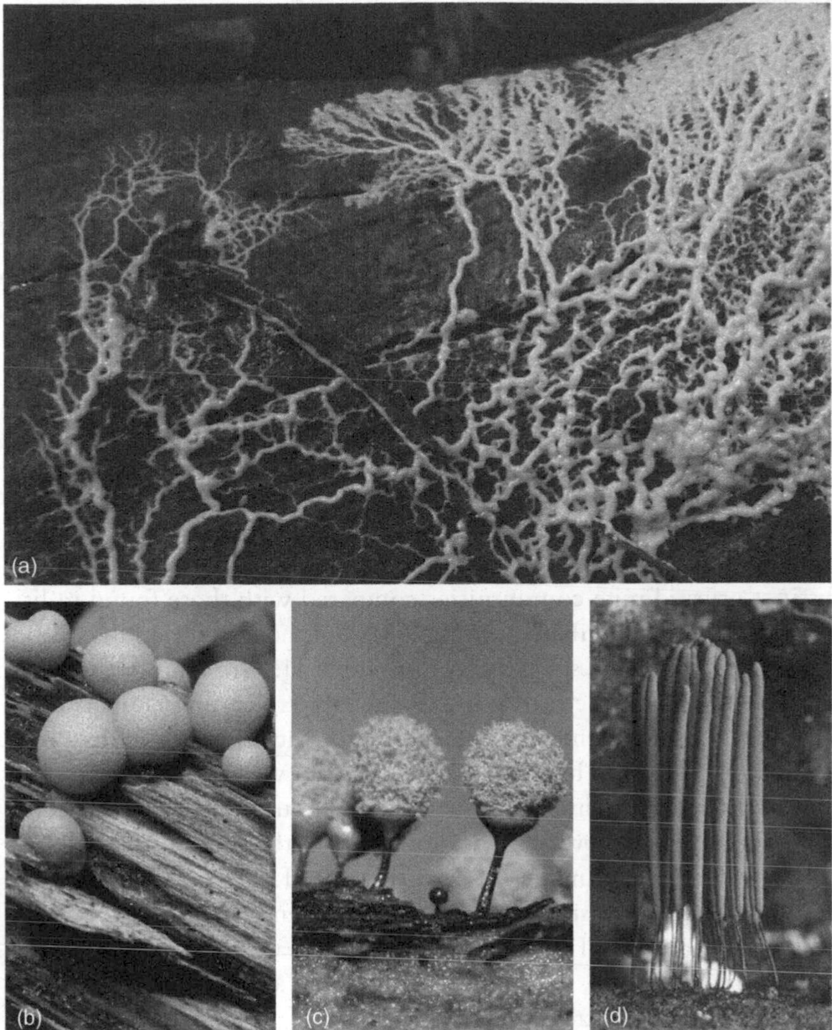

Figure 11.1. Different life-cycle stages of slime moulds: (a) the free-moving plasmodium stage of *Physarum polycephalum* (photo Ken Hickman); (b) sporocarps of *Lycogala epidendrum* (photo Luboš Čáp); (c) sporocarps of *Hemitrichia calyculata* (photo Kim Fleming); (d) sporocarps of *Stemonitis* sp. (photo Kim Fleming).

species are found in the orders Astigmata, Oribatida, Prostigmata and Mesostigmata, and in tens of different families. Mites are flightless, and many species living in freshly dead trees depend on insect vectors such as bark beetles for their dispersal (see Box 6.2). It is probable that other

Figure 11.2. (a) *Larca lata* is a pseudoscorpion that lives in old, large and hollow trees (photo John Hallmén); (b) the flightless pseudoscorpions need assistance to disperse between dead trees. Here an individual is attached to the leg of a saproxylic gnat (*Sylvicola* sp.) (photo Tom Murray).

saproxylic beetles can act as vectors for saproxylic mites living in more decayed wood and in other dead-wood microhabitats, such as polypore fruiting bodies. Decaying logs on the forest floor appear to host quite distinct oribatid mite communities compared with those found in litter and soil (Siira-Pietikäinen et al., 2008; Déchêne and Buddle, 2010). The numbers of mite species given in Table 11.1 are mainly based on data from Poland. The reason is that the saproxylic mite fauna is relatively poorly documented in the Nordic countries, whereas it is well studied in Poland. So by including Polish data we get a better impression of the species richness in this group, although some of the species occurring in Poland probably do not occur in the Nordic countries.

In spiders (Aranea) many species utilize dead trees for web-building or hunting, and some species occur most frequently on dead-wood surfaces (Buddle, 2001). However, there are only a few clearly saproxylic species. From Britain, Alexander (2002) listed only 10 species closely associated with dead trees.

Pseudoscorpions (Pseudoscorpionida) is a small arthropod order with rather few species in the Nordic countries. Some of these are obligate saproxylics that occur in hollow trees (Ranius and Wilander, 2000). Just like the mites, pseudoscorpions are flightless and depend upon other animals to colonize new dead wood units (Figure 11.2).

11.1.7 Beetles

The beetles are one of the three hyperdiverse insect orders occurring in decaying wood. Many beetle families include mainly detritivorous, wood-boring species such as the stag beetles (Lucanidae), longhorn beetles (Cerambycidae), deathwatch beetles (Anobiidae) and bark beetles (Scolytinae). Other beetle families include mainly fungivorous species such as the tree fungus beetles (Ciidae), erotylid beetles (Erotylidae), false darkling beetles (Melandryidae), hairy fungus beetles (Mycetophagidae) and mould beetles (Lathridiidae). In other beetle families most species are predatory, such as clown beetles (Histeridae), checkered beetles (Cleridae) and cylindrical bark beetles (Colydiidae). There are also some beetle families in which the majority of species are scavengers, such as the skin beetles (Dermestidae) and spider beetles (Ptinidae). Most beetles that occur in dead trees are obligate saproxylics, but in some families there are many facultative saproxylics, especially in rove beetles (Staphylinidae), silken fungus beetles (Cryptophagidae) and mould beetles (Lathridiidae).

The beetles are undoubtedly the best-known saproxylic insect group and they are frequently used as illustrative examples throughout this book. Also when it comes to our knowledge about individual species, the knowledge in the Nordic countries is very good. Key publications about saproxylic beetles include Saalas (1917, 1923), Palm (1951, 1959) and Ehnström and Axelsson (2002), and there are hundreds of publications on the ecology of individual species. We are quite sure that the total number of 1447 saproxylic species deviates from the true number by less than 10%.

11.1.8 Gnats and flies

Diptera is another mega-diverse insect order where a large number of species have larval development in different types of dead woody material. Many species are detritivores, as in the crane flies (Tipulidae) and hoverflies (Syrphidae). But even more species are fungivores, feeding either on fruiting bodies or mycelia inside decaying wood. The diverse group of fungus gnats (several related families placed in the superfamily Mycetophiloidea) includes mainly fungivorous species. Among the flies there are many predatory species, for example in the families robber flies (Asiliidae), awl-flies (Xylophagidae), muscid flies (Muscidae) and long-legged flies (Dolichopodidae). There is also a very diverse family of parasitic flies, the tachinids (Tachinidae).

The flies and gnats are more poorly documented than the beetles concerning their association with decaying wood. This is partly because fewer collectors have been interested in Diptera than in

Coleoptera, and partly due to different collecting traditions among dipterists, as they have done less larval rearing and detailed annotation of dead wood preferences. Furthermore, the larvae of small fungivorous gnats are morphologically simple and uniform and are therefore very difficult to identify. Despite a poorer knowledge base, it is becoming evident that there are more saproxylic species of Diptera than of Coleoptera in the Nordic countries. This situation is especially caused by an improvement in knowledge about larval ecology among the fungus gnats. During the last few years it has become evident that the majority of these species actually develop in decaying wood. We consider it likely that the number of saproxylic Diptera will continue to increase from the current number of 1550 species, maybe by 10–30%, as the larval ecology is documented for more species.

11.1.9 Wasps, including ants

The Hymenoptera is the third hyperdiverse saproxylic insect order. The ecology of these species falls into three distinctive categories. Woodwasps (Sciricidae and Xiphydriidae) constitute a rather small group of wood-boring species that feed on recently dead wood. The second, and most diverse, group of saproxylic hymenopterans are parasitoids on larvae of other saproxylic species. Large numbers of obligate saproxylic parasitoids are found in the families Braconidae, Eulophidae, Ichneumonidae and Pteromalidae. The third ecological group comprises species that are secondary hole-nesters in the galleries of other wood-boring insects (e.g. in the families Eumenidae, Sphecidae, Crabonidae and Megachilidae) or kleptoparasites on these secondary hole-nesters (particularly in the family Chrysidae). None of these species have decaying wood as their nutrition base, as the adult insect brings food to its larvae from the surroundings (see Chapter 4 for further details). Most ants that construct nests in wood represent another variety of this group, as they can also excavate galleries in the wood, although they do not feed on the wood itself.

The parasitic Hymenoptera is even more poorly documented than the Diptera concerning species-specific associations with decaying wood. There are many parasitic wasps that are obviously saproxylic, as they have frequently been observed on, or reared from, dead wood. But their host insects and other habitat requirements are often unknown. Hence, the number of roughly 800 saproxylic species (Table 11.1) is probably well below the real number of saproxylics.

It is difficult to assess the real number, but it could easily be at least 50% higher.

11.1.10 Other insect groups

The remaining insect orders known from the Nordic countries are quite modest with regard to their numbers of saproxylic species. The butterflies and moths (Lepidoptera), which is a mega-diverse insect order, only have about 65 species that are saproxylic. These are either fungivores (Tineidae and Oecophoridae) or wood-boring detritivores (Cossidae and Sesiidae). The snakeflies (Raphidioptera) is a small order of predatory insects and only a few species occur in the Nordic countries. The true bugs (Hemiptera) are represented by some species of which most are fungivores feeding on hyphae with their piercing–sucking mouthparts, although a few species that live in bark-beetle galleries are predatory. Also the thrips (Thysanoptera) have sucking mouthparts and are mainly fungivores. The tiny springtails (Collembola) are generally detritivores or fungivores, and it is likely that those occurring in decaying wood have the same feeding modes.

All these insect orders are rather well documented in the Nordic countries and we do not expect that the real numbers of saproxylic species are very much higher than those we have listed in the table. Collembola might represent an exception, as there are very few experts on this group, and these experts have traditionally investigated soil and litter strata rather than decaying wood when documenting the habitat preferences of springtails.

11.1.11 Nematodes

Hardly any serious inventories of nematodes have been made in any of the Nordic countries. Nematodes possibly have a large number of saproxylic species; almost 100 species associated with bark beetles in recently dead trees have been recorded in central Europe (Rühm, 1956). Saproxylic nematodes living in recently dead trees have no dispersal problems even though they are tiny, flightless animals. Most species have particular larval stages ('dauer larvae') which are adapted to dispersal using insect vectors. For instance, some nematode species are vectored by bark beetles (see Box 6.2), while dispersing larvae of other species crawl into the tracheae (respiratory tubes) of their saproxylic insect vectors. For instance, the pinewood nematode

(*Bursaphelenchus xylophilus*) is mainly transported by sawyer longhorn beetles (*Monochamus* spp.), and one newly hatched beetle may carry tens of thousands of nematode larvae (Fielding and Evans, 1996). We can only speculate (mainly based on Rühm 1956) that the number of saproxylic nematode species is most probably above 100.

11.1.12 Marine invertebrates

There are several marine invertebrates that are saproxylic, and some of them occur in the waters around the Nordic countries. These include seven species of shipworms (Teredinidae, Bivalvia) and one species of gribbles (Limnoriidae), which is an isopod crustacean. The shipworms bore into wood and digest woody material with the aid of symbiotic cellulose-degrading bacteria, while the gribbles seem to digest wood without the aid of microbes (see Chapter 9).

There could easily be additional marine saproxylic species (although probably not many) in the Nordic countries. For example, the small bivalves in the genus *Xylophaga* (see Chapter 9) might occur here, but we are not aware of any studies of these species living in sunken wood on deep ocean bottoms.

11.1.13 Vertebrates

The saproxylic vertebrates mainly comprise species that nest or hibernate in treeholes or decaying wood. Most of these species are birds, with 33 species (for a species list, see Esseen et al., 1992) and bats, with about 10 forest species, but there are also two salamander species that typically hibernate in decaying wood. There are only a few vertebrate species that use wood or wood-inhabiting invertebrates as a food source. These comprise six woodpecker species, which mainly forage on saproxylic insects, particularly during the winter.

All the terrestrial vertebrates are well known in the Nordic countries and the number of 45 saproxylic vertebrates is therefore very close to the actual number.

11.2 Additional saproxylic groups

The diversity of saproxylic species in the Nordic countries might seem impressive, but the global diversity is even wider. We now present some additional organisms that live in decaying wood in other parts of the world.

11.2.1 Termites

The termites are probably the saproxylic organisms that are treated most superficially in this book, considering their high species diversity and ecological importance in subtropical and tropical regions. The main reason for this meagre treatment is simply that there are no termites in northern Europe, and this book has a strong bias towards temperate and boreal forest ecosystems. However, the termites have been well studied from several perspectives, including symbiosis between termites and gut microorganisms, termite digestion, feeding ecology, impact on soils and ecosystem functioning, nest structure, social structure and caste systems, morphology, taxonomy and phylogeny. The book *Termites: Evolution, Sociality, Symbioses, Ecology* (Abe et al., 2000) is the best reference to consult for an overview of termites. The space available here does not permit the broad treatment of saproxylic termites, but we shall at least highlight a few facts about them.

Altogether there are about 2700 known species of termites. Roughly 1800 of these are strictly or predominantly wood-feeding, while the remaining species are humus- or soil-feeding (P. Eggleton, personal communication). A recent assessment suggests, however, that there might be as many as 4000 species (Chapman, 2009). It is reasonable to assume that the majority of the yet unknown termites primarily belong to the humus- or soil-feeding groups, due to their hidden nature.

Termites are widely distributed, being most abundant near the equator, and occur with the highest species richness in lowland wet and dry tropical forests, although they are also common in tropical and subtropical grasslands (Eggleton, 2000; Eggleton and Tayasu, 2001). In America and Asia, they occur regularly in the temperate zone, reaching southern Canada, central China and northern Japan. Termites are quite rare in Europe – with only a few species occurring in the Mediterranean area. Termites seem to be more widely distributed in the southern hemisphere, reaching as far as the southern parts of South America, Africa and Australia. Behind the global termite distribution, there are some very interesting patterns related to feeding and nesting habits (Eggleton and Tayasu, 2001). Termites nesting and feeding within single pieces of dry dead wood have the widest ranging distribution. Individual genera have quite large distribution areas and together they cover all continents and latitudes, even reaching distant oceanic islands. At the other extreme we find termites that nest in wet wood (typically large, partly decayed trunks lying on the forest floor). These have a scattered distribution in temperate forests, and individual genera have restricted distribution. Individual genera of soil-feeding termites also have a restricted distribution, but alternative genera occur in different areas and together the soil-feeding termites are widely distributed. Eggleton and Tayasu attributed these distribution patterns to the ecology and different dispersal potentials. Single-piece dry wood termites spend their whole colony cycle within a medium that floats and can therefore drift far away with rivers and ocean currents. This phenomenon of rafting is the most likely explanation for their occurrence on numerous oceanic islands. In addition, the fact that a colony consumes its own environment necessitates dispersal. The soil-feeding species, on the other hand, live in a long-term stable environment and thus have little stimulus to disperse.

11.2.2 Zoraptera

This is a very small order of tropical saproxylic insects with only about 40 known species altogether, belonging to only one genus, *Zorotypus* (Chapman, 2009). The members of this order are small insects, at most 3 mm in length, resembling tiny termites in appearance. They are gregarious and live in small colonies associated with rotting logs, where they feed on detritus and fungi.

11.2.3 Velvet worms

Velvet worms (Onychophora) make up the invertebrate phylum most closely related to arthropods. They have a worm-like segmented body with multiple pairs of conical legs with no joints, small eyes and antennae. About 200 species are known from the tropics and the temperate zone of the southern hemisphere. All the species are predatory and they have been described as hunting together like a pack of wolves. They prey on insects which they catch by squirting an adhesive slime over them. Velvet worms live in small groups of up to 15 females, males and young. These aggregations are not random assemblages but social groups of closely related individuals, organized in a hierarchy based on female dominance (Reinhard and Rowell, 2005).

The permeable skin of velvet worms makes them prone to desiccation; therefore they are nocturnal and prefer dark environments with high air humidity. Many species are associated with logs (Monge-Najera and Alfaro, 1995), and in drier regions, such as temperate Australia, most species are largely confined to the inside of rotting logs (Barclay et al., 2000; Yee et al., 2007).

11.2.4 Saproxylic curiosities among vertebrates

Few vertebrates, apart from the woodpeckers, have developed an intimate association with decaying wood and feed on the resources therein. As vertebrates are rarely studied in the context of saproxylic organisms, we will present some extraordinary species that demonstrate how the saproxylic lifestyle has evolved repeatedly in totally unrelated vertebrate groups.

The beaver can be considered as a saproxylic species for two reasons. Firstly, beavers are analogous to giant bark beetles in that they predominantly utilize the inner bark of recently dead trees. However, in contrast to bark beetles, beavers are not dependent on dead or weakened host trees but instead they fell vigorous trees for food (Vispo and Hume, 1995). From this point of view, they represent a borderline case outside the definition of a saproxylic species. But by killing the trees, the beavers are connected to the saproxylic food web. Secondly, beavers use dead wood for constructing their lodges.

Currently we have two beaver species (the North American *Castor canadensis* and the Eurasian *Castor fiber*) that are quite similar in size. But until 10 000 years ago there was also a giant beaver, *Castoroides ohioensis*, which lived in North America. This species was nearly 2.5 m long, or the size of a black bear, and weighed about 200 kg. There also existed another genus of woodcutting beavers (*Dipoides*). These beavers were smaller than the beavers of today. The habit of woodcutting probably evolved once within Castoridae and the hypothesized common ancestor lived at least 24 mya (Rybczynski, 2007).

Another use of decaying wood is illustrated by the aye-aye, *Daubentonia madagascariensis*, a lemur that lives in the tropical forests of Madagascar. With strong front teeth and a long, thin middle finger, the aye-aye seeks out food and fills the same ecological niche as the woodpeckers. They tap on the bark of old tree trunks and branches and listen for the sound of cavities where the larvae of wood-boring insects might occur. Then they rip open the bark and wood with their front teeth and search for their prey with their long finger. A short claw on the fingertip functions as a hook to pull out the larvae. Aye-ayes are not strictly confined to a saproxylic invertebrate diet, as they eat fungal fruiting bodies growing on the bark of trees as well as fruits, seeds and nectar (Andriamasimanana, 1994; Sterling, 1994).

The only other animal known to find food in the same way as the aye-aye is the striped possum *Dactylopsila trivirgata*, which lives in rain forests and eucalypt woodlands in New Guinea and tropical Australia.

This species is unique among the possums by having a thin, elongated fourth finger on each front leg, their tongue is unusually long, and their front teeth project forwards like chisels. Even though the striped possum and the aye-aye are totally unrelated, they have developed the same food-searching technique and morphological adaptation for extracting wood-boring insect larvae. Amazingly enough, there is also a fossil longfinger – *Heterohyus nanus*. This was a rat-sized vertebrate that lived 47 million years ago in southern Europe, which at that time was covered by tropical forest. *Heterohyus* was morphologically similar to the aye-aye, most notably with two very elongated fingers on each hand (Koenigswald, 1990). Also the general body form was similar, and it is likely that it searched for food in the same manner as the aye-aye.

Even among the fishes we find saproxylic species. In the Amazonian river system there are several species in the genera *Panaque* and *Cochliodon* (the catfish family, Loricariidae) that rasp at submerged wood to obtain their food. These fish have particular adaptations to their diet of wood, such as spoon-shaped teeth and highly angled jaws to chisel the wood. Nelson et al. (1999) reported that the panaque possesses gut bacteria that may allow the fish to digest the wood they consume. Recently, however, German and Bittong (2009) questioned the wood-degrading ability of the panaque gut bacteria. They suggested instead that the fish feed on wood already in the process of decomposition, and therefore it is most likely that it contains soluble organic compounds that the fish can assimilate directly.

It is perhaps not surprising to find that some dinosaurs also utilized decaying wood as a food source. The evidence comes from several records of fossilized faecal droppings found in 74–80 million-year-old strata in North America (Chin, 2007). The size of the droppings, nearby bone quarries and other clues indicate that the source animals were *Maiasaura* hadrosaurs. Most of these droppings were composed of fragmented conifer wood (up to 85%) with signs of fungal decay. Furthermore, the size of the wood fragments clearly indicated that they came from solid trunks and were not small twigs accidentally consumed while browsing foliage on live trees. The diet of the source animal was not completely composed of decaying wood, as other food items were present. It is clear, however, that the predominant consumption of decaying wood along with fungal mycelia and insect larvae was a feeding strategy that spanned at least 6 million years. Today, wood consumption is extremely rare among vertebrates. However, the selective degradation of lignin by white-rot fungi can make the cellulose in wood available for herbivorous vertebrates. This kind of naturally delignified wood (*palo podrido*) has proved to be an effective food source for cattle in Chile (González et al., 1989).

11.3 Why are there so many saproxylic species?

The great diversity of saproxylic organisms certainly calls for an explanation. How can there be so many different saproxylics? An answer to this question has several components, most notably niche differentiation between species, the huge energy base that woody material represents, and the coexistence of similar species. All these explanations represent general species richness hypotheses and have been extensively discussed in the ecological literature. Here we rephrase them in the context of decaying wood and show that they also have explanatory power here.

11.3.1 Niche differentiation between species

For a long time, ecologists have tried to develop a theoretical framework for the niche concept as a basis to explain the coexistence of species that utilize similar resources. Quite early on, Gause (1934) launched the idea that two competing species with identical niches cannot coexist indefinitely. The principle of competitive exclusion has subsequently been presented as a fundamental property and sometimes even as a law of nature. After Gause, the understanding of species coexistence and properties of community assemblages was almost exclusively framed in terms of competition for different resources. Gause's principle implied that there can only be a limited overlap in the usage of some shared resource for two species to coexist (Hutchinson, 1957). Hutchinson also extended the niche concept to a niche hyper-volume representing a multidimensional niche space of resources utilized by a species.

It is useful to consider niche differentiation to explain the number of saproxylic species. From this point of view, we can say that Chapters 5–8 describe different niche dimensions. That is, the host trees, decay stages, microhabitats, and size classes of wood represent resource dimensions where individual species utilize only a portion of each dimension. In addition, the way of utilizing this resource (Chapter 3–4) and the effect of the surrounding environment (Chapter 9) add further dimensions to the niche space of saproxylic species.

This mode of reasoning led Dahlberg and Stokland (2004) to calculate the potential number of niches as a basis for explaining the high number of saproxylic species in the Nordic countries. By multiplying the minimum number of different tree species (50), growth rates of the live tree (2), mortality factors (3), decay phases (4), microhabitats (6), diameter classes (3), surrounding environments (5) and other factors related to functional roles (10), they estimated that the number of possible combinations exceeded 1 million, thus allowing far more species to occur than the roughly 7500 saproxylic species known to exist in the area.

There are at least two weaknesses in this method of calculating the potential number of niches. First, the human classification of dead wood does not necessarily correspond to the way in which the saproxylic species experience this resource. For the majority of species, there is no significant difference between wood from different broadleaved trees. Similarly, while there are species that differentiate between wood in sun-exposed and shady positions, there are many others that do not respond to the degree of sun exposure (see Figure 9.4). The other weakness in the calculation is that the variations along different niche dimensions are not independent of each other. For example, the effect of different mortality factors is probably strong in the initial decomposition phase but weaker later in the decomposition (see Figure 6.4). Similarly, the difference between different host-tree species is significant for recently dead trees but almost negligible in strongly decayed wood (see Chapter 5). In addition, certain potential niche combinations simply do not exist. The microhabitat 'sap exudation' is only present in live trees, and the inner bark is almost invariably consumed early in the decomposition and it is no longer available in medium and strongly decayed trees. However, these limitations do not invalidate the fundamental principle, namely that saproxylic species subdivide the resource of decaying wood along multiple niche dimensions.

11.3.2 Coexistence of similar species

Research on niche differentiation, competition and number of coexisting species during the 1960s and 1970s was theoretically based on the mathematics of differential equations. This type of mathematics does not account for an important aspect of nature, namely spatial heterogeneity and habitat patchiness. The processes that take place in this mathematical model world implicitly occur in the same location.

During the 1980s and 1990s, the significance of spatially structured resources became widely acknowledged and explored. This led to the development of the metapopulation theory predicting that small and isolated habitat patches become more frequently empty for a particular species the smaller and more isolated the patches are (Hanski and Gilpin, 1991). In community ecology, it was realized that many competing species could coexist in a dynamic patchy environment if there is a trade-off between competitive and dispersal ability (Tilman, 1994). In other words, if increased competitive ability has a cost of reduced dispersal ability, then the inferior competitors survive together with stronger competitors because they are quicker to colonize empty patches. Soon thereafter, Hurtt and Pacala (1995) showed that Tilman's trade-off assumption was not necessary for explaining coexistence if dispersal and recruitment limitation were sufficiently strong. In this case, many sites become occupied by competitively inferior species because stronger competitors simply fail to reach all sites. This meant that Gause's competition exclusion principle is not generally true in spatial ecology.

Some years later, Hubbell (2001) took a further step by putting forward the 'neutral theory' in community ecology. Hubbell suggested that many observed patterns in ecological communities can be explained on the basis of species similarity rather than species differences. A cornerstone in this theory is the *functional equivalence* of species, namely that trophically similar species are (as a first approximation) demographically identical on a per capita basis in terms of birth-, death- and dispersal rates. In other words, the differences between trophically similar species are 'neutral', or irrelevant to their success. This similarity, combined with dispersal limitation and environmental heterogeneity, are the major factors allowing the coexistence of competing species (Hubbell, 2005). Hubbell further suggested that functionally similar species are likely to evolve in species-rich communities that are strongly dispersal and recruitment limited, because they converge on similar life-history strategies adapted to the most frequently encountered resources (Hubbell, 2006). This theory, and especially the assumption of functional equivalence as a basis to explain species richness, has been criticized by several prominent researchers (see Hubbell, 2006).

When we consider these general explanations for species coexistence, it becomes clear that decaying wood has exactly the properties that promote the coexistence of similar species. It is patchily distributed and temporally dynamic, as individual dead trees disappear due to decomposition and new dead trees appear as a result of natural mortality. It is therefore difficult for saproxylic species to colonize all the suitable dead wood units in this constantly varying environment. The effect of recruitment limitation is quite easy to observe when one investigates several dead trees with similar properties (same tree species, mortality cause, decay stage, dimension class, etc.). Even in the same local stand the species composition can be rather different on apparently similar dead trees. In a dataset based on observations on more than 4000 dead trees, the species frequency on similar dead logs was typically less than 10% for several hundred fungus species and hardly ever above 50% for the most common species (J. N. Stokland, unpublished data). In other words, the majority of suitable dead trees seem to be uncolonized even by the most frequent species.

We are not able to present hard evidence for the coexistence of functionally similar saproxylic species as we are unaware of such studies. But we strongly suspect that this is a common phenomenon among both fungi and insects. Thus, it is quite likely that this significantly contributes to the high species diversity in saproxylic communities.

11.3.3 The energy hypothesis

The species–energy hypothesis (Brown, 1981; Wright, 1983) proposes that the number of coexisting species in an area is limited by the amount of available energy. The available energy is partitioned between the local species, and the larger the energy base, the larger and more viable populations of each species can exist. It has long been acknowledged that geographical variations in species richness (e.g. from the equator towards the poles or from low to high altitudes) correlate positively with energy availability (Wallace, 1878; Hutchinson, 1959; Currie, 1991). But even if there it is a strong correlative pattern, it is not well understood which mechanisms cause this phenomenon.

It is intuitively logical to assume that the available energy limits population size, and there is plenty of evidence from invertebrates, birds and mammals showing that local population size increases with increasing food or energy availability (den Boer, 1996; Kaspari et al., 2000; Forsman and Mönkkönen, 2003). But the validity of the increased population size mechanism also rests on the assumption that additional energy becomes available across most of the resource range instead of being captured by some superior competitor (Evans et al., 2005). Few studies have explored these topics, but there is evidence suggesting that the relative proportion of energy available to different species does not change with total energy availability (Blackburn and Gaston, 1996). Thus, with an increasing amount of available energy, a wider range of resources (corresponding to potential niche positions) exceed the thresholds needed to support local populations. This is a related mechanism termed the 'niche position' mechanism (Evans et al., 2005).

We now turn our attention from these general energy considerations to the specifics of decaying wood. In a review of primary production and biomass distribution across major biomes and forest types, Rayner and Boddy (1988) calculated that wood makes up more than 90% of all standing biomass in forests and about 80% of all organic carbon in the biosphere. One should remember that the annual turnover (mortality and decomposition) of wood is much lower than for other plant material such as leaves and reproductive organs. But still, more than 50% of the annual net primary production in temperate trees is converted to trunk wood, branches or twigs, and less than 50% becomes leaf litter (Rayner and Boddy, 1988). It is therefore obvious that wood makes up a large proportion of all plant biomass. Furthermore, it has been around for at least 385 million years. Thus, wood has supported saproxylic populations and permitted the evolution of new species during all this time.

11.3.4 Summing up the species richness hypotheses

Above, we presented three key explanations for the great diversity of saproxylic species: niche differentiation, dead wood as a huge energy base, and the coexistence of similar species. These theories should not be considered as alternative competing hypotheses. Instead, they have most probably operated in concert and enabled many species to evolve during several hundred millions of years (Chapter 10). The large energy base of decaying wood has allowed a fine niche differentiation on very specific dead wood qualities. At the same time, the dynamic nature of this resource has promoted the coexistence of similar species, due to recruitment limitations and relaxed competition.

11.4 Global species richness of saproxylics

At the beginning of this chapter we cited studies indicating that total global biodiversity comprises around 10 million species. What would a corresponding figure for global saproxylic diversity look like? Right from the start we should emphasize the obvious: such calculations are surrounded by much uncertainty, often involving several steps of coarse ratio estimates and extrapolations.

11.4.1 Saproxylic diversity and tree diversity ratios

A very simplistic way of calculating the global number of saproxylic species is to consider the ratio between saproxylics and tree species. In the Nordic countries this ratio is roughly 170 : 1 (7500 saproxylic species and 43 indigenous tree species). If we assume a similar ratio on a global scale and use the low global estimate of 60 000 tree species, this would give a global figure of roughly 10 million saproxylic species. This calculation corresponds to the logic of Hawksworth (1991, 2001), who used the ratios between local fungal and plant diversity to calculate a global figure for fungal diversity. However, we do not consider this method of calculation to be valid for saproxylic species. The saproxylics are usually not host-plant-specific like many mycorrhizal, endophytic and pathogenic fungi are. As we have shown in Chapter 5, only a small proportion of the saproxylic species are strictly host-tree-specific.

It is worth noting that the high saproxylic–tree species ratio in the Nordic countries is the result of very low tree diversity. There is no empirical evidence suggesting a similar ratio in the tropics, where the tree diversity is very high. On the other hand, if we consider the few host-tree-specific saproxylics that do occur in the Nordic countries, there are at least five monophagous species, on average, on each tree species. This might indicate a global number of 300 000 host-tree-specific saproxylics plus all the other saproxylics that are less selective for host trees, although this way of extrapolating is quite unreliable. The main reason is that the most abundant tree genera, such as *Pinus*, *Picea*, *Populus*, *Betula* and *Quercus*, etc., are each represented by only one or two tree species in the Nordic countries. Thus, the species occurring on only one tree species in this area probably have a wider range of host trees elsewhere, where additional closely related tree species occur. So instead of counting the number of host-tree-specific saproxylics on individual tree species, one should instead relate host-specific species to the number of tree genera or higher taxonomic levels.

11.4.2 Saproxylic species as percentage of all species

Another way of assessing global saproxylic diversity is to use the percentage of saproxylic species with respect to all the species in a region. In the Nordic countries we know the number of saproxylic species, and also the total number of species, quite accurately. In both Norway and Sweden it has been estimated that the number of multicellular species is between 50 000 and 60 000 (ArtDatabanken, 2010a; 2010b). There is a large overlap in the species composition between these countries, but Sweden has more temperate species and Norway has a higher diversity of alpine and marine species. Finland and Denmark contribute relatively few species in addition to those that also occur in Norway and Sweden. Thus, it is likely that the total number of species could approach 70 000 in these countries. These numbers mean that saproxylic species represent about 10% of all species in the Nordic countries.

Are there good reasons to believe that this proportion is similar on a global scale? The answer is probably yes. We do not believe that the saproxylic species are better studied than other species groups in the Nordic countries. Furthermore, these countries have a wide ecosystem range and habitat diversity comparable to that found on a global scale: there are extensive areas of forests, arctic–alpine ecosystems, bare rocks and cliffs, mires and wetlands, freshwater systems, coastal habitats and marine environments, including coral reefs and depths of several thousand metres. What is missing are deserts and extensive grasslands. Furthermore, the forests are quite diverse, including coniferous and broadleaved forests across several climatic zones (subalpine, boreal, temperate). If the relative diversity of habitats should be lower anywhere, it is probably the forest systems of the Nordic countries, as there is no counterpart to mangrove forests, and no contrast like that between wet and dry tropical forests (although there is a tiny area of boreal 'rain forest' in Norway, but no saproxylic species are known to be specific to these wet forests). The potentially lower diversity of forest habitats in the north should make the 10% mentioned above a conservative value.

So if we accept that the proportion of saproxylic species is *c.* 10% of all species also on a global scale, this means that we have about a million saproxylic species worldwide. It is particularly interesting to consider the beetles in this context, as this is the most diverse organism group of all, with 350 000 known species, while the true species richness might be over 1 million species (Chapman, 2009). The 1447 saproxylic species in the Nordic countries make up 27% of the known 5403 beetle species in the area. It is possible that this proportion could be equally high in the tropics, where the beetle diversity is much higher. In a comprehensive study of beetles in a lowland tropical forest (about 5 km^2 in Sulawesi, Indonesia), over a million beetle specimens represented 3488 species, and 33% of these were saproxylic (Hanski and Hammond, 1995).

11.4.3 Nordic subset of the global saproxylic diversity

Still another way of approaching the question about global saproxylic diversity is to compare the number of Nordic saproxylic species with all saproxylic species in groups that are quite well studied on a global scale. There are not many such organism groups, but there are some. Among them we find three diverse beetle families (or subfamilies), woodpeckers, and the polyphyletic group known as polypores.

We shall first consider the insects, with 3946 known saproxylic species in the Nordic countries (Table 11.1). If we assume that the three beetle families in Table 11.2 reflect a representative sample of the ratio between global and Nordic diversity, this means that we should expect 72–253 times more saproxylic insect species on a global scale, i.e. 280 000–990 000 species. These figures might easily be conservative, as the global diversity is undoubtedly higher in these beetle families than the numbers in Table 11.2 indicate, because more species remain to be described globally (but hardly any in the Nordic countries). On the other hand, not all species in these families are saproxylic. In the Nordic countries, these proportions range from 84% to 100% (Table 11.2). So, if there are more new species to be described than there are non-saproxylic species in these families, the range calculated above should be conservative. If the calculation is restricted to beetles only, the range will be 105 000–365 000. We will not develop these calculations further for insects, but simply conclude that the number of saproxylic insect species must be well above 100 000, and probably also above 500 000, considering the large number of species that await description.

The saproxylic fungi are far more difficult to assess concerning their global diversity. To start with the empirical evidence, we know that there are at least 2350 saproxylic species of macrofungi in the Nordic countries (Table 11.1). Can we expect a multiplicative factor above

Table 11.2. *Number of known species of polypores and three groups of beetles on a global scale and in the Nordic countries.*

	Global diversity[a]	Nordic diversity[b]	Global : Nordic ratio	% saproxylic in Nordic countries[b]
Jewel beetles (Buprestidae)	14 700	57	253 : 1	84
Longhorn beetles (Cerambycidae)	35 000	158	221 : 1	93
Bark- and ambrosia beetles (Scolytinae and Platypodinae)	7 300	102	72 : 1	100
Woodpeckers	216	9	24 : 1	78
Polypores	1 200	214	6 : 1	95

[a] *Sources:* Buprestidae (Bellamy, 2008); Cerambycidae (Lawrence, 1982); bark beetles and ambrosia beetles (Wood and Bright, 1992; Bright and Skidmore, 1997); woodpeckers (Mikusinski, 2006); polypores (Mueller et al., 2007).
[b] *Source:* Nordic Saproxylic Database.

100, as for the beetles? The only well-documented saproxylic fungus group on a Nordic and also on a global scale, the polypores, suggests that the answer is no. The global : Nordic ratio for this group is 6 : 1. This ratio is most likely to be conservative, as the polypores are well known in the Nordic countries, while many species remain to be described elsewhere. So, what does this low ratio indicate? It is quite evident that most polypore species are widely distributed and the degree of local endemism is low. The overlap in species composition between Europe and East Asia is 80%, and between Europe and North America it is 70% (Mueller et al., 2007). Also, in the tropics, many polypore species are widely distributed, as the overlap between Africa and the Neotropics is 55% (Mueller et al., 2007). In another and more diverse group of saproxylic fungi, the Xylariaceae in the Ascomycetes, the degree of endemism seems to be much higher. In three areas of Central America (Venezuela, Caribbean islands and Mexico), there are about 500 known species, but the degree of overlap between each of these three areas is typically less than 50% (Mueller et al., 2007).

The figures above for fungi indicate a minimum number of 14 000 saproxylic species on a global scale by using the 2350 known species in the Nordic countries and a 6 : 1 ratio between global and Nordic diversity. But we have just argued that this polypore ratio is conservative and that other fungal groups could easily have higher ratios. Thus, it is likely that a global figure for saproxylic fungi is substantially larger. Moreover, the fungi we have considered so far are those we conventionally refer to as macrofungi, i.e. fungi with fruiting bodies that are visible to the human eye. These fungi potentially represent just 10% of all fungi (Rossman, 1994).

There are many additional saproxylic fungi among the microfungi, especially in the groups of true yeasts and ophiostomatoid fungi (bluestain and other sapstain fungi). But here we enter the huge and largely unexplored world of insect-associated fungi. We shall only point out that there are close associations between ophiostomatoid fungi and tree species as well as bark beetles (Kirisits, 2004) and that large numbers of yeast species new to science have recently been discovered as gut symbionts of wood-boring beetles (Suh et al., 2005). Thus, there is a large potential for additional saproxylic species among the microfungi. Unfortunately the knowledge base is still too weak to make a qualified assessment of the magnitude of saproxylic microfungal diversity.

11.4.4 Conclusions

We used three different approaches to calculate the global species richness of saproxylic species. In the first approach, we extrapolated from the ratio between the diversity of saproxylics and trees, but we consider this extrapolation to be so uncertain that we do not trust the result at all. In the second approach, we combined the Nordic ratio between saproxylic species and all multicellular species and the global estimate of all species. We then arrived at an estimate of about 1 million saproxylic species. In the third approach, we considered the diversity of well-studied saproxylic groups on a Nordic and global scale and arrived at numbers in the range of 300 000 to 1 million saproxylic insect species and a minimum of 14 000 macrofungal species (a group that represents about 10% of all fungi). If we consider these numbers in combination, it seems reasonable to assume that there could be about 0.4–1 million saproxylic species on a global scale.

12 · *Natural forest dynamics*

Bengt Gunnar Jonsson and Juha Siitonen

For millions of years, natural forest dynamics have created the variety of dead wood hosting the diversity of saproxylic life. This chapter describes the structure and natural dynamics of forests that develop without management or with negligible human interference. We will deal with stand-replacing dynamics driven by fire, storm events, or insect attacks; continuous-cover dynamics, including gap dynamics, in coniferous and broadleaved forests; and riparian dynamics caused by flooding and natural erosion. 'Parkland dynamics' in open wooded land maintained by large grazing herbivores will be presented in Chapter 16.

In this chapter, we also emphasize the abundance and variation in time and space of the qualities of dead wood. The patterns of habitat occurrence represent the environment to which saproxylic species are evolutionarily adapted. The consequences for their life strategies are discussed in Chapter 14.

The suitability of a dead tree as a habitat for a particular saproxylic species is partly related to the causes of its death. Different mortality factors open up different decomposition pathways, resulting in divergence in species composition during the decay succession (see Chapter 6). In natural forests, the full range of factors causing tree mortality is present, which means that a great variety of characteristics of dead trees becomes available. Thus, not only is the volume of dead wood higher in natural forests than in managed forests but also, and perhaps more importantly, the diversity of dead wood is much higher.

Yet another difference between natural and managed forests is that the recruitment of dead trees in natural forests occurs more evenly over time, since background mortality is higher and not just linked to single disturbance events (such as thinning and clear-cutting in managed forests). On the other hand, infrequent large-scale disturbances such as large wildfires can create huge concentrations of dead wood. This further implies that dead trees are created in more varied environments in natural forests, ranging from dead wood in open forests after stand-replacing disturbances to logs and snags in shaded conditions in late successional stages.

Finally, natural forests tend to consist of a mixture of different tree species and tree sizes, further increasing the variety of dead wood. Taking all these aspects into consideration (tree species, stand structure, mortality agents, tree sizes) gives an important explanation, in addition to the higher volume of dead wood, as to why natural forests are much more species-rich in saproxylic species than managed forests (Kirby et al., 1998; Grove, 2001; Siitonen, 2001; Jonsson et al., 2005). Below we describe in more detail the different natural disturbances that create dead wood.

12.1 Spatial and temporal variability in mortality

The term 'mortality factor' was introduced in Chapter 6 and defined as any agent causing the death of trees. Most mortality factors have distinct spatial patterns at the landscape scale (Franklin et al., 1987). Windthrow is most common on wet soils where rooting zones are restricted, and in particular topographic positions. Forest fires have clear patterns in relation to site type and topography. Upslope and dry sites tend to burn more frequently than downslope mesic or moist sites. Mortality caused by fluvial processes, such as bank erosion or flooding, obviously has very strong and predictable spatial patterns, as it only occurs adjacent to water bodies. Temporal variation of tree mortality is also large, both at the stand and landscape scales. Some mortality factors, such as competition, contribute to tree death continuously, while most exogenous factors cause irregular and episodic mortality. The spatial and temporal scales of mortality are interlinked: large-scale mortality episodes occur infrequently, while small-scale mortality of individual trees is a continuous process (Figure 12.1).

At the stand scale, both mortality rate and the relative importance of different mortality factors vary greatly during stand development (Table 12.1). After a stand-replacing disturbance (see below), four distinct successional stages have often been recognized (Peet and Christensen, 1987; Oliver and Larson, 1990; Harper et al., 2005; but see Franklin et al., 2002, for a refined classification: establishment, stem exclusion, transition and old-growth stages. Risk of death is highest during the establishment stage, when small seedlings with low carbohydrate

Table 12.1. *Prevailing mortality factors during forest successional stages in boreal forests (modified from Franklin et al., 1987).*

Establishment stage	Stem-exclusion stage	Transition stage	Old-growth stage
Environmental stress (e.g. drought, frost)	Competition	Fire	Heart rot + wind
Herbivory	Fungal diseases	Wind	Senescence
Fungal diseases	Snow load	Defoliation	Fire
		Competition	

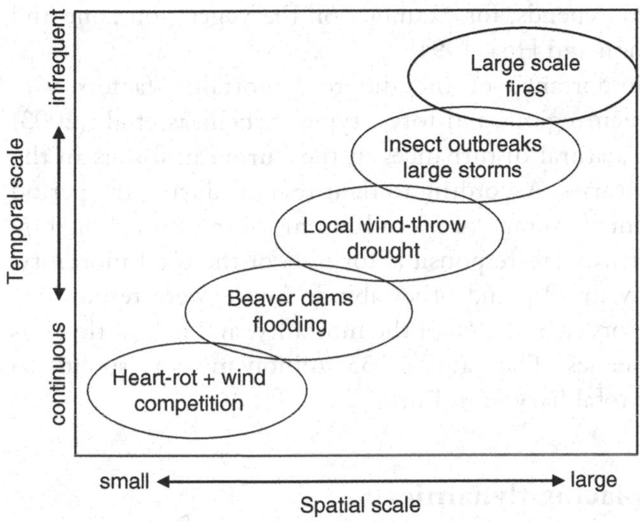

Figure 12.1. Spatial and temporal domains of some mortality factors at the landscape scale (modified from Kuuluvainen, 2002). Small-scale mortality, caused e.g. by competition, operates at the scale of individual trees continuously, whereas large-scale forest fires can cause tree mortality over large areas but are infrequent events.

reserves are susceptible to even minor environmental stresses, herbivory and diseases. The most important primary mortality factor in young stands entering the stem-exclusion stage is competition. All the growing space in the stand is occupied, and mortality is strongly density-dependent. The mortality rate is high and usually declines as the stand becomes older. The importance of competition decreases in the transition stage, and exogenous mortality factors become more important than competition. Mortality and growth may reach an approximate equilibrium in the old-growth stage.

The average annual background (non-catastrophic) mortality rate in the transition stage and old-growth natural forests ranges from about 0.2% to over 2%, being lowest in boreal and temperate coniferous forests and highest in tropical rain forests (Parker et al., 1985; Swaine et al., 1987; Ranius et al., 2004; Ozolincius et al., 2005; van Mantgem et al., 2009). However, large-scale disturbances (fires, storms, drought periods, etc.) may increase the mortality rate considerably. Viewed at the landscape scale and over a long time period, the continuous background mortality and the irregular large-scale disturbances create about equal amounts of dead wood, although this is a very rough generalization and depends, for example, on the vegetation zone and forest type (Harmon and Hua, 1991).

The relative importance of the different mortality factors varies strongly between regions and forest types. Schelhaas et al. (2003) compiled data on natural disturbances in the European forests in the 19th and 20th centuries. According to their review, during the period 1950–2000 an annual average of 35 million m^3 of trees was killed by disturbances. Storms were responsible for 53% of the total mortality, fire for 16%, snow for 3%, and other abiotic causes were responsible for 5%. Biotic factors caused 16% of the mortality, and half of this was caused by bark beetles. The value of 35 million m^3 corresponds to about 8.1% of the total harvest in Europe.

12.2 Stand-replacing dynamics

In many forest ecosystems, natural disturbances have the potential to kill almost all the trees within a stand during a short time period. These events are termed 'stand-replacing disturbances'. Obviously they provide a massive input of dead wood, but they also result in the occurrence of dead trees in open habitats as well as resetting the forest to early successional stages. These are often dominated by other tree species than old stands; therefore, stand-replacing disturbances contribute to the variation in types of dead wood over long time periods and large spatial scales.

12.2.1 Fire

Fire is the most important disturbance agent in many forested regions (Rowe and Scotter, 1973; Johnson and Miyanishi, 2001). In the boreal region, basically all forests are subject to fire, although in some humid regions and wetland sites, the fire return intervals are very long (e.g. Niklasson and Granström, 2000; Bergeron et al., 2004; Carcaillet et al., 2006). Ground fires can often develop into canopy fires that kill most of the trees (e.g. Turner and Romme, 1994; Kafka et al., 2001). In such situations, a large pulse of dead trees is obviously created. A well-documented pattern in the succession following fire is the decrease in CWD (coarse woody debris) volume over the course of time as the fire-generated dead wood is decomposed, and the young developing stand does not produce much new dead wood. After a low level at an intermediate time since fire, the volume then starts to increase as the stand gets older. This explains the U-shaped pattern of CWD abundance over time after a forest fire (Figure 12.2) found in both empirical and modelling studies (Harmon et al., 1986; Spies et al., 1988; Sturtevant et al., 1997; Siitonen, 2001; Harper et al., 2005; Brassard and Chen, 2008).

Natural fire regimes are characterized by three attributes: the fire return interval, fire severity, and size of fires (Johnson and Miyanishi, 2001). These attributes vary extensively among regions and forest types. The results of fire-ecological research highlight the importance

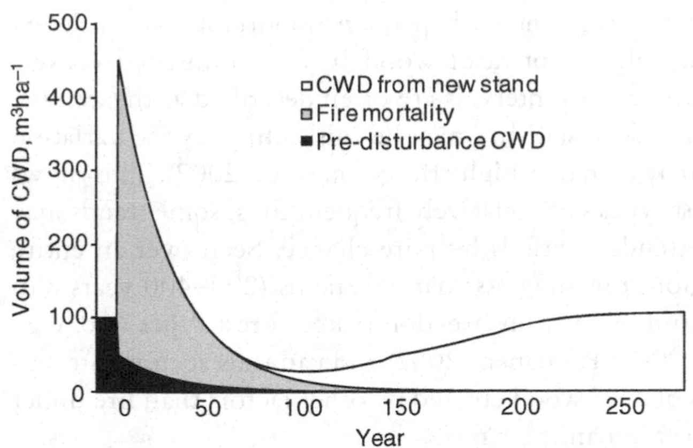

Figure 12.2. The availability of dead wood (CWD = coarse woody debris) after a stand-replacing disturbance generally follows a U-shaped pattern, based on a pulse of dead wood created by the disturbance followed by a slow accumulation of dead wood generated by the developing stand. The model (Siitonen, 2001) is based on data from natural spruce-dominated forests in southern Finland. The pattern is likely to be general, with the exception that the *x*- and *y*-axes (duration of stand development, volume of dead wood) are different depending on region, site type, and stand structure factors.

of variability, in itself, in understanding the effects of forest fires (Ryan, 2002). Fire return intervals correlate both with regional climatic conditions and local site factors. Thus, forests in the interior northwestern parts of North America have a higher fire frequency than forests in the more humid boreal forests of southeastern Canada (Johnson, 1992). Paralleling this pattern, fires are more frequent east of the Ural Mountains compared with more western parts of Eurasia (Bonan and Shugart, 1989). At a local scale, variation in site conditions and forest types has an important effect on fire frequency. In his classic work, Zackrisson (1977) showed that spruce-dominated stands in northern Sweden had a much longer fire return interval (average about 100 years) than pine-dominated stands on dry sites (50 years). Topography and forest vegetation add a component of systematic variability to fire frequencies (Gromtsev, 2002; Kuuluvainen, 2002). Dry, south-facing upslope areas tend to have a much higher fire frequency, while moist, north-facing areas have low fire frequency and may well escape fire for millennia, forming fire refugia. However, it should be stressed that the fire intensity may vary even within a single fire, leaving some smaller or larger patches more or less untouched. This implies that time since fire is variable over different scales, and most forest areas are a patchwork of varying times since last disturbance.

Given the variation of CWD abundance according to time since fire, the actual fire return intervals play an important role in determining the total volumes of dead wood in fire-dominated ecosystems. Although fire return intervals are often described as the average time between fires, such statistics may be misleading, as the variation between events may be rather high (Bergeron et al., 2002). This shows that even in forest types with relatively frequent fires, some stands may escape fire for extended periods by pure chance. Seen over an entire landscape or region, this suggests that old stands (200–400 years old) may be quite common even in fire-dominated forest types (see, e.g., Wimberly et al., 2000; Pennanen, 2002) and can thus accumulate significant volumes of dead wood created by other factors than fire under a continuous-cover dynamics regime.

The total area burned depends mainly on relatively infrequent but large fires. In Canada, 85% of the burned area results from fewer than 5% of the forest fires (Johnson et al., 1998; Stocks et al., 2003). During the period 1918–2005, approximately 1 million hectares per year burned in the Canadian forests. During extreme years (e.g. 1989 and 1995) up to almost 8 million hectares burned (Fauria and Johnson,

2008). In the other large boreal region, Russia, vast tracts of forest also burn annually. Recent record years are 2002 and 2003, when approximately 7.5 and 14.5 million hectares, respectively, burned. There are, however, indications that the majority of these fires are human-related (Achard et al., 2008). It is difficult to translate these area estimates into dead wood volumes created, but – given normal growing stocks – the rough estimates of dead wood added will be two orders of magnitude larger (in cubic metres) than the burned area (in hectares). That is, in Canadian and Russian forests, hundreds of millions of cubic metres of dead wood may be added through forest fires every year.

In addition to providing large amounts of dead wood, fires also add to the heterogeneity of dead wood types. This results from fire effects or scorching of dead trees already present in the stand as well as the creation of fire-killed trees. Some of the fire-created habitats are especially important for saproxylic species (see also Chapter 6). A fire may trigger strong resin production in living conifers; such trees will later form decay-resistant snags. Some trees become heavily scarred, creating habitat for specialized species. Furthermore, the fire may leave sun-exposed dead trees, favouring species that prefer warm habitats. For instance the large, long-lasting decorticated snags, so-called kelo trees (Leikola, 1969; Niemelä et al., 2002) often have a history of several fires, which caused the living trees to become impregnated with resins.

In terms of fire frequency, Australian eucalypt forests are among the most extreme forest types, where fires may occur with an interval of just a few years due to the flammability of the vegetation and dry weather conditions. In these conditions, the variation itself may be as important as the average fire frequency in maintaining the forest structure (Gill and McCarthy, 1998). In other dry regions, short fire return intervals are common too. In northwestern Mexico, in an area with natural fire regimes, the average fire return interval in *Pinus jeffreyi* forests was between 6 and 15 years (Stephens et al., 2003).

12.2.2 Wind

Severe storms or hurricanes may cause large-scale stand-replacing disturbances in some situations (e.g. Foster and Boose, 1992; Lässig and Mocalov, 2000; Hooper et al., 2001; Fischer et al., 2002). For instance, in natural boreal forest in Russia, wind-damaged forests which are several kilometres wide and up to more than 50 km long occur (Syrjänen et al., 1994). Another example is the winter storm in southern Sweden

Table 12.2. *Examples of large-scale storm events. Based on references in Fischer et al. (2002), Schütz et al. (2006) and Haanpää et al. (2006).*

Storm event	Region affected	Dead wood produced (million m^3)
Vivian/Wiebke, 28 Feb–1 March 1990	Central Europe	100–120
Lothar, 26 Dec 1999	Central Europe	*c.* 185
Gudrun, 7–9 Jan 2005	Scandinavia/Baltic States	*c.* 85

in January 2005 (see Haanpää et al., 2006), where a single storm felled around 75 million m^3. This corresponds to a full year's harvest in Sweden and caused dramatic changes in many coniferous stands. Despite the frequent occurrence of such events (Table 12.2), wind tends to be less important than fire in causing stand-replacing disturbances. In general, it seems that coniferous forests are more susceptible to severe blowdown than broadleaved stands (Foster and Boose, 1992; Baker et al., 2002), most probably due to higher crown fetch during winter storms and shallower root systems in conifers.

The degree of disturbance by wind is a function of wind speed. At wind speeds above 20 m/s the risk of major damages increases rapidly (Talkkari et al., 2000; Ancelin et al., 2004). The risk of windfall is also related to tree and stand characteristics (Peterson, 2000). Tall, slender trees have the highest risk of being felled by wind (Ancelin et al., 2004), while mixed stands are less wind-prone than monocultures (Schütz et al., 2006).

Compared with forest fires, wind disturbances provide a narrower range of dead-wood qualities and do not essentially influence the characteristics of pre-disturbance dead wood. It is mainly the full-grown (tall and large-diameter) trees that are felled, while smaller trees have a higher probability of survival. Wind disturbances mostly create downed woody debris and, to a lesser degree, standing dead wood (as a result of root breakage and subsequent weakening of trees).

12.2.3 Insects

There are also some insect species that can cause severe mortality and large-scale stand-replacing disturbances. Two main groups can be distinguished: defoliators and bark beetles. Defoliators include moth and sawfly species whose larvae feed on needles and leaves. Defoliators can feed on the foliage of completely healthy host trees, and the outbreaks are more connected with climatic factors and the population dynamics of the insects themselves than to reduced vigour of their hosts. On the other hand, bark beetles usually depend on reduced tree and stand vigour to cause extensive damage. Thus, there are clear interactions between abiotic disturbances and stress conditions of trees (drought, windblow, flooding, fire, etc.) and occasions when bark beetles cause stand-replacing disturbances (Parker et al., 2006; Gandhi et al., 2007).

Notorious examples of defoliator damage include the outbreaks of spruce bud worm (*Choristoneura fumiferana*) in balsam fir forests of eastern North America (Ghent et al., 1957; Blais, 1981; Bouchard et al., 2005). The humid boreal forest landscape is dominated by stand-replacing dynamics driven by insect defoliation. Three major spruce budworm outbreaks occurred in the 1900s, affecting a total forest area of almost 100 million hectares (Blais, 1983). Balsam fir often experiences over 80% mortality during an outbreak (e.g. Bouchard et al., 2005), and less preferred species (*Picea* spp.) are also affected (Hennigar et al., 2008). Depending on the actual mixture of tree species, the effect at the stand level may be either stand-replacing dieback or more limited disturbance, creating canopy gaps (McCarthy and Weetman, 2007).

There are also some defoliating species in Eurasia which may cause stand-level dieback and thus produce pulses of dead wood. The autumnal moth (*Epirrita autumnata*) is known to cause extensive mortality in mountain birch forests with about 10-year intervals (Tenow, 1972). The somewhat irregular cyclic outbreaks of this species have received great research interest and serve as a standard example for synchronous tree mortality over larger regions (e.g., Ruohomäki et al., 2000). Another species that can cause extensive mortality during population peaks is the Siberian moth (*Dendrolimus sibiricus*). The preferred host species are Siberian fir (*Abies sibirica*) and Siberian pine (*Pinus sibirica*), but during outbreaks, larches (*Larix*) and spruces (*Picea*) are also affected. It shows a strong cyclic behaviour with 10–11 year cycles, with peak populations generally lasting 2–3 years (Anon., 2005a). During peak years, the moth may affect vast areas of forests, measured on the scale of millions of hectares (Anon., 2005a; Kharuk et al., 2007). Such large-scale events are often followed by bark-beetle damage, thus providing situations with a huge input of dead wood, and may have strong effects on the population dynamics of secondary saproxylic species.

Bark beetle outbreaks often start after large-scale windthrow events, which provide ample amounts of breeding material, or drought combined with warm summers (Wermelinger, 2004; Raffa et al., 2008). Healthy conifers are generally resistant to bark-beetle attacks through their resin defence. However, if the defence system of the trees is weakened by drought or some other stress factor, or if there are sufficiently many adults boring into the trunk at the same time, the resin pressure is not sufficient to force them out (see Chapter 6). The mass-attacking capability is restricted to those bark-beetle species that have an aggregation pheromone system, which can attract thousands of individuals to the same tree.

All bark-beetle species are associated with blue-stain fungi and act as vectors for these fungi, carrying them from infested trees to uninfested ones. The virulence of different blue-stain fungus species varies, as does the virulence of different strains of the same fungus species. Most of the bark-beetle species that can cause large-scale epidemics are associated with pathogenic blue-stain fungi. The particular fungal strain carried by the bark beetles may determine the degree of stand mortality caused by the attack (Krokene and Solheim, 2001; Solheim et al., 2001, see also Box 6.2 for further details).

In western North America, bark beetles are a major mortality factor, often affecting larger areas annually than fire. Population eruptions are known to have occurred on numerous occasions during the last hundred years, causing up to 90% mortality of large trees over several million hectares (Romme et al., 1986; Berg et al., 2006; Raffa et al., 2008). However, during the last 10 years, almost 50 million hectares in different regions and conifer species have been affected (Raffa et al., 2008). The most important species among bark beetles belong to the genus *Dendroctonus*. The southern pine beetle (*D. frontalis*) and Mexican pine beetle (*D. mexicanus*) feed on southern pines (e.g. loblolly pine *Pinus taeda*, shortleaf pine *P. echinata*, and several other species) in the southern USA and Central America (Price et al., 1992), and cause cyclic epidemics with about 10-year intervals. The mountain pine beetle (*D. ponderosae*) lives on lodgepole pine (*Pinus contorta*) and causes outbreaks in the Rocky Mountain area in northwestern North America. Although native to the region, a series of mild winters and hot, dry summers has provided conditions leading to an unforeseen massive epidemic that started in the early 2000s (Lewis and Hrinkevich, 2008;

Raffa et al., 2008). In 2008, the outbreak area covered most of interior British Columbia, totalling 15 million hectares. The spruce bark beetle (*D. rufipennis*) is the most destructive of the spruce-inhabiting bark beetles in North America (Werner et al., 2006). Sporadic outbreaks following windblow and dry summers have occurred from Mexico to Alaska. A recent outbreak in Alaska has been estimated to amount to almost a million hectares.

In European and Asian boreal regions, the European spruce bark beetle (*Ips typographus*) is by far the most significant species, causing extensive mortality in Norway spruce forests (Weslien and Schröter, 1996; Wermelinger, 2004), especially during hot summers and favourable conditions. Other *Ips* species are also capable of causing stand-scale mortality; these include *I. sexdentatus* and *I. acuminatus* on Scots pine, as well as *I. cembrae* and *I. subelongatus* on larch species. Interestingly enough, unlike in North America, there are no *Dendroctonus* species in Eurasia that cause large-scale outbreaks in natural forests. The great spruce bark beetle (*Dendroctonus micans*) is capable of attacking seemingly vigorous trees, but it is generally rather rare and kills only individual trees.

12.2.4 Tree species composition during succession

A stand-replacing disturbance will also reset the succession. After forest fire, shade-intolerant, pioneer tree species with wind-dispersed seed and rapid growth will generally establish. Subsequent mortality will provide dead wood of different tree species than before the disturbance, which increases dead wood diversity significantly. Fairly distinct successional pathways following large-scale disturbances have been well described. In boreal Fennoscandia, a broadleaved stage with high abundance of birch, aspen and willows is common after forest fire (Esseen et al., 1997; Lilja et al., 2006). In Canadian boreal forests, broadleaved stages after forest fires may also occur as one of several potential successional pathways. For example, in Quebec, stands initially dominated by aspen and birch slowly developed into forests dominated by balsam fir (De Grandpré et al., 2000). Observed over longer time spans, the shift in dominant tree species also represents a shift in one of the most important habitat factors for saproxylic species, since the fungal and insect communities are distinctly different between coniferous and broadleaved forests (see Chapter 5). Natural succession therefore differs strikingly from the succession in managed forests, where variation in tree species composition over time is limited (Pedlar et al., 2002). Most protected forest areas are too small to include all the natural successional stages. Therefore, disturbances such as fire need to be reintroduced in order to reset the succession to earlier stages, and to provide the range of habitats used by saproxylic species (Linder et al., 1997; Kouki et al., 2004).

12.3 Continuous-cover dynamics

Stand-replacing disturbances are often spectacular events and receive much attention in forest ecology. For the most part, though, forest dynamics are less dramatic. In the absence of stand-replacing disturbances, stand development may proceed to a situation where shade-tolerant and long-lived species dominate. This is the prevailing type of forest dynamic throughout most of the Earth's forest biomes, from tropical through temperate to boreal forests. In such systems, the mortality of single trees represents an opportunity for other tree individuals to reach the canopy (Watt, 1947; Runkle, 1982; Kuuluvainen et al., 1998). Also, in forests where severe fire, windthrow or other large-scale disturbances frequently occur, it is a rule that small-scale tree mortality occurs in the periods between major disturbance events.

The term 'gap dynamics' has often been used to describe the mortality of individual trees or small groups of trees. Stand dynamics in natural forests are, however, complex and include other changes over time and space as well as those traditionally included in the term gap dynamics. The same factors which cause stand-replacing disturbances, such as fire, wind and insects, are capable of creating gaps. In addition, other factors such as fungi, drought, snow loads, competition and senescence of old trees also create canopy gaps (McCarthy, 2001; Worrall et al., 2005; Brassard and Chen, 2006). Many of these factors influence the colonization of saproxylic species and the subsequent succession (see Chapter 6).

A key aspect of continuous-cover dynamics is that most types of dead wood are produced more or less continuously on a relatively small spatial scale (i.e. within a few hectares only) for extended periods of time, sometimes for several hundreds or even thousands of years. Thus, the average distance to the nearest suitable dead wood unit is typically less than 50–100 m, and this situation prevails for long periods. This makes the temporal and spatial distribution of dead wood very different compared with that of stand-replacement dynamics, where the distance to the nearest suitable substrate is highly variable and the average distance is much longer. This obviously results in different selection pressure for dispersal abilities among species which live in forests characterized by continuous-cover or stand-replacement dynamics.

12.3.1 Surface fires

Not all forest fires are so intense that they kill most trees. Moderate surface fires may cause partial tree mortality, leaving a relatively large proportion of trees alive in burned areas (Figure 12.3). This seems to have been a common type of fire disturbance, for instance, in Fennoscandian boreal forests (Niklasson and Granström, 2000). Low-severity surface fires are common in other regions too (Heinselman, 1973; Ehle and Baker, 2003). Surface fires produce a pulse of dead wood, although less pronounced than with stand-replacing fires, thus contributing to a more continuous input of dead wood over time. Nevertheless, surface fires enhance the regeneration of different tree species and influence existing dead wood by surface charring, inflicting fire scars, and increasing the potential for fungal infections and insect attacks on living trees.

Figure 12.3. Low-intensity surface fires reset succession to earlier successional stages, and provide pulses of dead wood, thus influencing resident dead-wood and saproxylic species. These factors strongly influence the saproxylic community establishing after a forest fire. Note the large proportion of trees that have survived (photo B. G. Bengt Gunnar Jonsson).

In general, fires play a less important role in broadleaved forests in temperate and tropical regions. In the boreal region, broadleaved forests are often confined to moist ground conditions and the foliage has relatively higher water content than that of conifers. In addition, some broadleaved trees, such as oaks, have thick bark and are relatively resistant to high temperatures. By contrast, beech and other thin-barked species are very sensitive to high temperatures during a fire (Hengst and Dawson, 1994). It is the temperature in the cambium that is decisive for mortality in most tree species, and the insulating effect of bark is critical. In a study on tropical trees, the thickness of bark explained more than 50% of the variation in cambial temperature during a fire, while other factors (e.g. moisture content) had minor effects (Pinard and Huffman, 1997). Similarly, the resistance to fire varies among conifer species (Johnson and Miyanishi, 2001), with some (e.g. *Pinus* spp., *Sequoia* spp.) having evolved to survive fires, while other genera are sensitive to fire (e.g. *Picea* spp., *Abies* spp.).

12.3.2 Wind

Wind is the main driver in gap creation, and repeated wind events may have a strong influence on overall stand dynamics (McCarthy, 2001; Brassard and Chen, 2006). Wind itself may kill individual trees either by snapping or uprooting. However, it is well established that wind is often only a secondary reason for the death of individual trees. Fungal infection often contributes to windthrow (Hubert, 1918; Worrall et al., 2005; Lännenpää et al., 2008). Wind may be the dominant factor causing uprooting, but for snapped trees, fungi have normally worked on the tree before it is eventually felled by a storm event. In an old-growth forest in boreal Sweden, Edman et al. (2007) showed that although most trees had fall directions corresponding to the prevalent winds, a large proportion of Norway spruce trees that died during the study period were infected by the polypore species *Phellinus chrysoloma*, which causes extensive rot in live trees. Similarly, Worrall and Harrington (1988) noted that disease accounted for 66% of the mortality in low-elevation subalpine forests in New Hampshire, while windthrow was more important at higher elevations. Furthermore, severe wind may account for moderate-severity disturbances, where individual gaps coalesce such that 30–50% of the canopy is removed in a single event (Frelich and Lorimer, 1991; Hanson and Lorimer, 2007).

12.3.3 Insects

Insects also cause small-scale disturbances (Stewart et al., 1991; Filion et al., 2006; Fraver et al., 2007) and may interact with other disturbances (Lundquist, 1995). A vast number of species utilize living trees as a food resource, and some species may cause the death of individual trees. Species capable of causing stand-replacing disturbances through defoliation (see above) are also involved in small-scale gap creation.

In bark beetles and longhorn beetles, some species frequently kill suppressed or damaged trees, thus creating gaps, but never cause outbreaks. In northern Europe these include *Polygraphus poligraphus* (Scolytinae) and *Tetropium* species (Cerambycidae). Although defoliators are usually noticed when causing large-scale dieback, they also contribute to small-scale disturbances in natural forests. In the coniferous forests of eastern North America, the spruce budworm, *Choristoneura fumiferana*, causes periodical disturbances of varying intensity, thus being important in shaping stand structure (Fraver and White, 2005; Brassard and Chen, 2006), ranging from small-scale impacts to complete dieback of whole stands (see above).

12.3.4 Fungi

In natural old-growth forests, fungi are important tree killers and gap creators. In particular, trees weakened by other damages and stresses are often attacked by parasitic fungi. As noted above, the majority of dying trees may actually be colonized by heart-rot fungi in some forest types (Worrall and Harrington, 1988; Edman et al., 2007). Depending on which fungus species causes the eventual death of a tree, the following saproxylic succession is to some extent influenced by the primary parasitic fungi (Renvall, 1995). It is also likely that heart rot caused by parasitic fungi results in a more rapid decay of the wood as, for instance, in Norway spruce logs colonized by *Phellinus chrysoloma*.

One of the most widespread parasitic fungus species is the honey fungus (*Armillaria mellea*, which actually consists of a complex of sibling species). These fungi can attack many different tree species, including both coniferous and broadleaved, and in some forest types may cause extensive dieback, e.g. in mixed coniferous forests in the northern USA. Here it is a serious pathogen on some conifer genera (*Pseudotsuga, Tsuga* and *Abies* spp.), but it can affect many hundreds of woody plant species (Shaw and Kile, 1991). Parasitic fungi are not commonly associated with stand-replacing disturbances in natural forests. *Heterobasidion* spp. and *Phellinus weirii*, however, frequently cause expanding forest gaps due to spread via root grafts from infested to neighbouring healthy trees (Worrall, 1994; Hansen and Goheen, 2000).

12.3.5 Drought

Although different tree species have widely varying tolerance to drought, trees and whole stands growing at the margin of their environmental tolerance frequently become exposed to drought (Gitlin et al., 2006). For example, recently two consecutive years of drought in southern Europe led to extensive mortality in Scots pine populations (Martínez-Vilalta and Piñol, 2002). Due to the El Niño cycle, tropical areas in South America are subject to drought periods, and these also increase natural mortality rates significantly (Condit et al., 1995; Williamson et al., 2000). At the landscape scale, mortality is typically highest in the relatively most arid locations such as south-facing aspects, well-drained soils and ridge tops.

Drought interacts with several other disturbance agents. For instance, there is a clear relation between drought periods and fire risk, as well as insect outbreaks and fungal infestations. Drought associated with unusually warm weather will have an impact on insect population dynamics, including increased population growth rate, higher number of generations per year, lower winter mortality, and expanding geographical range (Ungerer et al., 1999; Logan et al., 2003; Jönsson A. M. et al., 2009).

12.3.6 Snow loads

Snow loads and avalanches may be important disturbance agents in forests at high altitudes and latitudes (see, e.g., Hesselman, 1912; Veblen et al., 1994; Shen et al., 2001). The impact may be significant; it has been estimated that between 300 and 500 tonnes of snow per hectare settled on the tree canopy of a Finnish northern boreal forest, causing severe stem breakage. At elevations of around 300 m above sea level, a significant proportion of spruce, pine and birch trees (maximum percentage 49%, 100% and 33%, respectively) had broken tops (Jalkanen and Konopka, 1998). In another study, top breaks were noted on 12% of spruce trees under natural conditions in northern Sweden (Fraver et al., 2008). These breaks also serve as infection courts for wood-decay fungi, which increases the mortality risk of the trees (Hennon and McClellan, 2003).

12.3.7 Competition

Self-thinning is a common phenomenon during the early successional stages after stand-replacing disturbances when the first cohort of trees emerges, and intense competition between stems causes mortality (Oliver and Larson, 1990). Self-thinning occurs inevitably in even-aged stands; an increase in the mean diameter of trees results in a decrease in tree density (Reineke, 1933; Westoby, 1984). The ultimate reason for this is that the stand has reached the upper boundary of leaf area or cross-sectional area of stems, limiting photosynthesis and water transportation (Pretzsch and Mette, 2008). Self-thinning boundaries may vary depending on the tree species, site type and environmental conditions (Hynynen, 1993; Pretzsch and Schütze, 2005; Pretzsch, 2006).

Self-thinning provides an input of small to medium-sized dead trees, which often die standing. It is also one reason for the U-shaped mortality pattern that has been reported in old-growth forests (Runkle, 2000; Lorimer et al., 2001; Busing, 2005; Fraver et al., 2008; Figure 12.4). These studies suggest that small-diameter trees have markedly higher mortality rates than medium-sized trees, as competition plays a greater

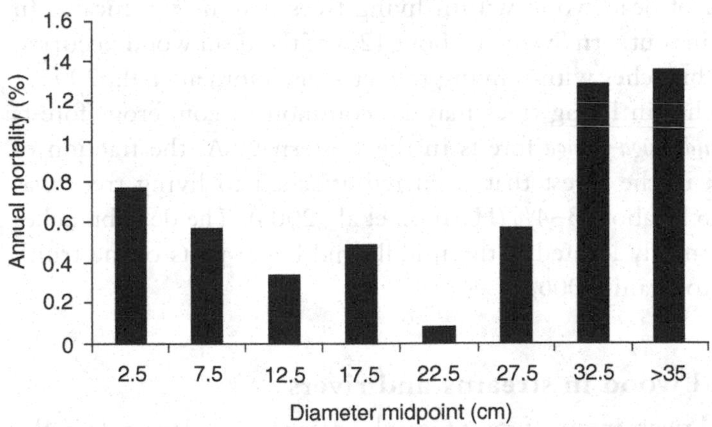

Figure 12.4. Mortality in old-growth forests often follows a U-shaped pattern as small trees die due to competition and old trees die due to senescence and pathogen loads. Data from a northern Swedish old-growth spruce forest (Fraver et al., 2008).

role at this stage, whereas large trees die because of external disturbance agents or senescence.

12.3.8 Senescence

When natural forests develop freely without major disturbance events, the growth rate of trees peaks at an intermediate age. Thereafter the growth rate decreases steadily for decades or even centuries, depending on tree species. Finally, the tree becomes senescent due to unfavourable photosynthetic balance and simply dies of old age. Old conifers have been shown to exhibit reduced growth efficiency (Seymour and Kenefic, 2002) and foliar efficiency (Day et al., 2001). It is unclear to what extent this alone causes tree mortality at an advanced age, since fungal infections and insect attacks most often distress the old, weakened trees. The probability of root and heart rot generally increases with the age and size of trees. All these factors contribute to significantly increased mortality rates in older trees, and senescence is a primary mode of tree death in many old-growth systems (Runkle, 1990; Krasny and Whitmore, 1992; Fraver et al., 2008). The combination of reduced physiological vigour and increased pathogen loads often results in higher mortality rates of old trees and contributes to the U-shaped mortality pattern (see Figure 12.4).

Before their death, old and large trees contribute a regular input of dead wood in the form of large dead branches. For some tree species, the volume of dead wood within living trees may be significant. In oak forests in southern Sweden, about 12% of the dead wood occurred as attached branches within living tree crowns (Nordén et al., 2004a). Dead branches in living trees may be common in coniferous forests too. In *Pseudotsuga–Tsuga* forests in the western USA, the fraction of dead wood in the forest that occurred attached to living trees was estimated to be about 3–4% (Harmon et al., 2004). The dead branches tend to be mostly located at the middle and lower parts of the trunk (Ishii and Kodatani, 2006).

12.4 Dead wood in streams and rivers

Streams and rivers intersect most forest landscapes and play a role in the natural dynamics of dead wood (Hassan et al., 2005). Alluvial forests along streams and lakes are special habitats, which are often productive and constitute sites with high biodiversity (Naiman and Décamps,

1997). The turnover in the tree layer is often high, and consequently the input of dead wood is also high. The dynamics differ from upland forests, as some of the dead wood enters the water system by being waterlogged, and sometimes it is transported downstream.

12.4.1 Input and effects

Floods, eroding river banks and simple tree falls all add dead wood to streams and rivers. The input rate is clearly related to the type of forest along the stream, but due to the slower decay rate in water (see, e.g., Hyatt and Naiman, 2001; see also Chapter 2), streams tend to accumulate more dead wood than the adjacent forest. Extreme values of dead wood in streams are reported from temperate rain forests, where volumes may be well above 1000 m^3/ha (Harmon et al., 1986). Stream size influences the amount of dead wood, as smaller streams have proportionally higher dead wood volume per stretch. This is due to closer proximity to the forest, but also to less forceful downstream transport. Thus, wood jams are more common in small streams (Harmon et al., 1986). Dead wood volumes in old-growth streams tend to be much higher than in streams in managed forest. Dahlström and Nilsson (2006) compared the two stream types and noted over three times (91 m^3/ha as compared with 26 m^3/ha) more dead wood in old-growth streams than in managed-forest streams.

The effects of dead wood in streams are many. Dead wood both plays an important role in the dynamics of river morphology and serves as a habitat for numerous saproxylic aquatic species (see Chapter 9). It adds significant heterogeneity to the aquatic habitat, and this is known to influence both the invertebrate community and the population sizes of fish. Despite its widely known positive effects on fish populations (see, e.g., Abbe and Montgomery, 1996), dead wood has been regularly removed from streams for flood control, navigation, and even to improve fishing (Sedell et al., 1984; Abbe and Montgomery, 1996). The main role of dead wood for fish communities is that it constitutes structures that help fish to avoid predators, influences sunlight and water velocities, as well as creates spawning sites (Crook and Robertson, 1999). Besides forming aquatic habitats, dead wood also strongly influences channel morphology and in this way influences the dynamics of riparian forests. Eroded materials directed and redirected by wood jams in streams are deposited at new sites and thus create starting points for forest succession along the stream.

12.4.2 Beaver dams

Beavers are considered as keystone species or ecological engineers, due to their strong influence on stream ecosystems (Naiman et al., 1986). They also exert a strong influence on forest dynamics. When a beaver dam is built, the flooding of riparian forests often results in high tree mortality, and hence pulses of dead wood (Figure 12.5). Upon abandonment of the dam, tree regeneration by pioneer species often takes place. Unfortunately it is difficult to estimate the historical impact of beavers, since both north European and North American beaver populations were extensively over-harvested during the 1800s, and the species almost went extinct in several regions (Jenkins and Busher, 1979). However, it has been estimated that beaver dams may occur with a frequency of up to 10 dams per kilometre of second- to fourth-order stream reaches (Naiman et al., 1986), roughly corresponding to streams up to 10 m width. Streams of these sizes may intersect most of the forest landscape, and beaver influence has probably been very large, at least in moist forests and swamp forests (Kuuluvainen, 2002).

Figure 12.5. Beaver dams may flood riparian forests, causing dieback of trees due to raised water levels. The fraction of forest land influenced by beavers was extensive in many boreal regions prior to the strong human influence on beaver populations (photo Mattias Edman).

12.4.3 Marine driftwood

Some of the dead wood transported by rivers may eventually reach marine waters. This represents a link between forested regions and coastal areas, where rivers play an important role in the transport. When reaching the mouth of the river, dead wood is redistributed by ocean currents, wind and ice. This is an important process, and marine driftwood, originating from natural forest landscapes, constitutes the habitat for numerous species in the marine environment. Nowadays, the marine saproxylic species are best known for their damage to wooden boats and other wood used as building material. But, for many species, the driftwood represents their original habitat. In Chapter 9 we describe such marine saproxylics in more detail.

Besides functioning as a habitat, driftwood may also facilitate dispersal for a wide variety of species. Rafting is a well-known phenomenon that refers to species being transported across oceans on various floating objects. Dead wood is the most important natural rafting agent and seems to play a key role in both northern and tropical waters (Thiel and Gutow, 2005). For example, the distribution of certain disjunct vascular plants between Greenland and Norway may have its explanation in transportation by driftwood from ice-free regions in Siberia and northwestern Russia during the final stages of the last glaciation (Johansen and Hytteborn, 2001).

Driftwood is also the major source of wood for people in Arctic regions, where cycles over longer time periods have most probably affected human settlement patterns (Alix, 2005). Current concern about these processes thus includes the effects of the ongoing climate change (Dyke et al., 1997).

12.5 Dead wood in natural forests

12.5.1 Volume

In the absence of stand-replacing disturbances, the average annual growth in a stand roughly determines the average levels of annual input of dead wood. Stand productivity is a function of climate, soil and water availability, in interaction with the tree species present. Thus, the global range of stemwood production is very large, ranging from below 0.1 m^3/ha to over 30 m^3/ha per year in temperate rain forest regions (see, e.g., Franklin and Dyrness, 1973). Somewhat surprisingly, closed tropical forests may not necessarily grow very fast, and old estimates of growth rates may be exaggerated (Vieira et al., 2005). This is further supported by relatively low mortality rates in certain tropical forests; for example, in Costa Rican rain forests, the largest trees have an annual mortality of only 0.6% (Clark and Clark, 1996).

Given a relatively continuous input of dead trees in old-growth forests, the local volume of dead wood is determined by stand productivity, mortality (dead wood input) rates and decay rates; these parameters serve as a basis for modelling dead-wood volumes during stand succession (Sollins, 1982; Harmon et al., 1986; Spies et al., 1988; Tyrrell and Crow, 1994; Siitonen, 2001; Ranius et al., 2004). We discussed decay rates in Chapter 6, and here only reiterate that several factors influence them. The decay rate constant in temperate and boreal forests is mostly within a relatively limited range (0.02–0.05; see Box 6.1 for details). This range corresponds to a 50% biomass loss during 15–35 years, and a 95% biomass loss during 60–150 years. In tropical regions, the decay is much faster (Chambers et al., 2000; Mackensen et al., 2003) because the decomposition takes place throughout the year, whereas in more northern latitudes the fungal decomposition activity ceases when the temperature inside the log drops below 4–5°C. Although annual tree mortality may vary widely, equilibrium volumes of dead wood can be modelled simply by combining information on average input rates (m^3/ha per year) and decay rates (Table 12.3). Therefore the same equilibrium volume can occur in forests with different mortality and decay rates.

A large number of studies have reported the volume of dead wood in different forest ecosystems. There are, however, several problems in comparing the results. Some studies report biomass and others report volume, which are not easily converted because density changes during decay (corresponding to biomass loss), while volume remains largely uninfluenced. Furthermore, estimates are highly sensitive to sample plot size (Woldendorp et al., 2004) and several examples of high reported volumes per hectare result from aggregated dead wood in very small (a few hundred m^2) sample plots (Linder et al., 1997; Gibb et al., 2005). In a similar way, volume estimates vary greatly depending on sampling transect length and number (Woldendorp et al., 2004). Finally, some bias can also be attributed to the mathematical formula used for estimating the volume of individual logs. In a comparison of six commonly used formulae, Fraver et al. (2007) showed that, depending on which formula is applied, estimates may vary up to 25%.

Table 12.3. *Dependence of the equilibrium volume of CWD (coarse woody debris) on average input and decay rates (after Siitonen, 2001). For details on decay rate models, see Chapter 6.*

Annual input rate (m^3/ha)	Decay rate constant, k						
	0.015	0.020	0.025	0.030	0.035	0.040	0.045
0.5	33	25	20	17	14	13	11
1.0	66	50	40	33	29	25	22
2.0	132	100	80	67	57	50	45
4.0	265	200	160	133	114	100	89
6.0	397	300	240	200	171	150	133
8.0	530	400	320	267	229	200	178
10.0	662	500	400	333	286	250	222

Bearing this uncertainty in mind, we present some summaries of dead wood volumes in natural forest ecosystems (Table 12.4). Siitonen (2001) presented a review of studies from boreal forests. Building on this, Hahn and Christensen (2004) compiled data from European boreal and temperate forest reserves. From North America, Stevens (1997) gave estimates for Canadian forests and Harmon et al. (1986) gave estimates for temperate forests. The data from the tropics are more scattered, but in Table 12.4 we include examples from tropical regions and from *Nothofagus* forests in southern latitudes.

Although there is considerable variation in many boreal and temperate natural forest ecosystems, the volume of dead wood typically appears to be around 100 m^3/ha, with lower values in northern latitudes and higher values in some areas of the temperate zone. Exceptions are found in tropical forests, where decay rates are very high and, as a consequence, relatively little dead wood may accumulate (Harmon et al., 1995; Grove, 2001). The average decay rate constant in the central Amazon forest is 0.17, corresponding to a 95% mass loss over 18 years (Chambers et al., 2000). The other extremes are found in temperate rain forests in northwest America. Here, very high productivity and growth of trees is combined with the dominance of rot-resistant, slowly decaying conifer species. Extreme values have been reported from *Pseudotsuga–Tsuga* forest in the Pacific Northwest, where volumes vary between 400 and 1000 m^3/ha (Harmon et al., 1986). Also, temperate beech forests (*Nothofagus* spp.) may have large

Table 12.4. *Compilation of a few selected published reviews reporting volumes of dead wood in natural forests.*

Forest type and region	Dead wood volume (m^3/ha)	References
Boreal Fennoscandia	20–120	Siitonen (2001)
Boreal Europe	60–80	Hahn and Christensen (2004)
Temperate Europe	130–250	Hahn and Christensen (2004)
Coniferous Pacific NW USA	60–1200	Harmon et al. (1986)
Tropical forests Mexico	40–120	Harmon et al. (1995)
Tropical forests Australia	20–45	Grove (2001)
Tropical forests Venezuela	5–80[a]	Delaney et al. (1998)
Nothofagus New Zealand	*c.* 100[a]	Hart et al. (2003)
Temperate *Nothofagus* forests	up to 800	Stewart and Burrows (1994)

[a] Recalculated from biomass estimates.

volumes, ranging from 100 to 800 m^3/ha (Stewart and Burrows, 1994; Hart et al., 2003).

The relation between dead volumes and living volumes varies between forest systems, but normally seems to fall within 10–40% of total volume or biomass. A general pattern seem to be that the fraction of dead wood is lower in tropical areas (see, e.g., Delaney et al., 1998; Houghton et al., 2001) than in boreal and temperate forests (see, e.g., Siitonen, 2001; Hahn and Christensen, 2004).

12.5.2 Spatial distribution of dead wood

The spatial distribution of dead trees represents a colonization challenge for saproxylic species. We discuss this in more detail in Chapter 14, when dealing with population dynamics. Here we stress that the abundance of dead wood may vary spatially, both within single stands and at the landscape scale, and that this variation influences the occurrence of saproxylic species.

The spatial variation within stands is highly dependent on the scale of resolution in the studies. In a Karelian (westernmost Russia) old-growth forest, the coefficient of variation decreased strongly when the size of sample plots increased from 0.01 to 0.20 ha, reflecting a small-scale variation in dead wood volumes (Karjalainen and Kuuluvainen, 2001). This is supported by other studies which also show that in both spruce and pine forests in boreal Fennoscandia, downed logs tend to be aggregated up to scales of 20–40 m (Edman and Jonsson, 2001; Rouvinen et al., 2002b). To what extent this small-scale aggregation may influence the colonization and establishment of saproxylic species is unclear, but results indicate that small-scale processes can indeed influence, at least, the colonization of wood fungi (Jönsson et al., 2008; see also Chapter 14). However, when increasing the scale to several hectares, the variation in most old-growth forests is relatively low, i.e. dead wood usually occurs abundantly at the scale of one to a few hectares under natural conditions (see, e.g., Jonsson, 2000).

At larger scales, between different forest stands, the variation in dead wood becomes higher. This relates to several interacting factors such as stand productivity, tree species composition, topographic position, successional stage and chance (Kennedy and Spies, 2007). To some extent, the average volumes are predictable from stand productivity, mortality rates and decay rates (see above). However, as the forest landscape is heterogeneous, the abundance of dead wood will vary across the landscape. In a chronosequence study of forest stands in Newfoundland, with ages ranging from 33 to 110 years, Sturtevant et al. (1997) estimated a range between 15 and up to almost 80 m^3/ha. Siitonen (2001), in a study based on modelling results, suggested a range between 30 and 600 m^3/ha in stands of different time since forest fire. This illustrates the fact that successional stages at the landscape scale may include variation in dead-wood volumes up to an order of magnitude. Accordingly, the presence of high-quality dead-wood habitats may vary strongly, even in natural landscapes.

12.5.3 Diversity of wood types

In previous chapters we described the intricate and complex relations between different types of dead wood and the associated saproxylic species. This emphasizes the importance of natural forests, as they maintain this diversity of wood types. Although forest harvesting creates pulses of dead wood and open habitats that are quite similar to those created by stand-replacing disturbances, it is not only the volume of dead wood that differs between managed forests and natural forests, but also the range of dead wood types. At least five important gradients of variation in wood types need to be considered.

1. In natural forests, dead wood input is relatively continuous, guaranteeing the local presence of dead trees at all stages of decay. Although

mortality rates vary depending on tree size and age (Fraver et al., 2008), natural forests tend to produce dead trees of all dimensions.

2. As discussed above, there are a variety of mortality agents involved in tree deaths, each providing specific types of substrates for the saproxylic species.
3. These substrates then undergo a decay succession, the different decay stages each hosting partly different communities of saproxylic species.
4. Natural forests often include a larger set of tree species compared with managed stands, which are often monocultures.

These four types of variation – *tree size*, *substrate type*, *decay stage* and *tree species* – constitute the main dead-wood diversity gradients. They may be viewed as a multidimensional space that sets the niches available (Figure 12.6).

5. On top of this variation is the role of environmental conditions, exposure to sun, humidity, etc., which typically show more spatial variation in natural forests than in managed forests.

In total, the ecological dimensions translate into an almost endless number of combinations and niches – probably a key driving force for the species diversity associated with dead wood (see Chapter 11

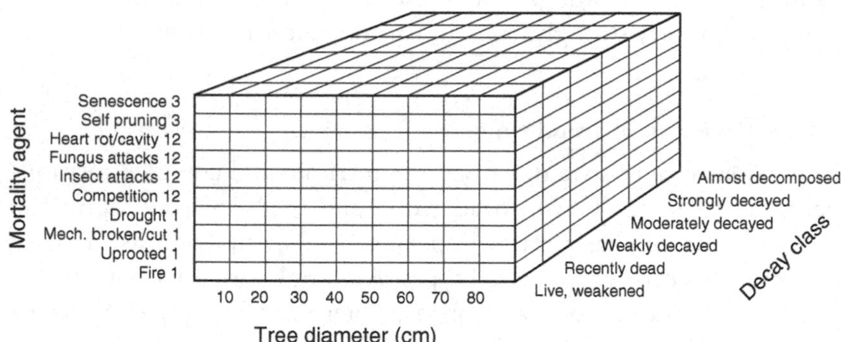

Figure 12.6. Diversity of dead wood categories, from a single tree species, as a combined result of diameter class, mortality agent and decay stage. The numbers following the mortality agents indicate (**1**) mainly abiotic factors (cutting by beavers or man are biotic, though), (**2**) interaction between abiotic weakening (most often drought) and subsequent biotic colonization, and (**3**) internal physiological processes in the tree. More than one number may apply to each agent.

for a detailed discussion). Thus, we strongly stress that not only are volumes of dead wood in natural forests higher than in managed stands, but perhaps even more important is the higher diversity of dead-wood types.

The level of naturalness in old-growth forests can be evaluated by the comprehensiveness of dead wood sizes and decay stages. Stokland (2001) suggested the use of the coarse woody profile, based on tree sizes and decay stages, as a means to evaluate naturalness and continuity in boreal forests. This approach represents an important starting point to quantify dead wood diversity. By including the additional dimensions of tree species, mortality agents and environmental conditions, a more comprehensive description of the dead wood resource is possible. A quantitative approach to describing dead wood diversity, focusing on substrate availability for saproxylic species, is given by Hottola et al. (2009).

13 · *Dead wood and sustainable forest management*

Bengt Gunnar Jonsson and Juha Siitonen

This chapter presents various forestry practices, from modest selective timber harvesting to plantation forestry. Particular attention is paid to differences in the amount, quality and dynamics of dead wood between managed and natural forests. The chapter also discusses management options and processes that might improve conditions for saproxylic species.

Many kinds of forestry practice exist, but it is beyond the scope of this chapter to give a full treatment of all aspects. Common to all types of forestry is the fact that trees are cut and removed from the forest. This clearly represents a situation of resource competition for the species dependent on dead wood. In many regions, clear-cutting and removing all the trees is the most common harvesting method and has obvious negative effects on many saproxylic species (Figure 13.1). But other management regimes also show negative effects on the dead wood biota. Balancing the extraction of wood with the demands of saproxylic species is a difficult task – a topic that is discussed in the second half of this chapter.

13.1 Amount, quality and dynamics of dead wood in managed forests

The most obvious effect of forestry is the extraction of trees for commercial purposes. This will result in a loss of dead wood resources for saproxylic species, in terms of quantity, quality and dynamics.

13.1.1 Dead wood volume

The volume of dead wood is generally lower in managed forests (Table 13.1) than in natural forests (Table 12.4). Typically, the volume in intensively managed forests is less than 10% of comparable types of natural forests (see Siitonen, 2001). This difference is linked to the

Figure 13.1. Well-managed forests are characterized by single cohorts of trees, evenly spaced and thinned to a density that allows maximum growth rates. Although economically valuable, these forests include only a small range of habitats for saproxylic species (photo Erkki Oksanen/Metla).

stand history as well as the intensity and type of current forest management. In regions with a long management history, the dead-wood volumes are generally lower than in regions where forestry has started more recently (Fridman and Walheim, 2000; Krankina et al., 2002; Rouvinen et al., 2002a; Webster and Jenkins, 2005). In the latter case, remnant snags and logs remaining from the unmanaged stage may be present to a larger extent, while these are lost due to the reduced input of new dead wood where forest management has a longer history. Furthermore, the differences in dead-wood volume between managed and natural forests depend also on the successional stage, being greatest early in the succession, after stand-replacing disturbances, and smallest in mature forest exceeding the normal rotation times (Duvall and Grigal, 1999).

Existing volumes also reflect management attitudes to dead wood. Management practice has gone through a number of stages in many regions. Initially, selective logging for particular tree species and qualities produced significant waste in the form of large-diameter logging residues and by increasing the mortality in the remnant stand. However, as the demand for timber and pulp increased, other resources also became valuable, which further intensified the extraction of wood

Table 13.1. *Compilation of selected published reviews providing average volumes of dead wood (with a minimum diameter of 10 cm) in managed forests.*

Region, type of forest	Dead wood volume (m^3/ha)	References
Sweden, boreal-temperate	6.1	Fridman and Walheim (2000)
Finland, boreal	5.4	Ihalainen and Mäkelä (2009)
Fennoscandia, boreal-temperate	4–10	Stokland et al. (2003)
Britain, temperate	< 20[a]	Kirby et al. (1998)
France, temperate	2.2	Vallauri et al. (2003)
France, maritime pine plantations	6.1	Brin et al. (2008)
Switzerland, temperate – oroboreal	8.9	Bretz Guby and Dobbertin (1996)
Switzerland, temperate – oroboreal	20–30[b]	Böhl and Brändli (2007)
Slovenia, temperate silver fir–beech forest (ecosystem-based management)	40–65 < 20% of natural stands	Debeljak (2006)
Pan-European assessment	normally < 25	Travaglini et al. (2007)
Russia, European parts and western Siberia, boreal-temperate	14–20	Krankina et al. (2002)
Australia, tropical lowland forests	20–30[c]	Grove (2001)

[a] Minimum diameter 5 cm, [b] minimum diameter 7 cm, [c] minimum diameter 7.5 cm.

and reduced the amount of structural legacies (Figure 13.2). In parallel with the view that dead wood was a wasted resource, another common idea was that dead trees represent a threat to the forest health. Dead wood was viewed as a sign of bad management and constituted a risk of pest outbreaks and other diseases. Thus, regulations are often in place, specifying how much dead wood is acceptable. Currently we have two opposing trends influencing dead wood volumes. On one hand, the appreciation of its importance is increasing, and the retention of dead wood and old trees is common. On the other hand, the extraction of biomass for energy is putting new pressure on the wood resources, and the harvesting of tops, branches and stumps is becoming more widespread.

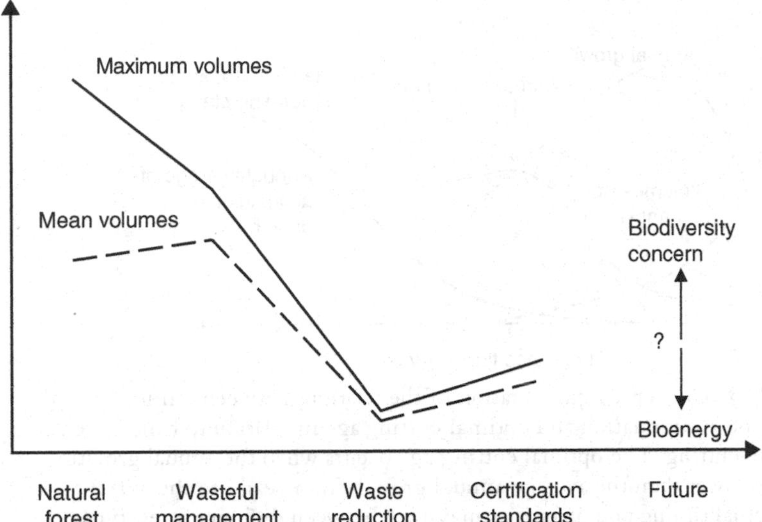

Figure 13.2. Changes in dead wood amounts during different stages of forest management history. During the early stages of forestry, only specific tree species and qualities of timber were harvested, which left relatively large amounts of dead wood as 'waste'. With time, this resource became usable raw material and was extracted either for economic purposes or to limit the risks associated with pest species. Current forest management and certification standards may include a higher degree of acceptance of dead wood, but still represent a major reduction compared with natural levels. In the future, the pressure for even further extraction may increase, as tops, branches and stumps are increasingly being used for bioenergy purposes (based on Harmon, 2001).

The combination of increased utilization, waste reduction, assumed pest management and biofuel harvesting has led to a situation with usually more than a 90% reduction in the available volume of dead wood. This is clearly not compatible with the long-term maintenance of the species richness associated with dead wood (see Box 15.1).

As well as an increasing fraction of trees being removed during logging, the rotation times in managed forests are significantly shorter than in natural forests. Forests are being cut at the stage where the annual volume increase of commercial timber is starting to level off. This age varies according to climatic region and site productivity, but is generally low compared with the forest age at which the accumulation of dead wood becomes significant. This means that trees are cut at the stage where the volume of dead wood is actually lowest during

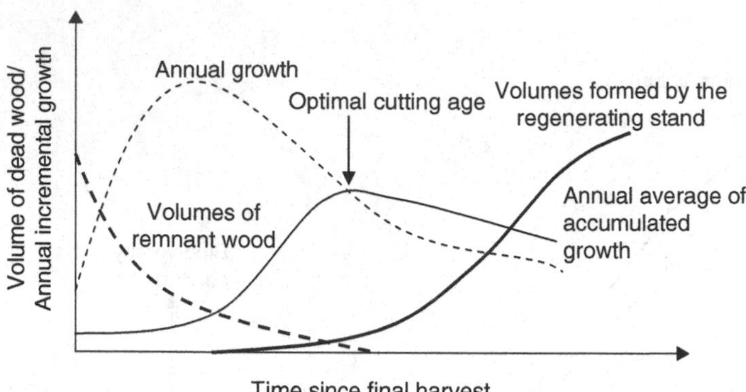

Figure 13.3. Conceptual presentation of the relations between remnant wood, dead wood accumulation and optimal cutting age in relation to time since last clear-cutting. The optimal cutting age occurs when the annual growth becomes lower than the average annual growth from year 1 to the current year. Actual timing and quantities may vary between different forest types, but generally logging maturation age, when stands are normally cut, is reached while stand mortality is still low and when most remnant logs have decayed.

the succession after a stand-replacing disturbance. Managed forests are normally not allowed to reach ages high enough for significant volumes of dead wood to have started to accumulate (Figure 13.3).

13.1.2 Dead wood quality

In addition to the sheer loss of volume, forest management often also brings about changes in the quality of available dead wood (Table 13.2). Decay stage distribution as well as size distribution is sensitive to cutting (Kruys et al., 1999; Storaunet et al., 2005). In particular, large-diameter logs and snags in advanced stages of decay are lacking in managed forests (Siitonen et al., 2000). As smaller logs tend to decay more rapidly (see Chapter 6) and become overgrown by ground vegetation, logs in later stages of decay also decrease. As we have shown earlier (Chapters 5, 9 and 12), many species require specific types of dead trees, and thus a given volume of dead wood in managed forests may not represent the full range of required qualities in the same way that a similar volume in a natural stand would.

Management implies even further impacts on the quality of dead wood. In many managed forests, production is favoured by selecting and planting tree species in monocultures. This causes less variation

Table 13.2. *Changes in wood quality lost in boreal managed forests as compared with natural forests.*

Degree if loss
Dead wood from over-mature trees
Large-diameter decorticated snags (kelo trees)
Severely reduced in abundance
Strongly decayed logs
Large-diameter snags and logs
Fire damaged and killed trees
Pre-rotted trees
Dead wood from slowly growing trees
Maintained with slight reduction or about the same abundance
Small-diameter, fine woody debris
Dead wood in open exposed habitats
Recently dead, medium-sized snags and logs

between sites in the quality of dead wood, and also frequently involves the introduction of exotic species which may not serve as habitat for the native biota. More subtle are the effects of management that attempt to increase stand production, which results in more fast-growing trees. As discussed in Chapter 9, wood density and the relation between early and late wood may affect the success of different saproxylic species. The effects of fast growth rates and well-maintained even-age stands are further magnified as trees are cut at a fairly early age. Old, damaged, senescent and slowly growing trees are thus not developed in managed forests. This contrasts with trees in natural forests, which demonstrate a much greater heterogeneity.

13.1.3 Dead wood dynamics and continuity

Finally, even when significant volumes of dead wood are allowed to accumulate in managed forests, previous management measures might have resulted in a break in the continuous supply of habitats for saproxylic species. There is an ongoing discussion on the importance of such continuity. The question is whether continuity *per se* is important or whether it is just currently available volumes that dictate the species richness and composition in a forest (Stokland, 2001; Rolstad et al., 2004). The answer to this question depends on the dispersal ability of the species in question. Given the large number of saproxylic species, it is reasonable to assume that some species do have some limited capability of long-distance dispersal. Such species may need local continuity of suitable substrates (see Chapter 14). The continuity of dead wood in a stand can be described either by detailed analysis of forest history or, more generally, by examining the abundance of dead wood at different stages of decay and of different sizes (Stokland, 2001).

13.2 Forest management regimes

Forests are managed in different ways, depending on the commercial demand for forest products and the ecological conditions in the forests in question. Below we present some aspects concerning broad scales of forest management systems. The focus is on the relation with dead wood abundance; it is not our aim to review forest management schemes in detail.

13.2.1 Selective cutting: continuous-cover forestry

Historically, commercial logging has been directed towards particular tree species, dimensions or other specific qualities which were merchantable for some specific purpose. In such systems, a large fraction of the stand was left intact and, in general, the forests maintained some of their natural characteristics and integrity. However, this harvest also had significant and long-lasting effects on the abundance of dead wood. For example, even 100–150 years after selective harvesting, dead wood volumes and composition have not fully recovered in Swedish boreal spruce forests (Jönsson M. T. et al., 2009), and in Norwegian spruce forests, current volume and quality of dead wood were significantly correlated with levels of harvest 50–100 years ago (Groven et al., 2002; Storaunet et al., 2005).

Currently, so-called 'continuous-cover forestry' is practised in many different forest types (Pommerening and Murphy, 2004; Raymond et al., 2009). This includes a range of different management systems ranging from selection of specific trees to systematic extraction of a fraction of all diameter classes. In some forest types, selective logging is considered to be an economically viable alternative to clear-cutting. Sometimes it is done to protect hydrology and prevent erosion and landslides. In other situations it is applied for aesthetic reasons and to promote recreational use of the forest. Here the value of the forest for people seeing and spending time in nature is valued more highly than the timber itself.

There is ongoing discussion concerning the economic potential of continuous-cover forestry as compared with traditional rotation management. Depending on the specific forest system and, perhaps more important, the valuation of different ecosystem services, the viability may vary. Sometimes even the removal of the largest most valuable stems can be economically feasible as well as representing sustainable forest harvesting. By assigning some economic value to the environmental benefits, this has been suggested as an alternative to clear-cutting, e.g. in British temperate forests. In a study of monocultures of Sitka spruce in Wales, Price and Price (2006) suggest that transformation into more complex stands was successfully achieved by selectively logging the largest and most valuable trees.

In boreal conifer forests, clear-cutting is the prevailing harvest method. To explore the alternatives, Lähde et al. (2002) performed a 12-year experimental study where 35–75% of the growing volume was harvested using six different cutting alternatives. The results showed that the diversity of dead wood was reduced only marginally compared with unmanaged old-growth forest and that the effect on the tree layer also was limited in some of the treatments (e.g. group selection and mixed tree shelterwood). It should be noted, though, that the study was relatively short term and thus did not address effects on the long-term input of dead wood.

The application of continuous-cover forestry in various forms seems to be growing. A feature common to all these management systems is that they maintain a closed tree canopy and thus a shaded and relatively moister microclimate than clear-cut stands. For drought-sensitive species this may be beneficial, as long as sufficient volumes of dead wood are retained. The shaded conditions will indeed favour some species groups such as lichens, bryophytes, fungi and some insect groups (see Chapter 9). However, as the richness of saproxylic species correlates primarily with the abundance of dead wood, the extensive extraction of trees, even when maintaining a closed canopy, implies loss of habitat for the saproxylic species. With repeated and efficient removal of trees, volumes of dead wood may be as low as – or even lower than – those following a stand-replacing harvest. In broad-scale analyses of the pros and cons of continuous-cover forestry in Sweden, a major conclusion was that, although beneficial for a range of biodiversity values (e.g.

mycorrhizal communities, epiphytic lichens, resident birds), the value for saproxylic species may be rather limited (Cedergren, 2008).

In tropical forests, selective logging of the most valuable trees is the rule rather than an exception. However, a major concern is that the logging activities cause extensive disturbance to the remaining tree canopy (Putz et al., 2000). Thus, the problems associated with selective logging in tropical areas are rather the opposite to that of coniferous forests – logging increases mortality and creates a surplus of dead wood, which is not compatible with long-term maintenance of biodiversity and forest integrity (Bawa and Seidler, 1998; Asner et al., 2006). This may result in secondary forests that are not suitable for the original saproxylics, despite the initial abundance of dead wood. So-called 'reduced impact logging' might offer a sustainable way of using forests in tropical regions. For instance, Keller et al. (2004) found that conventional logging produced 270% more dead wood than reduced impact logging approaches. But even with reduced impact harvest management, logged tropical forests have 50% more dead wood than undisturbed forests (Palace et al., 2007).

13.2.2 Clear-cutting

The dominant management regime in most forests today is clear-cutting, also called rotation management (Figure 13.4). Here the whole stand is harvested, while some limited retention of buffer zones or groups of trees may or may not occur. It is considered to be the economically most profitable management system, enabling efficient regeneration with selected tree species and a choice of suitable plant material. At the same time, there continues to be extensive criticism concerning the negative effects of clear-cutting on biodiversity. Again, there is an obvious trade-off between retaining trees and maintaining forest cover or extracting them for other uses. Any extraction, and especially clear-cutting, will lower dead wood volumes. Compared with other management options, however, clear-cutting is not necessarily the worst. The main issue is instead the level of retained trees, both living and dead.

There are potentially strong and positive effects of clear-cutting on the abundance of many insects and fungi associated with dead wood in open habitats. Clearly some species utilize the open, sun-exposed clear-cuts in the same way as the other early successional stages after a stand-replacing disturbance such as fire, provided that adequate qualities of

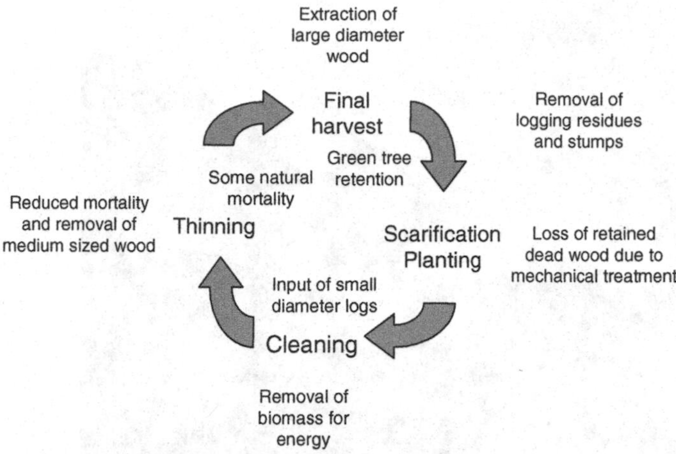

Figure 13.4. A generalized rotation management scheme showing the losses of dead wood outside the circle and additions of dead wood inside the circle.

dead wood are present. Such species may thus have their main distribution in the managed forest landscape on clear-cut areas. Especially in intensively managed landscapes with successful fire suppression, clear-cut areas may be the only early-successional forests available.

Although clear-cut areas can serve as habitat for potentially rather many saproxylic species, there are important limitations to consider. A forest fire does not normally consume more than 10% of the trees and available dead wood, while 95–98% of the volume is normally extracted during clear-cutting (Angelstam, 1996). Another problem is that some species clearly favour, or are even dependent on, the burnt wood itself (Chapter 6). Compared with fire, clear-cutting neither provides damaged trees, (e.g. fire-scarred pines) and, consequently, nor does it provide the full range of dead wood types typical after natural disturbances.

13.2.3 Whole-tree and logging residue harvesting

Global climate change is a major concern, and therefore bioenergy is increasingly being used to replace fossil fuels. This may result in the extraction of even more dead wood from forests (Rudolphi and Gustafsson, 2005; Framstad et al., 2009). Whole-tree harvesting (cutting and skidding the whole trees with their branches and leaves), as well as the harvesting of logging residues and stumps after clear-cutting, is

Figure 13.5. The need for alternatives to fossil fuels increases biomass extraction from forests. Harvest residues have thus become a commercial product, resulting in further losses of dead wood from the forest landscape (photo Erkki Oksanen/Metla).

increasing in many forest types. This means that not only the stems are removed during harvest, but also tree tops, branches, twigs and even stumps, all of which are left in traditional clear-cutting (Figure 13.5). This practice may have long-term effects on stand productivity and nutrient dynamics, although the effects vary between forest types (Egnell and Valinger, 2003; Mariani et al., 2006; McLaughlin and Phillips, 2006). Unfortunately, only limited knowledge is available so far on the effects on saproxylic species. Inventories show that, for instance, many saproxylic beetle species utilize these fractions of the tree, and some species tend to prefer the finer fractions of dead wood (Schiegg, 2001; Kappes and Topp, 2004; Jonsell et al., 2007). Among ascomycete fungi, the species richness is high on branches and twigs (Nordén et al., 2004b). Given the importance of tree tops, branches and twigs for many saproxylic species, the removal of these substrates may further decrease their populations. In addition, partial cutting combined with whole-tree harvesting of young trees reduces species richness of wood-decay fungi (Nordén et al., 2008).

Harvesting of cut stumps is also increasingly being practised. Cut stumps may be an even more important substrate for saproxylic species than logging slash in managed forests, since stumps are often the only large-diameter dead-wood substrates available. In Swedish pine forests, the remaining stumps after a final harvest may correspond to up to 20–25% of the volume harvested (Dahlberg et al., 2005). So far, rather limited knowledge is available on the role of these stumps for saproxylic diversity. Stumps have been shown to provide important habitats for rare epixylic lichens in managed forests (Caruso et al., 2008, Nascimbiene et al., 2008; Caruso and Rudolphi, 2009).

In a recent review, Framstad et al. (2009) point out some important implications of bioenergy harvesting for species associated with dead wood.

1. Whole-tree harvest further increases the contrast between managed forests and natural forests. This may restrict both the dispersal possibilities of species and the potential of managed forests to serve as permanent or transient habitats for saproxylic species.
2. Most studies addressing the effects of whole-tree harvest have been done in forests and landscapes that were strongly influenced by previous forest management. Thus the effects on especially demanding species may have been missed, since they were already lacking.
3. During the extraction of tree tops, branches, twigs and stumps, there is a risk that existing coarse dead wood may also be extracted or destroyed by machinery.

13.2.4 Salvage logging

Dead wood resulting from natural disturbances tends to be harvested, by virtue of perceptions of efficiency (not wasting resources) and for aesthetic reasons. So-called salvage logging is required by forest regulations and laws in many countries. Such waste reduction policies may be motivated by economic, social and safety reasons after large-scale disturbances. However, the ecological effects tend to be mostly negative (Lindenmayer and Noss, 2006; Lindenmayer et al., 2008). The salvage procedure in itself causes additional damage to the forest and may transform a relatively natural state (succession after a stand-replacing disturbance) to a heavily modified condition. Regarding the dead wood, salvaging obviously aims to reduce the volume below some predefined level. As we have shown earlier (Chapter 9), the lack of dead wood in open post-disturbance conditions is potentially a shortage that is as important as the loss of old-growth conditions.

Snags that are formed during large-scale disturbance events, and fresh snags in open habitats, often host a wide range of saproxylic species (see Chapter 9). This suggests a large negative impact of salvage logging of snags after fire and wind disturbances. As an example of the effects of salvage logging, Hutto and Gallo (2006) studied logging after a forest fire in Montana, USA. They showed that out of 18 cavity-nesting birds, only eight nested in the salvage-logged plots. They also showed that population densities of these eight species were generally higher in the unlogged fire areas. Similarly, Schroeder (2007) argues that the retention of snags created after wind storms and subsequent bark beetle attacks is a cost-efficient way of increasing dead wood volumes, as the risk of epidemic outbreaks of bark beetles is limited and the economic value of damaged trees is low. Müller et al. (2008a) showed that gaps created by a European spruce bark beetle (*Ips typographus*) outbreak had a higher species richness of saproxylic beetles and hosted more endangered saproxylic beetles than the adjacent closed forest.

13.2.5 Plantation forestry

Plantation forestry is the most intensive forest management system. It generally includes extensive site preparation, the use of exotic species or genetically improved plant material, the control of weeds, fertilizing, thinning, short rotations, and finally clear-cutting (sometimes whole-tree harvesting).

Compared with other management systems, plantation forestry entails truncated – or even eliminated – successional stages and significant changes in stand structure. As the focus is to maximize the production of commercial timber, such forests do not facilitate the natural death of trees and thus host few habitats for saproxylic species. Also the common use of exotic tree species implies that only generalist saproxylic species would be able to utilize the woody debris in plantations.

13.3 Sustainable forest management: background

Modern forestry increasingly recognizes that it has to deal with multiple values – not just economic but also social and ecological values. A central concept is sustainable forest management (Anon., 2003). This aims to ensure a long-term flow of societal values from forests by protecting ecosystem integrity, while also considering the ecological values that forests provide. It calls for balancing the need for raw materials to the forest industry with other values associated with forests, such as biodiversity and its protection. As a large proportion of the biodiversity in forests is associated with dead trees, the importance of considering this component is generally well recognized.

In the first half of this chapter we gave a broad overview of the main forest management systems and to what extent they are compatible with the protection of saproxylic species. Here we reiterate that the most critical aspects, regardless of management system, are

- the extent to which the existing dead wood remains, and new dead wood develops, after harvest;
- the extent to which the full range of different wood qualities is maintained.

As most management systems fail in some of these respects, this results in a lower and more uniform input of wood from one, or very few, tree species, and wood with a relatively high and similar growth rate. Much of the variation required by saproxylics (as depicted in Figure 12.6) is thus lost.

13.4 Disturbance regimes and forest management systems

A fundamental basis for forest biodiversity is the various disturbance regimes (see Chapter 12) that shape the structure of forest stands and landscapes in terms of tree species distribution and the occurrence of various successional stages. This implies that in order to be successful, sustainable forest management should to some extent mimic the patterns created by natural disturbances (Angelstam, 1996; Bergeron et al., 1999; Franklin et al., 2002; Kuuluvainen, 2002, 2009). In this context, it becomes important to distinguish broad types of disturbance regimes in forests and compare these with currently used management systems (Shorohova et al., 2009). At least three broad categories of dynamics can be recognized:

1. *successional dynamics* after a stand-replacing disturbance;
2. *cohort dynamics* in sites where old and mature trees survive fire;
3. *gap-phase dynamics*, when stand-replacing disturbances are rare.

Currently clear-cutting is the dominant forest management system which emulates, in some respects, successional dynamics after a stand-replacing disturbance. Cohort dynamics may be emulated by extensive green-tree retention at harvest, while gap-phase systems are not compatible with clear-cutting but can be emulated by various systems of selective cutting. Management systems emulating gap-phase dynamics must rely on continuous-cover forestry, and methods on how to transform even-aged stands to uneven-aged stands managed with continuous-cover forestry are being developed (e.g. Larsen and Nielsen, 2007).

The extent to which a better match between forest management systems and natural disturbance regimes translates to the protection of saproxylic species is, however, linked to levels of live tree and dead wood retention. A good match would imply environmental conditions (e.g. shade and moisture) suitable for the focal species. However, irrespective of the management system, if volumes of dead wood are kept low, saproxylic species will not benefit.

13.5 Retention

A first step to moderate the effects of forest harvesting is to leave at least some living and dead trees after harvest. The idea of structural retention, i.e. providing legacies from the old stand into the new one (Hansen et al., 1991; Franklin et al., 1997) has become a widespread conservation principle. Three main purposes of structural retention (including green-tree retention and the retention of dead trees) can be enumerated (Franklin et al., 1997).:

1. 'Life-boating' of species and processes over the regeneration phase; i.e. species present in the old stand do not disappear following logging, but at least part of their populations survive in the stand and can increase later in the course of succession.
2. Structural enrichment of the re-established stand with structural elements that would otherwise be absent.
3. Enhancing connectivity in the managed landscape.

Life-boating is directed to species that are already present in the old stand (continuous occupancy), whereas structural enrichment is meant to provide habitats for new species colonizing the re-established stand (often temporary occupancy). Structural enrichment can have both short-term and long-term effects. This practice has the potential to shortcut a critical phase during stand development when the new stand has not yet started to produce large-diameter dead wood (see Figure 13.3). From the point of view of saproxylic species, the important short-term effect is that disturbance-adapted species can colonize and breed in dead and dying trees that are left in clear-cut areas. This effect is very strongly substantiated with empirical studies, mainly from the Nordic countries. The combined effects of retention and prescribed burning have also been studied, and the results of these studies are described in Section 13.8.2 (Restoration fires). The long-term effects of structural legacies cannot be seen until some decades after the retention/cutting event, when the live trees retained after cutting have grown larger and eventually died, thus producing large-diameter dead wood that would otherwise be missing from the new stand. Because there are no experiments on green-tree retention that would be old enough, we do not yet have study results of the long-term effects of structural retention on biodiversity in general, and on saproxylic species in particular. There seem to be no empirical studies about the potential role of structural retention in enhancing the connectivity of managed landscapes either (Rosenvald and Lõhmus, 2008).

13.5.1 Green-tree retention

Leaving living trees at harvest is increasingly being adopted in clear-cut operations (Vanha-Majamaa and Jalonen, 2001; Rosenvald and Lõhmus, 2008). The retained trees may constitute habitats for species already present during the early stages after succession. This also includes saproxylic species, since mortality among the retained trees is relatively high. In two studies of retained aspens which had died during or after logging, it was found that large numbers of both saproxylic beetles (Martikainen, 2001) and polypore species (Junninen et al., 2007) utilized the retained trees, including a large set of red-listed species. These sun-exposed dead-wood habitats represented important structures that differed from dead aspens in closed forests.

The size of retention groups can vary widely, from single trees to groups of trees and small remnant forest stands (Franklin et al., 1997; Rosenvald and Lõhmus, 2008). Even single large, over-mature trees, such as old oaks, may have the potential to act as important life-boats for specialized saproxylic species (Ohsawa, 2007). The idea of saving patches of the old forest within the new stand is an extension of single-tree retention. The patches can provide a large range of important structures in the developing stand, including a variety of dead wood substrates. Their functionality in life-boating forest-interior species over the regeneration phase may be limited, since they are strongly influenced by edge effects (Vanha-Majamaa and Jalonen, 2001; Rosenvald and Lõhmus, 2008; Aubry et al., 2009). Note, however, that there appear to be no experimental results concerning the persistence of saproxylic species in retention patches. In a long-term study, Jönsson et al. (2007) showed that short-term tree mortality was very high in patches ranging from one-sixteenth of a hectare (0.06 ha) up to 1 ha. Therefore, such patches were all considered too small to maintain conditions for species demanding closed canopies, but they could be very useful for disturbance-adapted saproxylic species by providing an extended input of large-diameter dead trees in open conditions. In a North American study, about half of the retained living trees were alive and standing 10–18 years after harvest, while the other half had been topped or toppled by wind, or had become snags or logs (Busby et al., 2006). In boreal Canada, small forest remnants in clearcuts had a higher density of snags and higher species richness of saproxylic beetles than the interiors of large patches (Webb et al., 2008).

Early in the succession, the amount and quality of pre-existing dead wood is more important for saproxylic species than the level of green-tree retention (Jacobs et al., 2007; Junninen et al., 2008). In the longer term, it is crucial that the retained volumes are high enough to have a significant effect on the recruitment rate of large-diameter dead wood. This suggests that, to be efficient, the level of green-tree retention should be higher than what is normal (generally $<10\ m^3/ha$) in current forest management systems.

13.5.2 Snags and logs

The retention of already dead trees during forest harvest does not constitute a direct economic loss. With a growing awareness of the biodiversity value of dead wood, snags and logs are often left in harvested areas. An associated problem, however, is that a large proportion of the retained logs may be damaged during forest operations, both during harvest itself and during subsequent soil preparation. This loss may be significant, and Hautala et al. (2004) showed that as much as two-thirds of the initial pre-harvest dead wood was lost after soil scarification.

13.5.3 High stumps

During the last few decades, high stumps and snags have regularly been created during forest management. As visual representatives of new

Figure 13.6. Retention of high stumps during forest harvest is increasingly done in many regions and forest types. These stumps provide habitat for species adapted to open and warm microclimate, including many saproxylic beetles. As decay softens the wood, high stumps may become suitable for cavity-nesting birds (photo Erkki Oksanen/Metla).

ways of thinking, these stumps stand out like exclamation marks in harvested areas (Figure 13.6). Although they were initially mostly intended to provide nest sites for cavity-nesting birds (Hallett et al., 2001), their role as a habitat for other saproxylic species has been increasingly studied. Studies show that in many cases the high stumps actually do harbour significant numbers of saproxylic species that are otherwise rare in harvested areas (Lindhe and Lindelöw, 2004; Lindhe et al., 2005; Schroeder et al., 2006; Abrahamsson et al., 2009). In addition, threatened and red-listed species may utilize this partly artificial habitat. It has also been shown that high stumps host more species – and clearly different species assemblages – than ordinary cut stumps (Abrahamsson and Lindbladh, 2006). When high stumps are produced in harvesting, the whole top parts or pieces of the trunks can be left at the site as artificial logs. Such cut logs have been shown to host largely similar fungus assemblages as comparable windthrown trees (Lindhe et al., 2004).

13.6 Forest reserves

By setting aside forests for protection, the direct effect of forest harvesting is removed. Obviously this provides a situation where dead-

Figure 13.7. Large decorticated pine snag, a kelo tree, is an example of a dead wood substrate with a very long recruitment time. These are steadily declining and mostly occur in protected areas. Thus, they represent a challenge to both management and restoration (photo Bengt Gunnar Jonsson).

wood-associated species can thrive (Figure 13.7). In different countries, depending on the natural setting, history of forestry and local traditions, the system for area protection varies. In sparsely populated regions where modern forestry has not yet transformed the forest landscape, there might still exist large tracts of intact natural forests (Potapov et al., 2008). In these regions, such as parts of boreal Russia and Canada, it is possible to set aside entire landscapes of 10–1000 km^2

(DellaSala et al., 2001). Only in very large areas can natural forest dynamics and disturbance processes operate and provide all the needed variation in dead wood without human interference. Wind storms, forest fires and flooding (see Chapter 12) create pulses of dead wood and younger successional forests. Other areas might escape stand-replacing disturbances and develop into old-growth forests where old trees and trees dying through senescence are abundant.

In regions with a long land-use history, forest management has affected most of the landscape and only smaller areas (10–100 ha) of natural or close to natural forest fragments are left. Here protected areas serve an important role as well. Even if they individually are too small to include the full range of natural variability, they can collectively provide much of the needed variation in dead-wood habitats. Although traditionally left for free development, the protected area network in a human-dominated landscape requires stronger planning to ensure that all the important habitat types are represented, and active restoration measures may be needed.

13.7 Woodland key habitats

In addition to the reserve-sized areas, smaller patches (0.1–10 ha) of valuable habitats may occur embedded in managed forest landscapes. Such habitat patches are to an increasing degree set aside as so-called 'woodland key habitats' (WKHS), which are seen as important components of sustainable forest management in the Nordic and Baltic states (Timonen et al., 2010). They represent habitats where red-listed species can be expected to occur, and as such they include many sites rich in saproxylic species. It is, however, unclear to what extent such habitats maintain viable populations of late seral species. As WKHs are, on average, small forest stands, they are subject to edge effects (Aune et al., 2005). Furthermore, their small size means that they can host only small populations of most species, which implies a high risk of local extinction due to chance events (Hanski, 2005). WKHs have in many cases been subjected to previous forestry management, and thus are not true remnants of natural forests (Jönsson M. T. et al., 2009; Siitonen et al., 2009). Yet, in a situation where the remaining landscape is strongly modified and composed of young and middle-aged stands regenerated after forest harvesting, these sites may represent important hotspots of landscape biodiversity. Regarding saproxylic species, the fact that WKHs have a higher average volume of more variable dead wood than ordinary managed stands (Jönsson and Jonsson, 2007) is obviously an important factor contributing to the higher species richness in WKHs (Junninen and Kouki, 2006; Djupström et al., 2008; Hottola and Siitonen, 2008). Thus, WKHs may also be seen as cost-efficient starting points for further conservation or restoration efforts in predominantly managed landscapes and may serve as life-boats for certain species that are confined to older forest stands.

13.8 Restoration

In regions with a long history of forest management, the remaining habitats may be insufficient for the long-term survival of saproxylic species. This calls not only for setting aside valuable stands and adjusting management practices, but also for actively restoring habitats by manipulating the availability of dead wood.

In this respect two principles may be adopted. First, restoration may aim to restore sites back to their natural condition. A good example is the use of forest fire as a method of recreating multi-cohort pine stands with a high fraction of dead standing snags. Secondly, restoration can create habitats and substrates that are lacking in the larger landscape, and certain sites can be managed so as to increase their desirable qualities, regardless of what might be 'natural' for that particular site.

13.8.1 Felling of trees

An obvious way to increase saproxylic habitats is simply to kill trees. In situations where the current stand is too young to produce dead wood, active management may shortcut the long-term succession by increasing tree mortality. Thus, felling large trees and the artificial creation of snags may be done beyond what is considered natural. To be efficient, such management requires a well-planned choice of suitable sites, considering stand conditions and landscape context. Although potentially controversial, such management may greatly improve the situation for species facing an immediate risk of extinction.

Possibilities for creating dead wood vary considerably, depending on the kind of dead wood that is needed. Some types, such as fresh dead wood, can easily be created by cutting or girdling (Figure 13.8) or by applying prescribed fires. Other types, such as highly decayed logs and kelo trees (Figure 13.7) will take decades or even centuries to form, even if restorative management is carried out.

Figure 13.8. Active restoration in the form of a girdled pine tree. This will secure the presence of pine snags in relatively young and developing stands (photo Bengt Gunnar Jonsson).

There are also more drastic methods of killing trees. Blowing them up with dynamite to create fragmented snags is practised in several countries including Scotland (Abernethy Forest) and the USA (Oregon and Washington). This is not a completely new idea, since it has been practised to some extent for 30 years (Bull et al., 1981). Logs can be created, together with high stumps, by cutting, and snags can be created by girdling the tree at the base (Figure 13.8) or by topping the tree (Hallett et al., 2001), or even in combination with the inoculation of wood-decay fungi (Brandeis et al., 2002). Snags created in this way, by removing the tops of large conifers, provide nesting and foraging structures for cavity-nesting birds (Walter and Maguire, 2005) as well as for species utilizing the log on the ground.

13.8.2 Restoration fires

Controlled fire has been used as a forest management tool to reduce fire hazard risks and to improve regeneration. As clear-cutting does not capture all aspects of forest fires (McRae et al., 2001; Bergeron et al., 2002; Kuuluvainen, 2002), the need to reintroduce fire in managed forests is increasingly recognized. With a reasonable level of retention in combination with prescribed fires, several species groups may benefit. In a study on the importance of restoration fires and green-tree retention, Hyvärinen and co-authors showed that retention of as low as 10 m^3/ha had positive effects on the beetle fauna associated with dead wood. Two years after the prescribed fire, the numbers of all species (as well as red-listed and rare species) had increased about 50% compared with untreated controls, when tree retention was at least 10 m^3/ha (Hyvärinen et al., 2006, 2009). In a similar study, Toivanen and Kotiaho (2007b) also showed strong positive effects of prescribed fires in spruce forests on the abundance and richness of rare saproxylic beetles. Number of individuals was about three times larger on burnt sites compared with unburnt. Burned trees in fire areas also constitute the only suitable habitat for a group of pyrophilous species (see Chapter 6). Both experimental studies above showed that several pyrophilous species were able to utilize dead wood created at sites subjected to restoration fires, despite the fact that they were relatively small sites compared with most natural fires, and a large proportion of the trees were harvested before burning.

The previous studies only covered the short-term effects of fire and retention. The effects of different retention levels are likely to become evident later in the succession, when the retained trees produce varying amounts of coarse dead wood. In a retrospective study, Toivanen and Kotiaho (2007a) showed, firstly, that both the abundance and the species richness of saproxylic beetle species were positively affected by burning, and the effect of fire lasted about 20 years. Secondly, the difference between burned and unburned sites increased with the number of retention trees, and the effect of burning was not significant when there were fewer than about 15 trees per hectare left.

Concerning wood fungi, the effects of forest fires are more complex than for beetles. Immediately after a restoration fire, the number of fruiting species declines (Penttilä and Kotiranta, 1996; Olsson and Jonsson, 2010). The reasons behind this decline are not well known, but for many species it may simply be an effect of reduced fruiting during the first years after fire. In another study (Junninen et al., 2008), burning did not decrease the number of polypore species on pre-existing dead wood. At a longer time perspective, though, fire seems to promote the species richness of wood fungi. In a follow-up study after prescribed burning of two unlogged forest stands (100% retention), Penttilä (2004) showed that the number of species first decreased following the fire. It took 6 years for the species number to recover to the pre-disturbance level. Thirteen years after the fire, the number of species had increased over 40%, and the number of red-listed species had doubled.

Fires are also important for several groups of birds, and fire suppression is considered a threat to woodpecker occurrence in certain regions (Murphy and Lehnhausen, 1998). So far, few studies have analysed the specific effects of prescribed fires on woodpeckers (Pope et al., 2009). However, as fire increases the abundance of bark beetles and woodborers (Covert-Bratland et al., 2006) and the availability of tree cavities for cavity nesters (Hutto, 1995), prescribed burning should also improve habitat quality for cavity-nesting birds.

Although forest fires are generally positive for the biodiversity of dead wood, they also consume a portion of the dead wood resources that were available before the fire (Knapp et al., 2005). In natural conditions this is compensated for by the production of a large volume of dead trees during the fire. However, when prescribed burning is used after clear-cutting (to improve forest regeneration and decrease fire hazards), the consumption of available dead wood may represent a loss of habitat and an additional threat to saproxylic species. In a study of three different restoration fires addressing the consumption of pre-fire dead wood (Eriksson et al., submitted), the fires consumed between 20% and 45% of the existing volume of dead wood. There was also a strong relation to decay stage, as logs at higher decay stages were more consumed (Figure 13.9).

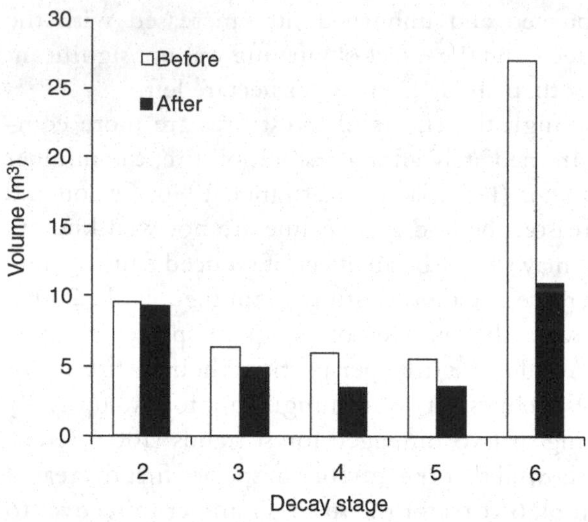

Figure 13.9. Example of consumption of dead wood during a prescribed fire in a boreal conifer forest (data from Eriksson et al., submitted). Decay stages range from recently dead trees to decay stage 6, where wood is soft, log outline is deformed and large pieces are starting to fall off. The consumption was on average 43% of the volume but as high as 59% at decay stage 6.

13.9 Management for dead wood

Forest management has been termed 'silviculture' when referring to tending the forest for the production of timber and pulp. As a parallel, the term 'morticulture' was introduced by Mark Harmon to denote management where the focus is also directed towards understanding its effects on dead wood. In his paper, Harmon (2001) states that

> As with silviculture, it [morticulture] would meet future needs, but instead of the type of logs to be harvested, it would deal with the methods to produce woody detritus structures for ecosystem function.

Furthermore, he stresses that its implementation should be in conjunction with silviculture and not in isolation.

It should be obvious from several of the chapters in this book that a key requirement for saproxylic diversity is linked to the diversity of dead wood (see Figure 12.6). Thus, morticulture needs to explicitly address not only volumes of dead wood but, perhaps even more importantly, its variety. In this respect, morticulture is perhaps even more demanding than traditional silviculture, which focuses on a very limited set of tree qualities (generally on the production of saw timber and pulpwood of a few preferred tree species).

Morticulture would need to include several components: (1) the links between living and dead trees, (2) the dynamics of the dead wood pool, (3) planning of the spatial distribution of dead wood both within stands and at the landscape scale, and (4) monitoring responses in the biota and ecosystem functions.

13.9.1 How much is enough?

A critical question for both retention and restoration is how much dead wood is needed. Although the question is simple to ask, the answer is very complex. The difficulty lays both in choosing the actual target (individual species, species richness or composition) and the complex relations between species occurrence and the different aspects of dead wood availability. Ideally, single threshold values should be available, but empirical as well as theoretical studies suggest that the use of single values may risk missing the demands of particular species (Ranius and Fahrig, 2006). In a recent study, Müller and Bütler (2010) reviewed 36 different European studies where thresholds for dead wood volumes were suggested. Their review showed that the derived threshold ranged between 10 and 80 m^3/ha in boreal and lowland forests and between 10 and 150 m^3/ha in mixed montane forests, but with common values in boreal forests of around 20–30 m^3/ha, in mixed montane forests around 30–40 m^3/ha, and in lowland oak–beech forests around 30–50 m^3/ha. Although these values are derived from relevant empirical studies, they do not necessarily include the full range of dead wood volumes and merely represent snapshots of presence/absence patterns. It is therefore difficult to provide definite recommendations based on available information, especially since the available studies mostly ignore the effect of landscape scale and the potential for transient dynamics, e.g. the extinction debt (see Chapters 14 and 15).

13.9.2 Stand models and dead wood dynamics

An obvious way to apply morticulture is to model dead wood dynamics, given different forest management systems. Stand development models aim to predict the growing volume of commercial timber and are used to select thinning and cutting regimes. By extending these models into taking the mortality of trees into account, allowing for the retention of trees and dead wood, and applying decay models (see Chapter 6) it is possible to construct integrated models that predict volumes of dead wood. Several such models have been developed (Tinker and Knight, 2001; Ranius et al., 2003; Wilhere, 2003; Ranius and Kindvall, 2004; Hynynen et al., 2005; Montes and Canellas, 2006). For example, Ranius et al. (2003) made predictions on the increase of dead wood in Swedish boreal forests managed according to forest certification criteria. They showed that volumes were expected to stabilize to around 12 m^3/ha viewed over the entire rotation period, and reaching up to 20 m^3/ha in mature stands before final cutting (Figure 13.10). This is about twice as much dead wood as these forests contain today.

These kinds of models have the potential not only to provide average values of dead wood volumes, but also to predict variation between stands and the distribution of dead wood in various size and decay classes. In addition, they enable the assessment of the effects of individual management prescriptions (such as green-tree retention) at varying intensities on dead wood. This opens up the possibility of modelling the availability of dead wood over time, taking different management options and their combinations into consideration. Such models provide an important tool for predicting habitat availability and for exploring the potential for saproxylic species to survive in man-

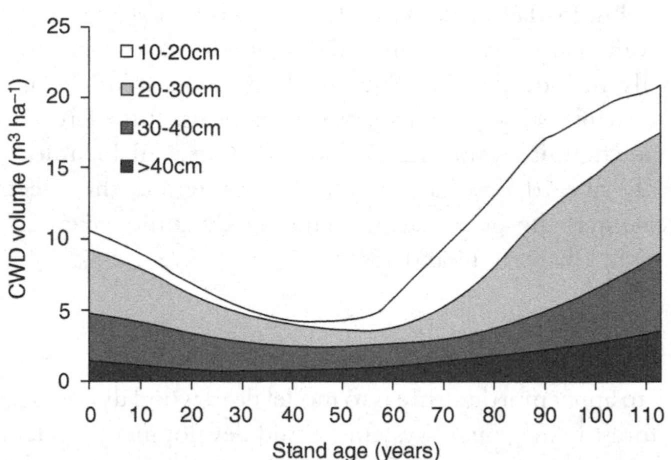

Figure 13.10. Result from a simulation model predicting dead-wood volumes (CWD = coarse woody debris) in managed forests. Figure shows average (1000 replicates) amount of CWD (m^3/ha of stems with a diameter larger than 10 cm), divided into different diameter classes. Parameter values are typical for biodiversity-oriented forestry (based on Ranius et al., 2003).

aged forest landscapes (Jonsson et al., 2005; Jonsson and Ranius, 2009; Sahlin and Ranius, 2009).

13.9.3 The landscape scale

Although individual species occur on dead trees within stands, their long-term viability is related to their population dynamics at larger scales. We explore the population dynamics in Chapter 14 in some detail, and suggest that forest management must consider the landscape in order to achieve long-term goals. By providing dead-wood substrates in managed forests, the contrast between protected areas and managed areas is decreased. This could help in mediating dispersal between natural forests as well as increasing population sizes of the species, i.e. increasing the metapopulation capacity of the landscape (Hanski and Ovaskainen, 2000). Clearly, activities carried out throughout the managed forest landscape will influence a much larger proportion of the forest area than those measures that are taken only in protected areas.

The dynamic nature of the habitats of saproxylic species suggests that many species may occur as metapopulations at the landscape scale (see Chapter 14). This may be especially true for those species which demand sites with higher volumes of dead wood, and which are thus restricted to patches of natural forests. This calls for modelling the long-term viability of species at the landscape scale. A further step in modelling is to integrate the dead-wood models with habitat suitability models of individual species, or with dynamic metapopulation models. Based on the latter approach, Schroeder et al. (2007) explored the population viability of a saproxylic beetle (*Harminius undulatus*) and showed the importance of understanding the dynamics of dead wood availability at the landscape level to correctly predict the occurrence and viability of the species. Some other attempts in this direction have also been made (Gu et al., 2002; Laaksonen et al., 2008). However, so far, a full application of metapopulation modelling to guide conservation management for saproxylic species seems to be lacking. In a recent review, Jonsson and Ranius (2009) listed data requirements for such an approach.:

- Preferably the species occurrences in the landscape should be known and provide the starting point for simulations of population development.
- Knowledge about which dead wood qualities the species utilize is required.

- For each forest stand in the landscape, a measure of habitat quality is needed. This measure should preferably be empirical data on species–habitat associations, but could also be a crude division of the forests into suitable and unsuitable habitats.
- To capture landscape dynamics, models that describe the change in habitat quality of individual stands over time must be available, including dead wood volume and quality, and other relevant stand characteristics (e.g. basal area, tree species composition).
- The species dispersal rate and distances, including its variability over time and among different sections of the landscape.
- Local persistence may be directly connected to habitat persistence, with local extinctions being deterministic when the habitat is lost. Alternatively, more complex models are needed that take into account local population dynamics, stochastic events, and rescue effects from neighbouring populations.
- Finally, the occurrence of time lags in colonization and age of first reproduction needs to be considered. This also includes survival and continued dispersal from patches no longer suitable for colonization. For long-lived species, these aspects may have a profound role to play in the dynamics at the landscape scale.

This is obviously data-demanding, but it should be possible to conduct such modelling for a set of well-chosen focal saproxylic species. This would allow the testing of various forest management scenarios at the landscape scale. As with most modelling exercises, the output may not provide exact predictions, but would facilitate exploring the relative benefits from different forest management options. Combined with other analyses, including economic, it may constitute part of a multi-objective planning system (Baskent, 2009).

13.9.4 Monitoring

Nature is unpredictable, and even the best management ideas, supported by empirical data and good models, may fail in practice (Villard and Jonsson, 2009b). This uncertainty should be addressed by an adaptive management system. Here, monitoring of the outcomes of the chosen management is a central activity. It is, of course, crucial to estimate the direct effects of the management on the availability for saproxylic habitats, i.e. the quantity and quality of dead wood should be measured. However, for a powerful morticulture, the effect of the chosen management system should be assessed on the saproxylic species themselves. Given the complexity of the landscape context and the potential effects of land-use history on the occurrence of saproxylic species (see Gu et al., 2002; Penttilä et al., 2006; Laaksonen et al., 2008), it is by no means certain that viable populations of target species will appear even when the right habitat quality is present.

13.10 Conservation goals and management standards

So where do we go from here? Above we have presented established and emerging management options that should be integrated with sustainable forest management. Although these approaches might help in maintaining saproxylic biodiversity, their incorporation into forest management rests upon other larger issues beyond pure biology and ecology. Below we discuss some of these issues that will support the implementation of sustainable forest management.

13.10.1 International policy

Global agreements to protect biodiversity are strong, as more than 190 countries worldwide have signed the Convention on Biological Diversity (CBD). The goals of the CBD were confirmed at the Johannesburg World Summit in 2002, where it was declared that the loss of biodiversity should be 'significantly reduced by 2010'. This ambition has been reiterated in national and regional strategies worldwide, and the new post-2010 targets established at the CBD meeting at Nagoya in October 2010 further emphasize the global commitment. In Europe, the ambition has been even more demanding, stating that the loss of biodiversity should be halted by 2010, and in accordance with this, the EU-level Malahide Declaration from 2004 declares the goal for European forests is 'to conserve and enhance biodiversity through sustainable forest management at national, regional and global levels'. This is further developed within the European Birds and Habitats Directive, which states that species and habitats listed should have 'favourable conservation status' (Anon., 1992). This clearly includes the importance of maintaining dead wood, as several of the listed species are saproxylic and because the occurrence of dead wood is a natural component of the forest habitats.

Other international processes consider forest biodiversity to a surprisingly small extent. In documents provided by United Nations Forum on Forests, biodiversity is hardly mentioned and only in connection with the CBD. This exemplifies alternative views of forests; some see forests as a sustainable source of resources, while others also see the problems associated with the large-scale conversion of natural forest ecosystems into silvicultural production systems.

This separation between the idea of forests as a primary resource base and the importance of protecting global biodiversity becomes obvious when studying economic incentives to forest management. There are many examples where national and international funding have stimulated activities that are harmful for biodiversity. These so-called 'perverse incentives' are an important topic for global discussion on biodiversity protection. In Fennoscandia there have been state subsidies for building roads into remote and low-productive forest areas. These roads have made it possible to harvest forests of high natural value: forests that otherwise would not have been economically profitable to harvest. In Norway, economic support is given to logging in steep terrain (cable crane extraction), which would be non-profitable without support.

13.10.2 Forest certification

Forest certification is based on the voluntary participation of forest owners in a system regulating how forestry is performed. Certification standards normally address all aspects of sustainable forestry (economic and social as well as ecological aspects), but here we only consider the extent to which they impact on dead wood.

Globally, the most renowned certification system is that of the Forest Stewardship Council (FSC; www.fsc.org). The FSC is an international organization which provides accreditation for national certification systems that comply with a general framework for sustainable forest management. These national standards are developed in a dialogue and negotiations between forest stakeholders, i.e. forest owners, environmental organizations, labour union representatives, local residents, indigenous people, etc. The standard is a consensus agreement and has, in many countries, influenced everyday forest practices. Dead wood is not mentioned explicitly in the general framework, but it is clearly an implicit aspect that should be considered (Box 13.1). Even though dead wood is not mentioned in the general framework, given that many saproxylic species are 'rare, threatened and endangered' and that they depend upon a proper management, many national standards address dead trees (see Box 13.1).

When properly enforced, a credible certification system is expected to provide forest owners with monetary compensation for their conservation efforts. On an open market, the certified products should be able to sell better and at a higher price. So far little effort has been made to evaluate the tangible ground-level effects in forests after the introduction of forest certification. To our knowledge, the only study that has evaluated forest management before and after the introduction of certification standards is that of Sverdrup-Thygeson et al. (2008). They concluded that tree retention levels have increased significantly, but that for many other field-level indicators (e.g. damage to existing logs, buffer zones along streams, and terrain damage) no improvement has been shown. It therefore remains to be seen how much voluntary approaches can improve habitat conditions for saproxylic and other forest species.

Box 13.1 Dead wood in forest certification standards

Below are selected sections from the certification standards set by the Forest Stewardship Council (FSC) that are specifically important for saproxylic species and dead wood.

The following text is included in Principle 6 of the FSC General Framework.

6.2. Safeguards shall exist which protect rare, threatened and endangered species and their habitats (e.g., nesting and feeding areas). Conservation zones and protection areas shall be established, appropriate to the scale and intensity of forest management and the uniqueness of the affected resources. Inappropriate hunting, fishing, trapping and collecting shall be controlled.

6.3. Ecological functions and values shall be maintained intact, enhanced, or restored, including: a) Forest regeneration and succession. b) Genetic, species, and ecosystem diversity. c) Natural cycles that affect the productivity of the forest ecosystem.

Box 13.1 *(continued)*

The UK FSC standard includes provisions for dead wood by stating:

In addition to these 'biodiversity areas', dead-wood habitats are being increased throughout the forest by:

- Retaining dead wood that reflects the sizes and species of tree present on the site and is matched to the requirements of species likely to be important on the site.
- Keeping 'snags', 'hulks', dead trees or those containing dead-wood habitats standing. (An average density of 3 standing and 3 fallen stems per hectare across the forest as a whole would be an appropriate minimum target.)
- Not harvesting windthrown stems, with the exception of logs of particularly high value or where >3m^3 per hectare is blown down.
- Retaining some fallen trunks or logs after each harvesting operation.
- Concentrating dead wood in areas where it is likely to be of greatest value – e.g. in shaded locations near to pre-existing dead wood.
- The amount, state and type of dead wood and lop and top retained is modified to accommodate any public safety or plant health constraints.

In Sweden the standard is even more detailed.

6.3.4S. Managers shall retain all snags, windthrows and other trees that have been dead for more than 1 year except when they:

a) constitute a safety risk for forestry workers or for the general public within recreation areas,
b) block up frequently used paths and roads,
c) constitute small-dimension felling residues,
d) constitute breeding substrate for pest insects in case there is a documented risk of mass propagation.

6.3.5S. Managers shall retain all snags, windthrows and other trees that have been dead less than 1 year:

a) that originate from trees with high biodiversity values (6.3.18) or other trees previously retained for nature conservation purposes,

b) in areas set aside for nature conservation, including care-demanding patches,
c) on low/non-productive forest land with an annual increment less than 1 cubic metre per hectare.

6.3.6S. Managers shall retain, on average, at least two coarse new windthrows per hectare when harvesting windthrown stems on final felled areas (in addition to windthrows addressed in 6.3.4S and 6.3.5S).

6.3.7S. Managers shall create, on average, at least three high stumps or girdled trees per hectare of areas harvested through regeneration felling and thick-stem thinning, striving to select for this purpose equal numbers of coarse pine, spruce, birch and aspen trees without high biodiversity values.

13.10.3 Acceptance and awareness

Unfortunately a number of myths about dead trees abound, which translate into threats. For extended periods, forest management has seen dead trees as a sanitary problem. One can still find reference to the old German expression '*Forsthygiene*' or its translations. The idea of being 'clean' has caused extensive efforts to remove wounded, dying and dead trees from the forest. Leaving these has been seen as bad forest management. Related to this is the worry about pests and diseases as a consequence of dead trees in forests. Fungi such as *Heterbasidion annosum*, *Armillaria* spp. and various bark beetles have been seen as serious threats to wood resources. Most often the outbreaks of these species have been associated with particular management regimes or serious disturbances that provide massive amounts of available habitat in restricted areas. Actually there are three main reasons for the spectacular outbreaks of pests in managed forests:

1. Management regimes creating optimal conditions for pests; e.g. the main reason for *I. typographus* outbreaks in central Europe is a consequence of the large-scale introduction of spruce into lowland areas that were formerly covered by beech-dominated forests;
2. Exceptional disturbances and/or drought conditions;
3. Introduced pest species, an example being the expanding dieback of pine forest in the Iberian Peninsula caused by the introduced pest, the pinewood nematode.

In natural forests, mass outbreaks of pest species are rare, and under normal conditions it is safe for forest owners to leave dead trees as part of their management strategy.

Another deeply rooted fear is the idea that dead trees pose a risk to visitors in protected areas. However, by keeping the main trails free from hanging dead wood, forest visitors should not be subject to risks. The potential to stumble, fall or have other accidents is, of course, increased whenever visitors take their own routes outside the trails, but this is hardly a reason to remove valuable components of forest reserves. Generally the greatest risk in visiting natural forests is connected to travelling to the area by car.

Dead trees have also been considered to increase fire hazards. This is clearly the case when considering slash after clear-cutting, where prescribed fires are commonly applied to reduce the risk of wildfire. In closed forests, however, the main fuel is in the ground vegetation and live trees rather than scattered dead logs and snags in the stand – forests burn under dry conditions, not due to dead wood.

Finally, dead wood is sometimes seen as a free resource to use as firewood, since it has no commercial value to the forest owner. Despite conservation intentions, dead wood may therefore be extracted by local residents as it is considered a free resource. This is definitely an issue in central Europe and elsewhere in densely populated areas, but also in other regions. For instance, the Australian Government (Department of the Environment and Water Resources) launched a programme in 2001 that aimed to inform local residents of the importance of dead wood for biodiversity. Their slogan was 'Logs have life inside – Are you burning their homes to warm yours?'. We acknowledge, however, that firewood extraction is an important part of subsistence in many rural Third World areas and needs special attention when addressing forest biodiversity conservation (see, e.g., Christensen and Heilmann-Clausen, 2009).

13.10.4 Multi-stakeholder involvement in forest management

A challenge for the future is to better integrate the multiple interests in our forests. Different groups, such as land-owners, local residents, tourist companies, NGOs, the forestry industry, all have their own opinions and interests in the forest and its resources. It is impossible to combine all these interests at every point in the forest landscape. Therefore, there is a need for a governance system that takes all these interests into account and, as far as possible, finds an optimal solution at the landscape scale. Traditionally it has been the domain of laws and official authorities to provide a legal and administrative framework for this. However, alternative governance models might be as effective, and more bottom-up control can potentially provide a better integration than traditional top-down control.

An interesting approach to a bottom-up governance model is represented by the increasing number of 'model forests' established around the world (www.imfn.net). A model forest represents an ongoing collaborative partnership between different stakeholders in a particular forest landscape. The aim is to develop a truly sustainable forest management, taking all stakeholders' interests into account. One example is the 800 000 ha Komi Model Forest in Russia (www.silvertaiga.ru/en/). Here the explicit goal is to improve the management of old-growth forests that are endangered by unsustainable forest practices, and also to encourage forest managers in regenerated sites to consider their environmental values and services.

14 · *Population dynamics and evolutionary strategies*

Bengt Gunnar Jonsson

A particular challenge for saproxylic species is that they live in ephemeral habitat patches ('sinking ships'): in dead trees that decompose and gradually vanish. The substrate units they inhabit will inevitably disappear, and they then need to colonize new suitable substrate units. This means that the reproductive success of an individual (and the fate of the local population) depends not only on its ability to reproduce in a suitable host tree now, but also on the availability of suitable host trees in the future. Different types of dead-wood substrates are highly variable with respect to their abundance and persistence time, and therefore contrasting life-history strategies and dispersal abilities have evolved among saproxylic species.

Using the arguments from the well-known evolutionary ecologist, Southwood, the conclusion is that the 'habitat is the life history template' (Southwood, 1977; Figure 14.1). This statement emphasizes the central problem that all species face: how to track suitable habitat given its distribution over time and space. In the most fundamental sense there are two basic questions: whether the individual should reproduce *here or somewhere else* and whether it should reproduce *now or later.* This connects strongly with change in habitat quality over time and space and how predictable this change is. Given the 'sinking ship' situation for saproxylic species, selection for efficient dispersal is obviously strong in most cases. In this chapter we briefly outline some of the basic issues on population dynamics, with a specific focus on the life histories of saproxylic species.

14.1 Life-history strategies

Many categorizations of life-history strategies have been proposed. One of the more well known refers to the trade-off between reproductive capacity and competitive ability, the so-called *r*- and *K*-selection strategies (Pianka, 1970). The former strategy concentrates on producing

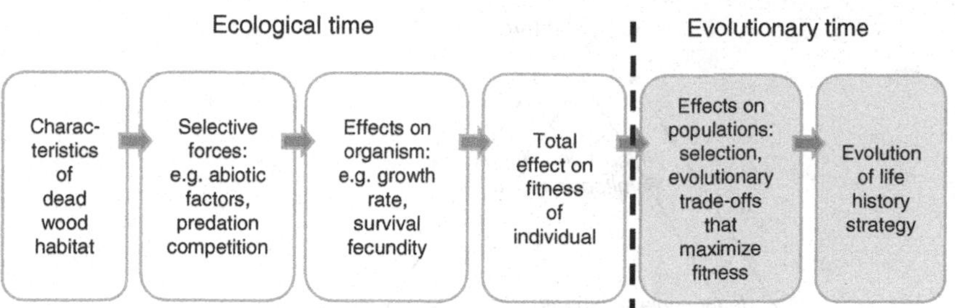

Figure 14.1. Depiction of the role of the habitat and interactions in shaping the ecological strategies of saproxylic species. Based on Southwood (1977).

abundant, small-sized offspring with high dispersal ability, whereas the latter strategy concentrates on producing fewer large-sized offspring with high competitive ability. The *r*-selection strategy is typical of species living in unpredictable, short-lived habitat patches in early successional stages, whereas the *K*-selection strategy is typical in stable conditions, often in late successional stages. In practice, there is a continuum of different life-history strategies between the archetypal *r*- and *K*-selection strategies.

Saproxylic succession normally follows a relatively predictable sequence (see Chapter 6). The species which establish early and slowly colonize the dead and dying tissues of still living trees face a stressful environment, with the tree defence mechanisms restricting access to resources. When the tree dies, ruderal species with good dispersal and establishment abilities are the first group of species to appear. Eventually more species colonize, resources become scarcer, and highly competitive species take over. Finally, when most of the easily decomposed fractions of the log have been consumed, again stress-tolerant species, although different from the initial set of species, become dominant.

A relevant classification was provided by Grime (1979) who focused on the relative importance of *stress*, *disturbance* and *competition*, resulting in three main categories of species. Although Grime originally presented his system with a focus on plants, it is also applicable to other groups of organisms. The idea is that in stressful environments, abiotic conditions limit growth and reproduction, while significant disturbance physically removes or kills individuals. No life strategy can cope with both severe stress and serious disturbance. By contrast, in stable environments with favourable abiotic conditions, competition will be more intense, selecting for competitive ability among the species. This

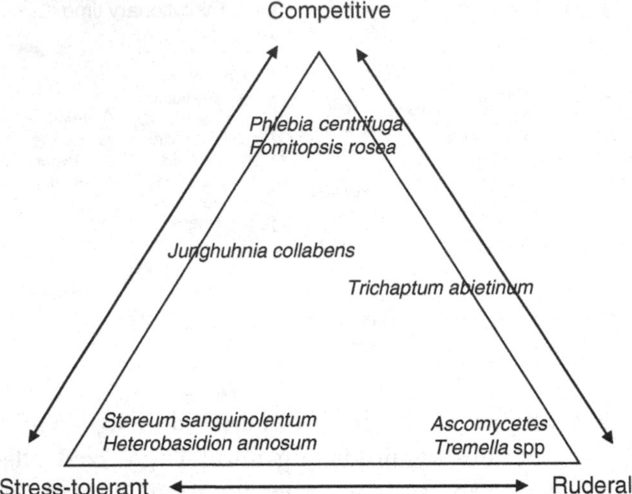

Figure 14.2. The three main strategies (competitive, stress-tolerant and ruderal) are best viewed as gradients between extremes. Fungal examples are based on the interpretation of experimental data and statements in Holmer (1996).

classification has been successfully applied to saproxylic fungi, which can broadly speaking be grouped into three main categories – stress-tolerant, ruderal and competitive species (Pugh, 1980; Rayner and Boddy, 1988; Boddy and Heilmann-Clausen, 2008; Figure 14.2).

Stress-tolerant fungi have evolved adaptations to adverse conditions and thus are able to utilize dead wood which, for various reasons, is uninhabitable for most other species. The adverse, stressful conditions in wood may include very dry wood or highly fluctuating moisture conditions, low nutrient content or resin-impregnated wood.

Ruderal fungi exploit newly formed, short-lived types of dead wood. Their adaptations allow them to be the first occupiers of the place and to utilize the empty and free habitat. This requires a strong investment in reproduction and dispersal. The newly dead tree is nutrient-rich and includes easily degradable organic compounds (see Chapter 3).

The competitive species are those that arrive at a later stage and may utilize less easily degraded organic compounds (i.e. cellulose, hemicellulose, lignin). As the first species lose some of their resource base, these later arrivals will slowly expand and eventually dominate the community. Competitive interactions between saproxylic fungi have often been demonstrated, and in several cases suggest that species form competitive hierarchies (Holmer and Stenlid, 1997; Wald et al., 2004b).

14.2 Factors affecting the population dynamics

14.2.1 Dispersal ability

As their habitat is transient in time, dispersal ability is a key factor in the population dynamics of saproxylic species. This includes short-distance dispersal between different dead trees within a single stand as well as dispersal between suitable stands within a more or less fragmented landscape. However, the relative importance of dispersal varies between species and seems to be linked to the temporal and spatial availability of their habitat. For example, among species living in very stable saproxylic habitats, such as hollows of living oaks (e.g. the beetle *Osmoderma eremita*), the majority of the individuals remain in their natal tree, and dispersal distances appear very short (Ranius, 2006; see Box 7.1). Fire-dependent insects such as the jewel beetle *Melanophila acuminata* are able to fly for extended time periods during the summer in search of fire areas suitable for regeneration (Wikars, 1997a). This species has large flight muscles and small ovaries and hence sacrifices reproductive potential for dispersal ability. By contrast, the closely related species *Phaenops formanecki* has small flight muscles but large ovaries. This species occurs in forested bogs and rarely has to move long distances to find a suitable breeding substrate (Wikars, 1997a). These species represent different solutions in a trade-off between investing in dispersal or reproduction and clearly show how the habitat dictates dispersal ability.

The variation in dispersal ability may also influence occurrence patterns and abundance of species. An illustrative example is the two beetle species *Bolitophagus reticulatus* and *Oplocephala haemorrhoidalis*. Both species live in the fruiting bodies of *Fomes fomentarius*, a common polypore on broadleaved trees. *O. haemorrhoidalis* is a rare and threatened species in northern Europe, while *B. reticulatus* is relatively common. In a study of their dispersal biology, Jonsson (2003) showed that although both species could potentially fly several kilometres, *B. reticulatus* flew for longer time periods due to its developed wing muscles, and in experiments it was more likely to take off. In addition, *O. haemorrhoidalis* produces fewer and larger eggs – a characteristic that limits its colonization potential.

Another example is provided by Jonsson and Nordlander (2006), who placed empty fruiting bodies at different distances from an old-growth forest rich in dead wood. The number of colonizing beetles was counted and identified. In general, most species showed no

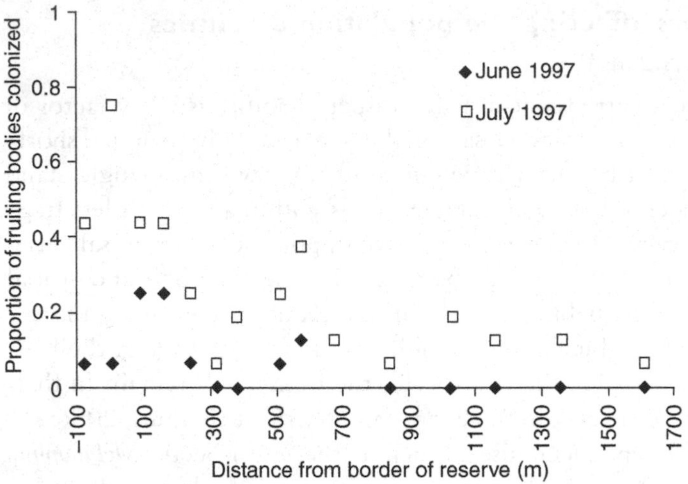

Figure 14.3. The proportion of *Fomitopsis pinicola* fruiting bodies colonized by the fungivorous beetle *Cis quadridens* at varying distances from an old-growth forest (after Jonsson and Nordlander, 2006).

marked decline in abundance at distances up to 1600 m from the old-growth forest. This indicates that many saproxylic beetles have good dispersal ability at scales of a few kilometres. However, for two species, the fungivorous beetle *Cis quadridens* (Figure 14.3) and the predatory fly *Medetera apicalis*, they observed a clear decline in abundance with increasing distance from the source habitat. *C. quadridens* is rare in managed forests and is included on the Swedish Red List (Gärdenfors, 2010). Both species are examples of species at higher trophic levels, which tend to have smaller populations than species at lower trophic levels. The results thus support other studies indicating that species at higher trophic levels are more sensitive to fragmentation and habitat loss than those utilizing the primary energy source, the wood itself (Komonen et al., 2000).

The range in spore sizes among fungi, bryophytes and lichens suggests large differences in dispersal potential. Wood-inhabiting fungi produce large numbers of very small spores and appear to have relatively good dispersal ability. Spores of old-growth species such as *Fomitopsis rosea* and *Phlebia centrifuga* also occur in significant numbers in landscapes with low amounts of old forests (Edman et al., 2004a). This indicates a potential for long-distance dispersal. Studies on long-distance dispersal for bryophytes and lichens are mostly lacking. However, given that these species often have significantly larger spores than wood-inhabiting fungi, and that some rely solely on large asexual diasporas, it is likely that some species may have limited dispersal ability. As an example, species in the lichen genus *Hypocenomyce* occur mainly on burned pine snags, a persistent substrate that may last for centuries. In this genus, several species rely mainly on large asexual soredia for dispersal, while spore production is very rare (Foucard, 2001).

Some species of epixylic lichens and bryophytes even show dual dispersal modes in order to balance the need for long-distance dispersal between forest stands and local dispersal between suitable dead trees within a stand. The hepatic *Anastrophyllum hellerianum* produces both sexual spores and larger asexual structures, gemmae. Gemmae may potentially contribute to long-distance dispersal but their main role is apparently to maintain local populations, while spores, besides providing sexual recombination, serve for long-distance dispersal (Pohjamo et al., 2006). It has also been shown that, among epixylic hepatics, species with predominantly asexual reproduction tend to occur in a more aggregated pattern within a stand than species with a predominance of spore dispersal (Laaka-Lindberg et al., 2006).

14.2.2 Establishment

The successful colonization of a dead tree includes both dispersal and establishment. Establishment relies on the location of a suitable substrate unit and gaining a foothold. This means that the character of the substrate, competitive interactions and abiotic conditions need to be considered. In general, we know surprisingly little about this. Occurrence patterns in terms of suitable decay stages, host-tree species and forest conditions have been documented (Chapters 5, 6 and 9). However, these relations are likely to indirectly reflect some more specific conditions that must be fulfilled, as – for example – a particular moisture availability, chemical composition of the wood, and temperature requirements.

To locate suitable dead trees, some insects use chemical compounds as cues, collectively called kairomones (Allison et al., 2004). The emission of ethanol and other volatile compounds from the dying cambium is a well-known attractor for bark beetles (Montgomery and Wargo, 1983; Schroeder and Lindelöw, 1989; Byers, 1995). Similarly the smoke from fire and the smell from charred trees are cues for several fire-adapted species (Evans, 1971). However, the pheromones emitted by many beetle species to attract individuals of the opposite sex are an even more intricate way to establish a population (Evans, 1971; Wood, 1982; Wertheim et al., 2005). This guarantees the efficient location of mates and subsequent reproduction in suitable substrates.

The establishment of some fungi is by specific vectors. For instance, the fungus *Amylostereum areolatum* is transferred to new host trees by woodwasps of the genus *Sirex*. A testimony to the efficiency of this interaction is the devastating effect of the fungus in *Pinus radiata* stands in Australia and New Zealand since the late 1940s, where it killed hundreds of thousands of hectares of pine plantations (Talbot, 1977). Other well-known examples of the importance of vectors for fungi establishment are the interactions between blue-stain fungi and bark beetles (see Box 6.2). Except for some of these well-documented cases, it is less clear to what extent beetles play a direct role in the colonization of wood-inhabiting fungi. The ability of insects to locate specific fungi have been tested in a few cases. This ability is in itself not conclusive evidence of their importance as dispersal vectors but represents an important prerequisite. The few available studies show contrasting results. Jonsell and Nordlander (1995) showed that insects associated with the fruiting bodies of bracket fungi are attracted by the odour emitted by specific fungal species. On the other hand, preference studies by Johansson et al. (2006), based on baited traps, have shown that only a limited number of beetles differentiated between empty control traps, traps with wood pieces, and traps with mycelium-infected wood.

In an intriguing set of papers, Robin Kimmerer (Kimmerer, 1993, 1994; Kimmerer and Young, 1995, 1996) tells the story of small-scale dynamics in the occurrence of epixylic bryophytes. She shows how the dynamics of the epixylic community are a miniature mirror of forest gap dynamics. Small-scale disturbance opens up the space for rapid colonizers, such as *Tetraphis pellucida* and *Dicranum flagellare*, which otherwise become overgrown and outcompeted by large carpet-forming mosses. These gap species rely on rapid colonization from large asexual propagules, gemmae, which are very efficient in establishing new populations when dispersal distances are short. Gaps in the system are created mainly from fragmentation of the log and the somewhat unexpected activity of banana slugs (*Arion* sp.).

Many species, however, rely on pure chance in their establishment, and cope with this life stage by depositing a vast number of dispersal propagules. This is especially evident among many fungi. Spores are produced at a rate of billions per day, and spore deposition in highly fragmented landscapes may range from tens of spores per m^2 per day up to 100 spores per m^2 per day, even for red-listed species restricted to old-growth forests, such as *Phlebia centrifuga* and *Fomitopsis rosea* (Edman et al., 2004a). In old-growth forests, spore deposition may be yet another order of magnitude higher, and for *F. rosea* the daily spore deposition could exceed 5000 spores per m^2 (Edman et al., 2004b; Jönsson et al., 2008). However, the actual level of spore deposition needed for successful colonization is still unknown. Empirical studies show that, despite a general high level of spore deposition, the colonization of logs is correlated with the availability of neighbouring occurrences of the species (Edman et al., 2004b; Jönsson et al., 2008). This suggests that very high spore deposition might be needed for successful establishment; this assumption is supported by the following case study.

To study the population dynamics of wood-inhabiting fungi, an 8.5 ha permanent plot in old-growth Norway spruce forest was established in a mountain forest of northern Sweden. Here all logs larger than 10 cm diameter ($N = 851$) were mapped in 1997 and the occurrence of fruiting bodies of all wood-inhabiting fungi recorded (Edman and Jonsson, 2001). The plot was re-inventoried 6 years later, providing spatial data on colonization and extinction events (Jönsson et al., 2008). To our knowledge, this is the only study following the spatial dynamics in established natural populations of wood-inhabiting fungi (but see Berglund et al., 2005, for fruiting-body dynamics).

The spatial analysis of the snapshot data in 1997 showed that only a few species occurred aggregated in the stand, suggesting that local dispersal did not strongly influence local distribution patterns. However, as the stand is an old-growth forest with abundant dead wood of all decay classes and long continuity, time may well have allowed dispersal-limited species to colonize the whole stand. The re-inventory allowed the study of actual colonization events in order to analyse whether these showed any spatial restriction. In five out of nine abundant species, there was a significant influence of distance to neighbouring occurrences on colonization (Jönsson et al., 2008). This shows that, even for abundant species, dispersal and establishment may be spatially restricted within stands.

14.2.3 Age of first reproduction, generation time and lifespan

After establishment, the next aspect to consider is the time a species requires for maturation, including the age of first reproduction and the generation time, i.e. the average age of reproducing individuals in the population. This is central for population growth and dynamics. The potential for rapid population growth is higher in species with a low age of first reproduction and short generation times. Due to the turnover of their substrate, most saproxylic species tend to have a short generation time, reproduce at an early age, and live for only a few years. The exceptions are species (mainly vertebrates) that move around and utilize several dead trees. The actual life-length is also correlated to the stage of decomposition of the dead tree in which the species occurs. Early, ruderal species (see above) tend to be short-lived, since their resource base is rapidly consumed and they often become outcompeted by later-arriving species. By contrast, species living in wood at later decay stages may well be present for many years on an individual substrate unit. For example, in the epixylic liverwort *Ptilidium pulcherrimum*, which utilizes logs at both early and intermediate decay stages, the development from establishment through colony expansion to sexual reproduction may take more than 10 years. An individual colony must reach a minimum size before male organs (antheridia) are formed. In a boreal region this takes approximately 4 years. In addition, 3–4 years are then required before the colony starts to produce sporophytes and spores. At this stage the colony often covers an area of several square decimetres and may continue to reproduce for several years (Jonsson and Söderström, 1989).

For most beetles, the larval stage is consistently the longest life stage and the adult stage is usually shorter and devoted to reproduction. Most species have an annual larval cycle. The exception are species utilizing fresh phloem (inner bark) resources, e.g. *Ips typographus* and other bark beetles that have larval development of just a few months and that may complete more than one generation per year. Large-sized species – such as many longhorn beetles and stag beetles – have a two-year or even longer development time. At the extreme end of the spectrum are species living in old, dry wood where larval development may take many years, or even decades, e.g. *Hylotrupes bajulus* and *Buprestis splendens* (Ehnström and Axelsson, 2002). For species with longer larval development time, the development seems to be rather variable and dependent on environmental conditions.

For wood-inhabiting fungi, relatively little is known about the length of time between establishment and reproduction. For early species it is evident that reproduction may take place almost immediately after establishment. It is less clear how long the species that reproduce during later decay stages might be present before they start to

Table 14.1. *Time window for fruiting of wood-inhabiting fungi living on Norway spruce. Data are based on extinctions from individual logs over a 6-year period (data from Jönsson et al., 2008). The species have been sorted according to average succession score of fruiting-body occurrences (see Chapter 6).*

Species	Succession score	Average no. of years fruiting
Stereum sanguinolentum	1.6	2.9
Trichaptum abietinum	2.0	4.8
Phellinus chrysoloma	2.2	4.9
Gloeophyllum sepiarium	2.4	4.1
Phellinus ferrugineofuscus	2.5	3.4
Phlebia centrifuga	2.5	3.6
Fomitopsis pinicola	2.5	5.8
Fomitopsis rosea	2.6	5.8
Antrodia serialis	2.9	4.7
Columnocystis abietina	2.9	5.7
Asterodon ferruginosus	3.1	5.6
Phellinus viticola	3.1	7.6
Phellinus nigrolimitatus	3.8	22.2

reproduce. An indication that many fungal species may be latent in logs comes from studies on DNA-identified mycelia taken from wood samples compared with fruiting species (Ovaskainen et al., 2010b). For instance, Gustafsson et al. (2002) found that only about one-third of the species present in *Picea abies* logs were represented by fruiting bodies. This indicates the presence of a latent fungal community and suggests that a significant time might elapse between colonization and reproduction in some species.

Persistence time within single substrate units is not well known for most wood-inhabiting fungi. In the study cited above, which monitored more than 800 logs over 6 years, local extinctions were mainly caused by the decay of logs (Jönsson et al. 2008). Depending on where in the successional sequence a species occurs, its 'time window' of appearance varies. Early species such as *Phellinus ferrugineofuscus* and *Stereum sanguinolentum* had a high annual extinction rate on individual logs (around 30%), while late species such as *P. nigrolimitatus* had a very low annual extinction rate (less than 5%). This translates into time windows of occurrence from a few years up to more than 20 years. In general, most species had fairly short periods of occurrence, normally less than 8 years (Table 14.1). In addition to decay stage, log size also influences extinction risks on individual logs. For most species, the annual mortality rate decreased with increasing log size.

14.3 Metapopulation dynamics

The field of metapopulation ecology has developed a great deal since its original foundation in the 1970s (Levins, 1970; Hanski and Gaggiotti, 2004). It depicts species as occurring in relatively discrete patches of suitable habitats and states that local populations interact by migration (Figure 14.4). Due to human land use, more and more saproxylic species occur in separated and relatively isolated populations, and their persistence at the landscape scale becomes a function of the dynamics of – and migration between – the remaining forest stands.

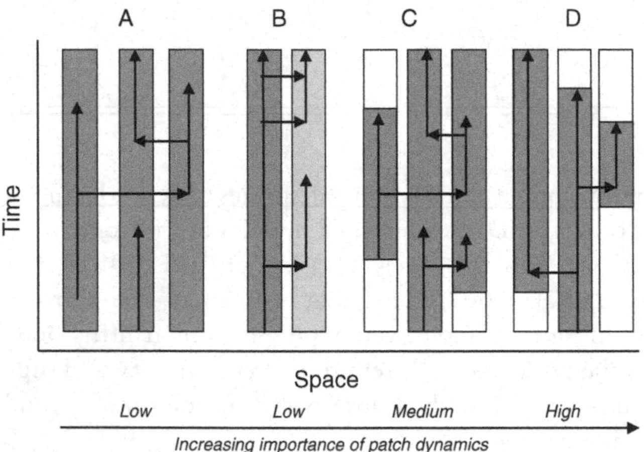

Figure 14.4. General models that describe the regional dynamics of species occurring on patchy habitats where **A** represents a classical metapopulation, **B** source–sink dynamics, **C** habitat-tracking dynamics and **D** patch-tracking dynamics. The axes describe time and space. Dark grey colour means that the patch is suitable for the species (light grey indicates suitable habitat patch of reduced quality), and the presence of species is indicated by arrows. The importance of habitat dynamics increases towards the models to the right. All models may be relevant for saproxylic species, but for individual dead wood units the habitat- and patch-tracking models are the most relevant. For larger forest stands, the lifespan of saproxylic species may also be potentially described as the classical or source–sink models. Illustration based on Snäll et al. (2003) and Eriksson (1996).

14.3.1 Types of metapopulation dynamics

In its original sense (see Hanski, 1999), a metapopulation is composed of a number of separate subpopulations, each living in discrete sites located within an inhospitable matrix. The local subpopulations may go extinct but the metapopulation persists if local extinctions are compensated by recolonization of temporarily empty habitat patches (Figure 14.4). This view of population dynamics focuses on dispersal, colonization and extinction events as the driving processes of population dynamics and population persistence. The influence of metapopulation theory on current population ecology is extensive. This is partly due to the integrity of the approach as such, but also because understanding metapopulation dynamics is becoming increasingly important as a consequence of the ongoing loss and fragmentation of natural habitats.

Similarly to classical metapopulations, source–sink dynamics mostly address the dynamics between separate habitat patches in the landscape. However, in this case it is recognized that the ability of individual patches to support local populations varies, and that some patches (sinks) are dependent on immigration from other patches (sources) to maintain their populations (see Figure 14.4B). This emphasizes that the importance of habitat patches cannot be deduced merely by the presence of the species, since sink habitats are clearly less important for the population than source habitats. It would, however, be erroneous to assume that sink habitats are unimportant for the viability of a metapopulation. These habitats might indeed be crucial for dispersal and migration between available habitat patches.

To further complicate the picture, but at the same time to increase realism in the saproxylic context, not only does habitat quality vary over time, but patches may also completely disappear in one place and new ones may appear elsewhere. This clearly concerns saproxylic species, as the individual decaying tree only exists for a limited time period. Regarding habitat-tracking metapopulations, the habitat dynamics are incorporated into the population models, and in the long run, a local population can be lost either by extinction due to stochastic events, or deterministically because the dead tree decays and becomes unsuitable as a habitat patch.

Patch-tracking metapopulation dynamics concerns a situation where it is only the habitat dynamics itself that limits the time during which a species can utilize the resource. Here the idea is that no extinctions occur due to the dynamics of the species in question, and only deterministic extinctions due to habitat dynamics take place (e.g. the decay of logs and the fall of snags).

As depicted in Figure 14.4 the role of habitat dynamics is the greatest in this kind of situation and understanding the input and loss of dead wood becomes crucial for understanding the dynamics of the associated populations.

Theoretical analyses suggest that, given a metapopulation system, a concentration of suitable habitats and resources is more efficient in maintaining declining populations than dispersing efforts evenly across a landscape (Hanski, 2000). In order to test the persistence of five different hypothetical saproxylic species, Ranius and Kindvall (2006) developed a metapopulation model for different landscape management scenarios. The 'species' represented a range of population turnovers and dispersal abilities. A so-called 'baseline species' was constructed, with realistic dispersal ability and population turnover rates (i.e. extinction and colonization rates). Four other 'species' were derived, representing combinations of high versus low turnover and short-distance versus long-distance dispersal ability. The management scenarios included only single large reserves, several smaller set-asides, or a high level of biodiversity concern in managed forest but with no reserves. Finally, a combined scenario including all three conservation efforts was considered. All four scenarios were constructed so that, in the long run, an equal amount of dead wood was available, and assuming that there is no difference in wood quality between managed and natural forests. Thus, the analysis considered the spatial configuration of the available habitat for the model species.

In a situation with constant conservation efforts at the landscape level, quite small conservation efforts were needed to maintain the species. One large reserve was better than a few small reserves, and to achieve the same outcome with no reserves but with the addition of dead wood across all the managed landscape, a slightly higher total volume of dead wood was required. Thus, this is consistent with the conventional wisdom that a concentration of conservation efforts in one space is better than more widespread efforts. However, when a bottleneck period was introduced (e.g. as a consequence of historical extensive clear-cutting) the results suggested that a larger set of small reserves was the most efficient to support the long-term persistence of a species at the landscape scale (Figure 14.5).

The study by Ranius and Kindvall nicely demonstrates the importance of understanding the consequences at the landscape scale for

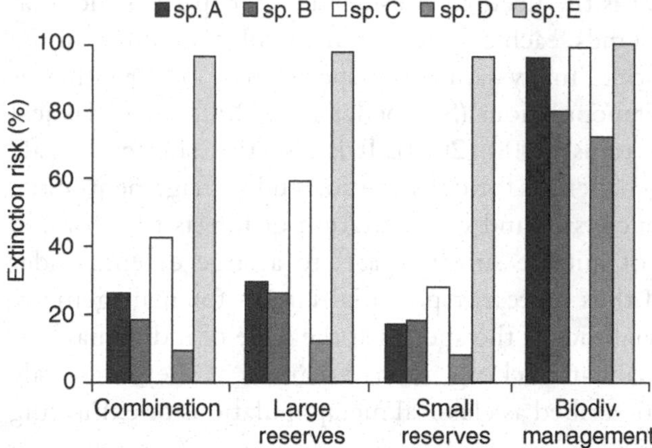

Figure 14.5. Extinction risk during a period of 100 years, according to simulation results for five model species representing baseline parameters (**A**); long-distance dispersal (**B**); short-distance dispersal (**C**); high population turnover rate (**D**); and low population turnover rate (**E**). The model scenario includes an initial bottleneck period with very low dead-wood volumes, and assumes that conservation efforts take place at the beginning of the 100-year period. Forest management scenarios include large reserves, small reserves, high level of biodiversity management, and a combination of all three. The four different management scenarios provide, in the long run, the same total amount of dead wood in the landscape (from Ranius and Kindvall, 2006).

species with different life strategies. It also stresses the need for better knowledge on the actual dispersal ability and population dynamics of saproxylic species. Although their study used relevant parameter estimates for the model, it is not known how many and which real species correspond to each of the hypothetical species.

14.3.2 Two scales of dynamics

From the above, it should be evident that the conceptual basis for understanding saproxylic population dynamics is rather well developed. It is, however, important to note that there are two distinct spatial scales that should be considered. To understand the population dynamics of a given saproxylic species, the occurrence pattern of its substrates, and the colonization and extinction rates on individual substrate units are central. This is the scale where the actual processes are taking place. For most saproxylic species, this small-scale dynamics is best described as a habitat-tracking or patch-tracking metapopulation (see Box 7.1 for an example).

Equally important is the need to consider a larger spatial scale: that of individual forest stands, each containing many substrate units, in the landscape. At this scale, many saproxylic species may occur either as habitat-tracking metapopulations (Schroeder et al., 2007) or as classical metapopulations (Ranius, 2000, 2001). It is also this landscape scale that constitutes the challenges for conservation and management of the species. The frequency, size and configuration of forests that contain sufficient amounts of suitable substrate are, to a large extent, under human control and thus represent potential targets for management. Even though the dynamics of the species at the scale of individual logs may be patch- or habitat-tracking, their dynamics at the stand scale may well be better described as classical metapopulations or exhibiting source–sink dynamics.

14.3.3 Extinction thresholds and extinction debts

Populations of saproxylic species survive in the long term when the average colonization rate of new trees compensates for the average extinction rate from old host trees. If habitat quality deteriorates and habitat loss becomes significant at the landscape scale, population growth may become negative, i.e. the average extinction rate exceeds the colonization rate. The minimum density of suitable habitat patches (at the level of either individual substrate units or forest stands, or both) below which the population cannot survive in the long term is known as the *extinction threshold* (see Chapter 15 for further details). Initially, the differences between extinction and colonization rates may be small and the negative trend in population size may be masked, e.g. by between-year variation. A continued negative population growth will, however, eventually result in population extinction. The time that this process takes may be considerable and has been called the *extinction debt* (see Chapter 15).

In a unique study, Gu et al. (2002) analysed a large landscape in eastern Finland with regard to the amount of old-growth forests during the last 50 years. The landscape had been exposed to major habitat loss, and old-growth stands had decreased by 75% between 1945 and 1995. During such a rapid fragmentation process, one may hypothesize that current population sizes and occurrences of species are not in equilibrium with the current landscape structure.

To analyse this, Gu and co-workers developed a spatially explicit metapopulation model (*sensu* Hanski, 1999) and used current as well as the historical landscape structure as predictors of the occurrence of four red-listed wood-inhabiting fungi. Their results showed that the current isolation of the forest stand from the nearest other old-growth spruce forest was not significantly correlated with the occurrence of any of the species. That is, one could potentially come to the conclusion that only stand-specific factors governed the occurrence of the species. However, when taking the historical landscape change into consideration, and thereby including the actual fragmentation process as a predictor, this explained a significant amount of the variation in the occurrence of three out of the four species. Thus, the study strongly supports the conclusion that the occurrence of wood-inhabiting fungi depends not only on the local stand-level factors (e.g. amount and quality of dead wood) but also on the structure of the surrounding landscape, affecting the long-term dispersal and colonization ability of the species.

14.4 The role of continuity

As shown earlier in this chapter, species differ widely in their dispersal ability and also in their degree of substrate specialization (see Chapters 5–8). Some species are adapted to woody substrates that are either locally common and/or have a long persistence time. For such species, the selection pressure for efficient dispersal may not be so strong under natural conditions. This suggests that all species are not sufficiently well adapted to track their habitats given the increased fragmentation and turnover of suitable stands in managed forest landscapes. Empirical studies also confirm that local continuity of substrates may be important to maintain the populations of some saproxylic species (Nilsson et al., 1995; Nilsson and Baranowski, 1997; Siitonen and Saaristo, 2000; Stokland and Kauserud, 2004). Based on these observations, it has been suggested that certain saproxylic species could be used as indicators of stand history and continuity (Nilsson et al., 1995; Bredesen et al., 1997; Alexander, 2004).

The role of continuity is most likely to be correlated with the dominant disturbance regime. In forest types where large-scale, stand-replacing disturbances are common, the selective forces for efficient dispersal of the associated species are higher. By contrast, in forest types dominated by gap-phase or small-scale disturbances, distances both in time and space between suitable habitat patches and substrate units will be much smaller. This is reflected, for instance, in two studies (Penttilä et al., 2006; Stokland and Larsson, 2011), where species associated with Norway spruce were more affected by habitat loss and fragmentation than species associated with Scots pine.

When studying the importance of continuity, a confounding factor is that long historical stand continuity tends to be strongly correlated with high current habitat quality. That is, in old-growth forests with long continuity, the abundance and quality of dead wood tend to be high too (Nordén and Appelqvist, 2001; Rolstad et al., 2002). Thus, it becomes difficult to separate the relative effect of current habitat conditions from the importance of historical continuity. Despite its potential importance, only a few studies have provided empirical data that supports the effect of continuity *per se*. In a study of the wood-decay fungi *Phellinus nigrolimitatus*, Stokland and Kauserud (2004) showed that the species was more frequent on suitable logs in stands that had been unmanaged for several tree generations than on similar logs in managed stands. The saproxylic beetle *Pytho kolwensis* is another example where continuity seems to play an important role in its occurrence. Siitonen and Saaristo (2000) studied the occurrence of the species in six spruce-mire forests in eastern Finland. The species has declined strongly during the 20th century (see Figure 15.2). All the studied spruce-mire forests represented sites with 170–300 years of continuity, and the long-term continuous availability of suitable host trees appeared to be the key factor explaining the occurrence of the species.

In addition to the problem with covariation between continuity and habitat quality, the general importance of continuity is challenged by some empirical studies. Groven et al. (2002) studied six different wood-decay fungi on sites where they had made a careful reconstruction of forest history. They came to the conclusion that a continuous supply of dead wood, at the scale of forest stands, was not crucial for the occurrence of the surveyed wood-decay fungi. Similarly, Rolstad et al. (2004) analysed whether the amount of new logs or old logs explained the occurrence probability of six red-listed and six common polypore species in a 200 ha large continuous forest. They found that spatio-temporal distribution (= continuity) of logs was of minor importance. Sverdrup-Thygeson and Lindenmayer (2003) studied *P. nigrolimitatus* and, contrary to Stokland and Kauserud (2004, see above), concluded that stand-level continuity was less important for the occurrence of the species than the proportion of old forests in the surrounding landscape. Common to all these three studies, however, was the fact that the plots sampled were generally small (0.2 ha, 1 ha and 0.16 ha, respectively) and the gradient in past forestry activity was rather limited in the first two studies, representing only forests with high natural values. The results are thus limited to a relatively small scale, and the influence of adjacent forest structure (within the same stand) may have obscured the relation between continuity and species occurrence. It therefore remains an important research challenge to understand the relative importance of continuity for the occurrence of saproxylic species.

15 · *Threatened saproxylic species*

Juha Siitonen

Saproxylic species are one of the most threatened organism groups. As with all forest species, they are suffering from the dwindling of forests. But the habitats of saproxylic species may also be decreasing in regions where, although the forest area is currently increasing, such as in Europe, practically all forests and other wooded areas have been taken into intensive economic use, resulting in a greatly reduced abundance of large over-mature trees and large-diameter dead wood (see Chapters 13 and 16). As a consequence, many saproxylic species dependent on these habitat structures have drastically declined and have become threatened.

In this chapter we examine threatened saproxylic species, their threat factors and the assessment of their threat status. The historical development which has led to so many saproxylic species becoming threatened is best known in Europe, and hence we first give a short account of the endangerment history of saproxylic species based on European examples. Next we list the current threat factors which reduce the number and extent of habitats for wood-inhabiting species. We also discuss the knowledge base, methods and criteria used in assessing the threat status of species. Knowledge on threatened species needs to be improved, and in the final section we examine survey methods. We have described, in other chapters, general measures that can be taken to maintain the overall diversity of saproxylic species in managed forests (Chapter 13) and in cultural habitats (Chapter 16). Most of these methods benefit threatened saproxylic species too.

15.1 Historical evidence for the decline of saproxylic species

15.1.1 Disappearance of the ancient-forest fauna

As Speight (1989) stated, 'a little-appreciated feature of the process of progressive eradication of Europe's saproxylic fauna is its great antiquity'. The large-scale modification of lowland forests had already caused the first regional extinctions of ancient-forest saproxylics thousands of years ago. By the Late Bronze Age, about 3000 years ago, pristine forests had disappeared from most cultivable lowland areas all over Europe, due to agricultural expansion. Forest clearance continued during the Iron Age and Medieval period. However, refugial areas for ancient-forest saproxylic species persisted in areas unsuitable for agriculture, such as mountain regions with rugged terrain and steep slopes, in wetland forests, and in low-productive rocky outcrops. In addition, wooded cultural habitats provided new environments for many saproxylic species adapted to open conditions (see Chapter 16).

Subfossil beetle faunas have been thoroughly studied in the British Isles. Many ancient-forest saproxylic beetle species in Britain disappear from the fossil record between 5000 and 3000 years ago, by the end of the Bronze Age. Some of the ancient-forest species, such as *Rhysodes sulcatus* (Rhysodidae), *Isorhipis melasoides* (Eucnemidae), *Pycnomerus terebrans* (Zopheridae) and *Prostomis mandibularis* (Prostomidae), which today have a scattered relict distribution in Europe (Figure 15.1), are frequently found in the subfossil data (Buckland and Dinnin, 1993; Whitehouse, 2006; Elias et al., 2009). The total number of regionally extinct saproxylic species found as subfossils in Britain has risen to at least 25 (Buckland, 2005; Whitehouse, 2006). Several other species which are common in the fossil record have not been found alive during the last 50 years (Warren and Key, 1991). At a single Late Bronze Age site in Thorne Moors, a total of 18 saproxylic beetle species extinct in Britain, as well as almost 50 threatened or rare ones, have been found. It seems that this wetland area, unsuitable for cultivation, may indeed have served as a refugium long after forest had been cleared elsewhere (Whitehouse, 2006). Some fire-associated beetle species have also been found only as subfossils in Britain (Whitehouse, 2006). These species, including *Peltis grossa* (Trogossitidae) and *Stagetus borealis* (Anobiidae), obviously benefited from the increased fire frequency caused by humans, and went regionally extinct after forest fires became rare.

In Sweden, *Sericoda bogemanni* (Carabidae), a ground-beetle species dwelling under the bark of fire-scorched trees, occurred through the Holocene (from about 10 000 years ago until about 1400–1850) based on peat deposits in a southern Swedish bog. The species is now probably extinct in the whole of Europe. Ancient-forest beetles recovered from the same site include *Prostomis mandibularis* (Prostomidae), *Isorhipis*

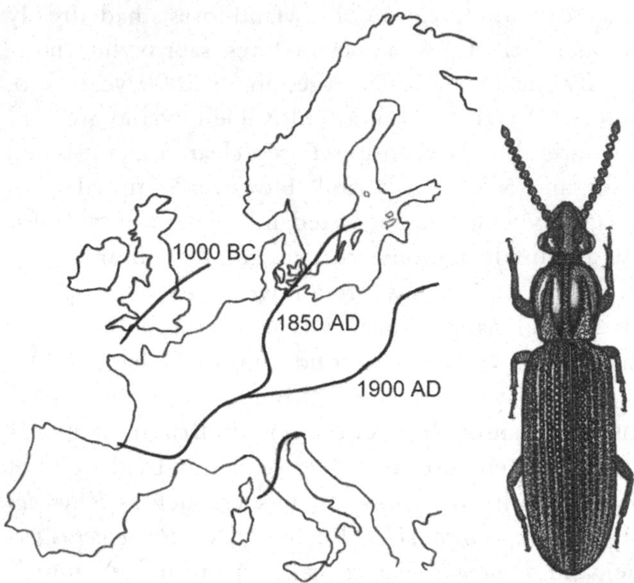

Figure 15.1. *Rhysodes sulcatus* is an archetypical ancient-forest saproxylic beetle that is becoming extinct in Europe. The map shows diagrammatically the progressive contraction of its range, starting with its disappearance from Great Britain some 3000 years ago. The species disappeared from its relict occurrences in western lowland Europe around 1850. At present it is known from some scattered localities in the remaining natural forest patches in mountainous regions. Redrawn after Speight (1989) and Buckland (2005).

marmottani (Eucnemidae), *Bothrideres contractus* (Bothrideridae) and *Elater ferrugineus* (Elateridae) (Olsson and Lemdahl, 2009).

15.1.2 Post-Linnean extinctions

The Swedish naturalist Carl von Linné developed the classification of species and the binary nomenclature used for describing species in the mid 1700s. When it became possible to identify and name species, knowledge about fauna and the number of regional faunistic studies increased rapidly in the late 1700s and the beginning of the 1800s. Old records can be compared with the present range of the same species, and thus it is possible to trace faunal changes during the last 200 years.

The early faunistic studies usually produced a list of species observed within a given area. No information on collecting methods, sampling

Table 15.1. *Collections made by Professor C. R. Sahlberg and his students during two excursions to Yläne, southwestern Finland (see Figure 15.1) in 1828 (Saalas, 1933). The current Red List category of each species in Finland (Rassi et al., 2010) is given according to the IUCN classification: CR = Critically Endangered, EN = Endangered, VU = Vulnerable, NT = Near Threatened. Rare – but not red-listed – species with at most 50 known occurrences in Finland are indicated with r. Number of individuals was recorded only on 19 May.*

Species	Category	19 May	27 May
Cucujus cinnaberinus	CR	5	x
Leptura thoracica	CR	–	x
Pytho kolwensis	EN	–	x
Dicerca alni	VU	–	x
Boros schneideri	VU	12	x
Platyrhinus resinosus	NT	2	x
Tropideres dorsalis	NT	–	x
Upis ceramboides	NT	20	–
Calitys scabra	r	–	x
Corticeus suturalis	r	2	x
Ipidia binotata	r	1	x
Lacon conspersus	r	–	x
Laemophloeus muticus	r	24	x
Mycetophagus fulvicollis	r	1	–
Mycetophagus quadripustulatus	r	–	x
Orchesia fasciata	r	–	x
Platynus mannerheimii	r	–	x
Platysoma deplanatum	r	–	x
Sacium pusillum	r	33	x
Zilora ferruginea	r	–	x

effort or abundances of species was generally provided. Nonetheless, some old publications strikingly reveal how much the saproxylic fauna has changed during the past two centuries. An illuminating example is the beetle material collected by the Finnish entomologist Professor C. R. Sahlberg and his students during two excursions in southwestern Finland in 1828 (Saalas, 1933). The number of rare and currently red-listed species collected in only two days is amazing (Table 15.1). Several of the observed species which occurred in southwestern Finland at that time have now disappeared from the region or whole of southern Finland. Many threatened saproxylic beetles still occur abundantly right across the eastern border of Finland in Russian Karelia, in areas where forestry has been much less intensive than in Finland until recently. In these areas it is still possible to observe large numbers of threatened saproxylic beetles within a couple of days – in much the same way as Professor Sahlberg did almost 200 years earlier (Siitonen and Martikainen, 1994; Siitonen et al., 1996).

From the first half of the 1900s, detailed faunistic publications are available from different parts of Europe. The term '*Urwaldrelikt*', ancient-forest relict, was coined by the German entomologist K. Dorn in 1935. It is possible to form a clear picture of the decline in many species which are currently threatened by simply plotting the old and new records on a map (Figure 15.2). It also becomes evident from the older faunistic works how saproxylic species which are at present regionally extinct or endangered may previously have been common in areas where suitable habitats still remained.

For example, the Swedish coleopterist Thure Palm explored saproxylic beetle fauna in the old-growth forests in the lower course of the Dalälven River, southern Sweden, in the 1930s. Unlogged forests had remained in this area because they were located on islands and peninsulas protected by rapids and bouldery channels of the branching river. A large number of ancient-forest relict species were found in the area (Palm, 1942). The present occurrence of some of the species found by Palm has more recently been surveyed within the same area (Eriksson, 2000). Some have most evidently disappeared (including the longhorn beetle species *Plagionotus detritus* and *Monochamus urussovii*) while several others are declining (for example, the flat bark beetle species *Cucujus cinnaberinus* and the stag beetle species *Ceruchus chrysomelinus*) in spite of the fact that the core area has been protected as a nature reserve. The white-backed woodpecker (*Dendrocopos leucotus*) is also about to become extinct in the area. It is dependent on large tracts of old deciduous forest with dying and dead trees. This kind of gradual disappearance from the remaining habitat patches can be an indication of extinction debt, which is discussed below.

The decline of saproxylic species associated with natural forests due to intensifying land use was already being noticed a long time ago. Entomologists and specialists of other groups were aware of this probably as early as the 1800s, but from the 1900s there are many published references documenting the decline of various species observed by devotees – long before the systematic threat status assessment of species began. The Finnish Professor of forest entomology, Esko Kangas,

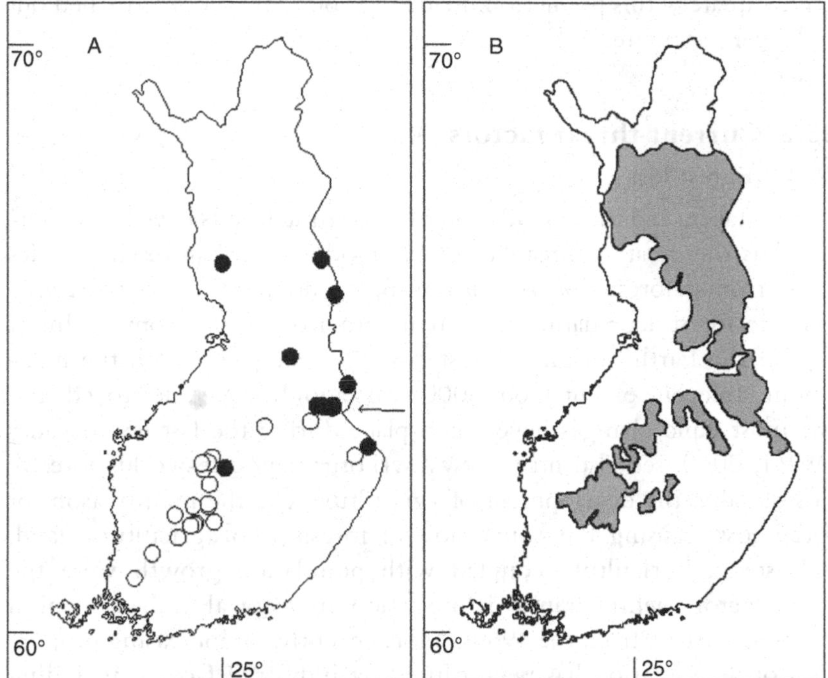

Figure 15.2. Records of the endangered old-growth beetle *Pytho kolwensis* in Finland (A). Open circles indicate records before 1960 and filled circles after 1960. The two most southwestern circles are the oldest records. The southwesternmost one is the type locality of the species, i.e. the species was described as new for science based on individuals collected in this place (Yläne, Kolwa) in 1828 (see Table 15.1). All the other records strikingly coincide with regions where forests were mostly unexploited in 1850 (B). It seems probable that this species, living in large fallen spruces as larvae, had become extinct in large parts of southern Finland, where slash and burn cultivation was intensive, as early as the 1800s or even earlier. Maps from Siitonen and Saaristo (2000).

described this development in the following way (Kangas, 1947; translated from Finnish):

> In particular, sick or stunted trees, snags, windthrows and snow-breaks as well as logs, in particular large ones, offer the only living chances to numerous curiosities and rarities of our beetle fauna. As the silviculture in our country develops and becomes more intensive, such tree individuals (or parts of them) disappear ever more thoroughly from our forests, and the beetles entirely dependent on them consequently become increasingly rare and withdraw to ever more restricted areas, until they possibly become extinct in our country.

Unfortunately, this prediction, made over 60 years ago, has turned out to be very accurate.

15.2 Current threat factors

15.2.1 Forest loss

Deforestation and the degradation of remaining forests have been identified as the greatest threats to global biodiversity. Saproxylic species suffer from deforestation as all forest species do, but they are more sensitive to forest degradation than many other organism groups. Almost half of the Earth's original forest cover (as compared with the maximum Holocene extent about 8000 years ago) has been destroyed, and the most rapid changes have taken place during the last few decades (WRI, 2000; see also http://www.wri.org/map/state-worlds-forests). Historically, the development of agriculture was the main reason for forest loss, causing the conversion of forests into agricultural land. Subsistence agriculture coupled with population growth were the main factors behind rapid deforestation in tropical and subtropical regions, starting from the 1960s. More recently, an increasing proportion of deforestation has been caused by industrial factors, including large-scale clearing for cropland and cattle ranching but also for timber extraction (Butler and Laurance, 2008). Industrial timber operations entail road expansion, which generally leads to further deforestation and forest degradation, in the tropics mainly because of human invasion and settlement (Laurance et al., 2009), but also in the boreal frontier forests because of human-induced fires (Achard et al., 2006).

At present, about 30% of the global land area is covered by forests, with a total area of 39 million km^2 (Schmitt et al., 2009). The overall deforestation rate has been 130 000 km^2 per year during the period 1990–2005 (FAO, 2006). The annual net loss of forest area is smaller, since deforestation is partly compensated for by reforestation (replanting of cleared forest) and afforestation (planting of treeless areas) in some regions. However, these plantation forests differ significantly in structure and species composition from primary forests. Primary forests have been estimated to account for one-third of the global forest area, but about 60 000 km^2 is lost or modified each year (FAO, 2006). The area of intact forest landscapes, i.e. unbroken large expanses of forest without signs of significant human activity, was recently estimated to be 13 million km^2, or 24% of the forested land area (Potapov et al., 2008; see also http://www.intactforests.org/).

15.2.2 Forest management

Different forest management practices simultaneously reduce the amount of dead wood at the levels of both individual stand and landscape (see Chapter 13). Clear-cutting followed by planting or natural regrowth are the prevailing harvesting methods over large regions in the boreal zone. The largest difference in dead wood between managed and natural stands occurs at the beginning of succession, immediately after a stand-replacing disturbance (Siitonen, 2001). In clear-cutting, most of the timber volume is extracted from the forest, in striking contrast to natural disturbances such as fire or windthrow, in which all the dead trees remain at the site. Mechanical harvesting and soil scarification, aiming to enhance regeneration, can destroy up to 80% of existing pre-treatment logs at the harvested site (Hautala et al., 2004). The complete depletion of trees in clear-cut harvesting both disrupts the continuous availability of dead trees at different stages of decay and reduces the initial volume and future accumulation of dead wood in the regenerating stand for at least a century.

Silvicultural thinning of stands reduces dead wood in the mid-successional stages, since dead, damaged and weakened trees are usually removed in thinnings. If large quantities of dead wood are created by natural disturbances such as forest fires and insect outbreaks, dead trees are often harvested through salvage or sanitation logging (DellaSala et al., 2006; Lindenmayer and Noss, 2006). This affects the amount of dead wood significantly and destroys habitats which are suitable for many saproxylic species adapted to disturbances (see Chapter 13). Short rotation times in managed forests truncate the stand development before the recruitment of large-diameter dead trees can even begin.

The use of exotic tree species can have a great adverse impact on saproxylic species. Most species are dependent on particular host-tree genera. If the original deciduous forests are replaced with plantations of conifers or distantly related deciduous species such as eucalypts, only the most generalist saproxylic species will be able to utilize the dead wood produced by the exotics. Furthermore, the use of exotic tree species is generally linked to intensive short-rotation forestry in which old trees and coarse dead wood are not formed at all.

15.2.3 Intensifying land use

The general intensification of land use and urbanization also reduce the number of over-mature living trees and amount of dead wood in non-forest environments. Open and semi-open grasslands and pasture woodlands with scattered old trees are the most important habitat for many saproxylic species. Following the industrial and agricultural revolutions, starting from the late 1700s, the density of ancient trees in cultural environments has been drastically reduced (see Chapter 16 for further details). The greatest changes have occurred after the 1950s following the large-scale conversion of traditional agricultural and grazing systems into agro-industrial systems.

15.2.4 Energy wood harvesting

Climatic change and the need to reduce CO_2 and other greenhouse gas emissions have rapidly increased the use of biofuels. Bioenergy production can be based on biomass plantations in agricultural or forest land, using various crop plants. Another source of bioenergy is forest biomass, including trees from thinning operations, and the use of logging residues (Berndes et al., 2003; Field et al., 2007).

Energy wood production and harvesting can have multiple negative effects on biodiversity in general, and on saproxylic species in particular. The increasing price of energy wood can make it profitable to convert marginal agricultural land to biomass production, increase harvesting from currently non-commercial woodlands, increase the intensity of forest management, etc. The harvesting of logging residues and stumps for fuel has direct effects on the amount of dead wood and subsequently on saproxylic species. Energy wood harvesting can further reduce the amount of dead wood in managed forests (Rudolphi and Gustafsson, 2005). Small-diameter logging residues (including twigs, branches and tree tops) and stumps, which have previously remained in regeneration areas after conventional clear-cutting, are removed. Hard logs still suitable for burning are regularly harvested along with the residues. Furthermore, mechanical harvesting involves several driving passages which can cause the unintentional destruction of existing decayed logs.

Logging residues and cut stumps have not generally been considered as important substrates for threatened saproxylic species. This is because these substrates are abundant in managed forest, and species which are able to utilize them are not likely to have become threatened. However, residues and stumps may have offered important substitutive substrates for many species which otherwise would have declined. Increased harvesting of residues and stumps may change the situation.

Recent studies have shown that large numbers of saproxylic invertebrate species, including several red-listed ones, can breed in logging slash. In southern Sweden, logging residues of different tree species hosted clearly different faunas, and oak and aspen, in particular, turned out to be important for red-listed species (Jonsell et al., 2007).

15.3 Effects of reduced dead-wood volume on saproxylic species

15.3.1 Species–area relationships

The data presented in the previous chapters (Chapters 12 and 13) indicate that the average volume of dead wood is considerably lower in managed forests than in natural forests, where dead wood is continuously formed by natural disturbances causing tree mortality. For instance, in the southern and middle boreal parts of Fennoscandia, the average volume of coarse dead wood at the landscape level has declined from about 60–90 m^3/ha to 2–5 m^3/ha, which means a reduction of 90–98%, depending on the region (Siitonen, 2001).

The effects of this reduction on the species richness of saproxylic species can be estimated on the basis of a general species–area relationship (Box 15.1). A decline of over 90% in available habitat, here in the amount of dead wood, can be expected to lead to the regional disappearance of more than 50% of the original saproxylic species from managed forests in the long term, assuming a typical relation ($z = 0.25$) between species number and amount of habitat. Assuming a conservative relation ($z = 0.1$), the minimum prediction for species loss following 90–98% habitat loss (the Fennoscandian example) is in any case 22–32% (Siitonen, 2001). If species suffer from habitat fragmentation in addition to habitat loss, the proportion of species expected to disappear can be even larger.

Box 15.1 Species–area relationship

It is considered an ecological law that the number of species increases when the habitat area increases and, vice versa, decreases when the habitat area decreases. Although several alternative models can be used to describe the functional form of this relationship (He and Legendre, 1996), the power function model has been widely used

Box 15.1 *(continued)*

and usually fits species–area data well (Connor and McCoy, 1979; Rosenzweig, 1995):

$$S = kA^z$$

where S represents the number of species, A is the available habitat area, and k and z are shape parameters. The relationship is easier to examine by taking a logarithm of both the number of species and the area which transforms the relationship into a straight line:

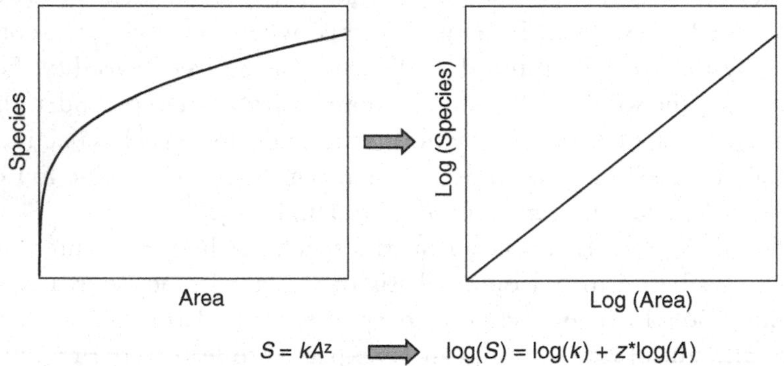

The species–area relationship represented by the power function in a non-transformed (left) and a log-transformed (right) version.

Considering saproxylic species, area can be substituted with dead wood volume, which better represents the amount of available habitat. An important aspect is that the z-value determines the rate at which species number decreases with decreasing amount of habitat, regardless of the initial amount. A certain percentage decrease in habitat is always associated with a certain percentage decrease in the number of species, represented by the z-value. The proportion of species lost as a function of suitable habitat remaining and the slope parameter z is:

$$f_e = 1 - (A_1/A_0)^z$$

where f_e stands for fraction extinct, A_0 is the original habitat area (or volume) and A_1 is the current habitat area (or volume) (Dial, 1995).

In their classic work, Connor and McCoy (1979) reviewed 100 studies reporting empirical species–area data; in 91% of these the z-value was at least 0.1, and typically within the interval of 0.2–0.4. However, if species–area relationships come from successively smaller parts of a region, the curves are nested and tend to have z-values within the range 0.1–0.2 (Leitner and Rosenzweig, 1997).

15.3.2 Extinction thresholds

Saproxylic species are living in a dynamic habitat, in dead trees that will disappear because of decomposition (see Chapter 14). To be able to survive in the long term, they must be able to colonize new suitable host trees at the same average rate as the old host trees become unsuitable. If the density of potential host trees is too low, the colonization rate of new host trees will not be sufficient to compensate for the local extinctions in old trees. This will lead to regional extinction of the species in the long run. The extinction threshold (Fahrig, 2002; Hanski, 2005) refers to the minimum density of habitats suitable for a species, below which the species cannot survive. Regarding saproxylics, the habitats can be either individual host trees or larger habitat fragments each containing several host trees. More generally, Hanski and Ovaskainen (2002) showed that a network of habitat fragments must satisfy certain necessary conditions in terms of number, size and spatial configuration of fragments to allow for the long-term persistence of the focal species. The extinction threshold distinguishes networks that satisfy these conditions from those that do not.

The species-specific extinction threshold depends mainly on two factors: the degree of specialization and dispersal power (Figure 15.3). Specialist species with strict host-tree requirements are more sensitive to habitat loss and fragmentation than generalist species which can utilize many kinds of dead-wood substrates. Similarly, species with low dispersal power are more sensitive to fragmentation than species with high dispersal power (Andrén, 1996).

The threshold concept is likely to be important for many threatened species, particularly those with strict host-tree requirements and poor dispersal ability, for defining the minimum density of suitable host trees at the stand or landscape level needed to sustain the species, and hence

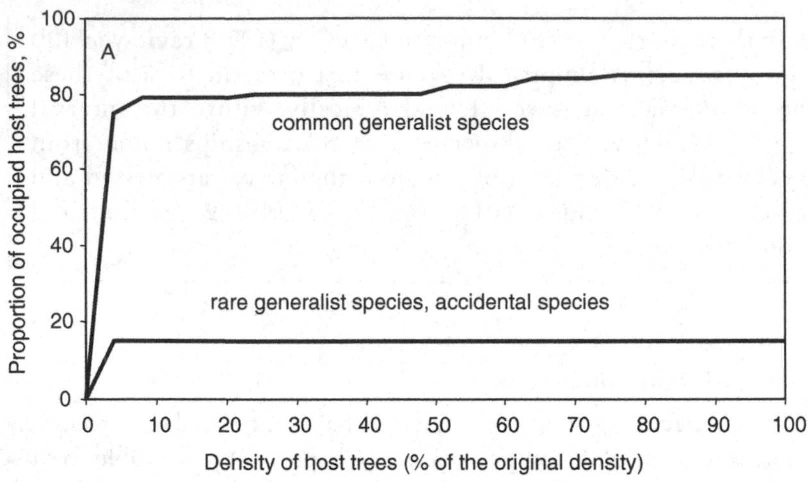

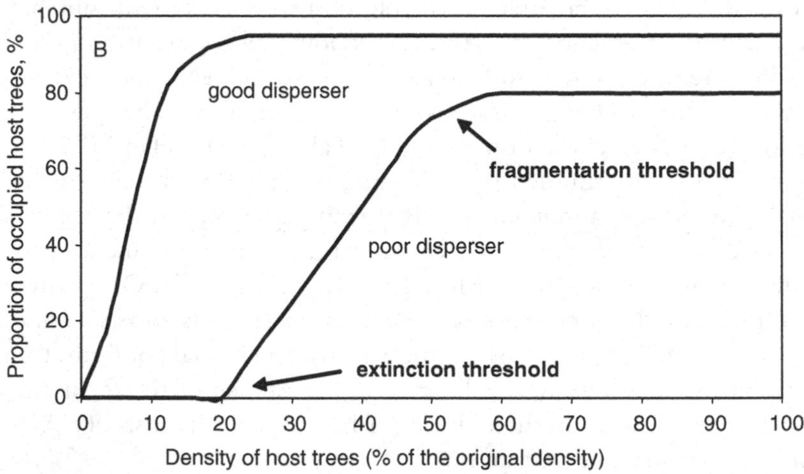

Figure 15.3. Responses of four hypothetical saproxylic species to reduction in the density of a specific host-tree type. For generalist species, which can utilize many different types of substrate, reduction in the specific host-tree type does not affect the average proportion of occupied trees, but common generalists inhabit a larger proportion of trees than rare or occasional ones (panel A). For specialist species, which are dependent on a particular host-tree type, the average proportion of occupied host trees first starts to decline when the density of suitable host trees is reduced below the so-called fragmentation threshold, and drops to zero below the extinction threshold density (panel B, modified from Andrén, 1996).

for setting management targets. However, since habitat requirements and ecological traits differ substantially between individual threatened saproxylic species, it is unlikely to be possible to define any single, general target of dead wood, in terms of cubic metres per hectare (m^3/ha), that would guarantee the survival of all threatened species in managed forests (Ranius and Jonsson, 2007). It is more likely that the number of saproxylic species per stand or per surface area follows the general species–area relationship to the volume of dead wood (see Martikainen et al., 2000).

Despite the potential usefulness of the threshold concept, there are only a few studies so far in which extinction thresholds for particular saproxylic species have been assessed. In their review, Müller and Bütler (2010) found published threshold values for 14 saproxylic species. The best examples so far concern woodpeckers, the top predators of the saproxylic food web. The probability of the presence of the three-toed woodpecker (*Picoides tridactylus*) dropped abruptly when the basal area of conifer snags was below 0.5 m^2/ha in Sweden or 1.3 m^2/ha in Switzerland (the latter figure corresponding to a snag volume of about 15 m^3/ha) (Bütler et al., 2004). The white-backed woodpecker (*Dendrocopos leucotus*) required landscapes where the volume of dead deciduous trees was at least 10–20 m^3/ha over a 100 ha area (Angelstam et al., 2003), or their basal area was at least 1.4 m^2/ha (Roberge et al., 2008). Another example concerns saproxylic beetles dependent on hollow oaks. Three threatened specialist species (*Tenebrio opacus*, *Elater ferrugineus*, *Osmoderma eremita*) were missing from stands where the density of hollow oaks was below 10 per stand (Ranius, 2002b).

15.3.3 Extinction debt

When a significant proportion of a particular habitat is lost, some of the species dependent on this habitat are also lost immediately. It is self-evident that if all the habitats of a given species are destroyed, the species will inevitably disappear too. The almost wholesale destruction of certain dead-wood habitats explains the documented regional extinctions of many saproxylic species mentioned above. However, it is important to note that when the amount of suitable habitat decreases, most species do not disappear immediately. The extinction of species caused by habitat loss and fragmentation generally takes place with shorter or longer time delays. Local populations which have been left in small and isolated habitat fragments can persist in that place for a long time but will eventually disappear.

Extinction debt is the number of extant species whose populations are deemed to go extinct gradually because of the environmental changes that have already taken place – even if the amount of suitable habitat is not reduced any more in the future (Tilman et al., 1994; Hanski and Ovaskainen, 2002; Kuussaari et al., 2009). The rate at which the local population will disappear depends on the magnitude of the habitat loss, the size and quality of the remaining habitat fragments, and the species' ecological characteristics such as generation time, degree of specialization, trophic level and dispersal ability. Extinction debt is especially high in habitats which have undergone major habitat loss only recently, and the time delay before extinction is especially long in those species which are just below their extinction threshold, in other words when the amount of suitable habitat is only slightly too small to allow long-term persistence (Hanski and Ovaskainen, 2002).

Many saproxylic species display rather fast population dynamics because they are able to live in the same host tree for only a limited time. An exception are species specialized in old hollow trees, thus comprising a group whose present occurrence does not necessarily reflect their long-term survival possibilities. This is due to the fact that old trees constitute long-lasting habitat patches which can remain suitable for many saproxylic species even for hundreds of years continuously (see Figure 16.3). The current occurrence of these species can reflect the density of suitable host trees in the past, and local extinctions are the prevailing feature of population dynamics. Consistently, the present occurrence of threatened decay fungi and epiphytic lichens specialized on old oaks in southern Sweden depends on both the current density of suitable host trees and the density of over-mature oaks in the 1830s (Ranius et al., 2008).

There are several ways to empirically explore the magnitude and progress of extinction debt (Kuussaari et al., 2009). Signs of extinction debt of threatened decay fungi and saproxylic beetles have been detected in several studies which either assessed the dependence of species richness on the present or previous amount of habitat, or compared species richness between habitats which were fragmented recently or long ago. In several cases, the current number of specialist species depended more on the historical rather than the current amount of suitable habitat (Gu et al., 2002; Paltto et al., 2006; Ranius et al., 2008). Similarly, the current number of specialist species was often found to be higher in recently isolated fragments of old-growth forest than in older fragments (Komonen et al., 2000; Berglund and Jonsson, 2005; Penttilä et al., 2006; Laaksonen et al., 2008).

15.4 Assessing the threat status of saproxylic species

15.4.1 Empirical studies showing the decline of saproxylics

How can we know for sure that the species richness of saproxylics has decreased because of intensive forest management and other forms of land use, and how do we know that certain saproxylic species are declining and have become threatened? There are a large number of studies that have investigated these questions: by comparing species richness and composition between natural and managed forests, and between landscapes or regions with different management histories, by comparing old and new records, etc. A thorough review is beyond the scope of this chapter, but our purpose is to highlight some consistent results which have emerged in many studies focusing on different saproxylic taxa, and across widely varying forest types (Table 15.2).

When managed forests have been compared with similar natural or semi-natural forests, species richness is invariably lower in managed stands. In cases where management history varies between stands or regions, species richness and the probability of occurrence of red-listed species generally decreases with increasing intensity and duration of management. Red-listed species are often confined to large-diameter, well-decayed logs and snags, i.e. on those dead wood qualities that have declined the most because of management.

In some studies, however, no connection was found between the past management intensity of a stand and the occurrence probability of late-successional species thought to indicate structural continuity, including red-listed wood-decaying fungi (Groven et al., 2002; Rolstad et al., 2004). These seemingly conflicting results can be explained by the different spatial scales used in the different studies. At very small scales (<1 ha) the continuity of logs is evidently not important for most species, and red-listed polypore species may be found in stands where the substrate continuity is unbroken in the surrounding landscape.

In contrast to most studies in Europe, some studies in North America did not discover significant differences in the species richness of saproxylic beetles between natural and managed stands (Zeran et al., 2006; Dollin et al., 2008; DeLancey et al., 2009). This is probably due to the relatively short management history of these stands, which had

Table 15.2. *Examples of studies that have documented the negative impacts of forest management on saproxylic species.*

Main findings	Species group (references)
• Species richness is lower in managed than natural forests • Species richness decreases with increasing management intensity	• Wood-decaying fungi (Bader et al., 1995; Sippola et al., 2001; Penttilä et al., 2004; Küffer and Senn-Irlet, 2005; Junninen et al., 2006) • Saproxylic beetles (Martikainen et al., 2000; Grove, 2002b; Maeto et al., 2002) • Saproxylic fungus gnats (Økland, 1994, 1996)
• Red-listed species are missing from stands exceeding a certain management intensity	• Wood-decaying fungi (Sippola et al., 2001, 2004; Penttilä et al., 2004; Junninen et al., 2006) • Saproxylic beetles (Müller et al., 2008; Brunet and Isacsson, 2009)
• Red-listed species are less abundant in regions with longer management history	• Wood-decaying fungi (Lindgren, 2001; Siitonen et al., 2001; Laaksonen et al., 2008) • Saproxylic beetles (Siitonen and Martikainen, 1994)
• Most occurrences of red-listed and rare species on large-diameter, well-decayed logs	• Wood-decaying fungi (Kruys et al., 1999; Stokland and Kauserud, 2004; Berglund et al., 2009; Stokland and Larsson, 2011) • Epixylic mosses and lichens (Söderström, 1988; Kruys et al., 1999; Berglund et al., 2009)

been subjected to their first management rotation and still had plenty of dead wood left as a legacy from the pre-harvest stand.

15.4.2 IUCN criteria and their application

Empirical studies such as those cited in the previous section can be used to identify species that are potentially threatened, even in circumstances where there is not enough background data to make exact assessments of the threat status of individual species. The Red List categories and criteria of the IUCN (International Union for the Conservation of Nature) have been developed to provide an explicit framework for the classification of species according to their extinction risk in an objective and comparable manner (IUCN, 2001). This system does not merely label the species as potentially threatened or not, but aims to assess the extinction risk of each species and subsequently to place all species into appropriate Red List categories (IUCN, 2001; Rodrigues et al., 2006). Application of the criteria is based on the available knowledge on the abundance, range size and changes in population size of each species. The term 'threatened species' includes the Red List categories critically endangered (CR), endangered (EN) and vulnerable (VU), while the term 'red-listed species' additionally includes near threatened (NT) species and those that have been classified as data deficient (DD).

Red Lists can be used, for example, to compare the threat status of different taxonomic groups. For instance, at the global scale, amphibians are more threatened than birds. Since saproxylic species are found in almost all taxonomic groups, it is not usually possible to directly extract the threat status estimates of saproxylics as one ecological group. The first international assessment that specifically concerns saproxylic species is the *European Red List of Saproxylic Beetles* (Nieto and Alexander, 2010). The assessment only covers a selection of 436 species belonging to some of the best-known families. Overall, nearly 11% of the assessed saproxylic beetles (46 species) were considered threatened and a further 13% (56 species) near threatened. Since the total number of saproxylic beetles in Europe is in the order of 10 000 species, and since the same threat factors affect both the evaluated and non-evaluated species, the real number of threatened species should be at least 20 times as high as the number of listed species, i.e. most probably over 1000 species.

All the Red List assessments that have been executed according to the IUCN instructions have documented the criteria on which the Red List category of each red-listed species is based. In some national assessments, the primary habitats and also the primary causes of threat have been documented for each species. If the reduction of dead wood has been used as a separate cause of threat, it is possible to examine the general threat status of saproxylics across taxonomic groups. For example, in the Finnish Red List assessment (*The 2010 Red List of Finnish Species*), the reduction in dead wood was distinguished as a separate threat factor (Rassi et al., 2010). A total of 814 forest-dwelling species across all the taxonomic groups were classified as threatened. Of these, the reduction in dead wood is one of the causes of threat for 279 species (34%). Although the Red List assessment in Finland is exceptionally comprehensive, about half of the species could not be evaluated regarding their extinction risk because of inadequate knowledge. Here also, the real number of threatened saproxylic species is likely to be about twice as high as the number of listed species, i.e. most probably over 500 species.

Komonen et al. (2008) reviewed the use of IUCN Red List criteria for saproxylic beetles in Sweden and Finland, and showed that there are some inherent problems in the application of these criteria to saproxylic invertebrates. The great majority of species (86%) had been listed based on criterion B (geographical range) and most of the remaining species (12%) were listed based on criterion D (very small population size). In contrast, criteria A (population reduction), C (small population size) and E (quantitative analyses) had not been applied at all or in few cases only. The main reason for this was that the data on population size and changes in population size (in terms of individuals or occupied area) of invertebrates and fungi are generally insufficient. Naturally, this problem is not restricted to saproxylics but concerns species living in other microhabitats as well. In addition, the threshold values for individuals apply mainly to vertebrate species. In the case of small saproxylic invertebrates, a single tree can host hundreds of individuals, but a viable local population still needs to inhabit sufficiently many host trees, and there needs to be continuous recruitment of new suitable host trees (Siitonen and Saaristo, 2000; Ranius, 2001). This means that the number of inhabited trees might be a better indicator of population size. Furthermore, the time window for assessing the population decline should also be related to the turnover time of the substrate. Hollow living trees have slow turnover rates, and several invertebrate generations can reproduce in the same tree.

To summarize, there are a large number of empirical studies showing the decline of common and medium-rare saproxylic species, while there is a lack of sufficiently detailed species-specific information for assessing the threat status according to the IUCN criteria. It is probable that the current population trend for many saproxylic species is a large-scale gradual decline.

15.4.3 Habitat requirements

As in any group of species, some saproxylic species are more susceptible to extinction than others. The extinction risk for each species depends both on its ecological traits, including habitat requirements, and on changes that have occurred in its habitats.

The most comprehensive analysis of the habitat requirements of red-listed saproxylic species to date was made by Jonsell et al. (1998).

They examined all the saproxylic invertebrates (542 species) which were red-listed in Sweden. Host-tree requirements were defined and classified according to tree species, decay stage, tree type (living, log, snag, stump), coarseness, part of the trunk, exposure and microhabitat. Some tree species or genera were far more important for threatened species than others. The number of threatened invertebrate species per tree genus varied from five species (*Sorbus*, *Malus*) to over 200 species (*Quercus*). Those tree genera that had the highest number of species also tended to have the highest number of monophagous species. However, almost all tree genera hosted at least some specialized monophagous species of their own. Faunas on conifers (*Picea*, *Pinus*) and deciduous trees differed the most from each other (about 5% shared species). Faunas on the boreal deciduous genera (*Betula*, *Alnus*, *Populus*) diverged from the faunas on temperate deciduous genera (at most 10% of common species), and in the latter group (*Quercus*, *Fagus*, *Acer*, *Ulmus*, *Tilia*, *Fraxinus*, *Carpinus*, *Corylus*) the similarity between communities varied (at most, 30% of common species). Faunas on different tree genera were thus complementary. The similarity of communities between tree species increased with the decay stage. The largest number of red-listed species (about half of the species) occurred in mid-decayed trunks, but each decay stage hosted some specialized species.

About equal numbers of species utilized living trees, snags and logs, whereas only about one-tenth of the species were able to utilize stumps, and none of the species were specialized on them. About one-third of the species were specialized on very coarse trunks (diameter, depending on the tree species, mostly 50–100 cm), but only a few species on thin trunks. A considerably larger proportion of species (about a quarter) preferred sun-exposed environments rather than shaded environments (about one-tenth of the species); the rest were indifferent to sun exposure (see Figure 9.4). The majority of red-listed species lived in wood or under the bark. About one-fifth of species occurred in hollow living trees, and slightly over one-tenth of species were specialized in this microhabitat.

Although the quantitative results referred to above concern red-listed saproxylic invertebrates occurring within northern Europe in Sweden, it is probable that the results can be qualitatively generalized to other regions as well, and can be summarized as follows. Most threatened species are specialized to use those types of dead wood that have declined the most as a result of forest management and other human impact, i.e. coarse, medium to very decayed snags and logs, as well as living hollow trees. Conversely, species that can utilize the types of dead wood that occur abundantly even in managed forests – freshly dead trees, small-diameter woody debris and cut stumps – are usually not under threat. Certain tree species are likely to be particularly important to threatened saproxylic species regionally; these include those tree species which originally hosted the largest numbers of specialized saproxylics, and in which the number of potential host trees had declined the most.

15.4.4 Ecological traits

The relationships between species' ecological traits and their extinction risk have been intensively studied (see, e.g., Fisher and Owens, 2004; Purvis et al., 2005). The main question is what differences exist between threatened and non-threatened species? The same traits that make species sensitive to habitat loss and fragmentation (Henle et al., 2004) also increase the extinction risk. Most of the studies concern vertebrates, but several ecological attributes, which are known to increase species' susceptibility to extinction, are likely to be applicable to saproxylics too.

The following attributes can be expected to be important in predicting the extinction vulnerability of saproxylics: population size, reproductive potential, dispersal power, length of reproductive cycle, ecological specialization, rarity, microhabitat and matrix use, disturbance and competition sensitiveness, and trophic position. Several of these traits are intercorrelated and related to other species characteristics. For instance, large body size (in proportion to taxonomically related species) can predict extinction vulnerability, because large organisms generally have lower population densities and slower reproductive cycles. Species possessing several predisposing traits are particularly vulnerable.

15.5 Survey methods and nature conservation evaluation

15.5.1 Survey methods

It has frequently been suggested that, due to their concealed lifestyle, saproxylic invertebrates are generally difficult to sample. This may be partly true, particularly as regards mass-trapping methods, but does not hold true for all saproxylics. As we shall see below, it is possible to develop efficient sampling methods specifically targeted towards surveying certain threatened species.

In general, it is far more difficult to gather sufficient quantitative materials of infrequent species than of common species. Making inventories of threatened saproxylic species may be particularly difficult because of their rarity, special microhabitat requirements and low abundance. A range of trapping methods have been developed for sampling saproxylic invertebrate species; these methods often yield large numbers of individuals and species with relatively little fieldwork effort. The most difficult and expensive phase in research is not the fieldwork (setting and emptying traps at frequent intervals) but sorting and identifying the samples afterwards.

Window traps and trunk-window traps have been the most frequently used sampling methods in quantitative studies involving saproxylic beetle species (Kaila, 1993; Siitonen, 1994; Sverdrup-Thygeson and Birkemoe, 2008). Pitfall traps buried in wood mould are a useful method for surveying beetle fauna in hollow trees; pitfalls collect specialized species which are rarely caught by window traps (Ranius and Jansson, 2002). A malaise trap is another type of flight interception trap which is suitable for quantitative studies on some other saproxylic insect groups such as Diptera and parasitic Hymenoptera (Økland, 1996; Darling and Packer, 1988).

If quantitative mass-trapping methods are used to determine how many red-listed species occur in different forest areas, the problem is that a very large sample size, in terms of individuals, is needed; otherwise red-listed species will be caught only accidentally. Martikainen and Kouki (2003) came to the conclusion that if red-listed saproxylic beetles are to be used for comparing the conservation value of boreal forest stands, samples which contain fewer than about 200 species, or 2000 individuals, are practically useless. The large sample sizes and daunting amount of identification work needed are likely to be an equal, or even worse, problem in the more species-rich temperate and tropical forests.

However, not all threatened saproxylic invertebrates are particularly difficult to discover when present. On the contrary, many species are very easy to find using direct searching if the suitable host-tree types and microhabitats of the species are known. A good example of the efficiency of microhabitat-based searching is the survey of certain threatened dipteran species in Scotland. When dipterists learned to recognize the microhabitats of these species (*Blera fallax* in wet rot holes in pine stumps, *Hammerschmidtia ferruginea* in the wet decaying layer under aspen bark, *Callicera rufa* in wet rot holes in conifers), it became possible to systematically survey the species in their potential habitats, and a large number of previously unknown localities were discovered (Rotheray and MacGowan, 2000). Systematic surveys and subsequent new records can lead to downgrading of threatened species, but equally well to upgrading if only a few habitats suitable for the species are found, or if the species is frequently missing from suitable microhabitats. The latter may indicate breaks in substrate continuity, habitat fragmentation, lack of recolonization sources, poor dispersal ability, etc.

A special feature shared by many saproxylic insects, particularly beetles, feeding on inner bark and wood is that they make characteristic galleries and exit holes which can enable a definite identification of species (Ehnström and Axelsson, 2002), sometimes even years after the adults have left the trees. Making inventories of threatened saproxylic insects based on searching galleries in suitable host trees is a potentially efficient method which, so far, has been used in only a few studies (Wikars, 2004; Buse et al., 2007; Hedgren, 2009).

Surveys of wood-decay fungi designed to compare species assemblages between different forest stands generally rely on sample plots of fixed size in which all dead trees larger than certain minimum dimensions are examined. Here also, adequate sample size may be a problem, because many threatened species and their host trees occur so sparsely – even in good sites – that they will only rarely fall within the sampling plots. In order to detect 80% of wood-inhabiting fungus species present in boreal old-growth forest, 300–600 logs, or an area of 2.5–5 ha, had to be sampled (Berglund et al., 2005). An additional problem and peculiarity of decay fungi is that most of the species have annual fruiting bodies and are observable only during a small part of the year, generally in autumn in boreal and temperate forests in the northern hemisphere. Most species produce fruiting bodies only during a few years on a single log (Berglund et al., 2005; Jönsson et al., 2008). As a consequence, there may be a hundredfold difference in detectability between different polypore species (Lõhmus, 2009).

In polypores a considerable number (about a quarter) of the species are perennial, i.e. the fruiting bodies last for several years. Perennial species are detectable throughout the year, easier to identify than annuals, and have much smaller year-to-year variation in abundance. These features make red-listed perennial polypores an attractive target group for ecological case studies concentrating on threatened wood-inhabiting fungi. For instance, the endangered polypore *Antrodia crassa* is probably the first fungus species for which an estimate of the national population size (*c.* 3000 inhabited trunks in Finland) was obtained, based on systematic searching in potential habitats covering a total forest area of 15 000 ha, and subsequent extrapolation using the total area of comparable forests (Junninen, 2009). The number of perennial species has been shown to give a good indication of the overall polypore species richness as well as the number of red-listed annual species, provided that sufficiently many records of perennials (at least 60–80) have been made (Halme et al., 2009).

15.5.2 Assessing nature conservation value based on threatened saproxylic species

Saproxylic species (including threatened ones) that are thought to indicate stand naturalness and long-term continuity of dead wood have been used to assess and rank the conservation value of woodlands. In Britain, the scoring and ranking systems have been developed based mainly on saproxylic beetles (see Chapter 16 for further details), whereas in the Nordic countries polypores have been widely used in inventories of old-growth forest (Karström, 1992; Niemelä, 2005). Even though the assumption that the presence of certain species indicates long continuity or the naturalness of a stand may be questioned (see, e.g., Nordén and Appelqvist, 2001; Rolstad et al., 2002), the presence of many threatened species is, in itself, a sign of high conservation value of the site. Recently, Jansson et al. (2009) developed a scoring system (the Conservation Priority Index) for old oak forests in southern Sweden based on red-listed saproxylic beetles and on a standardized trapping effort in the stands to be compared.

16 · *Dead wood in agricultural and urban habitats*

Juha Siitonen

Dead wood and saproxylic species do not occur only in forests. A rich saproxylic community also inhabits dead wood in habitats created by people, both in agricultural and urban landscapes, such as pasture woodlands and parks. Human-maintained habitats can provide important sites, or even the last footholds, for surprisingly many rare saproxylic species. Until recently, the biodiversity value and management of trees in agricultural and urban settings have not received the attention they deserve, despite the fact that these environments often contain greater concentrations of ancient and valuable trees than managed forests. In fact, most readers of this book are likely to find not only the closest populations of saproxylic species, but also the closest populations of threatened saproxylic species, only a few kilometres away, in the nearest park where old, hollow trees occur.

In this chapter, we deal with the occurrence, conservation and management of saproxylic species in different kinds of cultural environments, excluding forests managed primarily for wood production, which were treated in Chapter 13.

16.1 Cultural environments as habitats for saproxylic species

The importance of wooded cultural habitats for a number of saproxylic species can be explained as follows.

- These habitats are the products of traditional forms of land use, and some have a very long continuity, extending from several hundreds to even thousands of years back in time (Figure 16.1). The same kinds of habitats may have continuously occurred in the same regions for millennia, even though their amount and quality has fluctuated along with human populations and changes in land-use patterns.

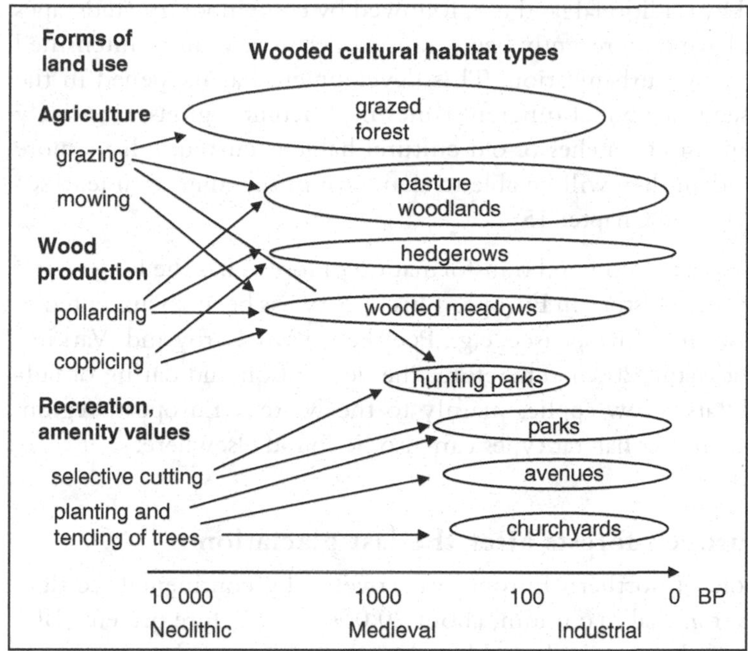

Figure 16.1. Cultural habitat types important for saproxylic species, the main forms of land use creating each habitat type, and their approximate ages and continuity in Europe.

- Human-modified landscapes can provide habitats which are analogous to natural ones, and which can accommodate perfect microhabitats for saproxylic species. Species adapted to naturally open, grazed or otherwise disturbed woodlands (see Chapter 12) can readily inhabit cultural woodlands, provided that suitable microhabitats are available.
- Open and warm conditions are favoured by many thermophilous species with southern or continental distribution in Europe. One can speculate that certain saproxylic species which are found almost exclusively in cultural habitats in western and northern Europe, such as the hermit beetle (*Osmoderma eremita*) and the great capricorn beetle (*Cerambyx cerdo*) (Ranius et al., 2005; Buse et al., 2007), have widened their distributions in the past through the novel habitats created by people.
- Human impact on saproxylic habitats constitutes a gradient, from pristine forests to landscapes first influenced and then completely

shaped by traditional land use, followed by contemporary landscapes which have only recently been transformed as a result of intensified land use and urbanization. This development has happened in the same sequence but at different times in different regions. It is likely that remaining patches of old cultural habitats currently have more species than they will be able to maintain in the long term (extinction debt, see Chapter 15).

Human-caused loss and transformation of forests has the longest and most pervasive history in Europe. This history has been documented in detail in western Europe (see, e.g., Peterken, 1996; Kirby and Watkins, 1998; Rackham, 2003). Therefore, the description and dating of cultural habitats below applies mainly to the western European region. However, similar habitat types can also be found elsewhere.

16.2 Europe's forests after the last glaciation

The whole of northern Europe was covered by continental ice during the last glacial maximum, about 20 000 years before present (BP). Because of the extremely cold and arid climate, treeless tundra or steppe vegetation covered northern Eurasia. Boreal coniferous and temperate deciduous forests were drastically reduced and displaced southwards (Prentice and Jolly, 2000; Tarasov et al., 2000). Both boreal and temperate forests (and hence all the saproxylic species) in Europe had retreated into small refugial areas, the most notable being in the Iberian, Apennine and Balkan peninsulas, and along the northern coast of the Black Sea (Bennet et al., 1991; Taberlet and Cheddadi, 2002). Recent studies have accumulated evidence that scattered populations of coniferous and some deciduous tree species survived in microenvironmentally favourable sites much farther north than previously assumed (Willis and Van Andel, 2004; Magri et al., 2006; Kullman, 2008; Svenning et al., 2008; Binney et al., 2009). Many other woodland species may also have survived in these cryptic refugia (Stewart and Lister, 2001; Provan and Bennett, 2008).

The climate had already started to warm up by the end of the Pleistocene, about 15 000 years BP, and the ranges of tree species began to expand from the refugial areas. The present interglacial period, the Holocene, started 11 500 years ago. The early Holocene forests were dominated by pine and birch, with hazel, alder and willow forming the bush layer. During the warm Atlantic period, 8000–5000 years

BP, most of the temperate deciduous tree species expanded their distribution and either reached or exceeded their present distribution limits (Prentice and Jolly, 2000; Brewer et al., 2002). In the early Atlantic period, the greater part of Europe was covered by continuous temperate forest in which large, over-mature tree individuals and coarse dead wood were major components (Box 16.1). Diversification of saproxylic fauna took place at the same time as the highly specialized ancient-forest species spread from their glacial refugia (Whitehouse, 2006).

Box 16.1 Primeval forests in Europe: open or closed?

The traditional view is that lowland temperate Europe was covered by closed-canopy old-growth forest before the onset of significant human influence some 5000 years ago. The *closed-forest hypothesis* was elaborated by vegetation scientists over 50 years ago, based on subfossil pollen deposits in lake sediments and peat, and has been widely accepted by forest ecologists and conservationists until lately. The closed-forest hypothesis was recently challenged by Vera (2000) who proposed that the natural vegetation is a mosaic of grasslands, scrubs, trees and groves, and that large mammals are the driving force determining forest structure by affecting tree regeneration.

The *wood-pasture hypothesis* has gained popularity among conservationists, lichenologists and invertebrate specialists. Both hypotheses need to be weighed against the available evidence. The existing palaeoecological data have been revisited by several plant ecologists to explore the likelihood that open parkland was the dominating vegetation type during the Holocene (Svenning, 2002; Bradshaw et al., 2003; Hodder et al., 2005; Mitchell, 2005). The palynological data indicate that open-canopy forests did not become common until during the last 3000 years following human settlement. An important piece of evidence comes from the subfossil insect assemblages. Abundant saproxylic but only scarce herb- and dung-dependent beetles have been recorded from most sites during the Atlantic period. Dung beetles – indicating the presence of large herbivores – as well as open-ground beetles generally appear and increase simultaneously with the spread of agriculture and human settlements (Buckland and Dinnin, 1993; Whitehouse, 2006; Elias et al., 2009). However, as pointed out by Alexander (2005), the subfossil saproxylic beetle fauna can more easily be interpreted to

Box 16.1 (*continued*)

match the wood-pasture hypothesis, with most species favouring open rather than closed-canopy conditions. It is evident that more integrated analysis is needed to resolve the question. Many sites from northwestern Europe show that widespread open vegetation indeed existed on river flood plains through the Holocene (Svenning, 2002). River dynamics created open vegetation that was maintained by grazing large herbivores.

The potential importance of large herbivores in shaping the structure of primeval forest was first raised by several authors already earlier. The *megaherbivore hypothesis* suggests that Eurasian vegetation evolved during the Late Tertiary period under the influence of large herbivores, most of which are extinct today (Owen-Smith, 1987; Andersson and Appelquist, 1990). Grasslands with solitary trees and open woodlands grazed by large herbivores developed and expanded during the Late Miocene and Pliocene periods about 10 million years ago, due to the massive mountain uplift in Europe, central Asia and western North America and the subsequent cooling and drying of continental climates (Potts and Behrensmeyer, 1992; Fortelius et al., 2002). At the same time, the area of closed forest declined. During the Pleistocene, between about 2.5 million and 12 000 years ago, repeated glaciations and severe climatic changes caused subsequent contractions and expansions of forests in all the continents. Closed forests versus open grassland-woodland types (tundra, steppe, savanna) have coexisted in highly variable proportions and configurations during the different glacial and interglacial periods throughout the Pleistocene. It seems reasonable to assume that these events have constituted a selection pressure favouring dispersal ability, and that many saproxylic taxa evolved to use scattered dead and ancient trees in open environments.

From the point of view of conservation and proper management of the remaining valuable woodland areas, the question about which of the two vegetation types, closed forest or open woodland, was the dominating one during the Early Holocene in western Europe seems to be rather irrelevant. Management targets in each individual case should be based primarily on the present structural features, occurrence of vulnerable species and their specific habitat requirements, and on the foreseen successional development with or without management interventions.

16.3 Prehistoric modification of forests by humans in Europe

16.3.1 Disappearance of natural forests

At the same time with the spread of temperate tree species during the climatically favourable Atlantic period, the Neolithic cultures based on agriculture started to expand rapidly. Clearance of forest into agricultural land began in Greece and the Balkans about 8000 years ago, and the Linear Pottery culture spread into central Europe along the major river valleys about 7500–6500 years ago. Ring-barking of trees and fire were used to clear the land for agriculture, but also to open up the forest and stimulate the growth of grasses in order to enhance grazing potential. By the Late Bronze Age, about 3000 years ago, natural forests had disappeared from most cultivable lowland areas all over Europe. Forest clearance continued in western Europe during the Iron Age from about 3000 to 1500 years ago, when more efficient axes and ploughs made of iron were brought into use. Woodland was reduced to what was necessary for the maintenance of communities, for producing timber for construction, firewood and fodder for livestock.

16.3.2 Pasture woodlands

Forests grazed by livestock and pasture woodlands (wood pastures) are ancient cultural habitats. Livestock grazing spread at the same time as agriculture and settlement. Starting from the Late Bronze Age, about 3000 years ago, until the 1800s, most of the land in temperate Europe was used for traditional animal husbandry. Cattle grazed almost everywhere in the forest. However, grazing pressure and woodland openness have varied both in time and space, from forests grazed only occasionally by cattle to permanently grazed grasslands with scattered trees (Rackham, 1986, 1998). In the boreal zone, cattle grazing in forests started later, during the Iron Age or in Medieval times.

Pasture woodland was defined by Kirby et al. (1995) as an area where trees coexist with grazing domestic or game animals, and where both are of value to people. The density of large herbivores is often higher than would normally occur in natural forest. Besides grazing, pasture woodlands have been used for wood production. Several different systems of combining grassland and trees have existed (Rackham, 2003). Poles used for construction and firewood were obtained by pollarding trees. In pollarding, branches were cut at a height of 2–3 m, which is sufficient to prevent browsing by cattle. The cut trunk starts to produce new branches which may then be cut at regular intervals as they become useful. Pollarding prolongs the lifespan on trees so that, for example, pollarded beeches commonly live 500 years, and pollarded oaks 800 years. The cut branch stubs on the trees provide access to wood that may be colonized by wood-decay fungi and saproxylic insects.

Traditionally managed pasture woodlands can be structurally very similar to pristine woodlands or savannas grazed by wild animals, and they often contain a rich and vulnerable flora and fauna including saproxylic species (Harding and Rose, 1986; Alexander, 1998). The presence of large and old broadleaved trees in open conditions is characteristic of these woodlands. Such living trees provide important dead wood and saproxylic microhabitats (see Chapter 7). Cavities and large dead branches occur on old living trees, and fallen branches can add a significant volume of dead wood to the ground. The impact of the grazing animals on the composition of the trees and shrubs should not be overlooked – the more palatable plants will be browsed preferentially and this may skew the balance between the tree species. Since some pasture woodlands are very old land-use systems, generations of suitable host trees have been available for saproxylic colonizers within their dispersal range for thousands of years, and some pasture woodlands may have direct links to the natural landscapes prior to human influence.

16.3.3 Wooded meadows

The wooded meadow is another type of habitat created by traditional animal husbandry. It is a mixture of low woodland and open meadow maintained by regular coppicing or pollarding of trees and mowing of hay (Häggström, 1998). Burning and seasonal grazing were also applied in some systems. In coppicing, small trunks are frequently harvested to produce firewood. Cutting close to the ground stimulates vegetative sprouting, and the end result is a group of vegetatively regenerated small stems. Certain tree genera such as *Alnus*, *Corylus*, *Carpinus* and *Fraxinus* are particularly suitable for coppicing, while most deciduous genera tolerate pollarding. With large herbivores excluded, trees and shrubs with more palatable foliage can thrive.

Wooded meadows were very common, or even the dominant habitat type, in the coastal areas around the Baltic Sea and in mountainous regions in central and southern Europe. For instance, in Estonia, wooded meadows still covered about 20% of the total land area at the beginning of the 1900s; of the original million hectares, only about a thousand hectares were traditionally managed by the beginning of the 2000s (Sammul et al., 2008). This is a habitat type extremely rich in vascular plants; up to 80 species have been recorded as occurring within one square metre. For saproxylic species in general, coppice meadows were probably not a particularly good habitat type because of the relatively low amount of dead wood present in the form of small stems. However, the multi-stemmed tree clones can be very old and continuously maintain stems at differed stages of decomposition.

16.3.4 Hedgerows

Hedgerows are managed rows of bushes and trees which are used to separate otherwise open fields and meadows. They are often bordered by strips of lower vegetation, and larger pollarded trees are also common in hedgerows in some regions. The structure of hedgerows is maintained by regular cutting. Traditionally, hedgerows were an important source of wood as well as other products (Baudry et al., 2000). The archaeological data show that hedgerows were already a permanent and widespread part of agricultural landscapes during Roman times (Rackham, 1986). Extensive networks of hedgerows are found in old agricultural landscapes in Europe and elsewhere (Baudry et al., 2000).

Obviously, hedgerows serve as extremely important habitats, refuges and corridors, maintaining biodiversity in the otherwise more intensively used agricultural environments. There are only a few studies concerning the potential significance of hedgerows to saproxylic species. Clements and Alexander (2009) studied the saproxylic beetle and dipteran species in hedgerows of different ages in southwestern England. Several ancient woodland indicator species were found, and the average species richness of both indicator and all saproxylic species was higher in hedgerows that were of ancient (pre-1100) or medieval origin. In general, hedgerows may be an optimal habitat for species that are specialized to use small-diameter dead wood in sunny conditions, and that need flowers for feeding as adults, such as many of the smaller jewel beetles (Buprestidae) and longhorn beetles (Cerambycidae). Furthermore, hedgerows containing large pollarded trees constitute vital habitat networks (Dubois et al., 2009).

16.4 Historic woodlands and parks

16.4.1 Hunting parks

A modern observer might think that medieval hunting parks cannot be much more than a curiosity as a saproxylic habitat. However, by isolating substantial areas of forest and parkland from changes in land use, royal families and the nobility initiated the first conservation measures for saproxylic species. These measures were of course incidental to their primary interests in hunting deer and other game, and creating attractive settings for their other recreational pursuits (McLean and Speight, 1993). The first hunting parks belonging to the Crown were already protected more than 1000 years ago. These hunting areas perpetuated habitats with ancient trees that might otherwise have been lost several hundred years earlier (Buckland and Dinnin, 1993). There is evidence that many medieval deer parks were formed from relict areas of wilderness still existing at that time, and probably containing fragments of primeval forests (Alexander, 2004).

Some of the most important sites for saproxylic species in Europe have a period in their history when they were used as hunting parks. These include the New Forest, Windsor, Epping and Sherwood forests in England (Buckland and Dinnin, 1993; Harding and Alexander, 1993), Fontainebleau in France (McLean and Speight, 1993) and Białowieza in Poland (Wesołowski, 2005).

16.4.2 Parks

The words 'park' and 'parkland' are often used to refer to open, grazed areas with some tree cover. Here we use 'park' in a more restricted sense to denote areas with trees that are managed primarily for their scenic value, including designed landscape parks. Trees have been preserved and planted for their amenity value from antiquity. However, from the point of view of saproxylic species, the essential history of parks starts from the Renaissance (about the 1400s to the 1700s).

Many deer parks of medieval origin and former pasture woodlands were landscaped in the 1700s to create landscape parks. These are modified woodlands that incorporate landscape features such as old, impressive trees already existing at the site. In contrast, artificial parks are based on planting of trees in originally unwooded or cleared sites. Site history has been shown to have important implications for saproxylic species. Saproxylic beetle faunas in post-medieval landscape parks that included earlier woodlands are strikingly richer than faunas in parks without such continuity (Harding and Alexander, 1994; Alexander, 1998).

But artificial parks can also be important if they contain some natural elements, if they are located close to potential source areas, or if they are old enough. For instance, parks that were established at the beginning of the 1700s would already have old, hollow trees by the 1850s, when the surrounding countryside was completely different from the present post-industrial landscape. At that time, most parks were connected to the contiguous landscape with traditionally managed woodlands and over-mature trees. Later on over-mature trees have been largely eliminated from the ordinary rural landscape but have persisted in parks.

The most valuable parks are located in the vicinity of castles and mansions (Figure 16.2), sometimes in old churchyards and graveyards. Surveys of saproxylic beetles and dipterans in older parks in central Europe (Franc, 1997), Britain (Denton and Chandler, 2005)

Figure 16.2. Hollow lime (*Tilia cordata*) in Träskända Mansion Park near Helsinki, southern Finland. The park contains tens of large hollow trees hosting endangered saproxylic beetles such as the false click beetle *Eucnemis capucina* (Eucnemidae) and the click beetle *Crepidoderus mutilatus* (Elateridae) (photo Juha Siitonen).

and northern Europe (Andersson, 1999; Biström et al., 2000; Jonsell, 2004a, 2008) have shown that rare and threatened species are regularly found in these habitats, which can act as substitutes for the natural habitats to some extent.

16.4.3 Avenues

Avenues and alleys are roads or walkways lined by trees. The term alley (from the French term allée) is usually confined to avenues planted in parks. Similar to parks, old avenues contain hollow trees and thus provide habitats for saproxylic species. The trees are also effectively open-grown and can develop the full range of saproxylic habitats, as in hunting parks or wood pasture. Avenues have some distinctive properties as a habitat. Trees are often continuously replaced, i.e. when old trees fall, new ones will be planted. This has guaranteed the temporal continuity of suitable host trees (Jonsell, 2004b). Furthermore, avenues have formed extensive networks in the countryside in many regions (Oleksa et al., 2006), increasing the spatial continuity of host trees. A linear habitat can enhance the dispersal of species with limited dispersal capacity, and avenues may have acted as dispersal corridors between woodland patches. Results from southern Sweden (Gerell, 2000) point to the importance of connectivity: most hollow trees in avenues close to woodlands with over-mature trees hosted red-listed beetles, while none were found in avenues further than 5 km from potential source areas.

16.4.4 Ancient and veteran trees

The terms 'ancient' and 'veteran' trees are often used interchangeably, but for some purposes it may be useful to distinguish between them (ATF, 2008). 'Ancient tree' has been used to describe tree individuals that have passed the mature stage and reached a stage where the annual rings have a declining cross-sectional area. Other typical characteristics of ancient trees are crown dieback which exceeds growth, and the loss of large branches or sections of the crown because the tree is not able to maintain functional cambium over the whole stem area as this expands with age – this process is termed canopy retrenchment. Younger trees may develop similar characteristics through damage, and these can be referred to as 'veteran trees'. In this terminology, ancient trees are a subset of veteran trees. Collectively, veteran trees can be defined as

(a)

(b)

Figure 16.3. Storkeegen (stork oak) in Jägerspris Nordskov, Denmark: (a) lithograph by Gurlitt (1839) portraying the giant hollow oak standing in an open pasture in the early 1800s; (b) photograph by Ole Martin (1974) of the same tree almost 150 years later, showing beech regrowth following the cessation of grazing. The tree is about 800–900 years old and still alive. Specialized saproxylic beetle species such as the click beetles *Ampedus cardinalis* and *A. hjorti* and the darkling beetle *Tenebrio opacus* were still living in the hollow (Ole Martin, personal communication).

being of interest biologically, culturally or aesthetically because of their age, size or condition (Read, 2000). Veteran trees with a large diameter maintain a number of important microhabitats, including trunk and branch cavities, sap exudations, barkless areas with exposed wood, large quantities of dead wood in the canopy, fruiting bodies of wood-decaying fungi, and large fallen branches (see Figure 7.1).

Veteran trees are the key feature of almost all the wooded cultural habitats described above. In addition, scattered individual veteran trees occur in the countryside in field margins, river banks, along old roads, etc. The potential importance of individual trees should not be underestimated. Some ancient trees have sustained microhabitats such as trunk cavities for hundreds of years (Figure 16.3), and they may host remnant populations of saproxylic species which were much more common and widespread in the past. It is possible, though not documented, that individual trees can provide links (stepping stones) between habitat patches that would otherwise be too isolated.

16.5 Urban forests and wooded ruderal areas

Urban forests constitute one of the most recent wooded cultural habitat types. Urban forests reserved mainly for recreation have emerged only during the last 50 years. It is a cultural habitat type that can be distinguished from urban parks, which tend to be more open. In addition, while parks have been artificially constructed (though they may contain natural parts), urban forests have often developed from former production forests or rural woodlands when urbanization has expanded and adjacent areas have become more important for recreation than for timber harvest or other commodity production. Urban forests and trees cover significant areas and their extent is constantly growing (Nowak et al., 2001). For instance, in Sweden, urban and peri-urban forests comprise an average of 20% of the area of cities, which makes up an area that is larger than the total extent of protected forests (Hedblom and Söderström, 2008).

The distinctive properties of urban forests include persistent small-scale disturbances and stress factors in the form of trampling, increased nitrogen deposition, aerial pollution, etc. Urban forests are more open and have more pioneer deciduous trees and dead wood (Hedblom and Söderström, 2008) than the average managed forest. Because wood production is not the primary goal, they tend to be less intensively managed and structurally more diverse than production forests. Vegetation and bird assemblages in urban forests have been intensively studied, but there is very little information available on saproxylic species. Nevertheless, the structural features of urban forests indicate that they are likely to contain suitable habitats for disturbance-adapted saproxylic species.

Wooded ruderal areas are formerly treeless areas of derelict land with varying origin, such as land excavation or landfill sites, roadsides and abandoned fields, which have become spontaneously afforested. These urban edgelands often remain neglected, resulting in natural-like early successional stages with deciduous dead wood created by competition. For instance, wooded ruderal areas in Helsinki Metropolitan Area, southern Finland, have mature goat willows (*Salix caprea*) at much higher densities than any natural woodland type. These trees host rare saproxylic specialist species of goat willow, such as the longhorn beetle *Aromia moschata* and the woodwasp *Xiphydria prolongata* (Juha Siitonen, personal observation). These areas are likely to be short-lived, as they have no protection from development or misguided tidying.

16.6 Conservation and management of dead wood in cultural environments

16.6.1 Threats to cultural habitats

Traditional cultural landscapes have dramatically changed during the last hundred years, and the rate of change has been ever-increasing during the last few decades (Antrop, 2005). The driving forces are population growth, rising standards of living, urbanization, globalization, and thus a growing demand for land and for high productivity per unit area. These pressures cause serious threats to the persistence and ecological integrity of cultural habitats (Table 16.1). Most threat factors can be placed into two broad categories: intensification of land use and abandonment, i.e. cessation of traditional land-use practices (see, e.g., Jongman, 2002; Plieninger et al., 2006).

Intensification of land use can lead to the total destruction of habitats, but it is equally common that, even if habitats have persisted, their essential structural features have been significantly reduced. Ancient trees have continuously decreased in agricultural landscapes, and in some regions their well-documented elimination had already started 200 years ago (Eliasson and Nilsson, 2002). Old trees were

Table 16.1. *Threats to wooded cultural habitats. The significance of the different threat factors varies according to the habitat type, dominant tree species, region, etc. Compiled from Harding and Alexander (1993), Key and Ball (1993), Alexander (1995), Kirby et al. (1995), Höjer and Hultengren (2004) and Nieto and Alexander (2010).*

Threat factors
Complete destruction of sites for urban development, agriculture or tree plantations
Pasture improvement in pasture woodlands, including felling of trees, ploughing and fertilizing, causing root damage and susceptibility to drought and pathogens
Overgrazing of pasture woodlands, causing tree mortality and lacking regeneration
Too much young growth shading out existing old trees
Mature trees to replace veteran trees are lacking
New plantings too dense for development of open-grown trees which would mature and become ancient trees
Introduced invasive shrub and tree species
Introduced pest and pathogen species
Climate change: increased mortality because of more frequent drought and storm events
Felling of over-mature trees for the sake of public safety or aesthetic value
Removal of dead trees, dead branches or dead wood for tidiness
Harmful arboricultural practices, such as emptying wood mould and filling of cavities
Planting of exotic tree species unsuitable as future host trees
Collecting of firewood

first removed, as they became an impediment to tillage when former pasture woodlands were converted into more productive agricultural systems.

Abandonment of formerly open habitats such as pasture woodlands leads to regrowth and closed-canopy thickets (Figure 16.3). Over-mature trees that used to grow in open conditions do not tolerate the shading of younger trees but will die because of competition for light and nutrients. Shading is harmful for those saproxylic species that prefer sun-exposed host trees (Ranius and Jansson, 2000; Nieto and Alexander, 2010).

In many cultural habitats, the age structure of the tree population is unbalanced so that mature tree cohorts are lacking, or insufficient to compensate for the continuing loss of old trees. Furthermore, it is unlikely that those trees that are currently mature would be allowed to grow until they reach the ancient tree stage. Cutting is the main mortality factor of veteran trees, but also other factors may increase their mortality rate. These include introduced invasive shrubs and trees, as well as non-native pests and pathogens. The latter have caused large-scale devastating mortality episodes of several tree species that used to be common in parks, avenues and other human-maintained habitats. Climate change can affect the mortality of old trees directly by increasing drought and storm events, as well as indirectly by enhancing the suitability of climate to invasive pest and pathogen species.

One of the most serious examples of introduced pathogens is Dutch elm disease, caused by two ascomycete species (*Ophiostoma ulmi* and *O. nova-ulmi*) and spread by elm bark beetles (*Scolytus multistriatus, S. scolytus* and *Hylurgopinus rufipes*) (Gibbs, 1978; Brasier and Kirk, 2010). The disease has wiped out most of the large elms (*Ulmus* spp.) both in North America and Europe; elms have been widely used as shade and street trees because of their tolerance to stress factors. Unfortunately, comparable new epidemics continue to threaten several other tree species. European ash (*Fraxinus excelsior*) is currently declining throughout its distribution range because of ash dieback caused by the fungus *Chalara fraxinea* (Bakys et al., 2009), and oriental plane (*Platanus orientalis*), a highly valued shade tree species planted from ancient times and distributed from the Mediterranean area to India, is threatened by a canker stain disease caused by the fungus *Ceratocystis platani* (Ocasio-Morales et al., 2007). Oak is also under severe threat from *Phytophthora* and other diseases causing oak dieback.

Particular pressures are connected with those cultural habitats that are used for recreation and that are open to public access. The main drivers here are more social than ecological or economic in nature (see Grimm et al., 2000), and include issues such as people's amenity perceptions and values, public opinion and demand for 'tidiness', public safety, and institutional constraints. Over-mature or dead trees are seen by many people as signs of deterioration, neglect and waste. Managers responsible for urban forests and parks may lack knowledge about their biodiversity value. Even if managing organizations acknowledge the value of dead wood for biodiversity in theory, they may be reluctant to promote a non-interference type of management in practice because this might reduce their resources and eventually threaten their existence. Furthermore, the legal liability of the

Figure 16.4. Hollow trees near dwellings may be felled for firewood. This felled giant beech in Denmark hosted several endangered saproxylic species, including the hermit beetle (*Osmoderma eremita*) and the click-beetle species *Athous mutilatus*, *Elater ferrugineus* and *Procraerus tibialis*. (Ole Martin, photo and personal communication).

land-owner or manager for hazardous trees (see, e.g., Mortimer and Kane, 2005) can directly force the managers to remove large trees with visible signs of decay. The collection of firewood is a major threat to saproxylic species in rural areas in the developing countries (Christensen et al., 2009). In western countries, having a fireplace is both fashionable and supposedly ecologically friendly, and in peri-urban residential areas the collecting of firewood efficiently destroys saproxylic habitats (Figure 16.4.)

16.6.2 Surveys of important sites

The first step in conservation and management of cultural habitats important for saproxylic species is to map the locations of potentially important sites, and then to survey their conservation value. Surveys can be directed towards habitat features, individual trees or species assemblages. At each hierarchical level, the accuracy of assessing the conservation value increases, but at the same time the survey costs and requirements for expert contribution increase too. Surveys of veteran trees have been performed on a large scale in Britain and Sweden, and inventory protocols for recording data on sites and individual trees have been developed.

In the most simple type of inventory, potential sites are outlined on a map and classified according to the habitat type. The usefulness of the inventory increases considerably if the numbers of veteran trees by tree species and age classes or diameter classes are recorded. Rapid non-specialist survey protocols which enable the evaluation of a site for its veteran tree interest, based on the size and quality of the veteran tree population *per se*, can be a useful conservation tool (Castle and Mileto, 2005). The surface area of a site and the number of veteran trees in it are usually good surrogates for the species richness of saproxylic species, and for the occurrence probability of threatened species (see Harding and Alexander, 1993).

The next level of accuracy entails mapping and measuring individual trees and their structural features. These include tree species, diameter or circumference, condition, and the occurrence of microhabitats such as cavities, sap exudations, dead attached branches, fruiting bodies of decay fungi, etc. (Fay and de Berger, 1997, 2003; Hultengren and Nitare, 1999; Sörensson, 2008). Nowadays it has become easy to record the exact location of trees by means of GPS devices. Registering the location and quality of individual trees enables both reliable assessment of the conservation value and long-term monitoring of sites. Accurate information on individual trees, site conditions, threats and management requirements is a prerequisite for successful management and restoration programmes.

A survey of saproxylic species assemblages with a reasonable coverage is the alternative which requires the largest effort. If the purpose of an inventory is to assess and compare the conservation values of sites, it should be based either on standardized trapping methods with sufficient effort, or on long-term collecting at the site using various methods so that most of the noteworthy species are likely to have been found.

Surveys of saproxylic beetles have been used in practice for assessing the conservation value of woodland sites mainly in Britain and Sweden (Harding and Rose, 1986; Harding and Alexander, 1994; Fowles et al., 1999; Alexander, 2004; Jansson et al., 2009). Conservation priority indexes have been constructed based on the presence of red-listed species, or species thought to indicate long-term woodland continuity. For instance, the index of ecological continuity (IEC) is based on the occurrence and scoring of 180 species out of the 700 British native saproxylic beetles (Alexander, 2004). Index values can then be used to rank sites, and to identify nationally or regionally important areas. An inherent problem of this approach is that sites that have been intensively studied receive higher scores than poorly studied, potentially valuable sites. Fowles et al. (1999) introduced a method that overcomes the problem of differences in sampling effort, provided that at least about 50 species are recorded. The saproxylic quality index (SQI) is based on assigning a score to each species (1 for very common species up to 32 for red-listed species), summing the scores, and dividing the sum by the total number of species; the index thus describes the average rarity of species in each locality. The IEC and SQI approaches are both useful and provide different viewpoints: the former about the whole fauna, the latter about the presence of rarities.

16.6.3 Management options

There are many possibilities for the positive management of old trees, dead wood and saproxylic species in cultural habitats. Veteran trees are the key feature of cultural environments; therefore particular attention should be focused on their preservation (Key and Ball, 1993; Read, 2000; Höjer and Hultengren, 2004). The principal management recommendation is always to retain valuable trees (veteran trees, hollow trees) whenever possible. The complete elimination of valuable trees should be the very last option when action is needed; there are many other options that should be considered first, and which permit the trees to be retained, at least to some extent (Figure 16.5). When possible, the lifespan of veteran trees should be prolonged by appropriate arboricultural methods such as the careful pruning of old pollards (see Read, 2000, for a detailed discussion). Clearing around large old trees may be needed when they have been absorbed by regrowth and are suffering from competition (Höjer and Hultengren, 2004).

It is not sufficient to concentrate on individual trees, but one should manage for the benefit of the population of trees in each area. Different mortality factors will inevitably reduce the numbers of veteran trees. It is therefore important to make sure that sufficient mature trees are maintained so that they can replace the oldest trees in the future. However, the formation of saproxylic microhabitats is a slow process, even when the starting point is biologically mature but healthy trees. In some cases it may be useful to accelerate the development of microhabitats by cutting large branches or by injuring the bark, which

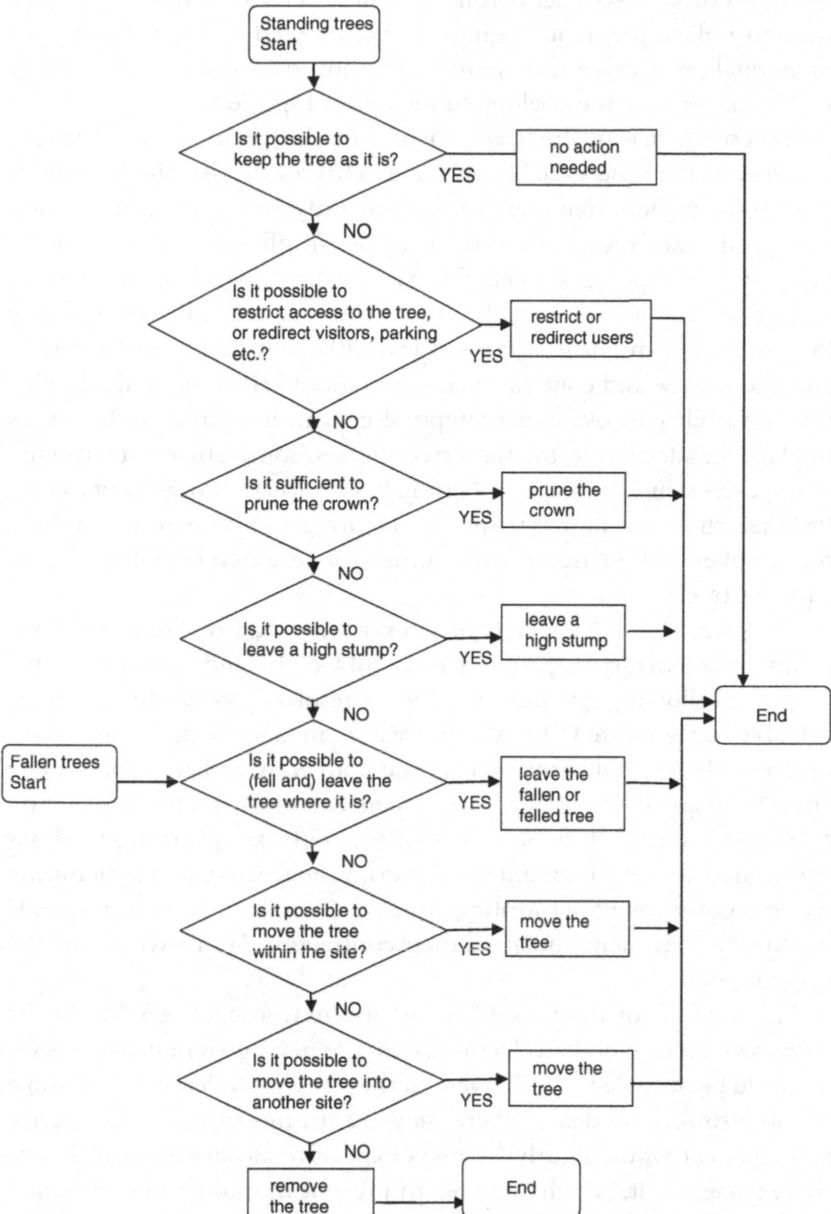

Figure 16.5. A flowchart of the management options for standing or fallen valuable trees in cultural habitats such as parks. Only the final option ('remove the tree': the last option at the bottom of the chart) leads to the complete destruction of biodiversity values.

promotes the access of decay fungi and insects into the tree. Damaging trees to induce premature aging can be directed to less valuable tree individuals, e.g. those that should be removed in any case, are not in visible places, or which belong to introduced species.

Sometimes mature trees are missing or they belong to a different tree species than the valuable old trees. This means that the long-term continuity of host trees can be secured only by planting new ones. In grazed areas, regeneration tends to be insufficient and mechanical protection of saplings is needed as well as planting. The transformation of saplings into trees with veteran tree characteristics takes a very long time. For instance, it takes at least 200 years before oaks start to become hollow and contain some wood mould (Ranius et al., 2009a). One possibility to overcome temporal gaps in host-tree availability is to plant, in addition to the focal tree species, some other fast-growing tree species which will age and produce substitutive microhabitats earlier than the main host tree species. Planting can be used to multiply the number of host trees in the future, or to extend small sites into adjacent treeless areas.

The successful management of veteran trees requires landscape-level planning and the participation of stakeholders. In landscape-level planning, the following questions need to be resolved: Where are the most valuable areas located? In which areas is management needed most urgently? Is it possible to increase the connectivity between separate areas by maintaining or creating green corridors, managing mature trees and planting? It is obvious that participatory planning methods are needed, and these should involve both the forestry and agricultural sectors, government authorities, district councils (or the corresponding local government units) and their contractors, land-owners and the general public.

The amount of dead wood in urban environments could also be increased. Since wood production is not the main goal in urban forests, it should be possible not to harvest all the trees but to leave at least some of them to die and decay where they are. In most cases, conservative management would clearly be a less expensive alternative than intensive management, which attempts to prevent mortality and efficiently removes all the dead trees or parts of them whenever they occur. The main hurdles here are the values and opinions of the public and managers with regard to the degree of management versus naturalness that is perceived as desirable for recreational purposes.

In general, increasing public awareness of the importance of dead wood to biodiversity is the key issue for sympathetic management. The conservation of saproxylic biodiversity in veteran trees should be connected to the existing interest that land-owners and the general public have in old trees, because of their aesthetic, cultural and historical values. Utilizing real examples of the often colourful and intriguing creatures that can be found in cultural habitats can also increase public interest in saproxylic species and in protecting their habitats.

17 · *The value and future of saproxylic diversity*

Bengt Gunnar Jonsson, Juha Siitonen and Jogeir N. Stokland

In this final chapter we address some issues that we consider essential for the future of dead wood and its biodiversity. Until now we have mainly focused on the biological aspects of dead wood and have only to a limited extent considered wider topics such as ecosystem functions, the future of forest biodiversity, and the need to disseminate information about the fascinating life in decaying wood.

17.1 Value of saproxylic diversity

We are convinced that the intrinsic value of the saproxylic species is a sufficient motivation for their protection. We can be fascinated by the peculiarity, strangeness and beauty of saproxylic species, and learn tremendously from the intricate interactions between them. But species living in dead wood are also more directly valuable, by providing products, ecological services and option values.

17.1.1 Ecological functions, services and resilience

The species that colonize and utilize dead trees provide a central ecosystem service, namely the decay of organic matter and the connected recycling of energy and nutrients. The decomposer community, mainly fungi but assisted by a multitude of invertebrates, performs this service at no cost, allowing the other living components of forest ecosystems to thrive. There is a growing awareness that the global loss of species threatens the provision of services such as decay and nutrient turnover. One might ask if the overwhelming variety of species living in wood is really necessary from this perspective. Maybe it would be enough that a few key species are present and that the vast majority of species are actually redundant. In any specific case there is probably no clear and simple answer, but a growing body of evidence suggests that a critical feature in ecosystems is the level of resilience (see, e.g., Rockström et al., 2009). It is highly unlikely that the exact number of 'necessary' species could be defined despite the fact that several species may perform the same function, rather than all of them having critical and unique functions.

The concept of resilience emphasizes that although several species may perform the same ecosystem function, the loss of species slowly degrades the ability of an ecosystem to withstand perturbation. Thus, a limited set of species may perform the ecosystem service under normal conditions, but when exposed to unexpected change, species-poor ecosystems may undergo drastic changes due to the absence of species which could replace the function of some declining species. The species richness in this case can be viewed as an insurance against the loss of key species, as there are alternative species that can step in and carry out the desired function.

In managed forests, insects such as bark beetles may cause significant economic damage. The end of epidemic outbreaks depends to an important extent on the population growth of bark beetle predators and parasitoids. These are also saproxylic species and thus rely on the presence of suitable dead wood habitats. This suggests that the ecosystem service provided by the natural enemies will be threatened when forest management does not allow the presence of sufficient amounts of dead wood. Currently, recommendations both in Fennoscandia and North America (Swedish Forest Agency and the 'Bugwood Network' supported by the US Forest Service) state that only dying and recently dead trees should be salvage logged during bark beetle outbreaks, and that trees from which bark beetles have already emerged should be maintained and protected to allow the build-up of predator and parasitoid populations.

We cannot argue that all saproxylic species are crucial for maintaining specific ecosystem functions in forests. It is likely that the loss of individual species may go unnoticed, but given the extensive decline in dead wood volumes in major forest ecosystems worldwide, and the consequent loss of saproxylic species, we are concerned that the resilience of these ecosystems is being eroded. This may be especially important in connection with ongoing climate change, which represents a so far unprecedented perturbation to forest ecosystems.

17.1.2 Industrial uses, food and medicines

Saproxylic species may also have direct economic value. Although this aspect remains to be explored further, there are already many good examples and even more interesting prospects.

Fungi are biochemical agents with numerous enzymes capable of degrading cellulose, lignin and secondary substances. These can be utilized through the so-called 'biopulping' process (Hatakka, 2001). Generally the separation of the valuable cellulose from lignin in making wood pulp is a process requiring either significant amounts of energy or the use of toxic chemicals. Therefore, the use of lignin-degrading fungi for the pre-treatment of wood is considered to be an extremely interesting option. By letting certain fungi do part of the job, it is possible both to save energy and to limit the use of chemicals. In a recent study, a research team investigated the lignin-degrading capacity of more than 300 previously untested white-rot fungi. Of these, the red-listed species *Physisporinus rivulosus* was found to be the most efficient lignin degrader (Hakala et al., 2004). This also exemplifies the value of less common species as potential sources for new products and services.

Many wood-decaying fungi are used as food all over the world, including the polypore species 'chicken of the woods' (*Laetiporus sulphureus*) and 'hen of the woods' (*Grifola frondosa*), the latter known as '*maitake*' in Japan. Some species, such as various oyster mushrooms (*Pleurotus* spp.) and shiitake (*Lentinula edodes*), can be cultivated and are produced on an industrial scale. There could probably be wide-ranging, yet still unexplored, possibilities for producing protein-rich food for humans from wood by cultivating wood-decay fungi in it – as termites and ambrosia beetles do.

Wood fungi also contain numerous compounds that may have medicinal uses (Zjawiony, 2004). Polypores are known to contain compounds that can enhance the immunological defences and suppress tumour growth and metastasis (Lin and Zhang, 2004; Zjawiony, 2004). The polypore *Fomitopsis officinalis*, nowadays rare in Europe and with its strongest footholds in the old-growth forests of the Pacific Northwest in the USA, is a renowned species in this respect (Weier, 2009). It has been found to selectively attack cowpox and vaccinia viruses, which are closely related to the smallpox virus (Weier, 2009). In addition, it also seems to provide a potential medicine against tuberculosis. The oyster fungus (*Pleurotus ostreatus*) is known for its medical properties, as it reduces cholesterol levels. This fungus has also been used for other purposes, as it may help in cleaning up petroleum wastes. This ability to degrade various pollutants may be of enormous value in the future (Stamets, 2005).

17.1.3 Tourism

The fastest-growing industry worldwide is tourism. This includes an increasing interest in ecotourism and outdoor activities. As national parks and nature reserves receive more and more visitors, the need to provide visitors with relevant information also increases. In many places, dead wood may be seen as an obstacle to a walk in the forest, and an old dying tree as a safety risk. However, with careful planning and relevant information, the presence of these structures may instead increase the value of the experience. Information centres are becoming more common outside of the larger protected areas, and should be utilized to inform visitors about the fascinating life in dead trees. On specific trails, short explanatory texts may be provided, to further inform the visitor about saproxylic biodiversity. Together with other information, it is likely that this will enrich the experience and encourage more people to explore nature. For instance, the majority of tourists visiting game reserves in South Africa responded positively to the idea of including information on invertebrates in ecotourism activities (Huntley et al., 2005). Jewel beetles (Buprestidae), including many saproxylic species, were listed among the potential insect groups to be included in such activities, because of the stunning, large, metallic-coloured adults that can easily be observed visiting flowers.

17.1.4 Option values

In Chapter 11 we made an attempt to calculate global saproxylic species richness. These numbers may be challenged, but whatever the true number is, in practice the list of species associated with dead wood is almost boundless. This means that there are also an endless number of potential economic values associated with these species. These so far unexplored values of species are often called their 'option value'. The notion is that although we see no direct value of these species today, they could prove to be valuable in the future. The option value could include anything from industrial products to food and medicines.

An interesting example of the great economic value of a saproxylic species is the dependence that we have on honey bees for food production. Originally the honey bee built its nests in hollow trees and thus used to be a saproxylic species. Over time, humans have learned how to manage bees and to provide artificial nesting sites: beehives. Although it is difficult to estimate the economic value of the pollinators, in the

USA alone it is estimated to be in the range of US$1.5–15 billion annually, of which the domesticated honey bee is the most significant species (Allsopp et al., 2008). There is currently a severe crisis in honey bee farming, as in many places the bee colonies have collapsed, usually for unknown reasons. Suggestions that this may be caused by widespread diseases or mortality due to pesticides have been put forward. As an example, this illustrates that the biodiversity of dead wood can provide huge economic value, and maybe the solution to the pollination crisis is to be found in other wild bee species still living in the dead trees in some remote forest.

17.2 Negative trends

Global biodiversity is declining and, at present, the species extinction rate has increased by as much as 1000 times over typical background rates during the planet's history (Anon., 2005b). The situation for saproxylic species is, in this respect, no better than for other species groups. Global forest cover is declining, and within regions with stable forest cover, ongoing forestry is seriously reducing the volumes of dead wood or keeping the dead wood volumes at very low levels. In the following sections, we specifically pinpoint the five aspects that we consider most serious for saproxylic biodiversity.

17.2.1 Loss of forest area

We mentioned the global loss of forest area as one of the major threats to saproxylic biodiversity in Chapter 15. The general relationship between habitat area and species richness suggest that the loss of tropical forests is the number one threat to saproxylic diversity. Therefore, as for most organism groups, preventing the loss of the tropical diversity hotspots is also a major goal for halting the global loss of saproxylic diversity.

The main cause behind the loss of forest area is its conversion to agricultural land and pastures. Between 2000 and 2005, the net loss of forests was estimated at 7.3 million hectares annually (FAO, 2006). The rate of loss is not equal across regions and, in particular, tropical areas have the highest rate of loss. By contrast, in some parts of Europe and China, the forest cover is actually increasing. However, this is mainly in the form of plantation forests with limited value for saproxylic species.

17.2.2 Expansion of forestry

The global demand for wood is predicted to increase according to various scenarios (Smeets and Faaij, 2007; Raunikar et al., 2010). This is mainly due to the predicted three- to six-fold increase in the use of fuelwood (Raunikar et al., 2010). A high fuelwood harvest would imply ecologically stressed forests in several countries, even under the minimum-demand scenarios.

In some regions, and particularly in boreal forests, the forest cover is not decreasing (FAO, 2006). Superficially, one might think that the threat to saproxylic species is less pronounced in these regions. However, in the boreal landscapes of Fennoscandia, in particular, but to an increasing extent also of Russia and Canada, large tracts of native and natural forest ecosystems have been transformed, and are currently being transformed, into even-aged planted forests, where the volumes of dead wood are down to a few percent of the natural levels (see Chapter 13). In effect, in these regions, modern forestry has brought about a profound ecosystem change that represents a major threat to the native saproxylic species (Chapter 15). Being residents of one of these regions, we have witnessed the loss of valuable habitats proceeding at a rate that has not been counterbalanced by the simultaneous increase in the area of protected forests or by the introduction of new biodiversity-oriented management methods starting in the 1990s (see Chapter 13). In this respect, we find that statistics indicating that forest conditions in northern Europe are in a good state disregard the decreased quality of habitats from the point of view of forest-dwelling species. When looking at these statistics, the old saying that we 'can't see the wood for the trees' becomes all too true.

17.2.3 Climate change

Climate change is the hottest environmental topic of all at present. The changes that are likely to be brought about by increasing global temperatures and the associated changes in other climatic factors (e.g. precipitation, storms, fires and drought) will also have strong and unpredictable effects on saproxylic diversity. It is an enormous challenge to predict how climate change will affect the biodiversity in dead wood. Models utilizing so-called climate envelopes (see, e.g., Thuiller et al., 2005) indicate how current climate regions and species distributions will be redistributed in the future. However, this is only part of a much more complex story. As species are differently sensitive to climate factors, and since they have different opportunities to migrate into new regions, we can assume that the composition of extant communities will break up and that the intricate relations between species, as depicted in the saproxylic food web (Chapter 3), will also break apart. As a consequence, new unforeseen effects on species interactions will appear. Although this might be interesting as a global experiment, the consequences for species diversity are most probably going to be negative. It is a safe – but pessimistic – prediction that climate change will speed up species extinctions among saproxylic species.

17.2.4 Bioenergy and carbon stores

An additional threat brought about by climate change is the steadily growing extraction of forest biomass for bioenergy. Previously unutilized biomass components, such as branches, tree tops, stumps and roots, are now being harvested to an increasing extent.

Through extensive lobbying, parts of the forest industry argue that by increasing wood production we can also increase carbon sequestration. It is obviously true that forest products are renewable and that sustainable forestry is possible when it is compatible with the conservation of forest biodiversity. However, the argument that old-growth forests should be harvested and converted into highly productive plantations in order to mediate the effects of climate change is seriously flawed. On the contrary, the transformation of old-growth forests, containing large stores of carbon in the living stand, dead wood and soil, into production forests actually constitutes a major source of greenhouse gases. This is generally understood in the case of tropical forests (see Chao et al., 2009), but applies equally to other, more northern, forest ecosystems (Harmon et al., 2000). Moreover, recent studies show that also old-growth forests continue to store significant volumes of carbon for centuries beyond the normal harvest age (Luyssaert et al., 2008).

A major problem in this respect seems to be the problem of understanding the temporal aspects. When an old-growth forest is harvested and replanted with young, fast-growing trees, the current net production will soon exceed that attained previously. However, the loss of the old-growth carbon stores, both in wood and in the soil, represents a carbon debt that must be paid. In most cases this takes many decades, and may even take centuries.

17.2.5 Invasive pathogens

Starting from the beginning of the 1900s, a number of invasive pathogens have spread into new regions. These include chestnut blight (*Cryphonectria* [= *Endothia*] *parasitica*), Dutch elm disease (*Ophiostoma ulmi*), ash wilt disease (*Chalara fraxinea*), the pinewood nematode (*Bursaphelenchus xylophilus*) and many more. The list is constantly increasing. Although such fungi and animals are a natural part of forest ecosystems within their natural distribution areas, they have spread with the help of us humans into entirely new regions. In the short term, the pathogens will, of course, increase the volume of dead wood of the affected tree species. However, in the long run, these tree species may be lost over large areas of their distribution, and eventually the dead wood of these species will become rare or absent. Given that many saproxylic species are host-specific (see Chapter 5), this clearly threatens their regional survival.

17.3 Research challenges

We are researchers and, as such, we of course are aware that many aspects of the biodiversity of saproxylic species remain to be explored. We acknowledge that the following list may be biased towards our interests, but it should still represent some of the main research challenges.

17.3.1 Functional diversity

As pointed out above, saproxylic organisms provide an important ecosystem service in the form of decomposing organic matter. It is still not well understood how sensitive this service is to the loss of saproxylic biodiversity. The complex interactions in the saproxylic food web suggest that the loss of basic decayers (representing specific decay pathways), as well as cascading effects on higher trophic levels, may influence key ecosystem processes. This calls for studies on the relations between the decomposer species, including studies on competition, facilitation and other kinds of interactions.

17.3.2 Alpha-taxonomy

Large numbers of saproxylic species remain to be identified and described. This branch of science is referred to as alpha-taxonomy, in contrast to those taxonomic branches that explore the evolutionary relationships between species groups (phylogeny). Even within regions where taxonomic knowledge is considered to be good, certain species groups are not well studied; this also includes groups in Fennoscandia. Based on recent inventories and collections, new species are constantly being found within groups such as fungus gnats (Mycetophilidae), mites, parasitoid wasps, and cryptic species among fungi.

Even more worrying is the huge number of undescribed saproxylic species in the tropics and other poorly investigated areas. Most groups of fungi have been inadequately and rather unsystematically collected in the tropics. The Xylariaceae, a diverse group of ascomycetes with many wood-decaying species, will serve as a good example. An expert on these fungi states, about the tropics, that

> almost every collecting expedition or box of specimens from correspondents reveal new taxa and other surprises (Rogers, 2000).

Even among the well-known and conspicuous polypores, there is a constant flow of descriptions of new species being published. Within a period of only three years, a group of Chinese mycologists have recently published details of 11 new polypore species from temperate and subtropical China (Dai et al., 2009, and references therein).

The challenge is even greater for invertebrates, since studies indicate that saproxylic species form a high proportion of the tropical forest insect fauna (Stork, 1987; Hammond, 1990). In a compelling study, Tavakilian et al. (1997) sampled longhorn beetles (Cerambycidae) in French Guyana. In this study, 348 longhorn beetle species were sampled from 690 felled trees and lianas, representing more than 200 plant species. Of these cerambycids, 90 species (26%) were undescribed. Considering that longhorn beetles are large, and are among the most popular beetle species to sample, this suggests that an even larger proportion of undescribed species can be expected in other insect families.

17.3.3 Surveys in frontier forests

Connected with the need for more alpha-taxonomic work is also a need for mapping saproxylic hotspot forests. These are likely to occur in tropical regions as well as along the so-called frontier forests. In interior areas of Canada and Russia there still exist tracts of unexploited boreal forests. The expansion of forestry is, however, slowly moving also into these remote areas, and the provision of baseline information on the location of the most valuable forests is lagging behind. Large-scale inventories are difficult to finance in these areas, but new and modern remote sensing techniques might help in providing a better understanding of the distribution of valuable forests. This requires better knowledge, with respect to these regions, on the relationship between forest types and saproxylic diversity.

17.3.4 Improving adaptive management

Awareness of the rich diversity associated with dead wood is increasing. This has resulted in changes in forest management in certain regions, aiming to improve the conditions for saproxylic species. Green-tree retention, preservation of valuable habitats, creation of high stumps, and sparing existing dead wood in logging are becoming standard procedures in many European and North American countries. From an economic perspective, these actions represent an additional cost for the land-owner. Yet the cost-efficiency of these methods is rarely evaluated. There is an obvious need to include a stronger component of adaptive management (Villard and Jonsson, 2009a). By setting explicit targets for actions and evaluating whether these are met with current management practices, we could move more quickly towards a truly sustainable forest management (Figure 17.1).

17.4 Knowledge synthesis and dissemination

It is our hope that the reader of this book should now be ready to acknowledge that dead wood is not a sign of poor forest health, but quite the opposite. The old notions of 'forest hygiene' and the overemphasized risk of pests and pathogens developing in dead wood are convincingly offset by our increasing knowledge about saproxylic diversity. We have written this book as a part of our own research interest in, and fascination with, the species living in dead trees. In parallel with our endeavours, north European researchers are gathering information on saproxylic species, with the aim of arranging this in a large database (Stokland and Meyke, 2008; http://www.saproxylic.org/). Here we hope that researchers, conservation managers and foresters will be able to find specific information about the habitat demands and distribution of north European saproxylic species, which will enable them to do a better job of protecting this crucial component of our forest biodiversity.

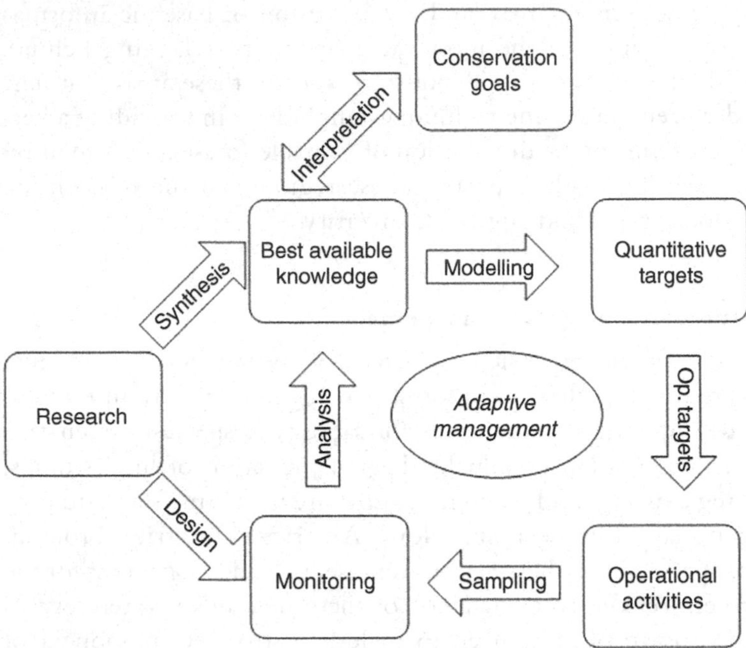

Figure 17.1. The diagram shows an adaptive management scheme, emphasizing the role of science and quantitative input into the process. Although the environmental goals are set at a policy level, the input from research to provide the best knowledge synthesis, to suggest modelling approaches, to identify targets, and to design and evaluate monitoring schemes, clearly shows that a value-free level of quantitative research is needed in the process (from Villard and Jonsson, 2009a).

To conclude, we would like to encourage and challenge you, as a reader, to tell friends, colleagues, conservation authorities and forest managers about this fascinating aspect of life on Earth. We feel that if people become aware of the diversity in decaying wood, this will serve to give a voice to those small but significant creatures dwelling in dead trees in our forests – they cannot speak for themselves, so we have to.

References

Abbe, T.B. & Montgomery, D.R. 1996. Large woody debris jams, channel hydraulics and habitat formation in large rivers. *Regulated Rivers: Research and Management*, **12**, 210–221.

Abe, T. & Higashi, M. 1991. Cellulose centered perspective on terrestrial community structure. *Oikos*, **60**, 127–133.

Abe, T., Bignell, D.E. & Higashi, M. 2000. *Termites: Evolution, Sociality, Symbioses, Ecology*. Dordrecht: Kluwer Academic Publishers.

Abe, Y. 1989. Effect of moisture on decay of wood by xylariaceous and diatrypaceous fungi and quantitative changes in the chemical components of decayed woods. *Transactions of the Mycological Society of Japan*, **30**, 169–181.

Abraham, L., Hoffman, B., Gao, Y. & Breuil, C. 1998. Action of *Ophiostoma piceae* proteinase and lipase on wood nutrients. *Canadian Journal of Microbiology*, **44**, 698–701.

Abrahamsson, M. & Lindbladh, M. 2006. A comparison of saproxylic beetle occurrence between man-made high- and low-stumps of spruce (*Picea abies*). *Forest Ecology and Management*, **226**, 230–237.

Abrahamsson, M., Jonsell, M., Niklasson, M. & Lindbladh, M. 2009. Saproxylic beetle assemblages in artificially created high stumps of spruce (*Picea abies*) and birch (*Betula pendula/pubescens*): does the surrounding landscape matter? *Insect Conservation and Diversity*, **2**, 284–294.

Achard, F., Mollicone, D., Stibig, H.-J., Aksenov, D., Laestadius, L., Li, Z., Potapov, P. & Yaroshenko, A. 2006. Areas of rapid forest-cover change in boreal Eurasia. *Forest Ecology and Management*, **237**, 322–334.

Achard, F., Eva, H.D., Mollicone, D. & Beuchle, R. 2008. The effect of climate anomalies and human ignition factor on wildfires in Russian boreal forests. *Philosophical Transactions of the Royal Society Series B*, **363**, 2331–2339.

Adams, A.S. & Six, D.L. 2007. Temporal variation in mycophagy and prevalence of fungi associated with developmental stages of *Dendroctonus ponderosae* (Coleoptera: Curculionidae). *Environmental Entomology*, **36**, 64–72.

Adams, A.S. & Six, D.L. 2008. Detection of host habitat by parasitoids using cues associated with mycangial fungi of the mountain pine beetle, *Dendroctonus ponderosae*. *Canadian Entomologist*, **140**, 124–127.

Adams, S.H. & Roth, L.F. 1969. Intraspecific competition among genotypes of *Fomes cajanderi* decaying young-growth Douglas-fir. *Forest Science*, **15**, 327–331.

Ahnlund, H. & Lindhe, A. 1992. Endangered wood-living insects in coniferous forests: some thoughts from studies of forest-fire sites, outcrops and clearcuttings in the province of Sörmland, Sweden [in Swedish with an English summary]. *Entomologisk Tidskrift*, **113**, 13–23.

Aime, M.C., Matheny, P.B., Henk, D.A., Frieders, E.M., Nilsson, R.H., Piepenbring, M., McLaughlin, D., Szabo, L.J., Begerow, D., Sampaio, J.P., Bauer, R., Weiss, M., Oberwinkler, F. & Hibbett, D.S. 2006. An overview of the higher level classification of *Pucciniomycotina* based on combined analyses of nuclear large and small subunit rDNA sequences. *Mycologia*, **98**, 896–905.

Aitken, K.E.H. & Martin, K. 2004. Nest cavity availability and selection in aspen-conifer groves in a grassland landscape. *Canadian Journal of Forest Research*, **34**, 2099–2109.

Aitken, K.E.H. & Martin, K. 2007. The importance of excavators in hole-nesting communities: availability and use of natural tree holes in old mixed forests of western Canada. *Journal of Ornithology*, **148** (Suppl. 2), 425–434.

Alamouti, S.M., Kim, J.-J., Humble, L.M., Uzunovik, A. & Breuil, C. 2007. Ophiostomatoid fungi associated with the northern spruce engraver, *Ips perturbatus,* in western Canada. *Antonie van Leeuwenhoek International Journal of General and Molecular Microbiology*, **91**, 19–34.

Alén, R., Kuoppala, E. & Oesch, P. 1996. Formation of the main degradation compound groups from wood and its components during pyrolysis. *Journal of Analytical and Applied Pyrolysis*, **36**, 137–148.

Alexander, K.N.A. 1995. Historic parks and pasture-woodlands: The National Trust resource and its conservation. *Biological Journal of the Linnean Society*, **56**, 155–175.

Alexander, K.N.A. 1998. The links between forest history and biodiversity: the invertebrate fauna of ancient pasture woodlands in Britain and its conservation. *In:* Kirby, K.J. & Watkins, C. (eds.) *The Ecological History of European Forests.* Wallingford: CAB International, 73–80.

Alexander, K.N.A. 2002. *The Invertebrates of Living and Decaying Timber in Britain and Ireland: A Provisional Annotated Checklist.* English Nature Research Reports No. 467.

Alexander, K.N.A. 2004. *Revision of the Index of Ecological Continuity as used for Saproxylic Beetles.* English Nature Research Reports No. 574.

Alexander, K.N.A. 2005. Wood decay, insects, palaeoecology, and woodland conservation policy and practice: breaking the halter. *Antenna*, **29**, 171–178.

Alexander, K.N.A. 2008. Tree biology and saproxylic Coleoptera: issues of definitions and conservation language. *Revue d'Ecologie (La Terre et la Vie)*, **63**, 1–6.

Alix, C. 2005. Deciphering the impact of change on the driftwood cycle: contribution to the study of human use of wood in the Arctic. *Global and Planetary Change*, **47**, 83–98.

Alkaslassy, E. 2005. Abundance of plethodontid salamanders in relation to coarse woody debris in a low-elevation mixed forest of the western cascades. *Northwest Science*, **79**, 156–163.

Allen, R.B., Buchanan, P.K., Clinton, P.W. & Cone, A.J. 2000. Composition and diversity of fungi on decaying logs in a New Zealand temperate beech (*Nothofagus*) forest. *Canadian Journal of Forest Research*, **30**, 1025–1033.

Allison, J. D., Borden, J. H. & Seybold, S. J. 2004. A review of the chemical ecology of the Cerambycidae (Coleoptera). *Chemoecology*, **14**, 123–150.

Allmér, J., Vasiliauskas, R., Ihrmark, K., Stenlid, J. & Dahlberg, A. 2006. Wood-inhabiting fungal communities in woody debris of Norway spruce (*Picea abies* (L.) Karst.), as reflected by sporocarps, mycelial isolations and T-RFLP identification. *FEMS Microbiology Ecology*, **55**, 57–67.

Allsopp, M. H., de Lange, W. J. & Veldtman, R. 2008. Valuing insect pollination services with cost of replacement. *PLoS ONE*, **3**, e3128.

Amezaga, I. & Rodríguez, M. A. 1998. Resource partitioning of four sympatric bark beetles depending on swarming dates and tree species. *Forest Ecology and Management*, **109**, 127–135.

Ancelin, P., Courbaud, B. & Fourcaud, T. 2004. Development of an individual tree-based mechanical model to predict wind damage within forest stands. *Forest Ecology and Management*, **203**, 101–121.

Andersson, H. 1999. Red-listed or rare invertebrates associated with hollow, rotting or sapping trees or polypores in the town of Lund [in Swedish with an English summary]. *Entomologisk Tidskrift*, **120**, 169–183.

Andersson, L. & Appelquist, T. 1990. The influence of the Pleistocene megafauna on the nemoral and the boreonemoral ecosystems: a hypothesis with implications for nature conservation strategy [in Swedish with an English summary]. *Svensk Botanisk Tidskrift*, **84**, 355–368.

Andersson, L. I. & Hytteborn, H. 1991. Bryophytes and decaying wood: a comparison between managed and natural forest. *Holarctic Ecology*, **14**, 121–130.

Andrén, H. 1996. Population responses to habitat fragmentation: statistical power and the random sample hypothesis. *Oikos*, **76**, 235–242.

Andriamasimanana, M. 1994. Ecoethological study of free-ranging aye-ayes (*Daubentonia madagascariensis*) in Madagascar. *Folia Primatologica*, **62**, 37–45.

Angelstam, P. 1996. The ghost of forest past: natural disturbance regimes as a basis for reconstruction of biologically diverse forests in Europe. *In:* Degraaf, R. M. & Miller, R. I. (eds.) *Conservation of Faunal Diversity in Forested Landscapes*. London: Chapman & Hall, 287–337.

Angelstam, P. K., Bütler, R., Lazdinis, M., Mikusiński, G. & Roberge, J.-M. 2003. Habitat thresholds for focal species at multiple scales and forest biodiversity conservation: dead wood as an example. *Annales Zoologici Fennici*, **40**, 473–482.

Annila, E. & Petäistö, R.-L. 1978. Insect attack on windthrown trees after the December 1975 storm in western Finland. *Communicationes Instituti Forestalis Fenniae*, **94**, 1–24.

Anon. 1992. *Council Directive 92/43/EEC of 21 May 1992 on the Conservation of Natural Habitats and of Wild Fauna and Flora*. The Council of the European Communities.

Anon. 2003. *Background Information for the Improved Pan-European Indicators for Sustainable Forest Management*. Vienna: Ministerial Conference on the Protection of European Forests.

Anon. 2005a. Data sheets on quarantine pests: *Dendrolimus sibiricus* and *Dendrolimus superans*. *Bulletin OEPP/EPPO Bulletin*, **35**, 390–395.

Anon. 2005b. *Millennium Ecosystem Assessment 2005: Ecosystems and Human Well-being.* Washington, DC: Island Press.

Antonsson, K., Hedin, J., Jansson, N., Nilsson, S.G. & Ranius, T. 2003. Occurrence of the hermit beetle (*Osmoderma eremita*) in Sweden. *Entomologisk Tidskrift*, **124**, 225–240.

Antrop, M. 2005. Why landscapes of the past are important for the future. *Landscape and Urban Planning*, **70**, 21–34.

Apolinário, F.E. & Martius, C. 2004. Ecological role of termites (Insecta, Isoptera) in tree trunks in central Amazonian rain forests. *Forest Ecology and Management*, **194**, 23–28.

ArtDatabanken 2010a. *Svenska artprojektet.* Homepage, accessed 8 January 2010.

ArtDatabanken 2010b. *Species in Norway.* Homepage, accessed 8 January 2010.

Ash, S.R. & Savidge, R.A. 2004. The bark of the late Triassic *Araucarioxylon arizonicum* tree from petrified forest national park, Arizona. *IAWA Journal*, **25**, 349–368.

Ashe, J.S. 1984. Major features in the evolution of relationships between gyrophaenine staphylinid beetles (Coleoptera: Staphylinidae: Aleocharinae) and fresh mushrooms. *In:* Wheeler, Q. & Blackwell, M. (eds.) *Fungus–Insect Relationships: Perspectives in Ecology and Evolution.* New York: Columbia University Press, 227–255.

Ashe, J.S. 1990. The larvae of *Placusa* Mannerheim (Coleoptera: Staphylinidae), with notes on the feeding habits. *Entomologica Scandinavica*, **21**, 477–485.

Askins, R.A. 1981. Survival in winter: the importance of roost holes to resident birds. *Loon*, **53**, 179–184.

Asner, G.P., Broadbent, E.N., Oliveira, P.J., Keller, M., Knapp, D.E. & Silva, J.N.M. 2006. Condition and fate of logged forests in the Brazilian Amazon. *Proceedings of the National Academy of Sciences of the USA*, **103**, 12947–12950.

Aspöck, H. 2002. The biology of Raphidioptera: a review of present knowledge. *Acta Zoologica Academiae Scientiarum Hungaricae*, **48** (Suppl. 2), 35–50.

ATF 2008. *Ancient Tree Guide No. 4: What are Ancient, Veteran and Other Trees of Special Interest?* Lincolnshire, UK: Woodland Trust.

Aubry, K.B., Jones, L.L.C. & Hall, P.A. 1988. Use of woody debris by plethodontid salamanders in Douglas-fir forests in Washington. *In:* Szaro, R.C., Severson, K.E. & Patton, D.R. (eds.) *Management of Amphibians, Reptiles, and Small Mammals in North America.* USDA Forest Service General Technical Report RM-166. Rocky Mountain Forest and Range Experimental Station, Fort Collins, CO.

Aubry, K.B., Halpern, C.B. & Peterson, C.E. 2009. Variable-retention harvests in the Pacific Northwest: a review of short-term findings from the DEMO study. *Forest Ecology and Management*, **258**, 398–408.

Audisio, P., Brustel, H., Carpaneto, G.M., Coletti, G., Mancini, E., Piattella, E., Trizzino, M., Dutto, M., Antonini, G. & Debiase, A. 2007. Updating the taxonomy and distribution of the European *Osmoderma*, and strategies for their conservation (Coleoptera, Scarabaeidae, Cetoniinae). *Fragmenta Entomologica*, **39**, 273–290.

Audisio, P., Brustel, H., Carpaneto, G.M., Coletti, G., Mancini, E., Trizzino, M., Antonini, G. & Debiase, A. 2009. Data on molecular taxonomy and genetic diversification of the European hermit beetles, a species complex of endangered insects (Coleoptera: Scarabaeidae, Cetoniinae, *Osmoderma*). *Journal of Zoological Systematics and Evolutionary Research*, **47**, 88–95.

Aune, K., Jonsson, B.G. & Moen, J. 2005. Isolation effects among woodland key habitats in Sweden: is forest policy promoting fragmentation? *Biological Conservation*, **124**, 89–95.

Ausmus, B.S. 1977. Regulation of wood decomposition rates by arthropod and annelid populations. *Ecological Bulletins*, **25**, 180–192.

Ayres, B.D., Ayres, M.P., Abrahamson, M.D. & Teale, S.A. 2001. Resource partitioning and overlap in three sympatric species of *Ips* bark beetles. *Oecologia*, **128**, 443–453.

Bader, P., Jansson, S. & Jonsson, B.G. 1995. Wood-inhabiting fungi and substratum decline in selectively logged boreal spruce forests. *Biological Conservation*, **72**, 355–362.

Bai, M.L., Wichmann, F. & Muehlenberg, M. 2003. The abundance of tree holes and their utilization by hole-nesting birds in a primeval boreal forest of Mongolia. *Acta Ornithologica*, **38**, 95–102.

Baker, W.L., Flaherty, P.H., Lindemann, J.D., Veblen, T.T., Eisenhart, K.S. & Kulakowski, D.W. 2002. Effect of vegetation on the impact of a severe blowdown in the southern Rocky Mountains, USA. *Forest Ecology and Management*, **168**, 63–75.

Bakys, R., Vasaitis, R., Barklund, P., Ihrmar, K. & Stenlid, J. 2009. Investigations concerning the role of *Chalara fraxinea* in declining *Fraxinus excelsior*. *Plant Pathology*, **58**, 284–292.

Baldrian, P. 2008. Enzymes of saprotrophic basidiomycetes. *In:* Boddy, L., Frankland, J.C. & van West, P. (eds.) *Ecology of Saprotrophic Basidiomycetes*. London: Academic Press/Elsevier, 19–41.

Baldrian, P. & Valášková, V. 2008. Degradation of cellulose by basidiomycetous fungi. *FEMS Microbiology Reviews*, **32**, 501–521.

Baranowski, R. 1985. Central and Northern European *Dorcatoma* (Coleoptera: Anobiidae), with a key and description of a new species. *Entomologica Scandinavica*, **16**, 203–207.

Barclay, R.M.R. & Kurta, A. 2007. Ecology and behaviour of bats roosting in tree cavities and under bark. *In:* Lacki, M.J., Kurta, A. & Hayes, J.P. (eds.) *Conservation and Management of Bats in Forests*. Baltimore, MD: Johns Hopkins University Press, 15–59.

Barclay, S.D., Ash, J.E. & Rowell, D.M. 2000. Environmental factors influencing the presence and abundance of a log-dwelling invertebrate, *Euperipatoides rowelli* (Onychophora). *Journal of Zoology*, **250**, 425–436.

Barker, J.S. 2008. Decomposition of Douglas-fir coarse woody debris in response to differing moisture content and initial heterotrophic colonization. *Forest Ecology and Management*, **255**, 598–604.

Barnett, H.L. & Binder, F.L. 1973. The fungal host–parasite relationship. *Annual Review of Phytopathology*, **11**, 273–292.

Barrera, R. 1996. Species occurrence and the structure of a community of aquatic insects in tree holes. *Journal of Vector Biology*, **21**, 66–80.

Barron, G.L. 2003. Predatory fungi, wood decay, and the carbon cycle. *Biodiversity*, **4**, 3–9.

Baskent, E.Z. 2009. Forest landscape modeling as a tool to develop conservation targets. *In:* Villard, M.A. & Jonsson, B.G. (eds.) *Setting Conservation Targets for Managed Forest Landscapes*. Cambridge: Cambridge University Press, 304–327.

Baudry, J., Bunce, R.G.H. & Burel, F. 2000. Hedgerows: an international perspective on their origin, function and management. *Journal of Environmental Management*, **60**, 7–22.

Bauer, R. & Oberwinkler, F. 1991. The colacosomes: new structures at the host–parasite interface of a mycoparasitic basidiomycete. *Botanica Acta*, **104**, 53–57.

Bawa, K.S. & Seidler, R. 1998. Natural forest management and conservation of biodiversity in tropical forests. *Conservation Biology*, **12**, 46–55.

Bayon, C. 1981. Modifications ultrastructurales des parois végétales dans le tube digestif d'une larva xylophage *Oryctes nasicornis* (Coleoptera, Scarabaeidae): role des bactéries. *Canadian Journal of Zoology*, **59**, 2020–2029.

Beaver, R.A. 1989. Insect–fungus relationships in the bark and ambrosia beetles. *In:* Wilding, N., Collins, N.M., Hammond, P.M. & Webber, J.F. (eds.) *Insect–Fungus Interactions*. London: Academic Press, 121–143.

Beck, C.B. 1960. Connection between *Archaeopteris* and *Callixylon*. *Science*, **131**, 1524–1525.

Begon, M., Townsend, C.R. & Harper, J.L. 2006. *Ecology: From Individuals to Ecosystems*. Oxford: Blackwell Publishing.

Bellamy, C.L. 2008. *A World Catalogue and Bibliography of the Jewel Beetles (Coleoptera: Buprestoidea), Volume 1: Introduction; Fossil Taxa; Schizopodidae; Buprestidae: Julodinae–Chrysochroinae: Poecilonotini*. Pensoft Series Faunistica, No. 76, 625 pp.

Benick, L. 1952. Pilzkäfer und Käferpilze: ökologische und statistische Untersuchungen. *Acta Zoologica Fennica*, **70**, 1–250.

Bennet, K.D., Tzedakis, P.C. & Willis, K.J. 1991. Quaternary refugia of north European trees. *Journal of Biogeography*, **18**, 103–115.

Berbee, M.L. & Taylor, J.W. 1993. Dating the evolutionary radiations of the true fungi. *Canadian Journal of Botany*, **71**, 1114–1127.

Berg, E.E., Henry, J.D., Fastie, C.L., De Volder, A.D. & Matsuoka, S.M. 2006. Spruce beetle outbreaks on the Kenai Peninsula, Alaska, and Kluane National Park and Reserve, Yukon Territory: relationship to summer temperatures and regional differences in disturbance regimes. *Forest Ecology and Management*, **227**, 219–232.

Bergeron, Y., Harvey, B., Leduc, A. & Gauthier, S. 1999. Forest management strategies based on the dynamics of natural disturbances: considerations and a proposal for a model allowing an even-management approach. *Forest Chronicle*, **75**, 55–61.

Bergeron, Y., Leduc, A., Harvey, B.D. & Gauthier, S. 2002. Natural fire regime: a guide for sustainable management of the Canadian boreal forests. *Silva Fennica*, **36**, 81–95.

Bergeron, Y., Flannigan, M., Gauthier, S., Leduc, A. & Lefort, P. 2004. Past, current and future fire frequency in the Canadian boreal forest: implications for sustainable forest management. *Ambio*, **33**, 356–360.

Berglund, H. & Jonsson, B. G. 2005. Verifying an extinction debt among lichens and fungi in northern Swedish boreal forests. *Conservation Biology*, **19**, 338–348.

Berglund, H., Edman, M. & Ericson, L. 2005. Temporal variation of wood-fungi diversity in boreal old-growth forests: implications for monitoring. *Ecological Applications*, **15**, 970–982.

Berglund, H., O'Hara, R. B. & Jonsson, B. G. 2009. Quantifying habitat requirements of tree-living species in fragmented boreal forests with Bayesian methods. *Conservation Biology*, **23**, 1127–1137.

Berkov, A., Feinstein, J., Centeno, P., Small, J. & Nkamany, M. 2007. Yeasts isolated from Neotropical wood-boring beetles in SE Peru. *Biotropica*, **39**, 530–538.

Berndes, G., Hoogwijk, M. & van den Broek, R. 2003. The contribution of biomass in the future global energy supply: a review of 17 studies. *Biomass and Bioenergy*, **25**, 1–28.

Berryman, A. A. 1989. Adaptive pathways in scolytid–fungus associations. *In:* Wilding, N., Collins, N. M., Hammond, P. M. & Webber, J. F. (eds.) *Insect–Fungus Interactions*. London: Academic Press, 145–159.

Bertone, M. A., Courtney, G. W. & Wiegmann, B. M. 2008. Phylogenetics and temporal diversification of the earliest true flies (Insecta: Diptera) based on multiple nuclear genes. *Systematic Entomology*, **33**, 668–687.

Biggs, A. R. 1992a. Anatomical and physiological responses of bark tissues to mechanical injury. *In:* Blanchette, R. A. & Biggs, A. R. (eds.) *Defence Mechanisms of Woody Plants Against Fungi*. Berlin: Springer-Verlag, 13–40.

Biggs, A. R. 1992b. Responses of angiosperm bark tissues to fungi causing cankers and canker rots. *In:* Blanchette, R. A. & Biggs, A. R. (eds.) *Defense Mechanisms of Woody Plants Against Fungi*. Berlin: Springer-Verlag, 41–61.

Bignell, D. E. 1977. An experimental study of cellulose and hemicellulose degradation in the alimentary canal of the American cockroach. *Canadian Journal of Zoology*, **55**, 579–589.

Binney, H. A., Willis, K. J., Edwards, M. E., Bhagwat, S. A., Anderson, P. M., Andreev, A. A., Blaauw, M., Damblon, F., Haesaerts, P., Kienast, F., Kremenetski, K. V., Krivonogov, S. K., Lozhkin, A. V., Macdonald, G. M., Novenko, E. Y., Oksanen, P., Sapelko, T. V., Väliranta, M. & Vazhenina, L. 2009. The distribution of late-Quaternary woody taxa in northern Eurasia: evidence from a new macrofossil database. *Quaternary Science Reviews*, **28**, 2445–2464.

Biström, O., Kaila, L. & Kullberg, J. 2000. Survey of tree-living Coleoptera in Herttoniemi manor-park, southern Finland [in Finnish with an English summary]. *Sahlbergia*, **5**, 14–20.

Bjurman, J. & Viitanen, H. 1996. Effects of wet storage on subsequent colonization and decay by *Coniophora puteana* at different moisture contents. *Material und Organismen*, **30**, 259–277.

Blackburn, T.M. & Gaston, K.J. 1996. A sideways look at patterns in species richness, or why there are so few species outside the tropics. *Biodiversity Letters*, **3**, 44–53.

Blackwell, M. 1984. Myxomycetes and their arthropod associates. *In:* Wheeler, Q. & Blackwell, M. (eds.) *Fungus–Insect Relationships: Perspectives in Ecology and Evolution*. New York: Columbia University Press, 67–90.

Blais, J.R. 1981. Mortality of balsam fir and white spruce following a spruce budworm outbreak in the Ottawa River watershed in Quebec. *Canadian Journal of Forest Research*, **11**, 620–629.

Blais, J.R. 1983. Trends in the frequency, extent and severity of spruce budworm outbreaks in eastern Canada. *Canadian Journal of Forest Research*, **13**, 539–547.

Blakely, T.J., Jellyman, P.G., Holdaway, R.J., Young, L., Burrows, B., Duncan, P., Thirkettle, D., Simpson, J., Ewers, R.M. & Didham, R.K. 2008. The abundance, distribution and structural characteristics of tree-holes in *Notophagus* forest, New Zealand. *Austral Ecology*, **33**, 963–974.

Blanc, L.A. & Walters, J.R. 2008. Cavity-nest webs in a longleaf pine ecosystem. *Condor*, **110**, 80–92.

Blanchette, R.A. 1984. Screening wood decayed by white rot fungi for preferential lignin degradation. *Applied and Environmental Microbiology*, **48**, 647–653.

Blanchette, R.A. 1991. Delignification by wood-decay fungi. *Annual Review of Phytopathology*, **29**, 381–398.

Blanchette, R.A. & Biggs, A.R. 1992. *Defense Mechanisms of Woody Plants against Fungi*. Berlin: Springer-Verlag.

Bleiker, K.P. & Six, D.L. 2007. Dietary benefits of fungal associates to an eruptive herbivore: potential implications of multiple associates on host population dynamics. *Environmental Entomology*, **36**, 1384–1396.

Bleiker, K.P. & Six, D.L. 2009. Competition and coexistence in a multi-partner mutualism: interactions between two fungal symbionts of the mountain pine beetle in beetle-attacked trees. *Microbial Ecology*, **57**, 191–202.

Block, W. 1991. To freeze or not to freeze? Invertebrate survival of sub-zero temperatures. *Functional Ecology*, **5**, 284–290.

Bobiec, A., Gutowski, J.M., Zub, K., Pawlaczyk, P. & Laudenslayer, W.F. 2005. *The Afterlife of a Tree*. WWF Poland.

Boddy, L. 1983. The effect of temperature and water potential on growth rates of wood-rotting basidiomycetes. *Transactions of the British Mycological Society*, **80**, 141–149.

Boddy, L. 1992. Development and function of fungal communities in decomposing wood. *In:* Wicklow, D.T. & Carroll, C.G. (eds.) *The Fungal Community: Its Organization and Role in the Ecosystem*. New York: Marcel Dekker, 749–782.

Boddy, L. 1993. Saprotrophic cord-forming fungi: warfare strategies and other ecological aspects. *Mycological Research*, **97**, 641–655.

Boddy, L. 1994. Latent decay fungi: the hidden foe? *Arboricultural Journal*, **18**, 113–135.

Boddy, L. 1999. Saprotrophic cord-forming fungi: meeting the challenge of heterogeneous environments. *Mycologia*, **91**, 13–32.

Boddy, L. 2001. Fungal community ecology and wood decomposition processes in angiosperms: from standing tree to complete decay of coarse woody debris. *Ecological Bulletins*, **49**, 43–56.

Boddy, L. & Heilmann-Clausen, J. 2008. Basidiomycete community development in temperate angiosperm wood. *In:* Boddy, L., Frankland, J.C. & van West, P. (eds.) *Ecology of Saprotrophic Basidiomycetes*. London: Academic Press/Elsevier, 211–237.

Boddy, L. & Jones, T.H. 2008. Interactions between Basidiomycota and invertebrates. *In:* Boddy, L., Frankland, J.C. & van West, P. (eds.) *Ecology of Saprotrophic Basidiomycetes*. London: Academic Press/Elsevier, 155–179.

Boddy, L. & Rayner, A.D.M. 1981. Fungal communities and formation of heartwood wings in attached oak branches undergoing decay. *Annals of Botany*, **47**, 271–274.

Boddy, L. & Rayner, A.D.M. 1983a. Ecological roles of basidiomycetes forming decay communities in attached oak branches. *New Phytologist*, **93**, 77–88.

Boddy, L. & Rayner, A.D.M. 1983b. Origins of decay in living deciduous trees: the role of moisture content and a re-appraisal of the expanded concept of tree decay. *New Phytologist*, **94**, 623–641.

Boddy, L., Bardsley, D.W. & Gibbon, O.M. 1987. Fungal communities in attached ash branches. *New Phytologist*, **107**, 143–154.

Boddy, L., Frankland, J.C. & van West, P. (eds.) 2008. *Ecology of Saprotrophic Basidiomycetes*. London: Academic Press/Elsevier.

Böhl, J. & Brändli, U.-B. 2007. Deadwood volume assessment in the third Swiss National Forest Inventory: methods and first results. *European Journal of Forest Research*, **126**, 449–457.

Bois, E. & Lieutier, F. 1997. Phenolic response of Scots pine clones to inoculation with *Leptographium wingfieldii,* a fungus associated with *Tomicus piniperda. Plant Physiology and Biochemistry*, **35**, 819–825.

Bonan, G.B. & Shugart, H.H. 1989. Environmental factors and ecological processes in boreal forests. *Annual Review of Ecology and Systematics*, **20**, 1–28.

Bonar, R.L. 2000. Availability of pileated woodpecker cavities and use by other species. *Journal of Wildlife Management*, **64**, 52–59.

Bonello, P. & Blodgett, J.T. 2003. *Pinus nigra–Sphaeropsis sapinea* as a model pathosystem to investigate local and systemic effects of fungal infection of pines. *Physiological and Molecular Plant Pathology*, **63**, 249–261.

Boone, C.K., Six, D.L., Zheng, Y.B. & Raffa, K.F. 2008. Parasitoids and dipteran predators exploit volatiles from microbial symbionts to locate bark beetles. *Environmental Entomology*, **37**, 150–161.

Borger, G.A. 1973. Development and shedding of bark. *In:* Kozlowski, T.T. (ed.) *Shedding of Plant Parts*. New York and London: Academic Press, 205–236.

Bouchard, M., Kneeshaw, D. & Bergeron, Y. 2005. Mortality and stand renewal patterns following the last spruce budworm outbreak in the Ottawa River watershed in Quebec. *Forest Ecology and Management*, **204**, 291–313.

Bouget, C. & Duelli, P. 2004. The effects of windthrow on forest insect communities: a literature review. *Biological Conservation*, **118**, 281–299.

Bouget, C., Brin, A. & Brustel, H. 2011. Exploring the 'last biotic frontier': are temperate forest canopies special for saproxylic beetles? *Forest Ecology and Management*, **261**, 211–220.

Boulanger, Y. & Sirois, L. 2007. Postfire succession of saproxylic arthropods, with emphasis on Coleoptera, in the north boreal forest of Quebec. *Environmental Entomology*, **36**, 128–141.

Bowles, J. M. & Lachance, M.-A. 1983. Patterns of variation in the yeast florae of exudates in an oak community. *Canadian Journal of Botany*, **61**, 2984–2995.

Boyle, W. A., Ganong, C. N., Clark, D. B. & Hast, M. A. 2008. Density, distribution, and attributes of tree cavities in an old-growth tropical rain forest. *Biotropica*, **40**, 241–245.

Bradshaw, R. H. W., Hannon, G. E. & Lister, A. M. 2003. A long-term perspective on ungulate–vegetation interactions. *Forest Ecology and Management*, **181**, 267–280.

Brandeis, T. J., Newton, M., Filip, G. M. & Cole, E. C. 2002. Cavity-nester habitat development in artificially made Douglas-fir snags. *Journal of Wildlife Management*, **66**, 625–633.

Brasier, C. M. & Kirk, S. A. 2010. Rapid emergence of hybrids between the two subspecies of *Ophiostoma novo-ulmi* with a high level of pathogenic fitness. *Plant Pathology*, **59**, 186–199.

Brassard, B. W. & Chen, H. Y. H. 2006. Stand structural dynamics of North American boreal forests. *Critical Reviews in Plant Sciences*, **25**, 115–137.

Brassard, B. W. & Chen, H. Y. H. 2008. Effects of forest type and disturbance on diversity of coarse woody debris in boreal forest. *Ecosystems*, **11**, 1078–1090.

Bréda, N., Huc, R., Granier, A. & Dreyer, E. 2006. Temperate forest trees and stands under severe drought: a review of ecophysiological responses, adaptation processes and long-term consequences. *Annals of Forest Science*, **63**, 625–644.

Bredesen, B., Haugan, R., Aanderaa, R., Lindblad, I., Økland, B. & Røsok, Ø. 1997. Wood-inhabiting fungi as indicator species of ecological continuity in southeast Norwegian spruce forests [in Norwegian with an English summary]. *Blyttia*, **54**, 131–140.

Bretz Guby, N. A. & Dobbertin, M. 1996. Quantitative estimates of coarse woody debris and standing trees in selected Swiss forests. *Global Ecology and Biography Letters*, **5**, 327–341.

Brewer, S., Cheddadi, R., de Beaulieu, J. L. & Reille, M. 2002. The spread of deciduous *Quercus* throughout Europe since the last glacial period. *Forest Ecology and Management*, **156**, 27–48.

Breznak, J. A. & Brune, A. 1994. Role of microorganisms in the digestion of lignocellulose by termites. *Annual Review of Entomology*, **39**, 453–487.

Bridges, J. R. 1987. Effects of terpenoid compound on growth of symbiotic fungi associated with southern pine beetle. *Phytopathology*, **77**, 83–85.

Bridges, J. R. & Moser, J. C. 1986. Relationship of phoretic mites (Acari: Tarsonemidae) to the bluestaining fungus, *Ceratocystis minor,* in trees infested by southern pine beetle (Coleoptera: Scolytidae). *Environmental Entomology*, **15**, 951–953.

Bright, D. E. & Skidmore, R. E. 1997. *A Catalog of Scolytidae and Platypodidae (Coleoptera), Supplement 1 (1990–1994)*. Ottawa: NRC Research Press.

Brightsmith, D. J. 2005a. Competition, predation and nest shifts among tropical cavity nesters: ecological evidence. *Journal of Avian Biology*, **36**, 74–83.

Brightsmith, D.J. 2005b. Parrot nesting in southeastern Peru: seasonal patterns and keystone trees. *Wilson Bulletin*, **117**, 296–305.

Brignolas, F., Lacroix, B., Lieutier, F., Sauvard, D., Drouet, A., Claudot, A.C., Yart, A., Berryman, A.A. & Christiansen, E. 1995. Induced responses in phenolic metabolism in two Norway spruce clones after wounding and inoculations with *Ophiostoma polonicum*, a bark beetle-associated fungus. *Plant Physiology*, **109**, 821–827.

Brin, A., Meredieu, C., Piou, D., Brutsel, H. & Jactel, H. 2008. Change in quantitative patterns of dead wood in maritime pine plantations over time. *Forest Ecology and Management*, **256**, 913–921.

Brown, A.V. & Brasier, C.M. 2007. Colonization of tree xylem by *Phytophthora ramorum*, *P. kernoviae* and other *Phytophthora* species. *Plant Pathology*, **56**, 227–241.

Brown, J.H. 1981. Two decades of homage to Santa Rosalia: toward a general theory of diversity. *American Zoologist*, **21**, 877–888.

Brunet, J. & Isacsson, G. 2009. Restoration of beech forest for saproxylic beetles: effects of habitat fragmentation and substrate density on species diversity and distribution. *Biodiversity and Conservation*, **18**, 2387–2404.

Brunner, A. & Kimmins, J.P. 2003. Nitrogen fixation in coarse woody debris of *Thuja plicata* and *Tsuga heterophylla* forests on northern Vancouver Island. *Canadian Journal of Forest Research*, **33**, 1670–1682.

Büche, B. & Lundberg, S. 2002. A new species of deathwatch beetle (Coleoptera: Anobiidae) discovered in Europe. *Entomologica Fennica*, **13**, 79–84.

Buckland, P.C. 2005. Palaeoecological evidence for the Vera hypothesis? *In:* Hodder, K.H., Bullock, J.M., Buckland, P.C. & Kirby, K.J. (eds.) *Large Herbivores in the Wildwood and Modern Naturalistic Grazing Systems*. English Nature Research Report No. 648, 62–116.

Buckland, P.C. & Dinnin, M.H. 1993. Holocene woodlands, the fossil evidence. *In:* Kirby, K.J. & Drake, C.M. (eds.) *Deadwood Matters: The Ecology and Conservation of Saproxylic Invertebrates in Britain*. Peterborough, UK: English Nature, 6–20.

Buddle, C.M. 2001. Spiders (Aranea) associated with downed woody material in a deciduous forest in central Alberta, Canada. *Agricultural and Forest Entomology*, **3**, 241–251.

Bull, E.L., Partridge, A.D. & Williams, W.G. 1981. *Creating Snags with Explosives*. Research Note PNW-393. Pacific Northwest Forest and Range Experiment Station: USDA.

Bull, E.L., Nielsen-Pincus, N., Wales, B.C. & Hayes, J.L. 2007. The influence of disturbance events on pileated woodpeckers in Northeastern Oregon. *Forest Ecology and Management*, **243**, 320–329.

Burke, R.M. & Cairney, J.W.G. 2002. Laccases and other polyphenol oxidases in ecto- and ericoid mycorrhizal fungi. *Mycorrhiza*, **12**, 105–116.

Burtin, P., Jay-Allemand, C., Charpentier, J.P. & Janin, G. 1998. Natural wood colouring process in *Juglans* sp. (*J. nigra*, *J. regia* and hybrid *J. nigra* × *J. regia*) depends on native phenolic compounds accumulated in the transition zone between sapwood and heartwood. *Trees*, **12**, 258–264.

Busby, P.E., Adler, P., Warren, T.L. & Swandon, F.J. 2006. Fates of live trees retained in forest cutting units, western Cascade Range, Oregon. *Canadian Journal of Forest Research*, **36**, 2550–2560.

Buse, J., Schröder, B. & Assmann, T. 2007. Modelling habitat and spatial distribution of an endangered longhorn beetle: a case study for saproxylic insect conservation. *Biological Conservation*, **137**, 372–381.

Busing, R.T. 2005. Tree mortality, canopy turnover, and woody detritus in old cove forests of the southern Appalachians. *Ecology*, **86**, 73–84.

Buswell, J.A. 1991. Fungal degradation of lignin. *In:* Arora, D.K., Rai, B., Mukerji, K.G. & Knudsen, G.R. (eds.) *Handbook of Applied Mycology, Vol. 1.* New York: Marcel Dekker, 425–480.

Butin, H. 1995. *Tree Diseases and Disorders: Causes, Biology, and Control in Forest and Amenity Trees.* Oxford, New York, Tokyo: Oxford University Press.

Butin, H. & Kowalski, T. 1983a. Die natürliche Astreinigung und ihre biologischen Voraussetzungen. I. Die Pilzflora der Buche (*Fagus sylvatica* L.). *European Journal of Forest Pathology*, **13**, 322–334.

Butin, H. & Kowalski, T. 1983b. Die natürliche Astreinigung und ihre biologischen Voraussetzungen. II. Die Pilzflora der Stieleiche (*Quercus robur* L.). *European Journal of Forest Pathology*, **13**, 428–439.

Butin, H. & Kowalski, T. 1986. Die natürliche Astreinigung und ihre biologischen Voraussetzungen. III. Die Pilzflora von Ahorn, Erle, Birke, Hainbuche und Esche. *European Journal of Forest Pathology*, **16**, 129–138.

Butin, H. & Kowalski, T. 1989. Die natürliche Astreinigung und ihre biologischen Voraussetzungen. IV. Die Pilzflora der Tanne (*Abies alba* Mill.). *Zeitschrift für Mykologie*, **55**, 189–196.

Butin, H. & Kowalski, T. 1990. Die natürliche Astreinigung und ihre biologischen Voraussetzungen. V. Die Pilzflora von Fichte, Kiefer und Lärche. *European Journal of Forest Pathology*, **20**, 44–54.

Bütler, R., Angelstam, P., Ekelund, P. & Schlaepfer, R. 2004. Dead wood threshold values for the three-toed woodpecker presence in boreal and sub-Alpine forest. *Biological Conservation*, **119**, 305–318.

Butler, R.A. & Laurance, F.L. 2008. New strategies for conserving tropical forests. *Trends in Ecology and Evolution*, **23**, 469–472.

Butovitsch, V. 1971. Untersuchen über das Auftreten von Forstschädlingen in den von Schnestürmen heimgesuchten Fichtenwälder des Küstgebiets der Provinz Västernorrland in den Jahren 1967–69 [in Swedish with a German summary]. *Institutionen för Skogszoologi Rapport och Uppsatser*, **8**, 1–204.

Byers, J.A. 1995. Host tree chemistry affecting colonization in bark beetles. *In:* Cardé, R.T. & Bell, W.J. (eds.) *Chemical Ecology of Insects, Vol. 2.* New York: Chapman & Hall, 154–213.

Byers, J.A. 2004. Chemical ecology of bark beetles in a complex olfactory landscape. *In:* Lieutier, F., Day, K.R., Battisti, A., Grégoire, J.-C. & Evans, H.F. (eds.) *Bark and Wood Boring Insects in Living Trees in Europe: A Synthesis.* Dordrecht, Boston, London: Kluwer Academic Publishers, 89–134.

Cairney, J.W.G. 2005. Basidiomycete mycelia in forest soils: dimensions, dynamics and roles in nutrient distribution. *Mycological Research*, **109**, 7–20.

Cairney, J. W. G., Taylor, A. F. S. & Burke, R. M. 2003. No evidence for lignin peroxidase genes in ectomycorrhizal fungi. *New Phytologist*, **160**, 461–462.

Carcaillet, C., Bergman, I., Delorme, S., Hörnberg, G. & Zackrisson, O. 2006. Long-term fire frequency not linked to prehistoric occupations in northern Swedish boreal forests. *Ecology*, **88**, 465–477.

Cardoza, Y. J., Klepzig, K. D. & Raffa, K. F. 2006a. Bacteria in oral secretions of an endophytic insect inhibit antagonistic fungi. *Ecological Entomology*, **31**, 636–645.

Cardoza, Y. J., Paskewitz, S. & Raffa, K. F. 2006b. Travelling through time and space on wings of beetles: a tripartite insect–fungi–nematode association. *Symbiosis*, **41**, 71–79.

Cardoza, Y. J., Moser, J. C., Klepzig, K. D. & Raffa, K. F. 2008. Multipartite symbioses among fungi, mites, nematodes, and the spruce beetle, *Dendroctonus rufipennis*. *Environmental Entomology*, **37**, 956–963.

Carlson, A., Sandström, U. & Olsson, K. 1998. Availability and use of natural tree holes by cavity nesting birds in a Swedish deciduous forest. *Ardea*, **86**, 109–119.

Carlsson, F., Edman, M., Holm, S., Eriksson, A.-M. & Jonsson, B. G. 2011. Increased heat resistance in mycelia from wood fungi prevalent in forests characterized by fire: a possible adaptation to forest fire. *Fungal Biology*, in press.

Carpenter, S. R. 1983. Resource limitation of larval treehole mosquitoes subsisting on beech detritus. *Ecology*, **64**, 219–223.

Cartwright, K. T. St. G. 1937. A reinvestigation into the cause of 'brown oak', *Fistulina hepatica* (huds.) fr. *Transactions of the British Mycological Society*, **21**, 68–83.

Caruso, A. & Rudolphi, J. 2009. Influence of substrate age and quality on species diversity of lichens and bryophytes on stumps. *Bryologist*, **112**, 520–531.

Caruso, A., Rudolphi, J. & Thor, G. 2008. Lichen species diversity and substrate amounts in young planted boreal forests: a comparison between slash and stumps of *Picea abies*. *Biological Conservation*, **141**, 47–55.

Castello, J. D., Shaw, C. G. & Furniss, M. M. 1976. Isolation of *Cryptoporus volvatus* and *Fomes pinicola* from *Dendroctonus pseudotsugae*. *Phytopathology*, **66**, 1431–1434.

Castle, G. & Mileto, R. 2005. *Development of Veteran Tree Site Assessment Protocol*. English Nature Research Reports 628.

Cedergren, J. 2008. Kontinuitetskogar och hyggesfritt skogsbruk. Skogsstyrelsen.

Chambers, J. Q., Higuchi, N., Schimel, J. P., Ferreira, L. V. & Melack, J. M. 2000. Decomposition and carbon cycling of dead trees in tropical forests of the central Amazon. *Oecologia*, **122**, 380–388.

Chambers, S. M., Burke, R. M., Brooks, P. R. & Cairney, J. W. G. 1999. Molecular and biochemical evidence for manganese-dependent peroxidase activity in *Tylospora fibrillosa*. *Mycological Research*, **103**, 1098–1102.

Chandler, P. 2001. *The Flat-footed Flies (Diptera: Opetiidae and Platypezidae) of Europe*. Leiden, Boston, Köln: Brill.

Chao, K.-J., Phillips, O.L., Baker, T.R., Peacock, J., Lopez-Gonzalez, G., Vásquez Martinez, R., Montegudo, A. & Torres-Lezama, A. 2009. After trees die: quantities and determinants of necromass across Amazonia. *Biogeosciences*, **6**, 1615–1626.

Chapela, I.H. 1989. Fungi in healthy stems and branches of American beech and aspen: a comparative study. *New Phytologist*, **113**, 65–75.

Chapela, I.H. & Boddy, L. 1988. Fungal colonization of attached beech branches. II. Spatial and temporal organization of communities arising from latent invaders in bark and functional sapwood, under different moisture regimes. *New Phytologist*, **110**, 47–57.

Chapman, A.D. 2009. *Numbers of Living Species in Australia and the World*, 2nd edition. Canberra: Australian Biological Resources Study.

Chararas, C. & Koutroumpas, A. 1977. Etude comparée de l'equipement osidasique de 2 Lépidoptères Cossidae xylophages (*Cossus cossus* L. et *Zeuzera pyrina* L.) et de divers coléoptères xylophages. *Comptes Rendus de l'Academie des Sciences, Serie D*, **285**, 369–371.

Chen, D.M., Taylor, A.F.S., Burke, R.M. & Cairney, J.W.G. 2001. Identification of genes for lignin peroxidases and manganese peroxidases in ectomycorrhizal fungi. *New Phytologist*, **152**, 151–158.

Chen, N., Siegel, S.M. & Siegel, B.L. 1980. Gravity and land plant evolution: experimental induction of lignification by simulated hypergravity and water stress. *Life Sciences and Space Research*, **18**, 193–198.

Chin, K. 2007. The paleobiological implications of herbivorous dinosaur coprolites from the upper Cretaceous Two Medicine formation of Montana: why eat wood? *Palaios*, **22**, 554–566.

Christensen, M. & Heilmann-Clausen, J. 2009. Forest biodiversity gradients and the human impact in Annapurna Conservation Area, Nepal. *Biodiversity and Conservation*, **18**, 2205–2221.

Christensen, M., Rayamajhi, S. & Meilby, H. 2009. Balancing fuelwood and biodiversity concerns in rural Nepal. *Ecological Modelling*, **220**, 522–532.

Christiansen, E. 1992. After-effects of drought did not predispose young *Picea abies* to infection by bark beetle-transmitted blue-stain fungus *Ophiostoma polonicum*. *Scandinavian Journal of Forest Research*, **7**, 557–569.

Christiansen, E. & Solheim, H. 1990. The bark beetle-associated blue-stain fungus *Ophiostoma polonicum* can kill various spruces and Douglas fir. *European Journal of Forest Pathology*, **20**, 436–446.

Clark, D.B. & Clark, D.A. 1996. Abundance, growth and mortality of very large trees in neotropical lowland rain forest. *Forest Ecology and Management*, **80**, 235–244.

Clements, D.K. & Alexander, K.N.A. 2009. A comparative study of the invertebrate faunas of hedgerows of differing ages, with particular reference to indicators of ancient woodland and 'old growth'. *Journal of Practical Ecology and Conservation*, **8**, 7–27.

Cleveland, L.R. 1924. The physiological and symbiotic relationships between the intestinal protozoa of termites and their host, with special reference to *Reticulitermes flavipes* Kollar. *Biological Bulletin*, **46**, 203–227.

Cleveland, L.R., Hall, S.R., Sanders, E.P. & Collier, J. 1934. *The Wood-feeding Roach Cryptocerus, its Protozoa, and the Symbiosis between Protozoa and Roach.* Memoirs of the American Academy of Arts and Sciences, Vol. 17. Menasha, WI: George Banta Publishing Co., 185–342.

Cline, A.R. & Leschen, R.A.B. 2005. Coleoptera associated with the oyster mushroom, *Pleurotus ostreatus* Fries, in North America. *Southeastern Naturalist,* **4**, 409–420.

Cobb, F.W., Jr., Krstic, M., Zavarin, E. & Barber, H.W., Jr. 1968. Inhibitory effects of volatile oleoresin components on *Fomes annosus* and four *Ceratocystis* species. *Phytopathology,* **58**, 1327–1335.

Cockle, K., Martin, K. & Wiebe, K. 2008. Availability of cavities for nesting birds in the Atlantic forest, Argentina. *Ornitologia Neotropical,* **19** (Suppl.), 269–278.

Cognato, A.I. & Grimaldi, D. 2009. 100 million years of morphological conservation in bark beetles (Coleoptera: Curculionidae: Scolytinae). *Systematic Entomology,* **34**, 93–100.

Cohen, J.E., Briand, F. & Newman, C.M. 1990. *Community Food Webs: Data and Theory.* Berlin, Springer-Verlag.

Condit, R., Hubbel, S.P. & Foster, R.B. 1995. Mortality rates of 205 neotropical tree and shrub species and the impact of a severe drought. *Ecological Monographs,* **65**, 419–439.

Conner, R.N. & Locke, B.A. 1982. Fungi and red-cockaded woodpecker cavity trees. *Wilson Bulletin,* **94**, 64–70.

Conner, R.N., Miller, O.K.J. & Adkisson, C.S. 1976. Woodpecker dependence on trees infected by fungal heart rots. *Wilson Bulletin,* **88**, 575–581.

Connor, E.F. & McCoy, E.D. 1979. The statistics and biology of the species–area relationship. *American Naturalist,* **113**, 791–833.

Cooper, S.J. 1999. The thermal and energetic significance of cavity roosting in mountain chickadees and juniper titmice. *Condor,* **101**, 863–866.

Cornelius, C., Cockle, K., Politi, N., Berkunsky, I., Sandoval, L., Ojeda, V., Rivera, L., Hunter, M.J. & Martin, K. 2008. Cavity-nesting birds in Neotropical forests: cavities as a potentially limiting resource. *Ornitologia Neotropical,* **19** (Suppl.), 253–268.

Covert-Bratland, K.A., Block, W.M. & Theimer, T.C. 2006. Hairy woodpecker winter ecology in ponderosa pine forests representing different ages since wildfire. *Journal of Wildlife Management,* **70**, 1379–1392.

Craighead, F.C. 1928. Interrelation of tree killing bark beetles (*Dendroctonus*) and blue stain. *Journal of Forestry,* **26**, 886–887.

Crane, P.R., Herendeen, P. & Friis, E.M. 2004. Fossils and plant phylogeny. *American Journal of Botany,* **91**, 1683–1699.

Crawford, D.L. & Sutherland, J.B. 1979. The role of Actinomycetes in the decomposition of lignocellulose. *Developments in Industrial Microbiology,* **20**, 143–151.

Croisé, L., Lieutier, F., Cochard, H. & Dreyer, E. 2001. Effects of drought stress and high density stem inoculations with *Leptographium wingfieldii* on hydraulic properties of young Scots pine trees. *Tree Physiology,* **21**, 427–436.

Crook, D.A. & Robertson, A.I. 1999. Relationships between riverine fish and woody debris: implications for lowland rivers. *Marine and Freshwater Research,* **50**, 941–953.

Crowson, R.A. 1981. *The Biology of the Coleoptera*. London: Academic Press.

Crowson, R.A. 1984. The associations of Coleoptera with Ascomycetes. *In:* Wheeler, Q. & Blackwell, M. (eds.) *Fungus–Insect Relationships: Perspectives in Ecology and Evolution*. New York: Columbia University Press, 256–285.

Cruden, D.L. & Markovetz, A.J. 1979. Carboxy-methylcellulase decomposition by intestinal bacteria of cockroaches. *Applied and Environmental Microbiology*, **38**, 369–372.

Cullen, D. 1997. Recent advances on the molecular genetics of lignolytic fungi. *Journal of Biotechnology*, **53**, 273–289.

Currie, D.J. 1991. Energy and large-scale patterns of animal and plant species richness. *American Naturalist*, **137**, 27–49.

Dahlberg, A. & Stokland, J.N. 2004. *Substrate Requirements of Wood-inhabiting Species: A Compilation and Analysis of 3600 Species* [in Swedish with an English summary]. Skogsstyrelsen Rapport 2004: 7, 75 pp.

Dahlberg, A., Allmér, J., Kruys, N., Nyström, K., Hyvönen, R., Ågren, G. & Madji, H. 2005. Carbon availability in litter for saprotrophic fungi in Norway spruce forests: a modelling approach of mass and flux of dead plant matter from the tree-, field-, and bottom-layer. *In:* Allmér, J. (ed.) *Fungal Communities in Branch Litter of Norway Spruce: Dead Wood Dynamics, Species Detection and Substrate Preferences*. Uppsala: SLU.

Dahlsten, D.L. & Stephen, F.M. 1974. Natural enemies and insect associates of the mountain pine beetle, *Dendroctonus ponderosae* (Coleoptera: Scolytidae), in sugar pine. *Canadian Entomologist*, **106**, 1211–1217.

Dahlström, N. & Nilsson, C. 2006. The dynamics of coarse woody debris in boreal Swedish forests are similar between stream channels and adjacent riparian forests. *Canadian Journal of Forest Research*, **36**, 1139–1148.

Dai, Y.-C., Cui, B.-K. & Yuan, H.-S. 2009. *Trichaptum* (Basidiomycota, Hymenochaetales) from China with description of three new species. *Mycological Progress*, **8**, 281–287.

Daily, G.C., Ehrlich, P.R. & Haddad, N.M. 1993. Double keystone bird in a keystone species complex. *Proceedings of the National Academy of Sciences of the USA*, **90**, 592–594.

Dajoz, R. 1966. Ecologie et biologie des coléoptères xylophages de la hêtraie. *Vie et Milieu*, **17**, 525–736.

Dajoz, R. 1977. Les biocénoses de coléoptères terricoles et xylophages de la Haute Vallée d'Aure et du Massif de Néouvielle (Hautes-Pyrénées). *Bulletin des Naturalistes Parisiens, Nouvelle Série*, **31**, 1–36.

Dajoz, R. 2000. *Insects and Forests: The Role and Diversity of Insects in the Forest Environment*. Paris: Intercept.

Darling, D.C. & Packer, L. 1988. Effectiveness of Malaise traps in collecting Hymenoptera: the influence of trap design, mesh size, and location. *Canadian Entomologist*, **120**, 787–796.

Day, M.E., Greenwood, M.S. & White, A.S. 2001. Age-related changes in foliar morphology and physiology in red spruce and their influence on declining photosynthetic rates and productivity with tree age. *Tree Physiology*, **21**, 1195–1204.

De Belie, N., Richardson, M., Braam, C.R., Svennerstedt, B., Lenehan, J.J. & Sonck, J.J. 2000. Durability of building materials and components in the agricultural environment. Part I. The agricultural environment and timber structures. *Journal of Agricultural Engineering Research*, **75**, 225–241.

De Grandpré, L., Morissette, J. & Gauthier, S. 2000. Long-term post-fire changes in the northeastern boreal forest of Quebec. *Journal of Vegetation Science*, **11**, 791–800.

Debeljak, M. 2006. Coarse woody debris in virgin and managed forest. *Ecological Indicators*, **6**, 733–742.

Déchêne, A.D. & Buddle, C.M. 2010. Decomposing logs increase oribatid mite assemblage diversity in mixed-wood boreal forest. *Biodiversity and Conservation*, **19**, 237–256.

Deflorio, G., Barry, K.M., Johnson, C. & Mohammed, C.L. 2007. The influence of wound location on decay extent in plantation-grown *Eucalyptus globulus* and *Eucalyptus nitens*. *Forest Ecology and Management*, **242**, 353–362.

Delancey, J.B., Majka, C.G., Bondrup-Nielsen, S. & Peck, S.B. 2009. Deadwood and saproxylic beetle diversity in naturally disturbed and managed spruce forests in Nova Scotia. *ZooKeys*, **22**, 309–340.

Delaney, M., Brown, S., Lugo, A.E., Torres-Lezama, A. & Quintero, N.B. 1998. The quantity and turnover of dead wood in permanent forest plots in six life zones of Venezuela. *Biotropica*, **30**, 2–11.

DellaSala, D.A., Staus, N.L., Strittholt, J.R., Hackman, A. & Jacobelli, A. 2001. An updated protected areas database for the United States and Canada. *Natural Areas Journal*, **21**, 124–135.

DellaSala, D.A., Karr, J.R., Schoenagel, T., Perry, D., Noss, R.F., Lindenmayer, D.B., Beschta, R., Hutto, R.L., Swanson, M.E. & Evans, J. 2006. Post-fire logging debate ignores many issues. *Science*, **314**, 51.

den Boer, P.J. 1996. *Regulation and Stabilization Paradigms in Population Ecology*. Population and Community Biology Series, No. 16. London: Chapman & Hall.

Dennis, R.L. 1970. A middle Pennsylvanian basidiomycete with clamp connections. *Mycologia*, **62**, 578–584.

Denton, J. & Chandler, P. 2005. Rotherfield Park, North Hampshire: an important site for saproxylic Coleoptera, Diptera and other insects. *British Journal of Entomology and Natural History*, **18**, 9–15.

Desprez-Lousteau, M.-L., Marçais, B., Nageleisen, L.-M., Piou, D. & Vannini, A. 2006. Interactive effects of drought and pathogens in forest trees. *Annals of Forest Science*, **63**, 595–610.

Dial, R.J. 1995. Species-area curves and Koopowitz et al.'s simulation on stochastic extinctions. *Conservation Biology*, **9**, 960–961.

Distel, D.L. & Roberts, S.J. 1997. Bacterial endosymbionts in the gills of the deep-sea wood-boring bivalves *Xylophaga atlantica* and *Xylophaga washingtona*. *Biological Bulletin*, **192**, 253–261.

Dix, N.J. 1985. Changes in relationship between water content and water potential after decay and its significance for fungal successions. *Transactions of the British Mycological Society*, **85**, 649–653.

Djupström, L. B., Weslien, J. & Schroeder, M. K. 2008. Dead wood and saproxylic beetles in set-aside and non-set-aside forest in a boreal region. *Forest Ecology and Management,* **255**, 3340–3350.

Dollin, P. E., Majka, C. G. & Duinker, P. N. 2008. Saproxylic beetle (Coleoptera) communities and forest management practices in coniferous stands in southwestern Nova Scotia, Canada. *ZooKeys,* **2**, 291–336.

Donisthorpe, H. 1935. The British fungicolous Coleoptera. *Entomologist's Monthly Magazine,* **71**, 21–31.

Donnelly, D. P. & Boddy, L. 1998. Developmental and morphological responses of mycelial systems of *Stropharia caerulea* and *Phanerochaete velutina* to soil nutrient enrichment. *New Phytologist,* **138**, 519–531.

Dorado, J., Claassen, F. W., van Beek, T. A., Lenon, G., Wijnberg, B. P. A. & Sierra-Alvarez, R. 2000. Elimination and detoxification of softwood extractives by white-rot fungi. *Journal of Biotechnology,* **80**, 231–240.

Doyle, J. A. 2008. Integrating molecular phylogenetic and paleobotanical evidence on origin of the flower. *International Journal of Plant Sciences,* **169**, 816–843.

Drossel, B. & McCane, A. J. 2003. Modelling food webs. *In:* Bornholdt, S. & Schuster, H. G. (eds.) *Handbook of Graphs and Networks.* Berlin: Wiley-VCH, 218–247.

Dubois, G. & Vignon, V. 2008. First results of radio-tracking of *Osmoderma eremita* (Coleoptera: Cetoniidae) in French chestnut orchards. *Revue d'Ecologie: La Terre et la Vie,* **63**, 123–130.

Dubois, G. F., Vignon, V., Delettre, Y. R., Rantier, Y., Vernon, P. & Burel, F. 2009. Factors affecting the occurrence of the endangered saproxylic beetle *Osmoderma eremita* (Scopoli, 1763) (Coleoptera: Cetoniidae) in an agricultural landscape. *Landscape and Urban Planning,* **91**, 152–159.

Dubois, G. F., Le Gouar, P. J., Delettre, Y. R., Brustel, H. & Vernon, P. 2010. Sex-biased and body condition dependent dispersal capacity in the endangered saproxylic beetle *Osmoderma eremita* (Coleoptera: Cetoniidae). *Journal of Insect Conservation,* **14**, 679–687.

Dudley, T. L. & Anderson, N. H. 1982. A survey of invertebrates associated with wood debris in aquatic habitats. *Melanderia,* **39**, 1–21.

Dunn, J. P. & Lorio, P. L. 1993. Modified water regimes affect photosynthesis, xylem water potential, cambial growth, and resistance of juvenile *Pinus taeda* L. to *Dendroctonus frontalis,* Coleoptera, Scolytidae. *Environmental Entomology,* **22**, 948–957.

Duvall, M. D. & Grigal, D. F. 1999. Effects of timber harvest on coarse woody debris in red pine forests across the Great Lakes states, USA. *Canadian Journal of Forest Research,* **29**, 1926–1934.

Dybas, H. S. 1956. A new genus of minute fungus-pore beetles from Oregon (Coleoptera: Ptiliidae). *Fieldiana Zoology,* **34**, 441–448.

Dybas, H. S. 1976. The larval characters of featherwing and limulodid beetles and their family relationships in the Staphylinoidea (Coleoptera: Ptiliidae and Limulodidae). *Fieldiana Zoology,* **70**, 29–78.

Dyke, A. S., England, J., Reimnitz, E. & Jette, H. 1997. Changes in driftwood delivery to the Canadian Arctic archipelago: the hypothesis of postglacial oscillations of the transpolar drift. *Arctic,* **50**, 1–16.

Edman, M. & Jonsson, B. G. 2001. Spatial pattern of downed logs and wood-decaying fungi in an old-growth *Picea abies* forest. *Journal of Vegetation Science*, **12**, 609–620.

Edman, M., Gustafsson, M., Stenlid, J., Jonsson, B. G. & Ericson, L. 2004a. Spore deposition of wood-decaying fungi: importance of landscape composition. *Ecography*, **27**, 103–111.

Edman, M., Kruys, N. & Jonsson, B. G. 2004b. Local dispersal sources strongly affect colonisation patterns of wood-decaying fungi on experimental logs. *Ecological Applications*, **14**, 893–901.

Edman, M., Möller, R. & Ericson, L. 2006. Effects of enhanced tree growth rate on the decay capacities of three saprotrophic wood-fungi. *Forest Ecology and Management*, **232**, 12–18.

Edman, M., Jönsson, M. & Jonsson, B. G. 2007. Small-scale fungal- and wind-mediated disturbances strongly influence the temporal availability of logs in an old-growth *Picea abies* forest. *Ecological Applications*, **17**, 482–490.

Eggleton, P. 2000. Global patterns of termite diversity. *In:* Abe, T., Bignell, D. E. & Higashi, M. (eds.) *Termites: Evolution, Sociality, Symbioses, Ecology.* Dordrecht: Kluwer Academic Publishers, 25–51.

Eggleton, P. & Belshaw, R. 1992. Insect parasitoids: an evolutionary overview. *Philosophical Transactions of the Royal Society of London, Series B*, **337**, 1–20.

Eggleton, P. & Tayasu, I. 2001. Feeding groups, lifetypes and the global ecology of termites. *Ecological Research*, **16**, 941–960.

Egnell, G. & Valinger, E. 2003. Survival, growth, and growth allocation of planted Scots pine trees after different levels of biomass removal in clear-felling. *Forest Ecology and Management*, **177**, 65–74.

Ehle, D. S. & Baker, W. L. 2003. Disturbance and stand dynamics in ponderosa pine forests in Rocky Mountain National Park, USA. *Ecological Monographs*, **73**, 543–566.

Ehnström, B. 1983. Faunistic notes on tree-living beetles (Coleoptera) [in Swedish with an English summary]. *Entomologisk Tidskrift*, **104**, 75–79.

Ehnström, B. & Axelsson, R. 2002. *Insektsgnag i bark och ved* [*Insect Galleries in Bark and Wood, in Swedish*]. Uppsala: ArtDatabanken, SLU.

Ehnström, B. & Waldén, H. W. 1986. *Faunavård i skogsbruket – den lägre faunan* [Fauna Management in Forestry - the Invertebrate Fauna, in Swedish]. Jönköping: Skogstyrelsen.

Elias, S. A., Webster, L. & Amer, M. 2009. A beetle's eye view of London from the Mesolithic to Late Bronze Age. *Geological Journal*, **44**, 537–567.

Eliasson, P. & Nilsson, S. G. 2002. 'You should hate young oaks and young noblemen': the environmental history of oaks in eighteenth- and nineteenth-century Sweden. *Environmental History*, **7**, 659–677.

Ellis, A. M., Lounibos, L. P. & Holyoak, M. 2006. Evaluating the long-term metacommunity dynamics of tree hole mosquitoes. *Ecology*, **87**, 2582–2590.

Engel, M. S., Grimaldi, D. A. & Krishna, K. 2009. Termites (Isoptera): their phylogeny, classification, and rise to ecological dominance. *American Museum Novitates*, **3650**, 1–27.

Eriksson, A.-M., Edman, M., Ohlsson, J., Toivanen, S. & Jonsson, B. G. submitted. Restoration fires as a conservation tool: effects on deadwood heterogeneity and availability.

Eriksson, K.-E., Blanchette, R.A. & Ander, P. 1990. *Microbial and Enzymatic Degradation of Wood and Wood Components*. Berlin: Springer-Verlag.

Eriksson, O. 1996. Regional dynamics of plants: a review of evidence for remnant, source–sink and metapopulations. *Oikos*, **77**, 248–258.

Eriksson, P. 2000. Long term variation in population densities of saproxylic Coleoptera species at the river of Dalälven, Sweden [in Swedish with an English summary]. *Entomologisk Tidskrift*, **121**, 119–135.

Erwin, T.L. 1982. Tropical forests: their richness in Coleoptera and other arthropod species. *Coleopterists Bulletin*, **36**, 74–75.

Esseen, P.-A., Ehnström, B., Ericson, L. & Sjöberg, K. 1992. Boreal forests: the focal habitats of Fennoscandia. *In:* Hanson, L. (ed.) *Ecological Principles of Nature Conservation*. London: Elsevier, 252–323.

Esseen, P.-A., Ehnström, B., Ericson, L. & Sjöberg, K. 1997. Boreal forests. *Ecological Bulletins*, **46**, 16–47.

Evans, K.L., Warren, P.H. & Gaston, K.J. 2005. Species–energy relationships at the macroecological scale: a review of the mechanisms. *Biological Reviews*, **80**, 1–25.

Evans, W.G. 1966. Perception of infrared radiation from forest fires by *Melanophila acuminata* DeGeer (Buprestidae, Coleoptera). *Ecology*, **47**, 1061–1065.

Evans, W.G. 1971. The attraction of insects to forest fires. *In:* Komarek, E.V. (ed.) *Proceedings of the Tall Timbers Conference on Ecological Animal Control by Habitat Management, Volume 3*. FL-Tallahassee: University of Florida, Department of Entomology, 115–127.

Evenhuis, N.L. 2006. *Catalog of the Keroplatidae of the World (Insecta: Diptera)*. Bishop Museum Bulletin in Entomology, No. 13.

Fahrig, L. 2002. Effect of habitat fragmentation on the extinction threshold: a synthesis. *Ecological Applications*, **12**, 346–353.

Fäldt, J., Jonsell, M., Nordlander, G. & Borg-Karlsson, A.K. 1999. Volatiles of the bracket fungi *Fomitopsis pinicola* and *Fomes fomentarius* and their functions as insect attractants. *Journal of Chemical Ecology*, **25**, 567–590.

FAO 2002. *International Standards for Phytosanitary Measures: Guidelines for Regulating Wood Packaging Material in International Trade*. Rome: FAO of the UN.

FAO 2006. *Global Forest Resources Assessment 2005: Progress towards Sustainable Forest Management*. Rome: FAO.

Farjon, A. 2003. The remaining diversity of conifers. *Acta Horticulturae*, **615**, 75–89.

Farrell, B.D., Sequeira, A.S., O'Meara, B.C., Normark, B.B., Chung, J.H. & Jordal, B.H. 2001. The evolution of agriculture in beetles (Curculionidae: Scolytinae and Platypodinae). *Evolution*, **55**, 2011–2027.

Farris, K.L., Huss, M.J. & Zack, S. 2004. The role of foraging woodpeckers in the decomposition of ponderosa pine snags. *Condor*, **106**, 50–59.

Fauria, M.M. & Johnson, E.A. 2008. Climate and wildfires in the North American boreal forest. *Philosophical Transactions of the Royal Society, Series B*, **363**, 2317–2329.

Fay, N. & de Berger, N. 1997. *Veteran Trees Initiative: Specialist Survey Method*. Peterborough, UK: English Nature.

Fay, N. & de Berger, N. 2003. *Evaluation of the Specialist Survey Method for Veteran Tree Recording.* English Nature Research Reports 529.

Ferguson, B.A., Dreisbach, T.A., Parks, C.G., Filip, G.M. & Schmitt, C.L. 2003. Coarse-scale population structure of pathogenic *Armillaria* species in a mixed-conifer forest in the Blue Mountains of northeast Oregon. *Canadian Journal of Forest Research*, **33**, 612–623.

Ferrar, P.A. 1987. *A Guide to the Breeding Habits and Immature Stages of Diptera Cyclorrhapha.* Entomograph.

Field, C.B., Campbell, J.E. & Lobell, D.B. 2007. Biomass energy: the scale of the potential resource. *Trends in Ecology and Evolution*, **23**, 65–72.

Fielder, H.J. & Hunger, W. 1963. Über den Einfluss einer Kalkdüngung auf Vorkommen, Wachstum und Nährelementgehalt höherer Pilze im Fichtenbestand. *Archiv für das Forstwesen*, **12**, 936–962.

Fielding, N.J. & Evans, H.F. 1996. The pine wood nematode *Bursaphelenchus xylohilus* (Steiner and Buhrer) Nickle (= *B. lignicolus* Mamiya and Kiohara): an assessment of the current position. *Forestry*, **69**, 36–46.

Filion, L., Payette, S., Robert, E.C., Delwaide, E. C. & Lemieux, C. 2006. Insect-induced tree dieback and mortality gaps in high-altitude balsam fir forests of northern New England and adjacent areas. *Ecoscience*, **13**, 275–287.

Fischer, A., Lindner, M., Abs, C. & Lasch, P. 2002. Vegetation dynamics in central European forest ecosystems (near-natural as well as managed) after storm events. *Folia Geobotanica*, **37**, 17–32.

Fisher, D.O. & Owens, I.P.F. 2004. The comparative method in conservation biology. *Trends in Ecology and Evolution*, **19**, 391–398.

Flodin, K. & Fries, N. 1978. Studies on volatile compounds from *Pinus silvestris* and their effect on wood-decomposing fungi. II. Effects of some volatile compounds on fungal growth. *European Journal of Forest Pathology*, **8**, 300–310.

Flores, G. & Hubbes, M. 1980. The nature and role of phytoalexin produced by aspen (*Populus tremuloides* Michx). *European Journal of Forest Pathology*, **10**, 95–103.

Foit, J. 2010. Distribution of early-arriving saproxylic beetles on standing dead Scots pine trees. *Agricultural and Forest Entomology*, **12**, 133–141.

Forsman, J.T. & Mönkkönen, M. 2003. The role of climate in limiting European resident bird populations. *Journal of Biogeography*, **30**, 55–70.

Fortelius, M.J., Eronen, J., Jernvall, J., Liu, L., Pushkina, J., Rinne, A., Tesakov, I., Vislobokova, Z., Zhang, Z. & Zhou, L. 2002. Fossil mammals resolve regional patterns of Eurasian climate change over 20 million years. *Evolutionary Ecology Research*, **4**, 1005–1016.

Fossli, T.E. & Andersen, J. 1998. Host preference of Cisidae (Coleoptera) on tree-inhabiting fungi in northern Norway. *Entomologica Fennica*, **9**, 65–78.

Foster, D.R. & Boose, E.R. 1992. Patterns of forest damage resulting from catastrophic wind in central New England, USA. *Journal of Ecology*, **80**, 79–98.

Foster, R. W. & Kurta, A. 1999. Roosting ecology of the northern bat (*Myotis septentrionalis*) and comparisons with the endangered Indiana bat (*Myotis sodalis*). *Journal of Mammalogy*, **80**, 659–672.

Foucard, T. 2001. *Svenska skorplavar och svampar som växer på dem.* [Swedish Crustose Lichens and Fungi that Grow on them, in Swedish] Stockholm: Interpublishing.

Fowles, A.P., Alexander, K.N.A. & Key, R.S. 1999. The saproxylic quality index: evaluating wooded habitats for the conservation of dead-wood Coleoptera. *Coleopterist*, **8**, 121–141.

Fox, J.C., Hamilton, F. & Ades, P.K. 2008. Models of tree-level hollow incidence in Victorian State forests. *Forest Ecology and Management*, **255**, 2846–2857.

Framstad, E., Berglund, H., Gundersen, V., Heikkilä, R., Lankinen, N., Peltola, T., Risbøl, O. & Weih, M. 2009. *Increased Biomass Harvesting for Bioenergy: Effects on Biodiversity, Landscape Amenities and Cultural Heritage Values.* Copenhagen: Nordic Council of Ministers.

Franc, V. 1997. Old trees in urban environments: refugia for rare and endangered beetles (Coleoptera). *Acta Universitatis Carolinae Biologica*, **41**, 273–283.

Franceschi, V.R., Krokene, P., Christiansen, E. & Krekling, T. 2005. Anatomical and chemical defenses of conifer bark against bark beetles and other pests. *New Phytologist*, **167**, 353–376.

Franklin, J.F. & Dyrness, C.T. 1973. *Natural Vegetation of Oregon and Washington.* USDA Forest Service, Pacific Northwest Forest and Range Experiment Station.

Franklin, J.F., Shugart, H.H. & Harmon, M.E. 1987. Tree death as an ecological process: the causes, consequences, and variability of tree mortality. *BioScience*, **37**, 550–556.

Franklin, J.F., Berg, D.R., Thornburgh, D.A. & Tappeiner, J.C. 1997. Alternative silvicultural approaches to timber harvesting: variable retention harvest systems. *In:* Kohm, K.A. & Franklin, J.F. (eds.) *Creating a Forestry for the 21st Century.* Washington, DC: Island Press, 111–140.

Franklin, J.F., Spies, T.A., Van Pelt, R., Carey, A.B., Thornburgh, D.A., Berg, D.R., Lindenmayer, D.B., Harmon, M.E., Keeton, W.S., Shaw, D.C., Bible, K. & Chen, J.Q. 2002. Disturbances and structural development of natural forest ecosystems with silvicultural implications, using Douglas-fir forests as an example. *Forest Ecology and Management*, **155**, 399–423.

Fraver, S. & White, A.S. 2005. Disturbance dynamics of old-growth *Picea rubens* forests of northern Maine. *Journal of Vegetation Science*, **16**, 597–610.

Fraver, S., Seymour, R.S., Speer, J.H. & White, A.S. 2007. Dendrochronological reconstruction of spruce budworm outbreaks in northern Maine, USA. *Canadian Journal of Forest Research*, **37**, 523–529.

Fraver, S., Jonsson, B.G., Jönsson, M. & Esseen, P.-A. 2008. Demographics and disturbance history of a boreal old-growth *Picea abies* forest. *Journal of Vegetation Science*, **19**, 789–798.

Frelich, L.E. & Lorimer, C.G. 1991. Natural disturbance regimes in hemlock-hardwood forests of the upper Great Lakes region. *Ecological Monographs*, **61**, 145–164.

Fridman, J. & Walheim, M. 2000. Amount, structure and dynamics of dead wood on managed forestland in Sweden. *Forest Ecology and Management*, **131**, 23–26.

Fukasawa, Y., Osono, T. & Takeda, H. 2009. Dynamics of physiochemical properties and occurrence of fungal fruit bodies during decomposition of coarse woody debris of *Fagus crenata*. *Journal of Forest Research*, **14**, 20–29.

Gandhi, K.J.K., Gilmore, D.W., Katovich, S.A., Mattson, W.J., Spence, J.R. & Seybold, S.J. 2007. Physical effects of weather events on the abundance and diversity of insects in North American forests. *Environmental Reviews*, **15**, 113–152.

Ganter, P.F., Starmer, W.T., Lachance, M.-A. & Pfaff, H.J. 1986. Yeast communities from host plants and associated *Drosophila* in southern Arizona: new isolations and analysis of the relative importance of hosts as vectors on community composition. *Oecologia*, **70**, 386–392.

Gara, R.I., Werner, R.A., Whitmore, M.C. & Holsten, E.H. 1995. Arthropod associates of the spruce beetle *Dendroctonus rufipennis* (Kirby) (Col., Scolytidae) in spruce stands of south-central and interior Alaska. *Journal of Applied Entomology*, **119**, 585–590.

Garbaye, J., Kabre, A., Le Tacon, F., Moussain, D. & Piou, D. 1979. Fertilization minérale et fructification des champignons supérieurs en hětraie. *Annales des Sciences Forestieres*, **36**, 151–164.

Gärdenfors, U. 2010. *The 2010 Red List of Swedish Species*. Uppsala: ArtDatabanken, SLU.

Gärdenfors, U. & Baranowski, R. 1992. Beetles living in open deciduous forests prefer different tree species than those living in dense forests [in Swedish with an English summary]. *Entomologisk Tidskrift*, **113**, 1–11.

Gardiner, L.M. 1957. Deterioration of fire-killed pine in Ontario and the causal wood-boring beetles. *Canadian Entomologist*, **89**, 241–263.

Garrett, S.D. 1951. Ecological groups of soil fungi: a survey of substrate relationships. *New Phytologist*, **50**, 149–166.

Gause, I. 1934. *The Struggle for Existence*. Baltimore, MD: Williams & Wilkins.

Gebauer, G. & Taylor, A.F.S. 1999. ^{15}N natural abundance in fruit bodies of different functional groups of fungi in relation to substrate utilization. *New Phytologist*, **142**, 93–101.

Gerell, R. 2000. The importance of avenues for threatened saproxylic beetles [in Swedish with an English summary]. *Entomologisk Tidskrift*, **121**, 59–66.

German, D.P. & Bittong, R.A. 2009. Digestive enzyme activity and gastrointestinal fermentation in wood-eating catfishes. *Journal of Comparative Physiology, B*, **179**, 1025–1042.

Ghent, A.W., Fraser, D.A. & Thomas, J.B. 1957. Studies of regeneration in forest stands devastated by the spruce budworm. *Forest Science*, **3**, 184–208.

Gibb, H., Ball, J.P., Johansson, T., Atlegrim, O., Hjältén, J. & Danell, K. 2005. Effects of management on coarse woody debris volume and composition in boreal forests in northern Sweden. *Scandinavian Journal of Forest Research*, **20**, 213–222.

Gibbons, P. & Lindenmayer, D.B. 2002. *Tree Hollows and Wildlife Conservation in Australia*. Melbourne: CSIRO Publishing.

Gibbons, P., Lindenmayer, D.B., Barry, S.C. & Tanton, M.T. 2002. Hollow selection by vertebrate fauna in forests of southeastern Australia and implications for forest management. *Biological Conservation*, **103**, 1–12.

Gibbs, J.N. 1978. Intercontinental epidemiology of Dutch elm disease. *Annual Review of Phytopathology*, **16**, 287–307.

Gibbs, J.N. 1993. The biology of ophiostomatoid fungi causing sapstain in trees and freshly cut logs. *In:* Wingfield, M.J., Seifert, K.A. & Webber, J.F. (eds.)

Ceratocystis and Ophiostoma: Taxonomy, Ecology and Pathogenicity. St. Paul, MN: APS Press, 153–160.

Gibbs, J.P., Hunter, M.L. & Melvin, S.M. 1993. Snag availability and communities of cavity-nesting birds in tropical versus temperate forests. *Biotropica*, **25**, 236–241.

Gierlinger, N., Jacques, D., Schwanninger, M., Wimmer, R. & Paques, L.E. 2004. Heartwood extractives and lignin content of different larch species (*Larix* sp.) and relationships to brown-rot decay-resistance. *Trees: Structure and Function*, **18**, 230–236.

Gilbert, G.S. & Sousa, W.P. 2002. Host specialization among wood-decay polypore fungi in a Caribbean mangrove forest. *Biotropica*, **34**, 396–404.

Gilbert, G.S., Gorospe, J. & Ryvarden, L. 2008. Host and habitat preferences of polypore fungi in Micronesian tropical flooded forests. *Mycological Research*, **112**, 674–680.

Gilbertson, R.L. 1980. Wood-rotting fungi of North America. *Mycologia*, **72**, 1–49.

Gill, A.M. & McCarthy, M.A. 1998. Intervals between prescribed fires in Australia: what intrinsic variation should apply? *Biological Conservation*, **85**, 161–169.

Gitlin, A.R., Sthultz, C.M., Bowker, M.A., Stumpf, S., Paxton, C.L., Kennedy, K., Muños, A., Bailey, J.K. & Whitham, T.G. 2006. Mortality gradients within and among dominant plant populations as barometers of ecosystem change during extreme drought. *Conservation Biology*, **20**, 1477–1486.

Gjerdrum, P. 2003. Heartwood in relation to age and growth rate in *Pinus sylvestris* L. in Scandinavia. *Forestry*, **76**, 413–424.

Glenn, J.K., Morgan, M.A., Mayfield, M.B., Kuwahara, M. & Gold, M.H. 1983. An extracellular H_2O_2-requiring enzyme preparation involved in lignin biodegradation by the white-rot basidiomycete *Phanerochaete chrysosporium*. *Biochemical and Biophysical Research Communications*, **114**, 1077–1083.

Glenz, C., Schlaepfer, R., Iorgulescu, I. & Kienast, F. 2006. Flooding tolerance of Central European tree and shrub species. *Forest Ecology and Management*, **235**, 1–13.

Gönczöl, J. & Revay, A. 2003. Treehole fungal communities: aquatic, aero-aquatic and dematiaceous hyphomycetes. *Fungal Diversity*, **12**, 19–34.

González, A.E., Martínez, A.T., Almendros, G. & Grinbergs, J. 1989. A study of yeasts during the delignification and fungal transformation of wood into cattle feed in Chilean rain forest. *Antonie van Leeuwenhoek*, **55**, 221–236.

Goodell, B. 2003. Brown-rot fungal degradation of wood: our evolving view. *In:* Goodell, B., Nicholas, D.D. & Schultz, T.P. (eds.) *Wood Deterioration and Preservation: Advances in our Changing World*. ACS Symposium Series, Vol. 845. Washington, DC: American Chemical Society, 97–118.

Goodell, B., Jellison, J., Liu, J., Daniel, G., Paszczynski, A., Fekete, F., Krishnamurthy, S., Jun, L. & Xu, G. 1997. Low molecular weight chelators and phenolic compounds isolated from wood decay fungi and their role in the fungal biodegradation of wood. *Journal of Biotechnology*, **53**, 133–162.

Graham, S.A. 1925. The felled tree trunk as an ecological unit. *Ecology*, **6**, 397–411.

Grandtner, M.M. 2005. *Elsevier's Dictionary of Trees, Volume 1: North America.* Amsterdam: Elsevier.

Graves, R.C. 1960. Ecological observations of the insects and other inhabitants of woody shelf fungi (Basidiomycetes: Polyporaceae) in the Chicago area. *Annals of the Entomological Society of America*, **53**, 61–78.

Greaves, H. 1971. The bacterial factor in wood decay. *Wood Science and Technology*, **5**, 6–16.

Griffith, G.S. & Boddy, L. 1990. Fungal decomposition of attached angiosperm twigs. I. Decay community development in ash, beech and oak. *New Phytologist*, **116**, 407–415.

Griffiths, B.S. & Cheshire, A.M.V. 1987. Digestion and excretion of nitrogen and carbohydrate by the cranefly larva *Tipula paludosa* (Diptera: Tipulidae). *Insect Biochemistry and Molecular Biology*, **17**, 277–282.

Grimaldi, D. & Engel, M.S. 2005. *Evolution of the Insects.* Cambridge: Cambridge University Press.

Grime, J.P. 1979. *Plant Strategies and Vegetation Processes.* Chichester, UK: John Wiley & Sons.

Grimm, N.B., Grove, J.M., Pickett, S.T.A. & Redman, C.L. 2000. Integrated approaches to long-term studies of urban ecological systems. *BioScience*, **50**, 571–584.

Gromtsev, A. 2002. Natural disturbance dynamics in the boreal forests of European Russia: a review. *Silva Fennica*, **36**, 41–55.

Groover, A.T. 2005. What genes make a tree a tree? *Trends in Plant Science*, **10**, 210–214.

Grove, S.J. 2001. Extent and composition of dead wood in Australian lowland tropical rainforest with different management histories. *Forest Ecology and Management*, **154**, 35–53.

Grove, S.J. 2002a. Saproxylic insect ecology and the sustainable management of forests. *Annual Review of Ecology and Systematics*, **33**, 1–23.

Grove, S.J. 2002b. The influence of forest management history on the integrity of the saproxylic beetle fauna in an Australian lowland tropical rainforest. *Biological Conservation*, **104**, 149–171.

Groven, R., Rolstad, J., Storaunet, K.-O. & Rolstad, E. 2002. Using forest stand reconstructions to assess the role of structural continuity for late-successional species. *Forest Ecology and Management*, **164**, 39–55.

Gu, W., Heikkilä, R. & Hanski, I. 2002. Estimating the consequences of habitat fragmentation on extinction risk in dynamic landscapes. *Landscape Ecology*, **17**, 699–710.

Guevara, R., Hutcheson, K.A., Mee, A.C., Rayner, A.D.M. & Reynolds, S.E. 2000a. Resource partitioning of the host fungus *Coriolus versicolor* by two ciid beetles: the role of odour compounds and host ageing. *Oikos*, **91**, 184–194.

Guevara, R., Rayner, A.D.M. & Reynolds, S.E. 2000b. Effects of fungivory by two specialist ciid beetles (*Octotemnus glabriculus* and *Cis boleti*) on the reproductive fitness of their host fungus, *Coriolous versicolor*. *New Phytologist*, **145**, 137–144.

Guevara, R., Rayner, A.D.M. & Reynolds, S.E. 2000c. Orientation of specialist and generalist fungivorous ciid beetles to host and non-host odours. *Physiological Entomology*, **25**, 288–295.

Gustafsson, M., Holmer, L. & Stenlid, J. 2002. Occurrence of fungal species in coarse logs of *Picea abies* in Sweden. *In:* Gustafsson, M. (ed.) *Distribution and Dispersal of Wood-decaying Fungi occurring on Spruce Logs*. Uppsala: Swedish University of Agricultural Sciences, 4.1–4.10.

Gutiérrez, A., del Río, J.C., Martínez, M.J. & Martínez, A.T. 1999. Fungal degradation of lipophilic extractives in *Eucalyptus globulus* wood. *Applied and Environmental Microbiology*, **65**, 1367–1371.

Gwiazdowicz, D.J. & Łakomy, P. 2002. Mites (Acari, Gamasida) occurring in fruiting bodies of Aphyllophorales. *Fragmenta Faunistica*, **45**, 81–89.

Haanpää, S., Lehtonen, S., Peltonen, L. & Talockaite, E. 2006. Impacts of winter storm Gudrun of 7th – 9th January 2005 and measures taken in Baltic Sea Region [available at http://www.mendeley.com/research/impacts-winter-storm-gudrun-7-th-9-th-january-2005-measures-taken-baltic-sea-region-1/].

Häggström, C.-A. 1998. Pollard meadows: multiple use of human-made nature. *In:* Kirby, K.J. & Watkins, C. (eds.) *The Ecological History of European Forests*. Wallingford, UK: CAB International, 33–41.

Hågvar, S. 1999. Saproxylic beetles visiting living sporocarps of *Fomitopsis pinicola* and *Fomes fomentarius*. *Norwegian Journal of Entomology*, **46**, 25–32.

Hågvar, S. & Økland, B. 1997. Saproxylic beetle fauna associated with living sporocarps of *Fomitopsis pinicola* (Fr.) Karst. in four spruce forests with different management histories. *Fauna Norvegica Serie B*, **44**, 95–105.

Hågvar, S., Hågvar, G. & Mønnes, E. 1990. Nest site selection in Norwegian woodpeckers. *Holarctic Ecology*, **13**, 156–165.

Hahn, K. & Christensen, M. 2004. Dead wood in European forest reserves: a reference for forest management. *In:* Marchetti, M. (ed.) *Monitoring and Indicators of Forest Biodiversity in Europe: From Ideas to Operationality*. EFI Proceedings No 51. Torikatu, Finland: European Forest Institute, 181–191.

Hakala, T.K., Maijala, P., Konn, J. & Hatakka, A. 2004. Evaluation of novel wood-rotting polypores and corticioid fungi for the decay and biopulping of Norway spruce (*Picea abies*) wood. *Enzyme and Microbial Technology*, **34**, 255–263.

Hall, S.J. & Raffaelli, D.G. 1993. Food webs: theory and reality. *Advances in Ecological Research*, **24**, 187–239.

Hall, W.E. 1999. Generic revision of the tribe Nanosellini (Coleoptera: Ptiliidae: Ptiliinae). *Transactions of the American Entomological Society*, **125**, 36–126.

Hallett, J.G., Lopez, T., O'Connell, M.A. & Borysewicz, M.A. 2001. Decay dynamics and avian use of artificially created snags. *Northwest Science*, **75**, 378–386.

Halme, P., Kotiaho, J.S., Ylisirniö, A.-L., Hottola, J., Junninen, K., Kouki, J., Lindgren, M., Mönkkönen, M., Penttilä, R., Renvall, P., Siitonen, J. & Similä, M. 2009. Perennial polypores as indicators of annual and red-listed polypores. *Ecological Indicators*, **9**, 256–266.

Hamilton, W.D. 1978. Evolution and diversity under bark. *In:* Mound, L.A. & Waloff, N. (eds.) *Diversity of Insect Faunas.* London: Royal Entomological Society, 154–175.

Hammond, H.E., Langor, D.W. & Spence, J. 2004. Saproxylic beetles (Coleoptera) using *Populus* in boreal aspen stands of western Canada: spatiotemporal variation and conservation of assemblages. *Canadian Journal of Forest Research,* **34**, 1–19.

Hammond, P.M. 1990. Insect abundance and diversity in the Dumoga-Bone National Park, N. Sulawesi, with special reference to the beetle fauna of lowland rainforest in the Toraut region. *In:* Knight, W.J. & Holloway, J.D. (eds.) *Insects and the Rain Forests of South East Asia (Wallacea).* London: Royal Entomological Society, 197–254.

Hammond, P.M. 1992. Species inventory. *In:* Groombridge, B. (ed.) *Global Diversity. Status of the Earth's Living Resources.* London: Chapman & Hall, 17–39.

Hansen, A.J., Spies, T.A., Swanson, F.J. & Ohmann, J.L. 1991. Conserving biodiversity in managed forests: lessons from natural forests. *BioScience,* **41**, 382–392.

Hansen, E.M. & Goheen, E.M. 2000. *Phellinus weirii* and other native root pathogens as determinants of forest structure and process in western North America. *Annual Review of Phytopathology,* **38**, 515–539.

Hanski, I. 1999. *Metapopulation Ecology.* Oxford: Oxford University Press.

Hanski, I. 2000. Extinction debt and species credit in boreal forests: modelling the consequences of different approaches to biodiversity conservation. *Annales Zoologici Fennici,* **37**, 271–280.

Hanski, I. 2005. *The Shrinking World: Ecological Consequences of Habitat Loss.* Oldendorf, Germany: International Ecological Institute.

Hanski, I. & Gaggiotti, O.E. 2004. *Ecology, Genetics, and Evolution of Metapopulations.* Amsterdam: Elsevier/Academic Press.

Hanski, I. & Gilpin, M.E. 1991. *Metapopulation Biology: Ecology, Genetics, and Evolution.* San Diego, CA: Academic Press.

Hanski, I. & Hammond, P. 1995. Biodiversity in boreal forests. *Trends in Ecology and Evolution,* **10**, 5–6.

Hanski, I. & Ovaskainen, O. 2000. The metapopulation capacity of a fragmented landscape. *Nature,* **404**, 755–758.

Hanski, I. & Ovaskainen, O. 2002. Extinction debt at extinction threshold. *Conservation Biology,* **16**, 666–673.

Hanson, J.J. & Lorimer, C.G. 2007. Forest structure and light regimes following moderate wind storms: implications for multi-cohort management. *Ecological Applications,* **17**, 1325–1340.

Hanula, J.L. 1996. Relationships of wood-feeding insects and coarse woody debris. *In: Biodiversity and Coarse Woody Debris in Southern Forests.* United States Department of Agriculture, Forest Service.

Harding, P.T. & Alexander, K.N.A. 1993. The saproxylic invertebrates of historic parklands: progress and problems. *In:* Kirby, K.J. & Drake, C.M. (eds.) *Dead Wood Matters: The Ecology and Conservation of Saproxylic Invertebrates in Britain.* Peterborough, UK: English Nature, 58–73.

Harding, P.T. & Alexander, K.N.A. 1994. The use of saproxylic invertebrates in the selection and evaluation of areas of relic forest in pasture-woodlands. *British Journal of Entomology and Natural History*, **7** (suppl. 1), 21–26.

Harding, P.T. & Rose, F. 1986. *Pasture-Woodlands in Lowland Britain: A Review of their Importance for Wildlife Conservation*. Huntingdon, UK: Natural Environment Research Council, Institute of Terrestrial Ecology.

Harju, A.M., Venäläinen, M., Anttonen, S., Viitanen, H., Kainulainen, P., Saranpää, P. & Vapaavuori, E. 2003. Chemical factors affecting the brown-rot decay resistance of Scots pine heartwood. *Trees – Structure and Function*, **17**, 263–268.

Harju, A.M., Venäläinen, M., Laakso, T. & Saranpää, P. 2009. Wounding response in xylem of Scots pine seedlings shows wide genetic variation and connection with the constitutive defence of heartwood. *Tree Physiology*, **29**, 19–25.

Härkönen, M., Ukkola, T. & Zeng, Z. 2004. Myxomycetes of the Hunan Province, China 2. *Systematics and Geography of Plants*, **74**, 199–208.

Harmon, M. 2001. Moving towards a new paradigm for woody detritus management. *Ecological Bulletins*, **49**, 269–278.

Harmon, M.E. & Hua, C. 1991. Coarse woody debris dynamics in two old-growth ecosystems: comparing a deciduous forest in China and a conifer forest in Oregon. *BioScience*, **41**, 604–610.

Harmon, M.E., Franklin, J.F., Swansson, F.J., Sollins, P., Gregory, S.V., Lattin, J.D., Anderson, N.H., Cline, S.P., Aumen, N.G., Sedell, J.R., Lienkamper, G.W., Cromack, K.J. & Cummins, K.W. 1986. Ecology of coarse woody debris in temperate ecosystems. *Advances in Ecological Research*, **15**, 133–302.

Harmon, M.E., Whigham, D.F., Sexton, J. & Olmsted, I. 1995. Decomposition and mass of woody detritus in the dry tropical forests of the northeastern Yucatan peninsula, Mexico. *Biotropica*, **27**, 305–316.

Harmon, M.E., Krankina, O.N. & Sexton, J. 2000. Decomposition vectors: a new approach to estimating woody detritus decomposition dynamics. *Canadian Journal of Forest Research*, **30**, 76–84.

Harmon, M.E., Bible, K., Ryan, M.G., Shaw, D.C., Chen, H., Klopatek, J. & Li, X. 2004. Production, respiration, and overall carbon balance in an old-growth *Pseudotsuga/Tsuga* forest ecosystem. *Ecosystems*, **7**, 498–512.

Harper, K.A., Bergeron, Y., Drapeau, P., Gauthier, S. & De Grandpré, L. 2005. Structural development following fire in black spruce boreal forest. *Forest Ecology and Management*, **206**, 293–306.

Harrington, T.C. 1993. Biology and taxonomy of fungi associated with bark beetles. *In:* Schowalter, T.D. & Filip, G.M. (eds.) *Beetle–Pathogen Interactions in Conifer Forests*. San Diego, CA: Academic Press, 37–51.

Hart, P.B.S., Clinton, P.W., Allen, R.B., Nordmeyer, A.H. & Evans, G. 2003. Biomass and macro-nutrients (above- and below-ground) in a New Zealand beech (*Nothofagus*) forest ecosystem: implications for carbon storage and sustainable forest management. *Forest Ecology and Management*, **174**, 281–294.

Harvey, A.E., Larsen, M.J. & Jurgensen, M.F. 1979. Comparative distribution of ectomycorrhizae in soils of three western Montana forest habitat types. *Forest Science*, **25**, 350–358.

Hassan, M.A., Hogan, D.L., Bird, S.A., May, C.L., Gomi, T. & Campbell, D. 2005. Spatial and temporal dynamics of wood in headwater streams of the Pacific Northwest. *Journal of the American Water Resources Association*, **41**, 899–919.

Hatakka, A. 2001. Biodegradation of lignin. *In:* Hofrichter, M. & Steinbüchel, A. (eds.) *Biopolymers: Vol. 1: Lignin, Humic Substances and Coal.* Weinheim, Germany: Wiley-VCH, 129–180.

Hautala, H., Jalonen, J., Laaka-Lindberg, S. & Vanha-Majamaa, I. 2004. Impacts of retention felling on coarse woody debris (CWD) in mature boreal spruce forests in Finland. *Biodiversity and Conservation*, **13**, 1541–1554.

Hawksworth, D.L. 1981. A survey of fungicolous fungi. *In:* Cole, G.T. (ed.) *Biology of Conidial Fungi.* New York: Academic Press, 171–244.

Hawksworth, D.L. 1991. The fungal dimension of biodiversity: magnitude, significance, and conservation. *Mycological Research*, **95**, 641–655.

Hawksworth, D.L. 2001. The magnitude of fungal diversity: the 1.5 million species estimate revisited. *Mycological Research*, **105**, 1422–1432.

Hayslett, M., Juzwik, J. & Moltzan, B. 2008. Three *Colopterus* beetle species carry the oak wilt fungus to fresh wounds on red oak in Missouri. *Plant Disease*, **92**, 270–275.

He, F. & Legendre, P. 1996. On species–area relations. *American Naturalist*, **148**, 719–737.

Hedblom, M. & Söderström, B. 2008. Woodlands across Swedish urban gradients: status, structure and management implications. *Landscape and Urban Planning*, **84**, 62–73.

Hedgren, O. 2009. Search for the rare buprestid beetle *Agrilus mendax* Mannerheim by its larval feeding marks [in Swedish with an English summary]. *Entomologisk Tidskrift*, **130**, 1–9.

Hedin, J., Ranius, T., Nilsson, S.G. & Smith, H.G. 2008. Restricted dispersal in a flying beetle assessed by telemetry. *Biodiversity and Conservation*, **17**, 675–684.

Heilmann-Clausen, J. 2001. A gradient analysis of communities of macrofungi and slime moulds on decaying beech logs. *Mycological Research*, **105**, 575–596.

Heilmann-Clausen, J. & Christensen, M. 2003. Fungal diversity on decaying beech logs – implications for sustainable forestry. *Biodiversity and Conservation*, **12**, 953–973.

Heilmann-Clausen, J. & Christensen, M. 2004. Does size matter? On the importance of various dead wood fractions for fungal diversity in Danish beech forests. *Forest Ecology and Management*, **201**, 105–117.

Heilmann-Clausen, J., Aude, E. & Christensen, M. 2005. Cryptogam communities on decaying deciduous wood: does tree species diversity matter? *Biodiversity and Conservation*, **14**, 2061–2078.

Heinselman, M.L. 1973. Fire in the virgin forests of the Boundary Waters Canoe Area, Minnesota. *Quaternary Research*, **3**, 329–382.

Hendrickson, O.Q. 1991. Abundance and activity of N_2-fixing bacteria in decaying wood. *Canadian Journal of Forest Research*, **21**, 1299–1304.

Hengst, G.E. & Dawson, J.O. 1994. Bark properties and fire resistance of selected tree species from the central hardwood region of North America. *Canadian Journal of Forest Research*, **24**, 688–696.

Henle, K., Davies, K. F., Kleyer, M., Margules, C. & Settele, J. 2004. Predictors of species sensitivity to fragmentation. *Biodiversity and Conservation*, **13**, 207–251.

Hennigar, C.R., MacLean, D.A., Quiring, D.T. & Kershaw, J.A. 2008. Differences in spruce budworm defoliation among balsam fir and white, red, and black spruce. *Forest Science*, **54**, 158–166.

Hennon, P.E. & McClellan, M.H. 2003. Tree mortality and forest structure in the temperate rain forests of southeast Alaska. *Canadian Journal of Forest Research*, **33**, 1621–1634.

Herard, F. & Mercadier, G. 1996. Natural enemies of *Tomicus piniperda* and *Ips acuminatus* (Col., Scolytidae) on *Pinus sylvestris* near Orléans, France. *Entomophaga*, **41**, 183–210.

Hernes, P.J. & Hedges, J. 2004. Tannin signatures of barks, needles, leaves, cones, and wood at the molecular level. *Geochimica et Cosmochimica Acta*, **68**, 1293–1307.

Hespenheide, H.A. 1976. Patterns in the use of single plant hosts by wood-boring beetles. *Oikos*, **27**, 161–164.

Hesselman, H. 1912. Om snöbrotten i norra Sverige vintern 1910–1911 [in Swedish]. *Skogsvårdsföreningens Tidskrift*, **10**, 145–172.

Hibbett, D. S. 2006. A phylogenetic overview of the Agaricomycotina. *Mycologia*, **98**, 917–925.

Hibbett, D. S. & Donoghue, M.J. 2001. Analysis of character correlations among wood decay mechanisms, mating systems, and substrate ranges in homobasidiomycetes. *Systematic Biology*, **50**, 215–242.

Hibbett, D.S., Donoghue, M.J. & Tomlinson, P.B. 1997a. Is *Phellinites digiustoi* the oldest homobasidiomycete? *American Journal of Botany*, **84**, 1005–1011.

Hibbett, D.S., Grimaldi, D. & Donoghue, M.J. 1997b. Fossil mushrooms from Miocene and Cretaceous ambers and the evolution of homobasidiomycetes. *American Journal of Botany*, **84**, 981–991.

Hibbett, D.S., Gilbert, L.-B. & Donoghue, M.J. 2000. Evolutionary instability of ectomycorrhizal symbioses in basidiomycetes. *Nature*, **407**, 506–508.

Hicks, E.A. 1971. Checklist and bibliography on the occurrence of insects in birds' nests. Supplement II. *Iowa State Journal of Science*, **46**, 123–338.

Hicks, W.T. & Harmon, M.E. 2002. Diffusion and seasonal dynamics of O_2 in woody debris from the Pacific Northwest, USA. *Plant and Soil*, **67**, 67–79.

Hillis, W.E. 1987. *Heartwood and Tree Exudates*. Berlin, New York: Springer-Verlag.

Hinds, T.E. 1972. Insect transmission of *Ceratocystis* species associated with aspen cankers. *Phytopathology*, **62**, 221–225.

Hinds, T.E. & Davidson, R.W. 1972. *Ceratocystis* species associated with the aspen ambrosia beetle. *Mycologia*, **64**, 405–409.

Hingley, M.R. 1971. The ascomycete fungus *Daldinia concentrica* as a habitat for animals. *Journal of Animal Ecology*, **40**, 17–32.

Hintikka, V. 1970. Selective effects of terpenes on wood-decomposing hymenomycetes. *Karstenia*, **11**, 28–32.

Hinton, H.E. 1948. On the origin and function of the pupal stage. *Transactions of the Royal Entomological Society of London*, **99**, 395–409.

Hodder, K.H., Bullock, J.M., Buckland, P.C. & Kirby, K.J. 2005. *Large Herbivores in the Wildwood and in Modern Naturalistic Grazing Systems.* English Nature Research Reports, No. 648. Peterborough, UK: English Nature.

Hoffmann, A. & Hering, D. 2000. Wood-associated macroinvertebrate fauna in central European streams. *International Review of Hydrobiology,* **85**, 25–48.

Hoffstetter, R.W., Cronin, J.T., Klepzig, K.D., Moser, J.C. & Ayres, M.P. 2006. Antagonisms, mutualisms and commensalisms affect outbreak dynamics of the southern pine beetle. *Oecologia,* **147**, 679–691.

Höjer, O. & Hultengren, S. 2004. Åtgärdsprogram för särskilt skyddsvärda träd i kulturlandskapet [in Swedish with an English summary]. Naturvårdsvärket.

Holmbom, B. 1998. Extractives. *In:* Sjöström, E. & Alén, R. (eds.) *Analytical Methods in Wood Chemistry, Pulping, and Papermaking.* Berlin: Springer-Verlag, 125–148.

Holmer, L. 1996. *Interspecific interactions between wood-inhabiting basidiomycetes in boreal forests.* PhD thesis, Swedish University of Agricultural Sciences.

Holmer, L. & Stenlid, J. 1997. Competitive hierarchies of wood decomposing basidiomycetes in artificial systems based on variable inoculum sizes. *Oikos,* **79**, 77–84.

Holmer, L., Renvall, P. & Stenlid, J. 1997. Selective replacement between species of wood-rotting basidiomycetes: a laboratory study. *Mycological Research,* **101**, 714–720.

Honigberg, B.M. 1970. Protozoa associated with termites and their role in digestion. *In:* Krishna, K. & Weesner, F.M. (eds.) *Biology of Termites, Vol. 2.* New York: Academic Press, 1–36.

Hooper, M.C., Arii, K. & Lechowicz, M.J. 2001. Impact of a major ice storm on an old-growth hardwood forest. *Canadian Journal of Forest Research,* **79**, 70–75.

Hopkins, A.J.M., Harrison, K.S., Grove, S.J., Wardlaw, T.J. & Mohammed, C.L. 2005. Wood-decay fungi and saproxylic beetles associated with living *Eucalyptus obliqua* trees: early results from studies at the Warra LTER Site, Tasmania. *Tasforests,* **16**, 111–125.

Hosoya, T., Kawamoto, H. & Saka, S. 2009. Solid/liquid- and vapor-phase interactions between cellulose- and lignin-derived pyrolysis products. *Journal of Analytic and Applied Pyrolysis,* **85**, 237–246.

Hottola, J. & Siitonen, J. 2008. Significance of woodland key habitats for polypore diversity and red-listed species in boreal forests. *Biodiversity Conservation,* **17**, 2559–2577.

Hottola, J., Ovaskainen, O. & Hanski, I. 2009. A unified measure of the number, volume and diversity of dead trees and the response of fungal communities. *Journal of Ecology,* **97**, 1320–1328.

Houghton, R.A., Lawrence, K.T., Hackler, J.L. & Brown, S. 2001. The spatial distribution of forest biomass in the Brazilian Amazon: a comparison of estimates. *Global Change Biology,* **7**, 731–746.

Hövenmeyer, K. & Schauermann, J. 2003. Succession of Diptera on dead beech wood: a 10-year study. *Pedobiologica,* **47**, 61–75.

Howden, H. F. & Vogt, G. B. 1951. Insect communities of standing dead pine (*Pinus virginiana* Mill.). *Annals of the Entomological Society of America*, **44**, 581–595.

Hsiau, P. T. W. & Harrington, T. C. 2003. Phylogenetics and adaptations of basidiomycetous fungi fed upon by bark beetles (Coleoptera: Scolytidae). *Symbiosis*, **34**, 111–131.

Hubbell, S. P. 2001. *The Unified Neutral Theory of Biodiversity and Biogeography.* Princeton, NJ: Princeton University Press.

Hubbell, S. P. 2005. Neutral theory in community ecology and the hypothesis of functional equivalence. *Functional Ecology*, **19**, 166–172.

Hubbell, S. P. 2006. Neutral theory and the evolution of ecological equivalence. *Ecology*, **87**, 1387–1398.

Hubert, E. E. 1918. Fungi as contributory causes of windfall in the Northwest. *Journal of Forestry*, **16**, 696–714.

Hudgins, J. W., Christiansen, E. & Franceschi, V. R. 2003. Methyl jasmonate induces changes mimicking anatomical and chemical defenses in diverse members of the Pinaceae. *Tree Physiology*, **23**, 361–371.

Hudson, H. J. 1968. The ecology of fungi on plant remains above the soil. *New Phytologist*, **67**, 837–874.

Hultengren, S. & Nitare, J. 1999. *Inventering av jätteträd: instruktion för inventering av grova lövträd i södra Sverige.* [Inventory of Giant Trees: Instructions for Surveying Coarse Broadleaved Trees in Southern Sweden, in Swedish] Skogstyrelsen.

Hunt, D. R. 1996. The genera of temperate broadleaved trees. *Broadleaves*, **2**, 4–5.

Hunt, T., Bergsten, J., Levkanicova, Z., Papadopoulou, A., St. John, O., Wild, R., Hammond, P. M., Ahrens, D., Balke, M., Caterino, M. S., Gómez-Zurita, J., Ribera, I., Barraclough, T. G., Bocakova, M., Bocak, L. & Vogler, A. P. 2007. A comprehensive phylogeny of beetles reveals the evolutionary origins of a superradiation. *Science*, **318**, 1913–1916.

Huntley, P. M., Van Noort, S. & Hamer, M. 2005. Giving increased value to invertebrates through ecotourism. *South African Journal of Wildlife Research*, **35**, 53–62.

Hurtt, G. C. & Pacala, S. W. 1995. The consequences of recruitment limitation: reconciling chance, history, and competitive differences between plants. *Journal of Theoretical Biology*, **176**, 1–12.

Hutchinson, G. E. 1957. Concluding remarks. *Cold Spring Harbor Symposium on Quantitative Biology*, **22**, 415–427.

Hutchinson, G. E. 1959. Homage to Santa Rosalia, or why there are so many kinds of animals. *American Naturalist*, **93**, 145–149.

Hutto, R. L. 1995. Composition of bird communities following stand-replacement fires in northern Rocky Mountain (U.S.A.) conifer forests. *Conservation Biology*, **9**, 1041–1058.

Hutto, R. L. & Gallo, S. M. 2006. The effects of postfire salvage logging on cavity-nesting birds. *Condor*, **108**, 817–831.

Hyatt, T. L. & Naiman, R. J. 2001. The residence time of large woody debris in the Queets River, Washington, USA. *Ecological Applications*, **11**, 191–202.

Hynynen, J. 1993. Self-thinning models for even-aged stands of *Pinus sylvestris*, *Picea abies*, and *Betula pendula*. *Scandinavian Journal of Forest Research*, **8**, 326–336.

Hynynen, J., Ahtikoski, A., Siitonen, J., Sievänen, R. & Liski, J. 2005. Applying the MOTTI simulator to analyse the effect of alternative management schedules on timber and non-timber production. *Forest Ecology and Management*, **207**, 5–18.

Hyvärinen, E., Kouki, J. & Martikainen, P. 2006. Fire and green-tree retention in conservation of red-listed and rare deadwood-dependent beetles in Finnish boreal forests. *Conservation Biology*, **20**, 1711–1719.

Hyvärinen, E., Kouki, J. & Martikainen, P. 2009. Prescribed burning and retention trees help to conserve beetle diversity in managed boreal forest despite their transient negative effects on some beetle groups. *Insect Conservation and Diversity*, **2**, 93–105.

Iablokoff, A. 1943. Ethologie de quelques Elaterides du massif de Fontainebleau. *Mémoires du Muséum National d'Histoire Naturelle, Nouvelle Sérié* **18**, 81–160.

Ihalainen, A. & Mäkelä, H. 2009. Kuolleen puuston määrä Etelä- ja Pohjois-Suomessa 2004–2007 [in Finnish]. *Metsätieteen aikakauskirja*, 35–56.

Ing, B. 1994. The phytosociology of myxomycetes. *New Phytologist*, **126**, 175–202.

Ingles, L. G. 1933. The succession of insects in tree trunks as shown by the collections from the various stages of decay. *Journal of Entomology and Zoology*, **25**, 57–59.

Ingold, C. T. 1954. Aquatic ascomycetes: discomycetes from lakes. *Transactions of the British Mycological Society*, **37**, 1–18.

Inward, D. J. G., Vogler, A. P. & Eggleton, P. 2007. A comprehensive phylogenetic analysis of termites (Isoptera) illuminates key aspects of their evolutionary biology. *Molecular Phylogenetics and Evolution*, **44**, 953–967.

Irmler, U., Heller, K. & Warning, J. 1996. Age and tree species as factors influencing the populations of insects living in dead wood (Coleoptera, Diptera: Sciaridae, Mycetophilidae). *Pedobiologia*, **40**, 134–148.

Ishii, H. & Kodatani, T. 2006. Biomass and dynamics of attached dead branches in the canopy of 450-year-old Douglas-fir trees. *Canadian Journal of Forest Research*, **36**, 378–389.

IUCN 2001. *IUCN Red List Categories and Criteria*. Version 3.1. Gland, Switzerland, and Cambridge, UK: IUCN.

Iwasa, M., Hori, K. & Aoki, N. 1995. Fly fauna of bird nests in Hokkaido, Japan (Diptera). *Canadian Entomologist*, **127**, 613–621.

Iwata, R., Maro, T., Yonezawa, Y., Yahagi, T. & Fujikawa, Y. 2007. Period of adult activity and response to wood moisture content as major segregating factors in the coexistence of two conifer longhorn beetles, *Callidiellum rufipenne* and *Semanotus bifasciatus* (Coleoptera: Cerambycidae). *European Journal of Entomology*, **104**, 341–345.

Jackson, J. A. 1977. Red-cockaded woodpeckers and pine red heart disease. *Auk*, **94**, 160–163.

Jackson, J. A. & Jackson, B. J. S. 2004. Ecological relationships between fungi and woodpecker cavity sites. *Condor*, **106**, 37–49.

Jacobs, J. M., Spence, J. R. & Langor, D. W. 2007. Variable retention harvest of white spruce stands and saproxylic beetle assemblages. *Canadian Journal of Forest Research*, **37**, 1631–1642.

Jahn, H. 1966. Pilzgesellschaften an *Populus tremula*. *Zeitschrift für Pilzkunde*, **32**, 26–42.

Jahn, H. 1968. Pilze an Weisstanne (*Abies alba*). *Westfälische Pilzebriefe*, **7**, 17–40.

Jakovlev, J. 1994. *Palearctic Diptera associated with Fungi and Myxomycetes* [in Russian with an English summary]. Petrozavodsk: Karelian Research Center, Russian Academy of Sciences, Forest Research Institute.

Jalkanen, R. & Konopka, B. 1998. Snow-packing as a potential harmful factor on *Picea abies*, *Pinus sylvestris* and *Betula pubescens* at high altitude in northern Finland. *European Journal of Forest Pathology*, **28**, 373–382.

James, T.Y., Kauff, F., Schoch, C.L., Matheny, P.B., Hofstetter, V., Cox, C.J., Celio, G., Gueidan, C., Fraker, E., Miadlikowska, J., Lumbsch, H.T., Rauhut, A., Reeb, V., Arnold, A.E., Amtoft, A., Stajich, J.E., Hosaka, K., Sung, G.-H., Johnson, D. & O'Rourke, B. 2006. Reconstructing the early evolution of fungi using a six-gene phylogeny. *Nature*, **443**, 818–822.

Jankowiak, R. 2005. Fungi associated with *Ips typographus* on *Picea abies* in southern Poland and their succession into the phloem and sapwood of beetle-infested trees and logs. *Forest Pathology*, **35**, 37–55.

Jansson, N. 1998. *Miljöövervakning av biotoper med gamla ekar i Östergötland* [Environmental Monitoring of Biotopes with Old Oaks in Östergötaland, in Swedish]. Linköping, Sweden: Länstyrelsen i Östergötlands Län.

Jansson, N., Bergman, K.-O., Jonsell, M. & Milberg, P. 2009. An indicator system for identification of sites of high conservation value for saproxylic oak (*Quercus* spp.) beetles in southern Sweden. *Journal of Insect Conservation*, **13**, 399–412.

Jeffries, P. 1995. Biology and ecology of mycoparasitism. *Canadian Journal of Botany*, **73** (Suppl. 1), S1284–S1290.

Jeffries, T.W. 1994. Biodegradation of lignin and hemicelluloses. *In:* Ratledge, C. (ed.) *Biochemistry of Microbial Degradation*. Dordrecht, The Netherlands: Kluwer Academic Publishers, 233–277.

Jenkins, B., Kitching, R.L. & Pimm, S.L. 1992. Productivity, disturbance and food web structure at a local spatial scale in experimental container habitats. *Oikos*, **65**, 249–255.

Jenkins, D.W. & Carpenter, S.J. 1946. Ecology of the tree hole breeding mosquitoes of Nearctic North America. *Ecological Monographs*, **16**, 31–47.

Jenkins, S.H. & Busher, P.E. 1979. *Castor canadiensis*. *Mammalian Species*, **120**, 1–9.

Jensen, W., Fremer, K.E., Sierilä, P. & Wartiowaara, V. 1963. The chemistry of bark. *In:* Browning, B.L. (ed.) *The Chemistry of Wood*. New York: Interscience Publishers, 587–666.

Johannesson, H., Vasiliauskas, R., Dahlberg, A., Penttilä, R. & Stenlid, J. 2001. Genetic differentiation in Eurasian populations of the postfire ascomycete *Daldinia loculata*. *Molecular Ecology*, **10**, 1665–1677.

Johansen, S. & Hytteborn, H. 2001. A contribution to the discussion of biota dispersal with drift ice and driftwood in the North Atlantic. *Journal of Biogeography*, **28**, 105–115.

Johansson, T., Olsson, J., Hjältén, J., Jonsson, B.G. & Ericson, L. 2006. Beetle attraction to sporocarps and mycelia of wood-decaying fungi. *Forest Ecology and Management*, **237**, 335–341.

Johnson, E.A. 1992. *Fire and Vegetation Dynamics: Studies from the North American Boreal Forest.* New York: Cambridge University Press.

Johnson, E.A. & Miyanishi, K. 2001. *Forest Fires: Behavior and Ecological Effects.* San Diego, CA: Associated Press.

Johnson, E.A., Miyanishi, K. & Weir, J.M.H. 1998. Wildfires in the western Canadian boreal forest: landscape patterns and ecosystem management. *Journal of Vegetation Science*, **9**, 603–610.

Johnsson, K., Nilsson, S.G. & Tjernberg, M. 1990. The black woodpecker: a key species in European forests. *In:* Carlsson, A. & Aulén, G. (eds.) *Conservation and Management of Woodpecker Populations.* Report 17. Swedish University of Agricultural Science, Department of Wildlife Ecology, 99–102.

Johnsson, K., Nilsson, S.G. & Tjernberg, M. 1993. Characteristics and utilization of old black woodpecker *Dryocopus martius* holes by hole-nesting species. *Ibis*, **135**, 410–416.

Jones, E.B.G. 1985. Wood-inhabiting fungi from San Juan Island, with special reference to ascospore appendages. *Botanical Journal of the Linnean Society*, **91**, 219–231.

Jongman, R.H.G. 2002. Homogenization and fragmentation of the European landscape: ecological consequences and solutions. *Landscape and Urban Planning*, **58**, 211–221.

Jonsell, M. 2004a. Old park trees: a highly desirable resource for both history and beetle diversity. *Journal of Arboriculture*, **30**, 238–243.

Jonsell, M. 2004b. Red-listed saproxylic beetles in the park at Skokloster Castle, Sweden [in Swedish with an English summary]. *Entomologisk Tidskrift*, **125**, 61–69.

Jonsell, M. 2008. Saproxylic beetles in the park at Drottningholm, Stockholm [in Swedish with an English summary]. *Entomologisk Tidskrift*, **129**, 103–120.

Jonsell, M. & Nordlander, G. 1995. Field attraction of Coleoptera to odours of the wood-decaying polypores *Fomitopsis pinicola* and *Fomes fomentarius*. *Annales Zoologici Fennici*, **32**, 391–402.

Jonsell, M. & Nordlander, G. 2004. Host selection patterns in insects breeding in bracket fungi. *Ecological Entomology*, **29**, 697–705.

Jonsell, M., Weslien, J. & Ehnström, B. 1998. Substrate requirements of red-listed saproxylic invertebrates in Sweden. *Biodiversity and Conservation*, **7**, 749–764.

Jonsell, M., Nordlander, G. & Jonsson, M. 1999. Colonization patterns of insects breeding in wood-decaying fungi. *Journal of Insect Conservation*, **3**, 145–161.

Jonsell, M., Nordlander, G. & Ehnström, B. 2001. Substrate associations of insects breeding in fruiting bodies of wood-decaying fungi. *Ecological Bulletins*, **49**, 173–194.

Jonsell, M., Hansson, J. & Wedmo, L. 2007. Diversity of saproxylic beetle species in logging residues in Sweden: comparisons between tree species and diameters. *Biological Conservation*, **138**, 89–99.

Jönsson, A.M., Appelberg, G., Harding, S. & Bärring, L. 2009. Spatio-temporal impact of climate change on the activity and voltinism of the spruce bark beetle, *Ips typographus*. *Global Change Biology*, **15**, 486–499.

Jonsson, B. G. 2000. Availability of coarse woody debris in an old-growth boreal spruce forest landscape. *Journal of Vegetation Science*, **11**, 51–56.

Jonsson, B. G. & Kruys, N. (eds.) 2001. *Ecology of Woody Debris in Boreal Forests*. Ecological Bulletins, No. 49. Oxford: Blackwell Science.

Jonsson, B. G. & Ranius, T. 2009. The temporal and spatial challenges of target setting for dynamic habitats: the case of dead wood in boreal forests. *In:* Villard, M. A. & Jonsson, B. G. (eds.) *Setting Conservation Targets for Managed Forest Landscapes*. Cambridge: Cambridge University Press.

Jonsson, B. G. & Söderström, L. 1989. Growth and reproduction in the leafy hepatic *Ptilidium pulcherrimum* (G. Web.) Vainio during a four year period. *Journal of Bryology*, **15**, 315–325.

Jonsson, B. G., Kruys, N. & Ranius, T. 2005. Ecology of species living on dead wood: lessons for dead wood management. *Silva Fennica*, **39**, 289–309.

Jonsson, M. 2003. Colonisation ability of the threatened tenebrionid beetle *Oplocephala haemorrhoidalis* and its common relative *Bolitophagus reticulatus*. *Ecological Entomology*, **28**, 159–167.

Jonsson, M. & Nordlander, G. 2006. Insect colonisation of fruiting bodies of the wood-decaying fungus *Fomitopsis pinicola* at different distances from an old-growth forest. *Biodiversity and Conservation*, **15**, 295–309.

Jonsson, M., Nordlander, G. & Jonsell, M. 1997. Pheromones affecting flying beetles colonizing the polypores *Fomes fomentarius* and *Fomitopsis pinicola*. *Entomologica Fennica*, **8**, 161–165.

Jonsson, M., Kindvall, O., Jonsell, M. & Nordlander, G. 2003. Modelling mating success of saproxylic beetles in relation to search behaviour, population density and substrate abundance. *Animal Behaviour*, **65**, 1069–1076.

Jönsson, M. T. & Jonsson, B. G. 2007. Assessing coarse woody debris in Swedish woodland key habitats: implications for conservation and management. *Forest Ecology and Management*, **242**, 363–373.

Jönsson, M. T., Fraver, S., Jonsson, B. G., Dynesius, M., Rydgård, M. & Esseen, P.-A. 2007. Eighteen years of tree mortality and structural change in an experimentally fragmented Norway spruce forest. *Forest Ecology and Management*, **242**, 306–313.

Jönsson, M. T., Edman, M. & Jonsson, B. G. 2008. Colonization and extinction patterns of wood-decaying fungi in a boreal old-growth *Picea abies* forest. *Journal of Ecology*, **96**, 1065–1075.

Jönsson, M. T., Fraver, S. & Jonsson, B. G. 2009. Forest history and the development of old-growth characteristics in fragmented boreal forests. *Journal of Vegetation Science*, **20**, 91–106.

Jönsson, N., Méndez, M. & Ranius, T. 2004. Nutrient richness of wood mould in tree hollows with the scarabaeid beetle *Osmoderma eremita*. *Animal Biodiversity and Conservation*, **27**, 79–82.

Judd, W. S., Campbell, C. S., Kellogg, E. A., Stevens, P. F. & Donoghue, M. J. 2002. *Plant Systematics: A Phylogenetic Approach*. Sunderland, MA: Sinauer Associates.

Junninen, K. 2009. Conservation of *Antrodia crassa* [in Finnish with an English summary]. *Metsähallituksen Luonnonsuojelujulkaisuja, Sarja A*, **182**, 51 pp.

Junninen, K. & Komonen, A. 2011. Conservation ecology of boreal polypores: a review. *Biological Conservation*, **144**, 11–20.

Junninen, K. & Kouki, J. 2006. Are woodland key habitats in Finland hotspots for polypores (Basidiomycota)? *Scandinavian Journal of Forest Research*, **21**, 32–40.

Junninen, K., Similä, M., Kouki, J. & Kotiranta, H. 2006. Assemblages of wood-inhabiting fungi along the gradient of succession and naturalness in boreal pine-dominated forests in Fennoscandia. *Ecography*, **29**, 75–83.

Junninen, K., Penttilä, R. & Martikainen, P. 2007. Fallen retention aspen trees on clear-cuts can be important habitats for red-listed polypores: a case study in Finland. *Biodiversity and Conservation*, **16**, 475–490.

Junninen, K., Kouki, J. & Renvall, P. 2008. Restoration of natural legacies of fire in European boreal forests: an experimental approach to the effects on wood-decaying fungi. *Canadian Journal of Forest Research*, **38**, 202–215.

Käärik, A. 1974. Decomposition of wood. *In:* Dickinson, C.H. & Pugh, G.J.F. (eds.) *Biology of Plant Litter Decomposition*. London: Academic Press.

Kafka, V., Gauthier, S. & Bergeron, Y. 2001. Fire impacts and crowning in the boreal forest: study of a large wildfire in western Quebec. *International Journal of Wildland Fire*, **10**, 119–127.

Kahler, H.A. & Andersson, J.T. 2006. Tree cavity resources for dependent cavity-using wildlife in West Virginia forests. *Northern Journal of Applied Forestry*, **23**, 114–121.

Kaila, L. 1993. A new method for collecting quantitative samples of insects associated with decaying wood or wood fungi. *Entomologica Fennica*, **4**, 21–23.

Kaila, L., Martikainen, P., Punttila, P. & Yakovlev, E. 1994. Saproxylic beetles (Coleoptera) on dead birch trunks decayed by different polypore species. *Annales Zoologici Fennici*, **31**, 97–107.

Kaila, L., Martikainen, P. & Punttila, P. 1997. Dead trees left in clear-cuts benefit saproxylic Coleoptera adapted to natural disturbances in boreal forest. *Biodiversity and Conservation*, **6**, 1–18.

Kalcounis-Ruppel, M.C., Psyllakis, J.M. & Brigham, R.M. 2005. Tree roost selection by bats: an empirical synthesis using meta-analysis. *Wildlife Society Bulletin*, **33**, 1123–1132.

Kangas, E. 1947. Kovakuoriaisfaunamme erikoisuuksia luonnonsuojelun kannalta. [Curiosities of our beetle fauna from the point of view of nature conservation, in Finnish]. *Suomen Luonto*, **6**, 45–55.

Kappeler, P.M. 1998. Nests, tree holes, and the evolution of primate life histories. *American Journal of Primatology*, **46**, 7–33.

Kappes, H. & Topp, W. 2004. Emergence of Coleoptera from deadwood in a managed broadleaved forest in central Europe. *Biodiversity and Conservation*, **13**, 1905–1924.

Karjalainen, L. & Kuuluvainen, T. 2001. Amount and diversity of coarse woody debris within a boreal forest landscape dominated by *Pinus sylvestris* in Vienansalo wilderness, eastern Fennoscandia. *Silva Fennica*, **36**, 147–167.

Karström, M. 1992. The project one step ahead: a presentation [in Swedish with an English summary]. *Svensk Botanisk Tidskrift*, **86**, 103–114.

Kaspari, M., O'Donnell, S. & Kercher, J.R. 2000. Energy, density and constraints to species richness, ant assemblages along a productivity gradient. *American Naturalist*, **155**, 280–293.

Kaufman, M.G., Pankratz, H.S. & Klug, M.J. 1986. Bacteria associated with the ectoperitrophic space in the midgut of the larva of the midge *Xylotopus par* (Diptera: Chironomidae). *Applied and Environmental Microbiology*, **51**, 657–660.

Keller, H. W. 2004. Tree canopy biodiversity: student research experiences in Great Smoky Mountains National Park. *Systematics and Geography of Plants*, **74**, 47–65.

Keller, H. W. & Braun, K. L. 1999. Myxomycetes of Ohio: their systematics, biology, and use in teaching. *Ohio Biological Survey, New Series*, **13**, 1–182.

Kellogg, D. W. & Taylor, E. L. 2004. Evidence of oribatid mite detritivory in Antarctica during the Late Paleozoic and Mesozoic. *Journal of Paleontology*, **78**, 1146–1153.

Kelner-Pillault, S. 1974. Étude écologique du peuplement entomologique des terreaux d'arbres creux (châtaigners et saules). *Bulletin d'Ecologie*, **5**, 123–156.

Kenis, M. & Hilszczanski, J. 2004. Natural enemies of Cerambycidae and Buprestidae infesting living trees. *In:* Lieutier, F., Day, K. R., Battisti, A., Grégoire, J.-C. & Evans, H. F. (eds.) *Bark and Wood Boring Insects in Living Trees in Europe: A Synthesis*. Dordrecht, Boston, London: Kluwer Academic Publishers, 475–498.

Kenis, M., Wegensteiner, R. & Griffin, C. T. 2004a. Parasitoids, predators, nematodes and pathogens associated with bark weevil pests. *In:* Lieutier, F., Day, K. R., Battisti, A., Grégoire, J.-C. & Evans, H. F. (eds.) *Bark and Wood Boring Insects in Living Trees in Europe: A Synthesis*. Dordrecht, Boston, London: Kluwer Academic Publishers, 395–414.

Kenis, M., Wermelinger, B. & Grégoire, J.-C. 2004b. Research on parasitoids and predators of Scolytidae: a review. *In:* Lieutier, F., Day, K. R., Battisti, A., Grégoire, J.-C. & Evans, H. F. (eds.) *Bark and Wood Boring Insects in Living Trees in Europe: A Synthesis*. Dordrecht, Boston, London: Kluwer Academic Publishers, 237–290.

Kennedy, R. S. H. & Spies, T. A. 2007. An assessment of dead wood patterns and their relationships with biophysical characteristics in two landscapes with different disturbance histories in coastal Oregon, USA. *Canadian Journal of Forest Research*, **37**, 940–956.

Kenrick, P. & Crane, P. R. 1997. The origin and early evolution of plants on land. *Nature*, **389**, 33–39.

Kerrigan, J., Smith, M. T., Rogers, J. D. & Poot, G. A. 2004. *Botryozoma mucatilis* sp. nov., an anamorphic ascomycetous yeast associated with nematodes in poplar slime flux. *FEMS Yeast Research*, **4**, 849–856.

Key, R. S. & Ball, S. G. 1993. Positive management for saproxylic invertebrates. *In:* Kirby, K. J. & Drake, C. M. (eds.) *Dead Wood Matters: The Ecology and Conservation of Saproxylic Invertebrates in Britain*. Peterborough, UK: English Nature, 89–101.

Kharuk, V. I., Ranson, K. J. & Fedotova, E. V. 2007. Spatial pattern of Siberian silk moth outbreak and taiga mortality. *Scandinavian Journal of Forest Research*, **22**, 531–536.

Kielczewski, B., Moser, J. C. & Wisniewski, J. 1983. Surveying the acarofauna associated with Polish Scolytidae. *Bulletin de la Societé des Amis des Sciences et des Lettres de Poznan, Serie D, Sciences Biologiques*, **22**, 151–159.

Kim, J.-J., Allen, E.A., Humble, L.M. & Breuil, C. 2005. Ophiostomatoid and basidiomycetous fungi associated with green, red, and grey lodgepole pines after mountain pine beetle (*Dendroctonus ponderosae*) infestation. *Canadian Journal of Forest Research*, **35**, 274–284.

Kim, Y.S. & Singh, A.P. 2000. Micromorphological characteristics of wood biodegradation in wet environments: a review. *IAWA Journal*, **21**, 135–155.

Kimmerer, R.W. 1993. Disturbance and dominance in *Tetraphis pellucida*: a model of disturbance frequency and reproductive mode. *Bryologist*, **96**, 73–79.

Kimmerer, R.W. 1994. Ecological consequences of sexual versus asexual reproduction in *Dicranum flagellare* and *Tetraphis pellucida*. *Bryologist*, **97**, 20–25.

Kimmerer, R.W. & Young, C.C. 1995. The role of slugs in dispersal of the asexual propagules of *Dicranum flagellare*. *Bryologist*, **98**, 149–163.

Kimmerer, R.W. & Young, C.C. 1996. Effect of gap size and regeneration niche on species coexistence in bryophyte communities. *Bulletin of the Torrey Botanical Club*, **123**, 16–24.

King, A.J., Cragg, S.M., Li, Y., Dymond, J., Guille, M.J., Bowles, D.J., Bruce, N.C., Graham, I.A. & McQueen-Mason, S.J. 2010. Molecular insight into lignocellulose digestion by a marine isopod in the absence of gut microbes. *PNAS*, **107**, 5345–5350.

Kirby, K.J. & Watkins, C. 1998. *The Ecological History of European Forests*. Wallingford, UK: CAB International.

Kirby, K.J., Thomas, R.C., Key, R.S., McLean, I.F.G. & Hodgetts, N. 1995. Pasture-woodland and its conservation in Britain. *Biological Journal of the Linnean Society*, **56** (Suppl.), 135–153.

Kirby, K.J., Reid, C.M., Thomas, R.C. & Goldsmith, F.B. 1998. Preliminary estimates of fallen dead wood and standing dead trees in managed and unmanaged forests in Britain. *Journal of Applied Ecology*, **35**, 148–155.

Kirisits, T. 2004. Fungal associates of European bark beetles with special emphasis on the ophiostomatoid fungi. *In:* Lieutier, F., Day, K.R., Battisti, A., Grégoire, J.-C. & Evans, H.F. (eds.) *Bark and Wood Boring Insects in Living Trees in Europe: A Synthesis*. Dordrecht, Boston, London: Kluwer Academic Publishers, 181–235.

Kirk, T.K. & Cullen, D. 1998. Enzymology and molecular genetics of wood degradation by white-rot fungi. *In:* Young, R.A. & Akhtar, M. (eds.) *Environmentally Friendly Technologies for Pulp and Paper Industries*. New York: John Wiley & Sons, 273–307.

Kirk, T.K. & Farrell, R.L. 1987. Enzymatic 'combustion': the microbial degradation of lignin. *Annual Review of Microbiology*, **41**, 465–505.

Kirschner, R. 2001. Diversity of filamentous fungi in bark beetle galleries in central Europe. *In:* Misra, J.K. & Horn, B.W. (eds.) *Trichomycetes and other Fungal Groups: Robert W. Lichtwardt Commemoration Volume*. Enfield, NH: Science Publishers, 175–196.

Kirschner, R., Bauer, R. & Oberwinkler, F. 1999. *Atractocolax*, a new heterobasidiomycetous genus based on a species vectored by conifericolous bark beetles. *Mycologia*, **91**, 538–543.

Kitching, R.L. 1971. Water-filled tree-holes and their position in the woodland ecosystem. *Journal of Animal Ecology*, **40**, 281–302.

Kitching, R.L. 2000. *Food Webs and Container Habitats: The Natural History and Ecology of Phytotelmata.* Cambridge: Cambridge University Press.

Kitching, R.L. 2001. Food webs in phytotelmata: 'bottom-up' and 'top-down' explanation for community structure. *Annual Review of Entomology*, **46**, 729–760.

Klimaszewski, J. & Peck, S.B. 1987. Succession and phenology of beetle faunas in the fungus *Polyporellus squamosus* (Huds: Fr.) Karst (Polyporaceae) in Silesia, Poland. *Canadian Journal of Zoology*, **65**, 542–550.

Knapp, E.E., Keeley, J.E., Ballenger, E.A. & Brennan, T.J. 2005. Fuel reduction and coarse woody debris dynamics with early season and late prescribed fire in a Sierra Nevada mixed conifer forest. *Forest Ecology and Management*, **208**, 383–397.

Knudsen, H. & Vesterholt, J. 2008. *Funga Nordica: Agaricoid, Boletoid and Cyphelloid Genera.* Nordsvamp.

Koch, A.J., Munks, S.A. & Kirkpatrick, J.B. 2008a. Does hollow occurrence vary with forest type? A case study in wet and dry *Eucalyptus obliqua* forest. *Forest Ecology and Management*, **255**, 3938–3951.

Koch, A.J., Munks, S.A. & Woehler, E.J. 2008b. Hollow-using vertebrate fauna of Tasmania: distribution, hollow requirements and conservation status. *Australian Journal of Zoology*, **56**, 323–349.

Koch, J. & Petersen, K.R.L. 1996. A check list of higher marine fungi on wood from Danish coasts. *Mycotaxon*, **60**, 397–414.

Koch, K. 1989. *Die Käfer Mitteleuropas. Ökologie, Vol. I–III.* Krefeld, Germany: Goecke & Evers Verlag.

Koenigswald, W. 1990. Die Paläobiologie der Apatemyiden (Insectivora s.l.) und die Ausdeutung der Skelettfunde von *Heterohyus nanus* aus dem Mittelleozän von Messel bei Darmstadt. *Palaeontographica Abteilung A*, **210**, 41–77.

Kohlmeyer, J., Bebout, B. & Volkmann-Kohlmeyer, B. 1995. Decomposition of mangrove wood by marine fungi and teredinids in Belize. *Marine Ecology*, **16**, 27–39.

Koide, R.T., Sharda, J.N., Herr, J.R. & Malcolm, G.M. 2008. Ectomycorrhizal fungi and the biotrophy–saprotrophy continuum. *New Phytologist*, **178**, 230–233.

Kolarik, M. & Hulcr, J. 2009. Mycobiota associated with the ambrosia beetle *Scolytodes unipunctatus* (Coleoptera: Curculionidae, Scolytinae). *Mycological Research*, **113**, 44–60.

Kolarik, M., Kubatova, A., Hulcr, J. & Pazoutova, S. 2008. *Geosmithia* fungi are highly diverse and consistent bark beetle associates: evidence from their community structure in temperate Europe. *Microbial Ecology*, **55**, 65–80.

Kolb, T.E., Guerard, N., Hofstetter, R.W. & Wagner, M.R. 2006. Attack preference of *Ips pini* on *Pinus ponderosa* in northern Arizona: tree size and bole position. *Agricultural and Forest Entomology*, **8**, 295–303.

Kõljalg, U., Dahlberg, A., Taylor, A.F.S., Larsson, E., Hallenberg, N., Stenlid, J., Larsson, K.-H., Fransson, P.M., Kåren, O. & Jonsson, L. 2000. Diversity and abundance of resupinate thelephoroid fungi as ectomycorrhizal symbionts in Swedish boreal forests. *Molecular Ecology*, **9**, 1985–1996.

Komonen, A. 2001. Structure of insect communities inhabiting two old-growth forest specialist bracket fungi. *Ecological Entomology*, **26**, 63–75.

Komonen, A. & Kouki, J. 2005. Occurrence and abundance of fungus-dwelling beetle species (Ciidae) in boreal forests and clearcuts: habitat associations at two spatial scales. *Animal Biodiversity and Conservation*, **28**, 137–147.

Komonen, A., Penttilä, R., Lindgren, M. & Hanski, I. 2000. Forest fragmentation truncates a food chain based on an old-growth forest bracket fungus. *Oikos*, **90**, 119–126.

Komonen, A., Jonsell, M. & Ranius, T. 2008. Red-listing saproxylic beetles in Fennoscandia: current status and future perspectives. *Endangered Species Research*, **6**, 149–154.

Kouki, J., Arnold, K. & Martikainen, P. 2004. Long-term persistence of aspen – a key host for many threatened species – is endangered in old-growth conservation areas in Finland. *Journal for Nature Conservation*, **12**, 41–52.

Kozlowski, T. T. 1997. Responses of woody plants to flooding and salinity. *Tree Physiology Monographs*, **1**, 1–29.

Kozlowski, T. T., Kramer, P. J. & Pallardy, S. G. 1991. *The Physiological Ecology of Woody Plants*. San Diego, CA: Academic Press.

Kramer, P. J. & Kozlowski, T. T. 1979. *Physiology of Woody Plants*. New York: Academic Press.

Krankina, O. N., Harmon, M. E. & Griazkin, A. V. 1999. Nutrient stores and dynamics of woody detritus in a boreal forest: modeling potential implications at the stand level. *Canadian Journal of Forest Research*, **29**, 20–32.

Krankina, O. N., Harmon, M. E., Kukuev, Y. A., Treyfeld, R. F., Kashpor, N. N., Kresnov, V. G., Skudin, V. M., Protasov, N. A., Yatskov, M., Spycher, G. & Povarov, E. D. 2002. Coarse woody debris in forest regions of Russia. *Canadian Journal of Forest Research*, **32**, 768–778.

Krasny, M. E. & Whitmore, M. C. 1992. Gradual and sudden forest canopy gaps in Allegheny northern hardwood forests. *Canadian Journal of Forest Research*, **22**, 139–143.

Krivosheina, M. G. 2006. Taxonomic composition of dendrobiontic Diptera and the main trends of their adaptive radiation. *Entomological Review*, **86**, 557–567.

Krivosheina, N. P. & Zaitzev, A. I. 2008. Trophic relationships and main trends in morphological adaptations of larval mouthparts in sciaroid dipterans (Diptera, Sciaroidea). *Biology Bulletin*, **35**, 606–614.

Krokene, P. & Solheim, H. 1997. Growth of four bark-beetle-associated blue-stain fungi in relation to the induced wound response in Norway spruce. *Canadian Journal of Botany*, **75**, 618–625.

Krokene, P. & Solheim, H. 2001. Loss of pathogenicity in the blue-stain fungus *Ceratocystis polonica*. *Plant Pathology*, **50**, 497–502.

Krombein, K. V. 1967. *Trap-nesting Wasps and Bees: Life Histories, Nests and Associates*. Washington, DC: Smithsonian Press.

Kropp, B. R. 1982a. Fungi from decayed wood as ectomycorrhizal symbionts of western hemlock. *Canadian Journal of Forest Research*, **12**, 36–39.

Kropp, B. R. 1982b. Rotten wood as mycorrhizal inoculum for containerized western hemlock. *Canadian Journal of Forest Research*, **12**, 428–431.

Kruys, N. & Jonsson, B. G. 1997. Insular patterns of calicioid lichens in a boreal old-growth forest-wetland mosaic. *Ecography*, **20**, 605–613.

Kruys, N. & Jonsson, B. G. 1999. Fine woody debris is important for species richness on logs in managed boreal spruce forests of northern Sweden. *Canadian Journal of Forest Research*, **29**, 1295–1299.

Kruys, N., Fries, C., Jonsson, B. G., Lämås, T. & Ståhl, G. 1999. Wood-inhabiting cryptogams on dead Norway spruce (*Picea abies*) trees in managed Swedish boreal forests. *Canadian Journal of Forest Research*, **29**, 178–186.

Küffer, N. & Senn-Irlet, B. 2005. Influence of forest management on the species richness and composition of wood-inhabiting basidiomycetes in Swiss forests. *Biodiversity and Conservation*, **14**, 2419–2435.

Kukor, J.J. & Martin, M.M. 1983. Acquisition of digestive enzymes by siricid woodwasps from their fungal symbiont. *Science*, **220**, 1161–1163.

Kukor, J.J. & Martin, M.M. 1986. The transformation of *Saperda calcarata* (Coleoptera: Cerambycidae) into a cellulose digester through the inclusion of fungal enzymes in its diet. *Oecologia*, **71**, 138–141.

Kukor, J.J., Cowan, D.P. & Martin, M.M. 1988. The role of ingested fungal enzymes in cellulose digestion in larvae of cerambycid beetles. *Physiological Zoology*, **61**, 364–371.

Kullman, L. 2008. Early postglacial appearance of tree species in northern Scandinavia: review and perspective. *Quaternary Science Reviews*, **27**, 2467–2472.

Kunz, T.H. & Lumsden, L.F. 2003. Ecology of cavity and foliage roosting bats. *In:* Kunz, T.H. & Fenton, M.B. (eds.) *Bat Ecology*. Chicago: University of Chicago Press, 2–90.

Kuranouchi, T., Nakamura, T., Shimamura, S., Kojima, H., Goka, K., Okabe, K. & Mochizuki, A. 2006. Nitrogen fixation in the stag beetle, *Dorcus (macrodorcus) rectus* (Motschulsky) (Col., Lucanidae). *Journal of Applied Entomology*, **130**, 471–472.

Kushnevskaya, H., Mirin, D. & Shorohova, E. 2007. Patterns of epixylic vegetation on spruce logs in late-successional boreal forests. *Forest Ecology and Management*, **250**, 25–33.

Kuuluvainen, T. 2002. Natural variability of forests as a reference for restoring and managing biological diversity in boreal Fennoscandia. *Silva Fennica*, **36**, 97–125.

Kuuluvainen, T. 2009. Forest management and biodiversity conservation based on natural ecosystem dynamics in northern Europe: the complexity challenge. *Ambio*, **38**, 309–315.

Kuuluvainen, T., Syrjänen, K. & Kalliola, R. 1998. Structure of a pristine *Picea abies* forest in northeastern Europe. *Journal of Vegetation Science*, **9**, 563–574.

Kuussaari, M., Bommarco, R., Heikkinen, R.K., Helm, A., Krauss, J., Lindborg, R., Öckinger, E., Pärtel, M., Pino, J., Rodà, F., Stefanescu, C., Teder, T., Zobel, M. & Steffan-Dewenter, I. 2009. Extinction debt: a challenge for biodiversity conservation. *Trends in Ecology and Evolution*, **24**, 564–570.

Laaka-Lindberg, S., Korpelainen, H. & Pohjamo, M. 2006. Spatial distribution of epixylic hepatics in relation to substrate in a boreal old-growth forest. *Journal of the Hattori Botanical Club*, **100**, 311–323.

Laaksonen, M., Peuhu, E., Várkonyi, G. & Siitonen, J. 2008. Effects of habitat quality and landscape structure on saproxylic species dwelling in boreal spruce-swamp forests. *Oikos*, **117**, 1098–1110.

Laaksonen, M., Murdoch, K., Siitonen, J. & Várkonyi, G. 2010. Habitat associations of *Agathidium pulchellum,* an endangered old-growth forest beetle species living on slime moulds. *Journal of Insect Conservation*, **14**, 89–98.

Labandeira, C. C. 1998. Early history of arthropod and vascular plant associations. *Annual Review of Earth and Planetary Sciences*, **26**, 329–377.

Labandeira, C. C., Phillips, T. L. & Norton, R. A. 1997. Oribatid mites and the decomposition of plant tissues in Paleozoic coal-swamp forests. *Palaios*, **12**, 319–353.

Labandeira, C. C., Lepage, B. A. & Johnson, A. H. 2001. A *Dendroctonus* bark engraving (Coleoptera: Scolytidae) from a Middle Eocene *Larix* (Coniferales: Pinaceae): early or delayed colonization? *American Journal of Botany*, **88**, 2026–2039.

Lachance, M.-A., Metcalf, B. J. & Starmer, W. J. 1982. Yeasts from exudates of *Quercus*, *Ulmus*, *Populus*, and *Pseudotsuga*: new isolations and elucidation of some factors affecting ecological specificity. *Microbial Ecology*, **8**, 191–198.

Lacy, R. C. 1984. Ecological and genetic responses to mycophagy in Drosophilidae (Diptera). *In:* Wheeler, Q. & Blackwell, M. (eds.) *Fungus–Insect Relationships*. New York: Columbia University Press, 286–301.

Lähde, E., Eskelinen, T. & Väänänen, A. 2002. Growth and diversity effects of silvicultural alternatives on an old-growth forest in Finland. *Forestry*, **75**, 395–400.

Laiho, R. & Prescott, C. E. 2004. Decay and nutrient dynamics of coarse woody debris in northern coniferous forests: a synthesis. *Canadian Journal of Forest Research*, **34**, 763–777.

Lange, M. 1992. Sequence of macromycetes on decaying beech logs. *Persoonia*, **14**, 449–456.

Langor, D. W. 1991. Arthropods and nematodes co-occurring with the eastern larch beetle, *Dendroctonus simplex* [Col.: Scolytidae], in Newfoundland. *Entomophaga*, **36**, 303–313.

Lännenpää, A., Aakala, T., Kauhanen, H. & Kuuluvainen, T. 2008. Tree mortality agents in pristine Norway spruce forests in Northern Fennoscandia. *Silva Fennica*, **42**, 151–163.

Larsen, J. B. & Nielsen, A. B. 2007. Nature-based forest management: where are we going? Elaborating forest development types in and with practice. *Forest Ecology and Management*, **238**, 107–117.

Larsson, M. C. & Svensson, G. P. 2009. Pheromone monitoring of rare and threatened insects: exploiting a pheromone–kairomone system to estimate prey and predator abundance. *Conservation Biology*, **23**, 1516–1525.

Larsson, M. C., Hedin, J., Svensson, G. P., Tolasch, T. & Francke, W. 2003. Characteristic odor of *Osmoderma eremita* identified as male-released pheromone. *Journal of Chemical Ecology*, **29**, 575–587.

Lasker, R. & Giese, A. C. 1956. Cellulose digestion in the silverfish *Ctenolepisma lineata*. *Journal of Experimental Biology*, **33**, 542–553.

Lässig, R. & Mocalov, S. 2000. Frequency and characteristics of severe storms in the Urals and their influence on the development, structure and management of the boreal forests. *Forest Ecology and Management*, **135**, 179–194.

Laurance, W.F., Goosem, M. & Laurance, S.G.W. 2009. Impacts of roads and linear clearings on tropical forests. *Trends in Ecology and Evolution*, **24**, 659–669.

Lawrence, J.F. 1973. Host preference in ciid beetles (Coleoptera: Ciidae) inhabiting the fruiting bodies of basidiomycetes in North America. *Bulletin of the Museum of Comparative Zoology*, **145**, 163–212.

Lawrence, J.F. 1982. Coleoptera. *In:* Parker, S.P. (ed.) *Synopsis and Classification of Living Organisms, Vol. 2*. New York: McGraw-Hill, 482–553.

Lawrence, J.F. 1989. Mycophagy in the Coleoptera: feeding strategies and morphological adaptations. *In:* Wilding, N., Collins, N.M., Hammon, P.M. & Webber, J.F. (eds.) *Insect–Fungus Interactions*. London: Academic Press, 1–123.

Lawrence, J.F. & Powell, J.A. 1969. Host relationships in North American fungus-feeding moths (Oecophoridae, Oinophilidae, Tineidae). *Bulletin of the Museum of Comparative Zoology*, **138**, 29–51.

Lee, Y.S. 2000. Qualitative evaluation of ligninolytic enzymes in xylariaceous fungi. *Journal of Microbiology and Biotechnology*, **10**, 462–469.

Lehtinen, R.M., Lannoo, M.J. & Wassersug, R.J. 2004. Phytotelm breeding anurans: past, present and future research. In: Lehtinen, R.M. (ed.) *Ecology and Evolution of Phytotelm-Breeding Anurans*. Miscellaneous Publications of the University of Michigan Museum of Zoology, No. 193, 1–9.

Leikola, M. 1969. On the termination of diameter growth of Scots pine in old age in northernmost Finnish Lapland [in Finnish with an English abstract]. *Silva Fennica*, **3**, 50–61.

Leitner, W.A. & Rosenzweig, M.L. 1997. Nested species–area curves and stochastic sampling: a new theory. *Oikos*, **79**, 503–512.

Lekander, B. 1955. Das Auftreten der Schadinsekten in den vom Januarsturm 1954 verheerten Wäldern [in Swedish with a German summary]. *Meddelanden från Statens Skogsforskningsinstitut*, **45**, 1–35.

Lekounougou, S., Mounguengui, S., Dumarçay, S., Rose, C., Courty, P.E., Garbaye, J., Gérardin, P., Jacquot, J.P. & Gelhaye, E. 2008. Initial stages of *Fagus sylvatica* wood colonization by the white-rot basidiomycete *Trametes versicolor:* enzymatic characterization. *International Biodeterioration and Biodegradation*, **61**, 287–293.

Leopold, A.C. 1980. Aging and senescence in plant development. *In:* Thimann, K.V. (ed.) *Senescence in Plants*. Boca Raton, FL: CRSC Press, 1–12.

Levins, R. 1970. Extinction. *Lecture Notes in Mathematics*, **2**, 75–107.

Levy, J.F. 1975. Bacteria associated with wood in ground contact. *In:* Liese, W. (ed.) *Biological Transformation of Wood by Microorganisms*. Berlin, Heidelberg, New York: Springer, 64–73.

Levy, J.F. 1982. The place of basidiomycetes in the decay of wood in contact with the ground. *In:* Frankland, J.C., Hedger, J.N. & Swift, M.J. (eds.) *Decomposer Basidiomycetes: Their Biology and Ecology*. Cambridge: Cambridge University Press, 161–178.

Levy, J.F. 1987. The natural history of the degradation of wood. *Philosophical Transactions of the Royal Society of London, A*, **321**, 423–433.

Lewis, K. & Hrinkevich, K. 2008. *Using Reconstructed Outbreak Histories of Mountain Pine Beetle, Fire and Climate to Predict the Risk of Future Outbreaks.* Natural Resources Canada, Canadian Forest Service, Pacific Forestry Centre, Victoria, BC.

Li, P.J. & Martin, T.E. 1991. Nest-site selection and nesting success of cavity-nesting birds in high elevation forest drainages. *Auk*, **108**, 405–418.

Lieutier, F., Day, K.R., Battisti, A., Gregoire, J.-C. & Evans, H.F. (eds.) 2004. *Bark and Wood Boring Insects in Living Trees in Europe: A Synthesis.* Dordrecht, Boston, London: Kluwer Academic Press.

Lieutier, F., Yart, A. & Salle, A. 2009. Stimulation of tree defenses by ophiostomatoid fungi can explain attack success of bark beetles on conifers. *Annals of Forest Science*, **66**, 801.

Lilja, S., Wallenius, T. & Kuuluvainen, T. 2006. Structure and development of old *Picea abies* forests in northern boreal Fennoscandia. *Ecoscience*, **13**, 181–192.

Lim, Y.W., Kim, J.-J., Lu, M. & Breuil, C. 2005. Determining fungal diversity on *Dendroctonus ponderosae* and *Ips pini* affecting lodgepole pine using cultural and molecular methods. *Fungal Diversity*, **19**, 79–94.

Lin, Z.B. & Zhang, H.N. 2004. Anti-tumor and immunoregulatory activities of *Ganoderma lucidum* and its possible mechanisms. *Acta Pharmacologica Sinica*, **25**, 1387–1395.

Lindahl, B., Stenlid, J., Olsson, S. & Finlay, R. 1999. Translocation of ^{32}P between interacting mycelia of a wood-decomposing fungus and ectomycorrhizal fungi in microcosm systems. *New Phytologist*, **143**, 183–193.

Lindahl, B.D. & Finlay, R.D. 2006. Activities of chitinolytic enzymes during primary and secondary colonization of wood by basidiomycetous fungi. *New Phytologist*, **169**, 389–397.

Lindblad, I. 1998. Wood-inhabiting fungi on fallen logs of Norway spruce: relations to forest management and substrate quality. *Nordic Journal of Botany*, **18**, 243–255.

Lindblad, I. 2000. Host specificity of some wood-inhabiting fungi in a tropical forest. *Mycologia*, **92**, 399–405.

Lindenmayer, D.B. & Noss, R.F. 2006. Salvage logging, ecosystem processes and biodiversity conservation. *Conservation Biology*, **20**, 949–958.

Lindenmayer, D.B., Cunningham, R.B., Tanton, M.T., Smith, A.P. & Nix, H.A. 1990. The conservation of arboreal marsupials in the montane ash forest in the central highlands of Victoria, southeast Australia. Part I. Factors influencing the occupancy of trees with hollows. *Biological Conservation*, **54**, 111–132.

Lindenmayer, D.B., Cunningham, R.B., Tanton, M.T., Smith, A.P. & Nix, H.A. 1991. Characteristics of hollow-bearing trees occupied by arboreal marsupials in the montane ash forests of the central highlands of Victoria, southeast Australia. *Forest Ecology and Management*, **40**, 289–308.

Lindenmayer, D.B., Welsh, A., Donnelly, C.F. & Cunningham, R.B. 1996. Use of nest trees by the mountain brushtail possum (*Trichosurus carinus*) (Phalangeridae: Marsupialia). Part II. Characteristics of occupied trees. *Wildlife Research*, **23**, 531–545.

Lindenmayer, D.B., Cunningham, R.B., Pope, M.L., Gibbons, P. & Donnelly, C.F. 2000. Cavity sizes and types in Australian eucalypts from wet and dry forest types: a simple of rule of thumb for estimating size and number of cavities. *Forest Ecology and Management*, **137**, 139–150.

Lindenmayer, D.B., Burton, P.J. & Franklin, J.F. 2008. *Salvage Logging and its Ecological Consequences*. Washington, DC: Island Press.

Linder, P., Elfving, B. & Zackrisson, O. 1997. Stand structure and successional trends in virgin boreal forest reserves in Sweden. *Forest Ecology and Management*, **98**, 17–33.

Lindgren, M. 2001. Polypore (Basidiomycetes) species richness and community structure in natural boreal forests of NW Russian Karelia and adjacent areas in Finland. *Acta Botanica Fennica*, **170**, 1–41.

Lindhe, A. & Lindelöw, Å. 2004. Cut high stumps of spruce birch, aspen and oak as breeding substrates for saproxylic beetles. *Forest Ecology and Management*, **203**, 1–20.

Lindhe, A., Åsenblad, N. & Toresson, H.-G. 2004. Cut logs and high stumps of spruce, birch aspen and oak: nine years of saproxylic fungi succession. *Biological Conservation*, **119**, 443–454.

Lindhe, A., Lindelöw, Å. & Åsenblad, N. 2005. Saproxylic beetles in standing dead wood: density in relation to substrate sun exposure and diameter. *Biodiversity and Conservation*, **14**, 3033–3053.

Loehle, C. 1988. Tree life history strategies: the role of defenses. *Canadian Journal of Forest Research*, **18**, 209–222.

Logan, J.A., Régnière, J. & Powell, J.A. 2003. Assessing the impact of global warming on forest pest dynamics. *Frontiers in Ecology and Environment*, **1**, 130–137.

Lõhmus, A. 2009. Factors of species-specific detectability in conservation assessments of poorly studied taxa: the case of polypore fungi. *Biological Conservation*, **142**, 2792–2796.

Lõhmus, A. & Remm, J. 2005. Nest quality limits the number of hole-nesting passerines in their natural cavity-rich habitat. *Acta Oecologica*, **27**, 125–128.

Lõhmus, P. & Lõhmus, A. 2001. Snags and their lichen flora in old Estonian peatland forests. *Annales Botanici Fennici*, **38**, 265–280.

Lombardero, M.J., Ayres, M.P., Lorio, P.L. & Ruel, J.J. 2000. Environmental effects on constitutive and inducible resin defences of *Pinus taeda*. *Ecology Letters*, **3**, 329–339.

Lomholdt, O. 1975. The Sphecidae (Hymenoptera) of Fennoscandia and Denmark. *Fauna Entomologica Scandinavica*, **4**(1), 1–224.

Lomholdt, O. 1976. The Sphecidae (Hymenoptera) of Fennoscandia and Denmark. *Fauna Entomologica Scandinavica*, **4**(2), 225–452.

Looy, C.V., Brugman, W.A., Dilcher, D.L. & Visscher, H. 1999. The delayed resurgence of equatorial forests after the Permian–Triassic ecologic crisis. *PNAS*, **96**, 13857–13862.

Lorimer, C.G., Dahir, S.E. & Nordheim, E.V. 2001. Tree mortality and longevity in mature and old-growth hemlock-hardwood forests. *Journal of Ecology*, **89**, 960–971.

Losin, N., Floyd, C.H., Shweitzer, T.E. & Keller, S.J. 2006. Relationship between aspen heartwood rot and the location of cavity excavation by primary cavity-nester, the red-naped sapsucker. *Condor*, **108**, 706–710.

Lundberg, S. 1966. *Eicolyctus brunneus* Gyll. (Coleoptera): något om bl.a. biologin. [in Swedish with an English summary] *Entomologisk Tidskrift*, **87**, 47–49.

Lundberg, S. 1993. *Phryganophilus ruficollis* (Fabricius) (Coleoptera, Melandryidae) in north Fennoscandia: habitat and developmental biology. *Entomologisk Tidskrift*, **114**, 13–18.

Lundquist, J.E. 1995. Pest interactions and canopy gaps in ponderosa pine stands in the Black Hills, South Dakota, USA. *Forest Ecology and Management*, **74**, 37–48.

Luo, W., Vrijmoed, L.P. & Jones, E.B.G. 2005. Screening of marine fungi for lignocellulose-degrading enzyme activities. *Botanica Marina*, **48**, 379–386.

Luschka, N. 1993. Die Pilze des Nationalparks Bayerischer Wald im bayerisch-böhmischen Grenzegebiete. *Hoppea*, **53**, 5–363.

Luyssaert, S., Schulze, E.-D., Börner, A., Knohl, A., Hessenmöller, D., Law, B.E., Grace, J. & Ciais, P. 2008. Old-growth forests as global carbon sinks. *Nature*, **455**, 213–215.

Lygis, V., Vasiliauskas, R., Stenlid, J. & Vasiliauskas, A. 2004. Silvicultural and pathological evaluation of Scots pine afforestations mixed with trees to reduce the infections by *Heterobasidion annosum*. *Forest Ecology and Management*, **201**, 275–285.

MacDonald, D. 2001. *The New Encyclopedia of Mammals*. Oxford: Oxford University Press.

MacGowan, I. & Rotheray, G. 2008. *British Lonchaeidae: Diptera, Cyclorrhapha, Acalyptratae*. Handbooks for the Identification of British Insects, Vol. 10, Part 15. London: Royal Entomological Society of London.

Mackensen, J., Bauhus, J. & Webber, E. 2003. Decomposition rates of coarse woody debris: a review with particular emphasis on Australian tree species. *Australian Journal of Botany*, **51**, 27–37.

Maddocks, R.F. & Steineck, P.L. 1987. Ostracoda from experimental wood-island habitats in the deep sea. *Micropaleontology*, **33**, 318–355.

Maeto, K., Sato, S. & Miyata, H. 2002. Species diversity of longicorn beetles in humid warm-temperate forests: the impact of forest management practices on old-growth forest species in southwestern Japan. *Biodiversity and Conservation*, **11**, 1919–1937.

Magallón, S. & Castillo, A. 2009. Angiosperm diversification through time. *American Journal of Botany*, **96**, 349–365.

Magallón, S. & Sanderson, M.J. 2005. Angiosperm divergence times: the effects of genes, codon positions, and time constraints. *Evolution*, **59**, 1653–1670.

Magan, N. 2008. Ecophysiology: impact of environment on growth, synthesis of compatible solutes and enzyme production. *In:* Boddy, L., Frankland, J.C. & van West, P. (eds.) *Ecology of Saprotrophic Basidiomycetes*. London: Academic Press/Elsevier, 63–78.

Magel, E. 2000. Biochemistry and physiology of heartwood formation. *In:* Savidge, R., Barnett, J. & Napier, R. (eds.) *Molecular and Cell Biology of Wood Formation*. Oxford: BIOS Scientific Publishers, 363–376.

Magel, E., Jay-Allemand, C. & Ziegler, H. 1994. Formation of heartwood substances in the stemwood of *Robinia pseudoacacia* L. II. Distribution of nonstructural carbohydrates and wood extractives across the trunk. *Trees*, **8**, 165–171.

Magri, D., Vendramin, G. G., Comps, B., Dupanloup, I., Geburek, T., Gömöry, D., Latałowa, M., Litt, T., Paule, L., Roure, J. M., Tantau, I., van der Knaap, W. O., Petit, R. J. & de Beaulieu, J.-L. 2006. A new scenario for the quaternary history of European beech populations: palaeobotanical evidence and genetic consequences. *New Phytologist*, **171**, 199–221.

Makarova, O. L. 2004. Gamasid mites (Parasitiformes, Mesostigmata), dwellers of bracket fungi, from the Petchora-Ilychskii Reserve, Republic of Komi. *Entomological Review*, **84**, 667–672.

Mäkinen, H., Saranpää, P. & Linder, S. 2002. Wood-density variation of Norway spruce in relation to nutrient optimization and fibre dimensions. *Canadian Journal of Forest Research*, **32**, 185–194.

Mäkinen, H., Hynynen, J., Siitonen, J. & Sievänen, R. 2006. Predicting the decomposition of Scots pine, Norway spruce, and birch stems in Finland. *Ecological Applications*, **16**, 1865–1879.

Malloch, D. & Blackwell, M. 1993. Dispersal biology of the ophiostomatoid fungi. *In:* Wingfield, M. J., Seifert, K. A. & Webber, J. F. (eds.) *Ceratocystis and Ophiostoma: Taxonomy, Ecology, and Pathogenicity.* St. Paul, MN: The American Phytopathological Society, 195–206.

Manion, P. D. 1991. *Tree Disease Concepts.* Englewood Cliffs, NJ: Prentice-Hall.

Mariani, L., Chang, S. X. & Kabzems, R. 2006. Effects of tree harvesting, forest floor removal, and compaction on soil microbial biomass, microbial respiration, and N availability in a boreal aspen forest in British Columbia. *Soil Biology and Biochemistry*, **38**, 1734–1744.

Markham, P. & Bazin, M. J. 1991. Decomposition of cellulose by fungi. *In:* Arora, D. K., Rai, B., Mukerji, K. G. & Knudsen, G. R. (eds.) *Handbook of Applied Mycology, Vol. 1.* New York: Marcel Dekker, 379–424.

Martikainen, P. 2001. Conservation of threatened saproxylic species: significance of retained aspen *Populus tremula* on clearcut areas. *Ecological Bulletins*, **49**, 205–218.

Martikainen, P. & Kouki, J. 2003. Sampling the rarest: threatened beetles in boreal forest biodiversity inventories. *Biodiversity and Conservation*, **12**, 1815–1831.

Martikainen, P., Siitonen, J., Punttila, P., Kaila, L. & Rauh, J. 2000. Species richness of Coleoptera in mature managed and old growth boreal forests in southern Finland. *Biological Conservation*, **94**, 199–209.

Martin, K. & Eadie, J. M. 1999. Nest webs: a community-wide approach to the management and conservation of cavity-nesting forest birds. *Forest Ecology and Management*, **115**, 243–257.

Martin, K., Aitken, K. E. H. & Wiebe, K. L. 2004. Nest sites and nest webs for cavity-nesting communities in interior British Columbia, Canada: nest characteristics and niche partitioning. *Condor*, **106**, 5–19.

Martin, M. M. 1979. Biochemical implications of insect mycophagy. *Biological Reviews*, **54**, 1–21.

Martin, M.M. 1991. The evolution of cellulose digestion in insects. *Philosophical Transactions of the Royal Society, B*, **333**, 281–288.

Martin, M.M. & Martin, J.S. 1978. Cellulose digestion in the midgut of the fungus-growing termite *Macrotermes natalensis*: the role of acquired digestive enzymes. *Science*, **199**, 1453–1455.

Martin, O. 1989. Click beetles (Coleoptera, Elateridae) from old deciduous forests in Denmark. [in Danish with an English summary] *Entomologiske Meddelelser*, **57**, 1–107.

Martínez, Á. T., Speranza, M., Ruiz-Dueñas, F.J., Ferreira, P., Camarero, S., Guillén, F., Martínez, M.J., Gutiérrez, A. & del Río, J.C. 2005. Biodegradation of lignocellulosics: microbial, chemical, and enzymatic aspects of the fungal attack of lignin. *International Microbiology*, **8**, 195–204.

Martínez-Vilalta, J. & Piñol, J. 2002. Drought-induced mortality and hydraulic architecture in pine populations of the NE Iberian Peninsula. *Forest Ecology and Management*, **161**, 247–256.

Mašán, P. & Walther, D.E. 2004. Description of the male of *Hoploseius mariae* (Acari, Mesostigmata), an European ascid mite associated with wood-destroying fungi, with key to *Hoploseius* species. *Biologia, Bratislava*, **59**, 527–532.

Masuya, H., Yamaoka, Y., Kaneko, S. & Yamaura, Y. 2009. Ophiostomatoid fungi isolated from Japanese red pine and their relationships with bark beetles. *Mycoscience*, **50**, 212–223.

Mathiesen, A. 1950. The nitrogen nutrition and vitamin requirement of *Ophiostoma pini*. *Physiologia Plantarum*, **3**, 93–102.

Mathiesen-Käärik, A. 1953. Eine Übersicht über die gewönlichsten mit Borkenkäfer assoziierten Bläuepilze in Schweden und einige für Schweden neue Bläuepilze. *Meddelanden från Statens Skogsforskningsinstitut*, **43**, 1–74.

Mathiesen-Käärik, A. 1960a. *Growth and Sporulation of* Ophiostoma *and Some Other Blueing Fungi on Synthetic Media*. Symbolae Botanicae Upsalienses, Vol. 16, Issue 8, 168 pp.

Mathiesen-Käärik, A. 1960b. Studies on the ecology, taxonomy and physiology of Swedish insect-associated blue stain fungi, especially the genus *Ceratocystis*. *Oikos*, **11**, 1–25.

Matthewman, W.G. & Pielou, D.P. 1971. Arthropods inhabiting the sporophores of *Fomes fomentarius* (Polyporacea) in Gatineau Park, Quebec. *Canadian Entomologist*, **103**, 775–847.

May, R.M. 1973. *Stability and Complexity in Model Ecosystems*. Princeton, NJ: Princeton University Press.

Mayer, A.M. & Staples, R.C. 2002. Laccase: new functions for an old enzyme. *Phytochemistry*, **60**, 551–565.

McCann, K.S. 2000. The diversity–stability debate. *Nature*, **405**, 228–233.

McCarthy, J.W. 2001. Gap dynamics of forest trees: a review with particular attention to boreal forests. *Environmental Review*, **9**, 1–59.

McCarthy, J.W. & Weetman, G. 2007. Stand structure and development of an insect-mediated boreal forest landscape. *Forest Ecology and Management*, **241**, 101–114.

McCay, T.S. 2000. Use of woody debris by cotton mice (*Pteromyscus gossypinus*) in a southeastern pine forest. *Journal of Mammalogy*, **81**, 527–535.

McComb, W. & Lindenmayer, D.B. 1999. Dying, dead, and down trees. *In:* Hunter, M.J.J. (ed.) *Maintaining Biodiversity in Forest Ecosystems.* Cambridge: Cambridge University Press, 335–372.

McDowell, N., Pockman, W.T., Allen, C.D., Breshears, D.D., Cobb, N., Kolb, T., Plaut, J., Sperry, J., West, A., Williams, D.G. & Yepez, E.A. 2008. Mechanisms of plant survival and mortality during drought: why do some plants survive while others succumb to drought? *New Phytologist*, **178**, 719–739.

McLaughlin, J.W. & Phillips, S.A. 2006. Soil carbon, nitrogen, and base cation cycling 17 years after whole-tree harvesting in a low-elevation red spruce (*Picea rubens*)–balsam fir (*Abies balsamea*) forested watershed in central Maine, USA. *Forest Ecology and Management*, **222**, 235–253.

McLean, I.F.G. & Speight, M.C.D. 1993. Saproxylic invertebrates: the European context. *In:* Kirby, K.J. & Drake, C.M. (eds.) *Dead Wood Matters: The Ecology and Conservation of Saproxylic Invertebrates in Britain.* Peterborough, UK: English Nature, 21–32.

McRae, D.J., Duchesne, L.C., Freedman, B., Lynham, T.J. & Woodley, S. 2001. Comparisons between wildfire and forest harvesting and their implications in forest management. *Environmental Reviews*, **9**, 223–260.

Merrill, W. & Cowling, E.B. 1966. Role of nitrogen in wood deterioration: amount and distribution of nitrogen in fungi. *Phytopathology*, **56**, 1083–1090.

Meyer-Berthaud, B., Scheckler, S.E. & Wendt, J. 1999. *Archaeopteris* is the earliest known modern tree. *Nature*, **398**, 700–701.

Midtgaard, F., Rukke, B.A. & Sverdrup-Thygeson, A. 1998. Habitat use of the fungivorous beetle *Bolitophagus reticulatus* (Coleoptera: Tenebrionidae): effects of basidiocarp size, humidity and competitors. *European Journal of Entomology*, **95**, 559–570.

Mikusinski, G. 2006. Woodpeckers: distribution, conservation, and research in a global perspective. *Annales Zoologici Fennici*, **43**, 86–95.

Miller, C.N. 1999. Implications of fossil conifers for the phylogenetic relationships of living families. *Botanical Review*, **65**, 239–277.

Miller, K.B. & Wheeler, Q.D. 2005. Asymmetrical male mandibular horns and mating behavior in *Agathidium* Panzer (Coleoptera: Leiodidae). *Journal of Natural History*, **39**, 779–792.

Mitchell, F.J.G. 2005. How open were European primeval forests? Hypothesis testing using palaeoecological data. *Journal of Ecology*, **93**, 168–177.

Monge-Najera, J. & Alfaro, J.P. 1995. Geographic variation of habitats in Costa Rican velvet worms (Onychophora: Peripatidae). *Biogeographica*, **71**, 97–108.

Monterrubio-Rico, T.C. & Escalante-Pliego, P. 2006. Richness, distribution and conservation status of cavity nesting birds in Mexico. *Biological Conservation*, **128**, 67–78.

Montes, F. & Canellas, I. 2006. Modelling coarse woody debris dynamics in even-aged Scots pine forests. *Forest Ecology and Management*, **221**, 220–232.

Montgomery, M.E. & Wargo, P.M. 1983. Ethanol and other host-derived volatiles as attractants to beetles that bore into hardwoods. *Journal of Chemical Ecology*, **9**, 181–190.

Morales-Jimenéz, J., Zúniga, G., Villa-Tanaca, L. & Hernándes-Rodriguez, C. 2009. Bacterial community and nitrogen fixation in the red turpentine beetle, *Dendroctonus valens* LeConte (Coleoptera: Curculionidae: Scolytinae). *Microbial Ecology*, **58**, 879–891.

Morgan, F.D. 1968. Bionomics of Siricidae. *Annual Review of Entomology*, **13**, 239–256.

Morgenstern, I.M., Klopman, S. & Hibbett, D.S. 2008. Molecular evolution and diversity of lignin-degrading heme peroxidases in the Agaricomycetes. *Journal of Molecular Evolution*, **66**, 243–257.

Mortimer, M.J. & Kane, B. 2005. Hazard tree liability in the United States: uncertain risks for owners and professionals. *Urban Forestry & Urban Greening*, **2**, 159–165.

Moser, J.C. 1985. Use of sporothecae by phoretic *Tarsonemus* mites to transport ascospores of coniferous blue-stain fungi. *Transactions of the British Mycological Society*, **84**, 750–753.

Moser, J.C. & Macias-Samano, J.E. 2000. Tarsonemid mite associates of *Dendroctonus frontalis* (Coleoptera: Scolytidae): implications for the historical biogeography of *D. frontalis*. *Canadian Entomologist*, **132**, 765–771.

Moser, J.C. & Roton, L.M. 1971. Mites associated with southern pine bark beetles in Allen Parish, Louisiana. *Canadian Entomologist*, **103**, 1775–1796.

Moser, J.C., Eidmann, H.H. & Regnander, J.R. 1989a. The mites associated with *Ips typographus* in Sweden. *Annales Entomologici Fennici*, **55**, 23–27.

Moser, J.C., Perry, T.J. & Solheim, H. 1989b. Ascospores hyperphoretic on mites associated with *Ips typographus*. *Mycological Research*, **93**, 513–517.

Mswaka, A.Y. & Magan, N. 1999. Temperature and water potential relations of tropical *Trametes* and other wood-decay fungi from the indigenous forests of Zimbabwe. *Mycological Research*, **103**, 1309–1317.

Mueller, G.M., Schmit, J.P., Leacock, P.R., Buyck, B., Cifuentes, J., Desjardin, D.E., Halling, R.E., Hjortstam, K., Iturriaga, T., Larsson, K.-H., Lodge, D.J., May, T.W., Minter, D., Rajchenberg, M., Redhead, S.A., Ryvarden, L., Trappe, J.M., Watling, R. & Wu, Q. 2007. Global diversity and distribution of macrofungi. *Biodiversity and Conservation*, **16**, 37–48.

Mueller, U.G., Gerardo, N.M., Aanen, D.K., Six, D.L. & Schultz, T.R. 2005. The evolution of agriculture in insects. *Annual Review of Ecology, Evolution and Systematics*, **36**, 563–595.

Muhle, H. & LeBlanc, F. 1975. Bryophyte and lichen succession on decaying logs. I. Analysis along an evaporational gradient in eastern Canada. *Journal of the Hattori Botanical Laboratory*, **39**, 1–33.

Müller, J. & Bütler, R. 2010. A review of habitat thresholds for dead wood: a baseline for management recommendations in European forests. *European Journal of Forest Research*, **129**, 981–992.

Müller, J., Bussler, H., Gossner, M., Rettelbach, T. & Duelli, P. 2008a. The European spruce bark beetle *Ips typographus* (L.) in a national park: from pest to keystone species. *Biodiversity and Conservation*, **17**, 2979–3001.

Müller, J., Bussler, H. & Kneib, T. 2008b. Saproxylic beetle assemblages related to silvicultural management intensity and stand structures in a boreal beech forest in South Germany. *Journal of Insect Conservation*, **12**, 107–124.

Muona, J. & Rutanen, I. 1994. The short-term impact of fire on the beetle fauna in boreal coniferous forest. *Annales Zoologici Fennici*, **31**, 109–121.

Murdoch, C. W. & Campana, R. J. 1983. Bacterial species associated with wetwood of elm. *Phytopathology*, **73**, 1270–1273.

Murdoch, C. W., Biermann, C. J. & Campana, R. J. 1983. Pressure and composition of intrastem gases produced in wetwood of American elm. *Plant Disease*, **67**, 74–76.

Murphy, E. C. & Lehnhausen, W. A. 1998. Density and foraging ecology of woodpeckers following a stand-replacement fire. *Journal of Wildlife Management*, **62**, 1359–1372.

Næsset, E. 1999. Relationship between relative wood density of *Picea abies* logs and simple classification systems of decayed coarse woody debris. *Scandinavian Journal of Forest Research*, **14**, 454–461.

Naiman, R. J. & Décamps, H. 1997. The ecology of interfaces: riparian zones. *Annual Review of Ecology and Systematics*, **28**, 621–658.

Naiman, R. J., Melillo, J. M. & Hobbie, J. E. 1986. Ecosystem alteration of boreal forest streams by beaver (*Castor canadensis*). *Ecology*, **67**, 1254–1269.

Nascimbiene, J., Marini, L., Caniglia, G., Cester, D. & Nimis, P. L. 2008. Lichen diversity on stumps in relation to wood decay in subalpine forests of northern Italy. *Biodiversity and Conservation*, **17**, 2661–2670.

Nelson, J. A., Wubah, D. A., Whitmer, M. E., Johnson, E. A. & Stewart, D. J. 1999. Wood-eating catfishes of the genus *Panaque:* gut microflora and cellulolytic enzyme activities. *Journal of Fish Biology*, **54**, 1069–1082.

Newton, A. F. 1984. Mycophagy in Staphylinoidea (Coleoptera). *In:* Wheeler, Q. & Blackwell, M. (eds.) *Fungus–Insect Relationships: Perspectives in Ecology and Evolution*. New York: Columbia University Press, 302–353.

Newton, I. 2003. The role of nest sites in limiting the numbers of hole-nesting birds: a review. *Biological Conservation*, **70**, 265–276.

Niemelä, T. 2005. *Käävät: puiden sienet [Polypores: Lignicolous Fungi*] [in Finnish with an English summary]. Norrlinia, No. 13. Helsinki: Museum of Natural History, 320 pp.

Niemelä, T., Renvall, P. & Penttilä, R. 1995. Interactions of fungi at late stages of wood decomposition. *Annales Botanici Fennici*, **32**, 141–152.

Niemelä, T., Wallenius, T. & Kotiranta, H. 2002. The kelo tree, a vanishing substrate of specified wood-inhabiting fungi. *Polish Botanical Journal*, **47**, 91–101.

Nieto, A. & Alexander, K. N. A. 2010. *European Red List of Saproxylic Beetles*. Luxembourg: Publications Office of the European Union.

Nikitsky, N. B. & Schigel, D. S. 2004. Beetles in polypores of the Moscow region: checklist and ecological notes. *Entomologica Fennica*, **15**, 6–22.

Niklasson, M. & Granström, A. 2000. Numbers and sizes of fires: long-term spatially explicit fire history in a Swedish boreal forest. *Ecology*, **81**, 1484–1499.

Nikolajev, G. V. 1992. Taxonomical features and composition of genera of Mesozoic scarab beetles (Coleoptera, Scarabaeidae). *Paleontological Journal*, **1**, 76–88.

Nilsson, S.G. 1984. The evolution of nest-site selection among hole-nesting birds: the importance of nest predation and competition. *Ornis Scandinavica*, **15**, 167–175.

Nilsson, S.G. & Baranowski, R. 1997. Habitat predictability and the occurrence of wood beetles in old-growth beech forests. *Ecography*, **20**, 491–498.

Nilsson, S.G., Arup, U., Baranowski, R. & Ekman, S. 1995. Tree-dependent lichens and beetles as indicators in conservation forests. *Conservation Biology*, **9**, 1208–1215.

Nilsson, T. 1997. Survival and habitat preferences of adult *Bolitophagus reticulatus*. *Ecological Entomology*, **22**, 82–89.

Nordén, B. 1997. Genetic variation within and among populations of *Fomitopsis pinicola* (Basidiomycetes). *Nordic Journal of Botany*, **17**, 319–329.

Nordén, B. & Appelqvist, T. 2001. Conceptual problems of ecological continuity and its bioindicators. *Biodiversity and Conservation*, **10**, 779–791.

Nordén, B., Appelqvist, T., Lindahl, B. & Henningson, M. 1999. Cubic rot fungi – corticioid fungi in highly brown rotted spruce stumps. *Mycologia Helvetica*, **10**, 13–24.

Nordén, B., Götmark, F., Tönnberg, M. & Ryberg, M. 2004a. Dead wood in semi-natural temperate broadleaved woodland: contribution of coarse and fine dead wood, attached dead wood and stumps. *Forest Ecology and Management*, **194**, 235–248.

Nordén, B., Ryberg, M., Götmark, F. & Olausson, B. 2004b. Relative importance of coarse and fine woody debris for the diversity of wood-inhabiting fungi in temperate broadleaf forests. *Biological Conservation*, **117**, 1–10.

Nordén, B., Götmark, F., Ryberg, M., Paltto, H. & Allmér, J. 2008. Partial cutting reduces species richness of fungi on woody debris in oak-rich forests. *Canadian Journal of Forest Research*, **38**, 1807–1816.

Norstog, K.J. & Nicholls, T.J. 1997. *The Biology of the Cycads*. Ithaca, NY: Cornell University Press.

Nowak, D.J., Noble, M.H., Sisinni, S.M. & Dwyer, J.F. 2001. Assessing the US urban forest resource. *Journal of Forestry*, **99**, 37–42.

Nuñez, M. 1996. Hanging in the air: a tough skin for a tough life. *Mycologist*, **10**, 15–17.

Nuorteva, M. 1956. Über den Fichtenstamm-Bastkäfer, *Hylurgops palliatus* Gyll., und seine Insektenfeinde. *Acta Entomologica Fennica*, **13**, 1–116.

Nzokou, P., Tourtellot, S. & Kamdem, D.P. 2008. Kiln and microwave heat treatment of logs infested by the emerald ash borer (*Agrilus planipennis* Fairmaire) (Coleoptera: Buprestidae). *Forest Products Journal*, **58**, 68–72.

Oberprieler, R.G., Marvaldi, A.E. & Anderson, R.S. 2007. Weevils, weevils, weevils everywhere. *Zootaxa*, **1668**, 491–520.

Oberwinkler, F., Bandoni, R.J., Bauer, R., Deml, G. & Mkisimova-Horovitz, L. 1984. The life-history of *Christiansenia pallida*, a dimorphic, mycoparasitic heterobasidiomycete. *Mycologia*, **76**, 9–22.

Ocasio-Morales, R.G., Tsopelas, P. & Harrington, T.C. 2007. Origin of *Ceratocystis platani* on native *Platanus orientalis* in Greece and its impact on natural forests. *Plant Disease*, **91**, 901–904.

Ódor, P., Heilmann-Clausen, J., Christensen, M., Aude, E., van Dort, K.W., Piltaver, A., Siller, I., Veerkamp, M.T., Walleyn, R., Standóvar, T., van Hees, A.F.M., Kosec, J., Matočec, N., Kraigher, H. & Grebenc, T. 2006. Diversity of dead wood inhabiting fungi and bryophytes in semi-natural beech forests in Europe. *Biological Conservation*, **131**, 58–71.

Oevering, P. & Pitman, A.J. 2002. Characteristics of attack of coastal timbers by *Pselaphus spadix* (Herbs) (Col: Curc.: Cossoninae) and investigations of its life history. *Holzforschung*, **56**, 335–340.

Ohsawa, M. 2007. The role of isolated old oak trees in maintaining beetle diversity within larch plantations in the central mountainous region of Japan. *Forest Ecology and Management*, **250**, 215–226.

Økland, B. 1994. Mycetophilidae (Diptera), an insect group vulnerable to forestry practices? A comparison of clearcut, managed and semi-natural spruce forests in southern Norway. *Biodiversity and Conservation*, **3**, 68–85.

Økland, B. 1995. Insect fauna compared between six polypore species in a southern Norwegian spruce forest. *Fauna Norvegica Serie B*, **42**, 21–26.

Økland, B. 1996. Unlogged forests: important sites for preserving the diversity of mycetophilids (Diptera: Sciaroidea). *Biological Conservation*, **76**, 297–310.

Økland, B. & Hågvar, S. 1994. The insect fauna associated with carpophores of the fungus *Fomitopsis pinicola* (Fr.) Karts. in a southern Norwegian spruce forest. *Fauna Norvegica Serie B*, **41**, 29–42.

Oldfield, S., Lusty, C. & MacKinven, A. 1998. *The World List of Threatened Trees*. Cambridge, UK: World Conservation Press.

Oleksa, A., Ulrich, W. & Gawroński, R. 2006. Occurrence of the marbled rose-chafer (*Protaetia lugubris* Herbst, Coleoptera, Cetoniidae) in rural avenues in northern Poland. *Journal of Insect Conservation*, **10**, 241–247.

Oleksa, A., Ulrich, W. & Gawroński, R. 2007. Host tree preferences of hermit beetles (*Osmoderma eremita* Scop., Coleoptera: Scarabaeidae) in a network of rural avenues in Poland. *Polish Journal of Ecology*, **55**, 315–323.

Oliver, C.D. & Larson, B.C. 1990. *Forest Stand Dynamics*. New York: McGraw-Hill.

Olsson, F. & Lemdahl, G. 2009. A continuous Holocene beetle record from the site Stavsåkra, southern Sweden: implications for the last 10,600 years of forest and land-use history. *Journal of Quaternary Science*, **24**, 612–626.

Olsson, J. & Jonsson, B.G. 2010. Restoration fire and wood-inhabiting fungi in a Swedish *Pinus sylvestris* forest. *Forest Ecology and Management*, **259**, 1971–1980.

O'Neill, K.M. 2001. *Solitary Wasps: Behavior and Natural History*. Ithaca, NY: Cornell University Press.

Orledge, G.M. & Reynolds, S.E. 2005. Fungivore host-use groups from cluster analysis: patterns of utilisation of fungal fruiting bodies by ciid beetles. *Ecological Entomology*, **30**, 620–641.

Orth, A., Royse, D. & Tien, M. 1993. Ubiquity of lignin-degrading peroxidases among various wood-degrading fungi. *Applied and Environmental Microbiology*, **59**, 4017–4023.

Otjen, L. & Blanchette, R.A. 1986. A discussion of microstructural changes in wood during decomposition by white rot basidiomycetes. *Canadian Journal of Botany*, **64**, 905–911.

Ovaskainen, O., Hottola, J. & Siitonen, J. 2010a. Modeling species co-occurrence by multivariate logistic regression generates new hypotheses on fungal interactions. *Ecology*, **91**, 2514–2521.

Ovaskainen, O., Nokso-Koivisto, J., Hottola, J., Rajala, T., Pennanen, T., Ali-Kovero, H., Miettinen, O., Oinonen, P., Auvinen, P., Paulin, L., Larsson, K.-H. & Mäkipää, R. 2010b. Identifying wood-inhabiting fungi with 454 sequencing: what is the probability that BLAST gives the correct species? *Fungal Ecology*, **3**, 274–283.

Owen-Smith, N. 1987. Pleistocene extinctions: the pivotal role of megaherbivores. *Paleobiology*, **13**, 351–362.

Ozolincius, R., Miksys, V. & Stakenas, V. 2005. Growth-independent mortality of Lithuanian forest tree species. *Scandinavian Journal of Forest Research*, **20** (Suppl. 6), 153–160.

Paclík, M. & Weidinger, K. 2007. Microclimate of tree cavities during winter nights: implications for roost site selection in birds. *International Journal of Biometeorology*, **51**, 287–293.

Paine, T.D., Birch, M.C. & Svirha, P. 1981. Niche breadth and resource partitioning by four sympatric species of bark beetles (Coleoptera: Scolytidae). *Oecologia*, **48**, 1–6.

Paine, T.D., Raffa, K.F. & Harrington, T.C. 1997. Interactions among scolytid bark beetles, their associated fungi, and live host conifers. *Annual Review of Entomology*, **42**, 179–206.

Palace, M., Keller, M., Asner, G.P., Silva, J.N.M. & Passos, C. 2007. Necromass in undisturbed and logged forests in the Brazilian Amazon. *Forest Ecology and Management*, **238**, 309–318.

Palm, T. 1942. Coleopterfaunan vid nedre Dalälven. [in Swedish] *Entomologisk Tidskrift*, **63**, 1–58.

Palm, T. 1951. *Die Holz- und Rinden-Käfer der nordschwedischen Laubbäume.* Meddelanden från Statens Skogsforskningsinstitut, No. 40, 242 pp.

Palm, T. 1959. *Die Holz- und Rinden-Käfer der süd- un mittelschwedischen Laubbäume.* Opuscula Entomologica, Supplementum XVI, 374 pp.

Paltto, H., Nordén, B., Götmark, F. & Franc, N. 2006. At which spatial and temporal scales does landscape context affect local density of Red Data Book and Indicator species? *Biological Conservation*, **133**, 442–454.

Pang, K.-L., Abdel-Wahab, M.A., Sivichai, S., El-Sharouney, H.M. & Jones, E.B.G. 2002. Jahnulales (Dothideomycetes, Ascomycota): a new order of lignicolous freshwater ascomycetes. *Mycological Research*, **106**, 1031–1042.

Paradise, C.J. 2004. Relationships of water and leaf litter variability to insects inhabiting treeholes. *Journal of North American Benthological Society*, **23**, 793–805.

Paradise, C.J. & Dunson, W.A. 1997. Insect species interactions and resource effects in treeholes: are helodid beetles bottom-up facilitators of midge populations? *Oecologia*, **109**, 303–312.

Park, D. 1968. The ecology of terrestrial fungi. *In:* Ainsworth, G.C. & Sussman, A.S. (eds.) *The Fungi: An Advanced Treatise.* New York: Academic Press, 5–39.

Park, O. & Auerbach, S. 1954. Further study of the tree-hole complex with emphasis on quantitative aspects of the fauna. *Ecology*, **35**, 208–222.

Park, O., Auerbach, S. & Corley, G. 1950. The tree-hole habitat with emphasis on the pselaphid beetle fauna. *Bulletin of the Chicago Academy of Sciences*, **9**, 19–56.

Parker, G.G. 1995. Structure and microclimate of forest canopies. *In:* Lowman, M.D. & Nadkarni, N.M. (eds.) *Forest Canopies: A Review of Research on a Biological Frontier.* San Diego, CA: Academic Press, 73–106.

Parker, G.R., Leopold, D.J. & Eichenberger, J.K. 1985. Tree dynamics in an old-growth deciduous forest. *Forest Ecology and Management*, **11**, 31–57.

Parker, T.J., Clancy, K.M. & Mathiasen, R.L. 2006. Interactions among fire, insects and pathogens in coniferous forests of the interior western United States and Canada. *Agricultural and Forest Entomology*, **8**, 167–189.

Pattanavibool, A. & Edge, W.D. 1996. Single-tree selection silviculture affects cavity resources in mixed deciduous forests in Thailand. *Journal of Wildlife Management*, **60**, 67–73.

Paviour-Smith, K. 1960. The fruiting-bodies of macrofungi as habitats for beetles of the family Ciidae (Coleoptera). *Oikos*, **11**, 43–71.

Paviour-Smith, K. 1964. Habitats, headquarters and distribution of *Tetratoma fungorum*. *Entomologist's Monthly Magazine*, **100**, 71–80.

Paviour-Smith, K. & Elbourn, C.A. 1993. A quantitative study of the fauna of small dead and dying wood in living trees in Wytham Woods, near Oxford. *In:* Kirby, K.J. & Drake, C.M. (eds.) *Dead Wood Matters: The Ecology and Conservation of Saproxylic Invertebrates in Britain*. Peterborough, UK: English Nature, 33–57.

Pearce, R.B. 1991. Reaction zone relics and the dynamics of fungal spread in the xylem of woody angiosperms. *Physiological and Molecular Plant Pathology*, **39**, 41–55.

Pearce, R.B. 1996. Antimicrobial defences in the wood of living trees. *New Phytologist*, **132**, 203–233.

Pedlar, J., Pearce, J.L., Venier, L.A. & McKenney, D.W. 2002. Coarse woody debris in relation to disturbance and forest type in boreal Canada. *Forest Ecology and Management*, **158**, 189–194.

Peet, R.K. & Christensen, N.L. 1987. Competition and tree death. *BioScience*, **37**, 586–595.

Pennanen, J. 2002. Forest age distribution under mixed-severity fire regimes: a simulation-based analysis for middle boreal Fennoscandia. *Silva Fennica*, **36**, 213–231.

Penttilä, R. 2004. *The impacts of forestry on polyporous fungi in boreal forests.* PhD thesis, University of Helsinki.

Penttilä, R. & Kotiranta, H. 1996. Short-term effects of prescribed burning on wood-rotting fungi. *Silva Fennica*, **30**, 399–419.

Penttilä, R., Siitonen, J. & Kuusinen, M. 2004. Polypore diversity in managed and old-growth boreal *Picea abies* forests in southern Finland. *Biological Conservation*, **117**, 271–283.

Penttilä, R., Lindgren, M., Miettinen, O., Rita, H. & Hanski, I. 2006. Consequences of forest fragmentation for polyporous fungi at two spatial scales. *Oikos*, **114**, 225–240.

Pérez, J., Muñoz-Dorado, J., de la Rubia, T. & Martínez, J. 2002. Biodegradation and biological treatments of cellulose, hemicellulose and lignin: an overview. *International Microbiology*, **5**, 53–63.

Peterken, G.F. 1996. *Natural Woodland: Ecology and Conservation in Northern Temperate Regions*. Cambridge: Cambridge University Press.

Peterson, C.J. 2000. Catastrophic wind damage to North American forests and the potential impact of climate change. *Science of the Total Environment*, **262**, 287–311.

Petit, J.R. & Hampe, A. 2006. Some evolutionary consequences of being a tree. *Annual Review of Ecology, Evolution and Systematics*, **37**, 187–214.

Pettersson, E.M., Sullivan, B.T., Anderson, P., Berisford, C.W. & Birgersson, G. 2000. Odor perception in the bark beetle parasitoid *Roptrocerus xylophagorum* exposed to host-associated volatiles. *Journal of Chemical Ecology*, **26**, 2507–2525.

Pfister, D.H. 1994. *Orbilia fimicola*, a nematophagous discomycete and its *Arthrobotrys* anamorph. *Mycologia*, **86**, 451–453.

Phaff, H.J. & Knapp, E.P. 1956. The taxonomy of yeasts found in exudates of certain trees and other natural breeding sites of some species of *Drosophila*. *Antonie van Leeuwenhoek*, **22**, 117–130.

Phillips, D.H. & Burdekin, D.A. 1982. *Diseases of Forest and Ornamental Trees*. London: MacMillan.

Pianka, E.R. 1970. On *r*- and *K*-selection. *American Naturalist*, **104**, 592–597.

Pielou, D.P. & Verma, A.N. 1968. The arthropod fauna associated with the birch bracket fungi, *Polyporus betulinus*, in eastern Canada. *Canadian Entomologist*, **100**, 1179–1199.

Pimm, S.L. 2002. *Food Webs*. Chicago: University of Chicago Press.

Pimm, S.L., Lawton, J.H. & Cohen, J.E. 1991. Food web patterns and their consequences. *Nature*, **350**, 669–674.

Pinard, M.A. & Huffman, J. 1997. Fire resistance and bark properties of trees in a seasonally dry forest in eastern Bolivia. *Journal of Tropical Ecology*, **13**, 727–740.

Plattner, A., Kim, J.-J., Diguistini, S. & Breuil, C. 2008. Variation in pathogenicity of a mountain pine beetle-associated blue-stain fungus, *Grosmannia clavigera*, on young lodgepole pine in British Columbia. *Canadian Journal of Plant Pathology*, **30**, 457–466.

Plieninger, T., Höchtl, F. & Spek, T. 2006. Traditional land-use and nature conservation in European rural landscapes. *Environmental Science & Policy*, **9**, 317–321.

Pochon, J. 1939. Flore bactérienne cellulolytique du tube digestif de larves xylophages. *Comptes Rendus de l'Académie des Sciences*, **208**, 1684–1686.

Pócs, T. 1982. Tropical forest bryophytes. *In:* Smith, A.J.E. (ed.) *Bryophyte Ecology*. London: Chapman & Hall, 59–104.

Pohjamo, M., Laaka-Lindberg, S., Ovaskainen, O. & Korpelainen, H. 2006. Dispersal potential of spores and asexual propagules in the epixylic hepatic *Anastrophyllum hellerianum*. *Evolutionary Ecology*, **20**, 415–430.

Pointing, S.B., Parungao, M.M. & Hyde, K.D. 2003. Production of wood-decay enzymes, mass loss and lignin solubilization in wood by tropical *Xylariaceae*. *Mycological Research*, **107**, 231–235.

Pommerening, A. & Murphy, S. T. 2004. A review of the history, definitions and methods of continuous cover forestry with special attention to afforestation and restocking. *Forestry*, **77**, 27–44.

Ponomarenko, A. G. 2003. Ecological evolution of beetles (Insecta: Coleoptera). *Acta Zoologica Cracoviensia*, **46**, 319–328.

Pope, T. L., Block, W. M. & Beier, P. 2009. Prescribed fire effects on wintering, barkforaging birds in northern Arizona. *Journal of Wildlife Management*, **73**, 695–700.

Pore, R. S. 1986. The association of *Prototheca* spp. with slime flux in *Ulmus americana* and other trees. *Mycopathologia*, **94**, 67–73.

Potapov, P., Yaroshenko, A., Turubanova, S., Dubinin, M., Laestadius, L., Thies, C., Aksenov, D., Egorov, A., Yesipova, Y., Glushkov, I., Karpachevskiy, M., Kostikova, A., Manisha, A., Tsybikova, E. & Zhuravleva, I. 2008. Mapping the world's intact forest landscapes by remote sensing. *Ecology and Society*, **13**, 51.

Potts, R. & Behrensmeyer, A. K. 1992. Late Cenozoic terrestrial ecosystems. *In:* Behrensmeyer, A. K., Damuth, J. D., DiMichele, W. A., Potts, R., Sues, H.-D. & Wing, S. L. (eds.) *Terrestrial Ecosystems Through Time: Evolutionary Paleoecology of Terrestrial Plants and Animals*. Chicago, IL: Chicago University Press, 419–541.

Prentice, I. C. & Jolly, D. 2000. Mid-Holocene and glacial-maximum vegetation geography of the northern continents and Africa. *Journal of Biogeography*, **27**, 507–519.

Pretzsch, H. 2006. Species-specific allometric scaling under self-thinning: evidence from long-term plots in forest stands. *Oecologia*, **146**, 572–583.

Pretzsch, H. & Mette, T. 2008. Linking stand-level self-thinning allometry to the tree-level leaf biomass allometry. *Trees*, **22**, 611–622.

Pretzsch, H. & Schütze, G. 2005. Crown allometry and growing space efficiency of Norway spruce (*Picea abies* (L.) Karst.) and European beech (*Fagus sylvatica* L.) in pure and mixed stands. *Plant Biology*, **7**, 628–639.

Price, M. & Price, C. 2006. Creaming the best, or creatively transforming? Might felling the biggest trees first be a win–win strategy? *Forest Ecology and Management*, **224**, 297–303.

Price, T. S., Doggett, C., Pye, J. L. & Holmes, T. P. 1992. *A History of Southern Pine Beetle Outbreaks in the Southeastern United States*. Macon, GA: The Georgia Forestry Commission.

Prospero, S., Holdenrieder, O. & Rigling, D. 2003. Primary resource capture in two sympatric *Armillaria* species in managed Norway spruce forests. *Mycological Research*, **107**, 329–338.

Provan, J. & Bennett, K. D. 2008. Phylogeographic insights into cryptic glacial refugia. *Trends in Ecology and Evolution*, **16**, 608–613.

Pugh, G. J. F. 1980. Strategies in fungal ecology. *Transactions of the British Mycological Society*, **75**, 1–14.

Purvis, A., Cardillo, M., Grenyer, R. & Collen, B. 2005. Correlates of extinction risk: phylogeny, biology, threat and scale. *In:* Purvis, A., Gittleman, J. L. & Brooks, T. M. (eds.) *Phylogeny and Conservation*. Cambridge: Cambridge University Press, 295–316.

Putz, F.E., Dykstra, D.P. & Heinrich, R. 2000. Why poor logging practices persist in the Tropics. *Conservation Biology*, **14**, 951–956.

Quinn, C.J. & Price, R.A. 2003. Phylogeny of the Southern Hemisphere conifers. *Acta Horticulturae*, **615**, 129–136.

Råberg, U., Brischke, C., Rapp, A.O., Högberg, N.O.S. & Land, C.J. 2007. External and internal fungal flora of pine sapwood (*Pinus sylvestris* L.) specimens in above-ground field tests at six different sites in southwest Germany. *Holzforschung*, **61**, 104–111.

Rackham, O. 1986. *The History of the Countryside*. London: J.M. Dent.

Rackham, O. 1998. Savanna in Europe. *In:* Kirby, K.J. & Watkins, C. (eds.) *The Ecological History of European Forests*. Wallingford, UK: CAB International, 1–24.

Rackham, O. 2003. *Ancient Woodland: Its History, Vegetation and Uses in England*. Dalbeattie, UK: Castlepoint Press.

Raffa, K.F. & Berryman, A.A. 1983. The role of host plant resistance in the colonization behavior and ecology of bark beetles. *Ecological Monographs*, **53**, 27–49.

Raffa, K.F., Aukema, B.H., Erbilgin, N., Köepzig, K.D. & Wallin, K.F. 2005. Interactions among conifer terpenoids and bark beetles across multiple levels of scale: an attempt to understand links between population patterns and physiological processes. *Recent Advances in Phytochemistry*, **39**, 79–118.

Raffa, K.F., Aukema, B.H., Bentz, B.J., Carroll, A.L., Hicke, J.A., Turner, M.G. & Romme, W.H. 2008. Cross-scale drivers of natural disturbances prone to anthropogenic amplification: the dynamics of bark beetle eruptions. *BioScience*, **58**, 501–517.

Ranius, T. 2000. Minimum viable metapopulation size of a beetle, *Osmoderma eremita*, living in tree hollows. *Animal Conservation*, **3**, 37–43.

Ranius, T. 2001. Constancy and asynchrony of *Osmoderma eremita* populations in tree hollows. *Oecologia*, **126**, 208–215.

Ranius, T. 2002a. Influence of stand size and quality of tree hollows on saproxylic beetles in Sweden. *Biological Conservation*, **103**, 85–91.

Ranius, T. 2002b. *Osmoderma eremita* as an indicator of species richness of beetles in tree hollows. *Biodiversity and Conservation*, **11**, 931–941.

Ranius, T. 2006. Measuring the dispersal of saproxylic insects: a key characteristic for their conservation. *Population Ecology*, **48**, 177–188.

Ranius, T. 2007. Extinction risks in metapopulations of a beetle inhabiting hollow trees predicted from time series. *Ecography*, **30**, 716–726.

Ranius, T. & Fahrig, L. 2006. Targets for maintenance of dead wood for biodiversity conservation based on extinction thresholds. *Scandinavian Journal of Forest Research*, **21**, 201–208.

Ranius, T. & Hedin, J. 2001. The dispersal rate of a beetle, *Osmoderma eremita*, living in tree hollows. *Oecologia*, **126**, 363–370.

Ranius, T. & Hedin, J. 2004. Hermit beetle (*Osmoderma eremita*) in a fragmented landscape: predicting occupancy patterns. *In:* Akçakaya, H.R., Burgman, M.A., Kindvall, O., Wood, C.C., Sjögren-Gulve, P., Hatfield, J.S. & McCarthy, M.A. (eds.) *Species Conservation and Management: Case Studies*. Oxford: Oxford University Press, 162–170.

Ranius, T. & Jansson, N. 2000. The influence of forest regrowth, original canopy cover and tree size on saproxylic beetles associated with old oaks. *Biological Conservation*, **95**, 85–94.

Ranius, T. & Jansson, N. 2002. A comparison of three methods to survey saproxylic beetles in hollow oaks. *Biodiversity and Conservation*, **11**, 1759–1771.

Ranius, T. & Jonsson, M. 2007. Theoretical expectations for thresholds in the relationship between number of wood-living species and amount of coarse woody debris: a study case in spruce forests. *Journal for Nature Conservation*, **15**, 120–130.

Ranius, T. & Kindvall, O. 2004. Modelling the amount of coarse woody debris produced by the new biodiversity-oriented silvicultural practices in Sweden. *Biological Conservation*, **119**, 51–59.

Ranius, T. & Kindvall, O. 2006. Extinction risk of wood-living model species in forest landscapes as related to forest history and conservation strategy. *Landscape Ecology*, **21**, 687–698.

Ranius, T. & Nilsson, S. G. 1997. Habitat of *Osmoderma eremita* Scop. (Coleoptera: Scarabaeidae), a beetle living in hollow trees. *Journal of Insect Conservation*, **1**, 193–204.

Ranius, T. & Wilander, P. 2000. Occurrence of *Larca lata* H.J. Hansen (Pseudoscorpionida: Garypidae) and *Allochernes wideri* C.L. Koch (Pseudoscorpionida: Chernetidae) in tree hollows in relation to habitat quality and density. *Journal of Insect Conservation*, **4**, 23–31.

Ranius, T., Kindvall, O., Kruys, N. & Jonsson, B.G. 2003. Modelling dead wood in Norway spruce stands subject to different management regimes. *Forest Ecology and Management*, **182**, 13–29.

Ranius, T., Kruys, N. & Jonsson, B.G. 2004. Modelling dead wood in Fennoscandian old-growth forests dominated by Norway spruce. *Canadian Journal of Forest Research*, **34**, 1025–1034.

Ranius, T., Aguado, O., Antonsson, K., Audisio, P., Ballerio, A., Carpaneto, G.M., Chobot, K., Gjurašin, B., Hanssen, O., Huijbregts, H., Lakatos, F., Martin, O., Neculisenanu, Z., Nikitsky, N.B., Paill, W., Pirnat, A., Rizun, V., Ruicănescu, A., Stegner, J., Süda, I., Szwałko, P., Tamutis, V., Telnov, D., Tsinkevich, V., Versteirt, V., Vignon, V., Vögeli, M. & Zach, P. 2005. *Osmoderma eremita* (Coleoptera, Scarabaeidae, Cetoniinae) in Europe. *Animal Biodiversity and Conservation*, **28**, 1–44.

Ranius, T., Eliasson, P. & Johansson, P. 2008. Large-scale occurrence patterns of red-listed lichens and fungi on old oaks are influenced both by current and historical habitat density. *Biodiversity and Conservation*, **17**, 2371–2381.

Ranius, T., Niklasson, M. & Berg, N. 2009a. Development of tree hollows in pedunculate oak (*Quercus robur*). *Forest Ecology and Management*, **257**, 303–310.

Ranius, T., Svensson, G.P., Berg, N., Niklasson, M. & Larsson, M.C. 2009b. The successional change of hollow oaks affects their suitability for an inhabiting beetle, *Osmoderma eremita*. *Annales Zoologici Fennici*, **46**, 205–216.

Rassi, P., Hyvärinen, E., Juslén, A. & Mannerkoski, I. 2010. *The 2010 Red List of Finnish Species*. Helsinki: Ministry of the Environment and Finnish Environment Institute.

Ratcliffe, B.C. 1970. Collecting slime flux feeding Coleoptera in Japan. *Entomological News*, **81**, 255–256.

Raunikar, R., Buongiorno, J., Turner, J. & Zhu, S. 2010. Global outlook for wood and forests with the bioenergy demand implied by scenarios of the Intergovernmental Panel on Climate Change. *Forest Policy and Economics*, **12**, 48–56.

Rawlins, J.E. 1984. Mycophagy in Lepidotera. *In:* Wheeler, Q. & Blackwell, M. (eds.) *Fungus–Insect Relationships: Perspectives in Ecology and Evolution*. New York: Columbia University Press, 382–423.

Raymond, P., Bédard, S., Roy, V., Larouche, C. & Tremblay, S. 2009. The irregular shelterwood system: review, classification, and potential application to forests affected by partial disturbance. *Journal of Forestry*, **107**, 405–413.

Rayner, A.D.M. & Boddy, L. 1988. *Fungal Decomposition of Wood: Its Biology and Ecology*. Chichester, UK: John Wiley & Sons.

Rayner, A.D.M., Boddy, L. & Dowson, C.G. 1987. Temporary parasitism of *Coriolus* spp. by *Lenzites betulina*: a strategy for domain capture in wood decay fungi. *FEMS Microbiology Letters*, **45**, 53–58.

Read, H. 2000. *Veteran Trees: A Guide to Good Management*. Peterborough, UK: English Nature.

Rees, G. & Jones, E.B.G. 1984. Observations on the attachment of spores of marine fungi. *Botanica Marina*, **7**, 145–160.

Reeve, J.R., Ayres, M.P. & Lorio, P.L. 1995. Host suitability, predation and bark beetle population dynamics. *In:* Cappucino, N. & Price, P. (eds.) *Population Dynamics: New Approaches and Synthesis*. San Diego, CA: Academic Press, 339–357.

Reibnitz, J. 1999. Verbreitung und Lebensräume der Baumschwammfresser Südwestdeutschlands (Coleoptera: Cisidae). *Mitteilungen Entomologischer Verein Stuttgart*, **34**, 1–76.

Reineke, L.H. 1933. Perfecting a stand-density index for even-aged forests. *Journal of Agricultural Research*, **46**, 627–638.

Reinhard, J. & Rowell, D.M. 2005. Social behaviour in an Australian velvet worm, *Euperipatoides rowelli* (Onychophora: Peripatopsidae). *Journal of Zoology*, **267**, 1–7.

Remm, J., Lohmus, A. & Remm, K. 2006. Tree cavities in riverine forests: what determines their occurrence and use by hole-nesting passerines? *Forest Ecology and Management*, **221**, 267–277.

Renvall, P. 1995. Community structure and dynamics of wood-rotting basidiomycetes on decomposing conifer trunks in northern Finland. *Karstenia*, **35**, 1–51.

Retallack, G.J., Veevers, J.J. & Morante, R. 1996. Global coal gap between Permian-Triassic extinction and Middle Triassic recovery of peat-forming plants. *GSA Bulletin*, **108**, 195–207.

Roberge, J.-M., Angelstam, P. & Villard, M.-A. 2008. Specialised woodpeckers and naturalness in hemiboreal forests: deriving quantitative targets for conservation planning. *Biological Conservation*, **141**, 997–1012.

Rock, J., Badeck, F.-W. & Harmon, M.E. 2008. Estimating decomposition rate constants for European tree species from literature sources. *European Journal of Forest Research*, **127**, 301–313.

Rockström, J., Steffen, W., Noone, K., Persson, Å., Chapin, F.S., Lambin, E.F., Lenton, T.M., Scheffer, M., Folke, C., Schellnhuber, H.J., Nykvist, B., de Wit, C.A., Hughes, T., van der Leeuw, S., Rodhe, H., Sörlin, S., Snyder, P.K., Costanza, R., Svedin, U., Falkenmark, M., Karlberg, L., Corell, R.W., Fabry, V.J., Hansen, J., Walker, B., Liverman, D., Richardson, K., Crutzen, P. & Foley, J.A. 2009. A safe operating space for humanity. *Nature*, **461**, 472–475.

Rodrigues, A.S.L., Pilgrim, J.D., Lamourex, J.F., Hoffman, M. & Brooks, T.M. 2006. The value of the IUCN Red List for conservation. *Trends in Ecology and Evolution*, **21**, 71–76.

Rogers, J.D. 2000. Thoughts and musings on tropical Xylariaceae. *Mycological Research*, **104**, 1412–1420.

Rohrmann, S. & Molitoris, H.P. 1992. Screening of wood-degrading enzymes in marine fungi. *Canadian Journal of Botany*, **70**, 2116–2123.

Rolstad, J., Gjerde, I., Gundersen, V.S. & Sætersdal, M. 2002. Use of indicator species to assess forest continuity: a critique. *Conservation Biology*, **16**, 253–257.

Rolstad, J., Sætersdal, M., Gjerde, I. & Storaunet, K.O. 2004. Wood-decaying fungi in boreal forest: are species richness and abundances influenced by small-scale spatiotemporal distribution of dead wood? *Biological Conservation*, **117**, 539–555.

Romme, W.H., Knight, D.H. & Yavitt, J.B. 1986. Mountain pine beetle outbreaks in the Rocky Mountains: regulators of primary productivity. *American Naturalist*, **127**, 484–494.

Rosenvald, P. & Lõhmus, A. 2008. For what, when, and where is green-tree retention better than clear-cutting? A review of the biodiversity aspects. *Forest Ecology and Management*, **255**, 1–15.

Rosenzweig, M.I. 1995. *Species Diversity in Space and Time*. Cambridge: Cambridge University Press.

Rossman, A. 1994. A strategy for an all-taxa inventory of fungal biodiversity. *In:* Peng, C.I. & Chou, C.H. (eds.) *Biodiversity and Terrestrial Ecosystems*. Taipei: Academia Sinica Monograph Series, No. 14, 169–194.

Rotheray, G.E. 1990. Larval and puparial records of some hoverflies associated with dead wood (Diptera, Syrphidae). *Dipterists Digest*, **7**, 2–7.

Rotheray, G.E. 1991. Larval stages of 17 rare and poorly known British hoverflies (Diptera: Syrphidae). *Journal of Natural History*, **25**, 945–969.

Rotheray, G.E. 1994. Colour guide to hoverfly larvae (Diptera, Syrphidae) in Britain and Europe. *Dipterists Digest*, **9**, 1–156.

Rotheray, G.E. & Gilbert, F. 1999. Phylogeny of Palaearctic Syrphidae (Diptera): evidence from larval stages. *Zoological Journal of the Linnean Society*, **127**, 1–112.

Rotheray, G.E. & MacGowan, I. 2000. Status and breeding sites of three presumed endangered Scottish saproxylic syrphids (Diptera, Syrphidae). *Journal of Insect Conservation*, **4**, 215–223.

Rotheray, G.E., Hancock, G., Hewitt, S., Horsfield, D., MacGowan, I., Robertson, D. & Watt, K. 2001. The biodiversity and conservation of saproxylic Diptera in Scotland. *Journal of Insect Conservation*, **5**, 77–85.

Roualt, G., Candau, J.-N., Lieutier, F., Nageleisen, L.-M., Martin, J.-C. & Warzée, N. 2006. Effects of drought and heat on forest insect populations in relation to the 2003 drought in Western Europe. *Annals of Forest Science*, **63**, 613–624.

Rouvinen, S., Kuuluvainen, T. & Karjalainen, L. 2002a. Coarse woody debris in old *Pinus sylvestris* dominated forests along a geographic and human impact gradient in boreal Fennoscandia. *Canadian Journal of Forest Research*, **32**, 2184–2200.

Rouvinen, S., Kuuluvainen, T. & Siitonen, J. 2002b. Tree mortality in a *Pinus sylvestris* dominated boreal forest landscape in Vienansalo wilderness, eastern Fennoscandia. *Silva Fennica*, **36**, 127–145.

Rovira, I., Berkov, A., Parkinson, A., Tavakilian, G., Mori, S. & Meurer-Grimes, B. 1999. Antimicrobial activity of Neotropical wood and bark extracts. *Pharmaceutical Biology*, **37**, 208–215.

Rowe, J.S. & Scotter, G.W. 1973. Fire in the boreal forest. *Quaternary Research*, **3**, 444–464.

Rudolphi, J. & Gustafsson, L. 2005. Effects of forest-fuel harvesting on the amount of deadwood on clear-cuts. *Scandinavian Journal of Forest Research*, **20**, 235–242.

Rühm, W. 1956. *Die Nematoden der Ipidien*. Jena, Germany: G. Fischer-Verlag.

Runkle, J.R. 1982. Patterns of disturbance in some old-growth mesic forests of eastern North America. *Ecology*, **63**, 1533–1546.

Runkle, J.R. 1990. Eight years change in an old *Tsuga canadensis* woods affected by beech bark disease. *Bulletin of the Torrey Botanical Club*, **117**, 409–419.

Runkle, J.R. 2000. Canopy tree turnover in old-growth mesic forests of eastern North America. *Ecology*, **81**, 554–567.

Ruohomäki, K., Tanhuanpää, M., Ayres, M.P., Kaitaniemi, P., Tammaru, T. & Haukioja, E. 2000. Causes of cyclicity of *Epirrita autumnata* (Lepidoptera, Geometridae): grandiose theory and tedious practice. *Population Ecology*, **42**, 211–223.

Ruschka, F. 1924. Kleine Beiträge zur Kenntnis der forstlichen Chalcididen und Proctotrupiden von Schweden. *Entomologisk Tidskrift*, **45**, 6–16.

Ryan, K.C. 2002. Dynamic interactions between forest structure and fire behavior in boreal ecosystems. *Silva Fennica*, **36**, 13–39.

Rybczynski, N. 2007. Castorid phylogenetics: implications for the evolution of swimming and tree-exploitation in beavers. *Journal of Mammalian Evolution*, **14**, 1–35.

Ryvarden, L. & Gilbertson, R.L. 1993. *European Polypores: Volumes 1–2*. Oslo: Fungiflora.

Ryvarden, L. & Nuñez, M. 1992. Basidiomycetes in the canopy of an African rain forest. *In:* Hallé, F. & Pascal, O. (eds.) *Biologie d'une canopée de fôret équatoriale: II*. Lyon, France: Pro-Natura International & Operation Canopée, 116–118.

Saalas, U. 1917. Die Fichtenkäfer Finnlands. I. Allgemeiner Teil und spezieller Teil 1. *Annales Academiae Scientiarium Fennicae, Serie A*, **8**, 1–547.

Saalas, U. 1923. Die Fichtenkäfer Finnlands. II. Spezieller Teil. *Annales Academiae Scientiarum Fennicae, Serie A*, **22**, 1–746.

Saalas, U. 1933. Anteckningar över tvenne excursioner i Kolva urskogar i Yläne socken mer än 100 år sedan. [in Swedish] *Notulae Entomologicae*, **13**, 47–49.

Sahlin, E. & Ranius, T. 2009. Habitat availability in forests and clearcuts for saproxylic beetles associated with aspen. *Biodiversity and Conservation*, **18**, 621–638.

Saint-Germain, M., Drapeau, P. & Hébert, C. 2004a. Comparison of Coleoptera assemblages from a recently burned and unburned black spruce forests of northeastern North America. *Biological Conservation*, **118**, 583–592.

Saint-Germain, M., Drapeau, P. & Hébert, C. 2004b. Xylophagous insect species composition and patterns of substratum use on fire-killed black spruce in central Quebec. *Canadian Journal of Forest Research*, **34**, 677–685.

Saint-Germain, M., Drapeau, P. & Buddle, C.M. 2007. Host-use patterns of saproxylic phloeophagous and xylophagous Coleoptera adults and larvae along the decay gradient in standing dead black spruce and aspen. *Ecography*, **30**, 737–748.

Saint-Germain, M., Drapeau, P. & Buddle, C.M. 2008. Persistence of pyrophilous insects in fire-driven boreal forests: population dynamics in burned and unburned habitats. *Diversity and Distributions*, **14**, 713–720.

Sakamoto, Y. & Atsushi, K. 2002. Some properties of the bacterial wetwood (watermark) in *Salix sachalinensis* caused by *Erwinia salicis*. *IAWA Journal*, **23**, 179–190.

Sammul, M., Kattai, K., Lanno, K., Meltsov, V., Otsus, M., Nõuakas, L., Kukk, D., Mesipuu, M., Kana, S. & Kukk, T. 2008. Wooded meadows of Estonia: conservation efforts for a traditional habitat. *Agricultural and Food Science*, **17**, 413–429.

Sandoval, L. & Barrantes, G. 2009. Relationships between species richness of excavator birds and cavity-adopters in seven tropical forests in Costa Rica. *Wilson Journal of Ornithology*, **121**, 75–81.

Sarén, M.-P., Serimaa, R., Andersson, S., Saranpää, P., Keckes, J. & Fratzl, P. 2004. Effect of growth rate on mean microfibril angle and cross-sectional shape of tracheids of Norway spruce. *Trees – Structure and Function*, **18**, 354–362.

Saunders, D.A., Smith, G.T. & Rowley, I. 1982. The availability and dimensions of tree hollows that provide nest sites for cockatoos (Psittaciformes) in Western Australia. *Australian Wildlife Research*, **9**, 541–556.

Savory, J.G. 1954. Breakdown of timber by ascomycetes and fungi imperfecti. *Annals of Applied Biology*, **41**, 336–347.

Scalbert, A. 1991. Antimicrobial properties of tannins. *Phytochemistry*, **30**, 3875–3883.

Schaffrath, U. 2003. Zu Lebensweise, Verbreitung und Gefährdung von *Osmoderma eremita* (Scopoli, 1763) (Coleoptera: Scarabaeoidea, Cetoniidae, Trichiinae). Teil 1. *Philippia*, **10**, 157–248.

Schedl, K.E. 1958. Breeding habits of arboricole insects in Central Africa. *In:* Becker, E.C. (ed.) *Proceedings of the 10th International Congress of Entomology, Montreal, August 17–25, 1956, Vol. 1.* Montreal: Mortimer, 185–197.

Scheerpeltz, O. & Höfler, K. 1948. *Käfer und Pilze.* Vienna: Verlag für Jugend und Volk.

Schelhaas, M.-J., Nabuurs, G.-J. & Schuck, A. 2003. Natural disturbances in the European forests in the 19th and 20th centuries. *Global Change Biology,* **9**, 1620–1633.

Schepps, J., Lohr, S. & Martin, T.E. 1999. Does tree hardness influence nest-tree selection by primary cavity nesters? *Auk,* **116**, 658–665.

Schiegg, K. 2001. Saproxylic insect diversity of beech: limbs are richer than trunks. *Forest Ecology and Management,* **149**, 295–304.

Schigel, D.S. 2007. Fleshy fungi of the genera *Armillaria, Pleurotus* and *Grifola* as habitats of Coleoptera. *Karstenia,* **47**, 37–48.

Schigel, D.S., Niemelä, T. & Kinnunen, J. 2006. Polypores of western Finnish Lapland and seasonal dynamics of polypore species. *Karstenia,* **46**, 37–64.

Schimitschek, E. 1953. Forstentomologische Studien im Urwald Rotwald. *Zeitschrift für angewandte Entomologie,* **34**, 178–215, 513–542.

Schimitschek, E. 1954. Forstentomologische Studien im Urwald Rotwald. *Zeitschrift für angewandte Entomologie,* **35**, 1–54.

Schink, B., Ward, J.F. & Zeikus, J.G. 1981. Microbiology of wetwood: role of anaerobic bacterial population in living trees. *Journal of General Microbiology,* **123**, 313–322.

Schlyter, F. & Anderbrant, O. 1993. Competition and niche separation between two bark beetles: existence and mechanisms. *Oikos,* **68**, 437–447.

Schlyter, F. & Löfqvist, J. 1990. Colonization pattern in the pine shoot beetle, *Tomicus piniperda*: effects of host declination, structure and presence of conspecifics. *Entomologia Experimentalis et Applicata,* **54**, 163–172.

Schmidl, J. & Bussler, H. 2004. Ökologische Gilden xylobionter Käfer Deutschlands. *Naturschutz und Landschaftsplanung,* **36**, 202–218.

Schmidl, J., Sulzer, P. & Kitching, R.L. 2008. The insect assemblage in water-filled tree-holes in a European temperate deciduous forest: community composition reflects structural, trophic and physiochemical factors. *Hydrobiologia,* **598**, 285–303.

Schmidt, C., Bernhard, D. & Arndt, E. 2007. Ecological examinations concerning xylobiontic Coleotera in the canopy of a *Quercus–Fraxinus* forest. *In:* Unterseher, M., Morawetz, W., Klotz, S. & Arndt, E. (eds.) *The Canopy of a Temperate Floodplain Forest: Results from Five Years of Research at the Leipzig Canopy Crane.* Leipzig, Germany: Universität Leipzig, 97–105.

Schmidt, O. 2006. *Wood and Tree Fungi: Biology, Damage, Protection and Use.* Berlin: Springer.

Schmidt, O. & Liese, W. 1994. Occurrence and significance of bacteria in wood. *Holzforschung,* **48**, 271–277.

Schmidt, O., Dujesiefken, D., Stobbe, H., Moreth, U., Kehr, R. & Schröder, T. 2008. *Pseudomonas syringae* pv. *aesculi* associated with horse chestnut bleeding canker in Germany. *Forest Pathology,* **38**, 124–128.

Schmit, J.P. 2005. Species richness of tropical wood-inhabiting macrofungi provides support for species-energy theory. *Mycologia*, **97**, 751–761.

Schmit, J.P. & Shearer, C.A. 2003. A checklist of mangrove-associated fungi, their geographical distribution and known host plants. *Mycotaxon*, **85**, 423–477.

Schmitt, C.B., Burgess, N.D., Coad, L., Belokurov, A., Besançon, C., Boisrobert, L., Campbell, A., Fish, L., Gliddon, D., Humphries, K., Kapos, V., Loucks, C., Lysenko, I., Miles, L., Mills, C., Minnemeyer, S., Pistorius, T., Ravilious, C., Steininger, M. & Winkel, G. 2009. Global analysis of the protection status of the world's forests. *Biological Conservation*, **142**, 2122–2130.

Schmitz, H., Schmitz, A. & Bleckmann, H. 2000. A new type of infrared organ in the Australian 'fire-beetle' *Merimna atrata* (Coleoptera: Buprestidae). *Naturwissenschaften*, **87**, 542–545.

Schnittler, M. & Novozhilov, Y. 1996. The myxomycetes of boreal woodlands in Russian northern Karelia: a preliminary report. *Karstenia*, **36**, 19–40.

Schönborn, W., Dorfelt, H., Foissner, W., Krienitz, L. & Schafer, U. 1999. A fossilized microcenosis in Triassic amber. *Journal of Eukaryotic Microbiology*, **46**, 571–584.

Schroeder, L.M. 2007. Retention or salvage logging of standing trees killed by the spruce bark beetle *Ips typographus*: consequences for dead wood dynamics and biodiversity. *Scandinavian Journal of Forest Research*, **22**, 524–530.

Schroeder, L.M. & Lindelöw, Å. 1989. Attraction of scolytids and associated beetles by different absolute amounts and proportions of α-pinene and ethanol. *Journal of Chemical Ecology*, **15**, 807–817.

Schroeder, L.M., Ranius, T., Ekbom, B. & Larsson, S. 2006. Recruitment of saproxylic beetles in high stumps created for maintaining biodiversity in a boreal forest landscape. *Canadian Journal of Forest Research*, **36**, 2168–2178.

Schroeder, L.M., Ranius, T., Ekbom, B. & Larsson, S. 2007. Spatial occurrence in a habitat-tracking metapopulation of a saproxylic beetle inhabiting a managed forest landscape. *Ecological Applications*, **17**, 900–909.

Schuck, H.J. 1982. The chemical composition of the monoterpene fraction in wounded wood of *Picea abies* and its significance for the resistance against wound infecting fungi. *European Journal of Forest Pathology*, **12**, 175–181.

Schütz, J.-P., Götz, M., Schmid, W. & Mandallaz, D. 2006. Vulnerability of spruce (*Picea abies*) and beech (*Fagus sylvatica*) forest stands to storms and consequences for silviculture. *European Journal of Forest Research*, **125**, 291–302.

Schütz, S., Weissbecker, B., Hummel, H.E., Apel, K.-H., Schmitz, H. & Bleckmann, H. 1999. Insect antenna as a smoke detector. *Nature*, **398**, 298–299.

Schwarze, F.W.M.R. & Baum, S. 2000. Mechanisms of reaction zone penetration by decay fungi in wood of beech (*Fagus sylvatica*). *New Phytologist*, **146**, 129–140.

Schwarze, F.W.M.R., Baum, S. & Fink, S. 2000a. Dual modes of degradation by *Fistulina hepatica* in xylem cell walls of *Quercus robur*. *Mycological Research*, **104**, 846–852.

Schwarze, F.W.M.R., Engels, J. & Mattheck, C. 2000b. *Fungal Strategies of Decay in Trees*. Berlin: Springer.

Schweingruber, F.H., Börner, A. & Schulze, E.-D. 2006. *Atlas of Woody Plant Stems: Evolution, Structure and Environmental Modifications.* Berlin: Springer.

Scott, A.C. 2000. The pre-Quaternary history of fire. *Palaeogeography, Palaeoclimatology, Palaeoecology,* **164**, 281–329.

Scott, A.C. 2009. Forest fire in the fossil record. *In:* Cerdà, A. & Robichaud, P.R. (eds.) *Fire Effects on Soils and Restoration Strategies.* Enfield, NH: Science Publishers, 1–37.

Scott, J.J., Oh, D.-C., Yuceer, M.C., Klepzig, K.D., Clardy, J. & Currie, C.R. 2008. Bacterial protection of beetle–fungus mutualism. *Science,* **322**, 63.

Scott, V.E., Evans, K.E., Patton, D.R. & Stone, C.P. 1977. *Cavity-nesting Birds of North American Forests.* USDA Forest Service Agriculture Handbook 511.

Sedell, J.R., Swanson, F.J. & Gregory, S.V. 1984. Evaluating fish response to woody debris. *In:* Hassler, T.J. (ed.) *Proceedings of the Pacific Northwest Streams Habitat Management Workshop.* Arcata, CA: American Fisheries Society, Humbolt State University, 191–221.

Sedgeley, J.A. 2001. Quality of cavity microclimate as a factor influencing selection of maternity roosts by a tree-dwelling bat, *Chalinolobus tuberculatus,* in New Zealand. *Journal of Applied Ecology,* **38**, 425–438.

Sedgeley, J.A. & O'Donnell, C.F.J. 1999. Roost selection by the long-tailed bat, *Chalinolobus tuberculatus,* in temperate New Zealand rainforest and its implications for the conservation of bats in managed forests. *Biological Conservation,* **88**, 261–276.

Seifert, K.A. 1993. Sapstain of commercial lumber by species of *Ophiostoma* and *Ceratocystis. In:* Wingfield, M.J., Seifert, K.A. & Webber, J.F. (eds.) *Ceratocystis and Ophiostoma: Taxonomy, Ecology and Pathogenicity.* St. Paul, MN: APS Press, 141–151.

Ševčík, J. 2006. Diptera associated with fungi in the Czech and Slovak Republics. *Časopis Slezkého Muzea Opava,* **55** (Suppl. 2), 1–84.

Seymour, R.S. & Kenefic, L.S. 2002. Influence of age on growth efficiency of *Tsuga canadensis* and *Picea rubens* trees in mixed-species, multiaged northern conifer stands. *Canadian Journal of Forest Research,* **32**, 2032–2042.

Shain, L. & Hillis, W.E. 1971. Phenolic extractives in Norway spruce and their effects on *Fomes annosus. Phytopathology,* **61**, 841–845.

Sharkey, M.J. 2007. Phylogeny and classification of Hymenoptera. *Zootaxa,* **1668**, 521–548.

Shaw, C.G. 1985. *In vitro* responses of different *Armillaria* taxa to gallic acid, tannic acid and ethanol. *Plant Pathology,* **34**, 594–602.

Shaw, C.G. & Kile, G.A. 1991. *Armillaria Root Disease.* Agriculture Handbook No. 691. Washington, DC: United States Department of Agriculture.

Shaw, M.R. 1997. *Rearing Parasitic Hymenoptera.* The Amateur Entomologist Series, No. 25. Amateur Entomologists' Society.

Shearer, C.A. 1992. The role of woody debris. *In:* Bärlocher, F. (ed.) *The Ecology of Aquatic Hyphomycetes.* Ecological Monographs 94. Heidelberg and New York: Springer-Verlag, 77–98.

Shearer, C.A., Descals, E., Kohlmeyer, B., Kohlmeyer, J., Marvanová, L., Padgett, D., Porter, D., Raja, H.A., Schmit, J.P., Thorton, H.A. & Voglymayr, H.

2007. Fungal biodiversity in aquatic habitats. *Biodiversity Conservation*, **16**, 49–67.

Shen, Z.-H., Fang, J.-Y., Liu, Z.-L. & Wu, J. 2001. Structure and dynamics of *Abies fabri* population near the alpine timberline in Hailuo Clough of Gongga Mountain. *Acta Botanica Sinica*, **43**, 1288–1293.

Sherwood, M.A. 1981. Convergent evolution in discomycetes from bark and wood. *Botanical Journal of the Linnean Society*, **82**, 15–34.

Shigo, A.L. 1985. How tree branches are attached to trunks. *Canadian Journal of Botany*, **63**, 1391–1401.

Shorohova, E., Kuuluvainen, T., Kangur, A. & Jõgiste, K. 2009. Natural stand structures, disturbance regimes and successional dynamics in the Eurasian boreal forests: a review with special reference to Russian studies. *Annals of Forest Science*, **66**, 201.

Shortle, W.C. 1990. Compartmentalization of decay red maple and hybrid poplar trees. *Phytopathology*, **69**, 410–413.

Shortle, W.C., Tattar, T.A. & Rich, A.E. 1971. Effects of some phenolic compounds on the growth of *Phialophora melinii* and *Fomes coniiatus*. *Phytopathology*, **61**, 552–555.

Shrimpton, D.M. & Whimey, H.S. 1968. Inhibition of growth of blue stain fungi by wood extractives. *Canadian Journal of Botany*, **46**, 757–761.

Siira-Pietikäinen, A., Penttinen, R. & Huhta, V. 2008. Oribatid mites (Acari: Oribatida) in boreal forest floor and decaying wood. *Pedobiologia*, **52**, 111–118.

Siitonen, J. 2001. Forest management, coarse woody debris and saproxylic organisms: Fennoscandian boreal forests as an example. *Ecological Bulletins*, **49**, 11–41.

Siitonen, J. 1994. Decaying wood and saproxylic Coleoptera in two old spruce forests: a comparison based on two sampling methods. *Annales Zoologici Fennici*, **31**, 89–95.

Siitonen, J. & Martikainen, P. 1994. Occurrence of rare and threatened insects living on decaying *Populus tremula:* a comparison between Finnish and Russian Karelia. *Scandinavian Journal of Forest Research*, **9**, 185–191.

Siitonen, J. & Saaristo, L. 2000. Habitat requirements and conservation of *Pytho kolwensis*, a beetle species of old-growth boreal forest. *Biological Conservation*, **94**, 211–220.

Siitonen, J., Martikainen, P., Kaila, L., Mannerkoski, I., Rassi, P. & Rutanen, I. 1996. New faunistic records of threatened saproxylic Coleoptera, Diptera, Heteroptera, Homoptera and Lepidoptera from the Republic of Karelia, Russia. *Entomologica Fennica*, **7**, 69–76.

Siitonen, J., Martikainen, P., Punttila, P. & Rauh, J. 2000. Coarse woody debris and stand characteristics in mature managed and old-growth boreal mesic forests in southern Finland. *Forest Ecology and Management*, **128**, 211–225.

Siitonen, J., Penttilä, R. & Kotiranta, H. 2001. Coarse woody debris, polyporous fungi and saproxylic insects in an old-growth spruce forest in Vodlozero National Park, Russian Karelia. *Ecological Bulletins*, **49**, 231–242.

Siitonen, J., Hottola, J. & Immonen, A. 2009. Differences in stand characteristics between brook-side key habitats and managed forests in southern Finland. *Silva Fennica*, **43**, 21–37.

Silvestri, F. 1913. Descripzione di un nuove ordine di insetti. *Bolletino del Laboratorio di Zoologia generale e agraria della R. Sciola superiore d'Agricola in Portici*, 192–209.

Sinclair, B.J. 1999. Insect cold tolerance: how many kinds of frozen? *European Journal of Entomology*, **96**, 157–164.

Sipe, A.R., Wilbur, A.E. & Cary, S.C. 2000. Bacterial symbiont transmission in the wood-boring shipworm *Bankia setacea* (Bivalvia: Teredinidae). *Applied and Environmental Microbiology*, **66**, 1685–1691.

Sippola, A.-L., Lehesvirta, T. & Renvall, P. 2001. Effects of selective logging on coarse woody debris and diversity of wood-decaying polypores in eastern Finland. *Ecological Bulletins*, **49**, 243–254.

Sippola, A.-L., Similä, M., Mönkkönen, M. & Jokimäki, J. 2004. Diversity of polyporous fungi (Polyporaceae) in northern boreal forests: effects of forest site type and logging intensity. *Scandinavian Journal of Forest Research*, **19**, 152–163.

Sirén, G. 1961. Skogsgränstallen som indicator för klimatfluktuationerna i norra Fennoskandien under historisk tid [in Swedish]. *Communicationes Instituti Forestalis Fenniae*, **54**, 1–66.

Six, D.L. 2003. A comparison of mycangial and phoretic fungi of individual mountain pine beetles. *Canadian Journal of Forest Research*, **33**, 1331–1334.

Six, D.L. & Paine, T.D. 1998. Effects of mycangial fungi and host tree species on progeny survival and emergence of *Dendroctonus ponderosae* (Coleoptera: Scolytidae). *Environmental Entomology*, **27**, 1393–1401.

Sjöström, E. & Westermark, U. 1998. Chemical composition of wood and pulps: basic components and their distribution. *In:* Sjöström, E. & Alén, R. (eds.) *Analytical Methods in Wood Chemistry, Pulping, and Papermaking.* Berlin: Springer-Verlag, 1–35.

Smeets, E.M.W. & Faaij, A.P.C. 2007. Bioenergy potentials from forestry in 2050: an assessment of the drivers that determine the potentials. *Climatic Change*, **81**, 353–390.

Smith, D.B. & Sears, M.K. 1982. Mandibular structure and feeding habits of three morphologically similar coleopterous larvae: *Cucujus clavipes* (Cucujidae), *Dendroides canadensis* (Pyrochroidae), and *Pytho depressus* (Salpingidae). *Canadian Entomologist*, **114**, 173–175.

Snäll, T., Ribeiro, P.J. & Rydin, H. 2003. Spatial occurrence and colonisations in patch-tracking metapopulations: local conditions versus dispersal. *Oikos*, **103**, 566–578.

Söderström, L. 1988a. Sequence of bryophytes and lichens in relation to substrate variables of decaying coniferous wood in northern Sweden. *Nordic Journal of Botany*, **8**, 89–97.

Söderström, L. 1988b. Substrate preference in some forest bryophytes: a quantitative study. *Lindbergia*, **18**, 98–103.

Sokoloff, A. 1964. Studies on the ecology of *Drosophila* in the Yosemite region of California: a preliminary survey of species associated with *D. pseudobscura* and *D. persimilis* at slime fluxes and banana traps. *Pan-Pacific Entomologist*, **40**, 203–218.

Solheim, H. & Långström, B. 1991. Blue-stain fungi associated with *Tomicus piniperda* in Sweden and preliminary observations on their pathogenicity. *Annales des Sciences Forestières*, **48**, 149–156.

Solheim, H. & Saffranyik, L. 1997. Pathogenicity to Sitka spruce of *Ceratocystis rufipennis* and *Leptographium abietinum,* blue-stain fungi associated with the spruce beetle. *Canadian Journal of Forest Research*, **27**, 1336–1341.

Solheim, H., Krokene, P. & Långström, B. 2001. Effects of growth and virulence of associated blue-stain fungi on host colonization behaviour of the pine shoot beetles *Tomicus minor* and *T. piniperda. Plant Pathology*, **50**, 111–116.

Sollins, P. 1982. Input and decay of coarse woody debris in coniferous stands in western Oregon and Washington. *Canadian Journal of Forest Research*, **12**, 18–28.

Sollins, P., Cline, S.P., Verhoeven, T., Sachs, D. & Spycher, G. 1987. Patterns of log decay in old-growth Douglas-fir forests. *Canadian Journal of Forest Research*, **17**, 1585–1595.

Sörensson, M. 1997. Morphological and taxonomical novelties in the world's smallest beetles, and the first Old World record of *Nanosellini* (Coleoptera: Ptiliidae). *Systematic Entomology*, **22**, 257–283.

Sörensson, M. 2008. AHA: a simple method for evaluating conservation priorities of trees in South Swedish parks and urban areas from an entomosaproxylic viewpoint [in Swedish with an English summary]. *Entomologisk Tidskrift*, **129**, 81–90.

Southwood, T.R.E. 1977. Habitat, the templet for ecological strategies? *Journal of Animal Ecology*, **46**, 337–365.

Spatafora, J.W. & Blackwell, M. 1993. The polyphyletic origins of ophiostomatoid fungi. *Mycological Research*, **98**, 1–9.

Speight, M.C.D. 1989. *Saproxylic Invertebrates and their Conservation*. Strasbourg: Council of Europe, Publications and Documents Division.

Spies, T.A., Franklin, J.F. & Thomas, T.B. 1988. Coarse woody debris in Douglas-fir forests of western Oregon and Washington. *Ecology*, **69**, 1689–1702.

Spribille, T., Thor, G., Bunnell, F.L., Goward, T. & Björk, C.R. 2008. Lichens on dead wood: species–substrate relationships in the epiphytic lichen floras of the Pacific Northwest and Fennoscandia. *Ecography*, **31**, 741–750.

Srivastava, D. 2005. Do local processes scale to global patterns? The role of drought and the species pool in determining treehole insect diversity. *Oecologia*, **145**, 205–215.

Stamets, P. 2005. *Mycelium Running: How Mushrooms Can Help Save the World*. Berkeley, CA: Ten Speed Press.

Stanosz, G.R. & Patton, R.F. 1991. Quantification of *Armillaria* rhizomorphs in Wisconsin aspen sucker stands. *European Journal of Forest Pathology*, **21**, 5–16.

Stein, W. E., Mannolini, F., Hernick, L. V., Landing, E. & Berry, C. M. 2007. Giant cladoxylopsid trees resolve the enigma of the Earth's earliest forest stumps at Gilboa. *Nature*, **446**, 904–907.

Stenlid, J. & Johansson, M. 1987. Infection of roots of Norway spruce (*Picea abies*) by *Heterobasidion annosum*. II. Early changes in phenolic content and toxicity. *European Journal of Forest Pathology*, **17**, 217–226.

Stenlid, J., Penttilä, R. & Dahlberg, A. 2008. Wood-decay basidiomycetes in boreal forests: distribution and community development. *In:* Boddy, L., Frankland, J.C. & van West, P. (eds.) *Ecology of Saprotrophic Basidiomycetes*. London: Academic Press/Elsevier, 239–262.

Stephens, S.L., Skinner, C.N. & Gill, S.J. 2003. Dendrochronology-based fire history of Jeffrey pine–mixed conifer forests in the Sierra San Pedro Martir, Mexico. *Canadian Journal of Forest Research*, **33**, 1090–1101.

Stephenson, S.L. 1988. Distribution and ecology of myxomycetes in temperate forests. I. Patterns of occurrence in the upland forests of southwestern Virginia. *Canadian Journal of Botany*, **66**, 2187–2207.

Stephenson, S.L. & Stempen, H. 1994. *Myxomycetes: A Handbook of Slime Molds*. Portland, OR: Timber Press.

Stephenson, S.L., Wheeler, Q.D., McHugh, J.V. & Fraissinet, P.R. 1994. New North American associations of Coleoptera with Myxomycetes. *Journal of Natural History*, **28**, 921–936.

Sterling, E. 1994. Aye-ayes: specialists on structurally defended resources. *Folia Primatologica*, **62**, 142–154.

Stevens, V. 1997. *The Ecological Role of Coarse Woody Debris: An Overview of the Ecological Importance of CWD in BC Forests*. Working Paper 30/97. Victoria, BC, Canada: Research Program, BC Ministry of Forests.

Stewart, G.H. & Burrows, L.E. 1994. Coarse woody debris in old-growth temperate beech (*Nothofagus*) forests of New Zealand. *Canadian Journal of Forest Research*, **24**, 1989–1996.

Stewart, G.H., Rose, A.B. & Veblen, T.T. 1991. Forest development in canopy gaps in old-growth beech (*Nothofagus*) forests, New Zealand. *Journal of Vegetation Science*, **2**, 679–690.

Stewart, J.R. & Lister, A.M. 2001. Cryptic northern refugia and the origins of the modern biota. *Trends in Ecology and Evolution*, **16**, 608–613.

Stireman, J.O., O'Hara, J.E. & Wood, D.M. 2006. Tachinidae: evolution, behavior, and ecology. *Annual Review of Entomology*, **51**, 525–555.

Stocks, B.J., Mason, J.A., Todd, J.B., Bosch, E.M., Wotton, B.M., Amiro, B.D., Flannigan, M.D., Hirsch, K.G., Logan, K.A., Martell, D.L. & Skinner, W.R. 2003. Large forest fires in Canada, 1959–1997. *Journal of Geophysical Research – Atmospheres*, **107**, 8149, doi:10.1029/2001JD000484.

Stokland, J.N. 2001. The coarse woody debris profile: an archive of recent forest history and an important biodiversity indicator. *Ecological Bulletins*, **49**, 71–83.

Stokland, J.N. & Kauserud, H. 2004. *Phellinus nigrolimitatus*: a wood-decomposing fungus highly influenced by forestry. *Forest Ecology and Management*, **187**, 333–343.

Stokland, J.N. & Larsson, K.-H. 2011. Legacies from natural forest dynamics: different effects of forest management on wood-inhabiting fungi in pine and spruce forests. *Forest Ecology and Management*, **261**, 1701–1721.

Stokland, J.N. & Meyke, E. 2008. The saproxylic database: an emerging overview of the biological diversity in dead wood. *Revue d'Ecologie (Terre Vie)*, **63**, 29–40.

Stokland, J.N., Eriksen, R., Tomter, S.M., Korhonen, K., Tomppo, E., Rajaniemi, S., Söderberg, U., Toet, H. & Riis-Nielsen, T. 2003. *Forest Biodiversity Indicators in the Nordic Countries: Status Based on National Forest Inventories*. Report No. 2003:514. Denmark: TemaNord, 106 pp.

Stokland, J.N., Tomter, S.M. & Söderberg, U. 2004. Development of dead wood indicators for biodiversity monitoring: experiences from Scandinavia. *In:*

Marchetti, M. (ed.) *Monitoring and Indicators of Forest Biodiversity in Europe: From Ideas to Operationality.* EFI Proceedings No. 51. Tonkatu, Finland: European Forest Institute, 207–226.

Storaunet, K.O., Rolstad, J., Gjerde, I. & Gundersen, V.S. 2005. Historical logging, productivity, and structural characteristics of boreal coniferous forests in Norway. *Silva Fennica,* **39**, 429–442.

Stork, N.E. 1987. Guild structure of arthropods from Bornean rain forest trees. *Ecological Entomology,* **12**, 69–80.

Stork, N.E., Hammond, P.M., Russel, B.L. & Hadwen, W.L. 2001. The spatial distribution of beetles within the canopies of oak trees in Richmond Park, UK. *Ecological Entomology,* **26**, 302–311.

Strand, M.R. & Pech, L.L. 1995. Immunological basis for compatibility in parasitoid–host relationships. *Annual Review of Entomology,* **40**, 31–56.

Stubblefield, S.P., Taylor, T.N. & Beck, C.B. 1985. Studies of Paleozoic fungi. IV. Wood-decaying fungi in *Callixylon newberryi* from the Upper Devonian. *American Journal of Botany,* **72**, 1765–1774.

Sturtevant, B.R., Bissonette, J.A., Long, J.N. & Roberts, D.W. 1997. Coarse woody debris as a function of age, stand structure, and disturbance in boreal Newfoundland. *Ecological Applications,* **7**, 702–712.

Suckling, D.M., Gibb, A.R., Daly, J.M., Chen, X. & Brockerhoff, E.G. 2001. Behavioral and electrophysiological responses of *Arhopalus tristis* to burnt pine and other stimuli. *Journal of Chemical Ecology,* **27**, 1091–1104.

Süda, I. & Nagirnyi, V. 2002. The *Dorcatoma* Herbst, 1792 (Coleoptera: Anobiidae) species of Estonia. *Entomologica Fennica,* **13**, 116–122.

Suh, S.-O., Marshall, C., McHugh, J.V. & Blackwell, M. 2003. Wood ingestion by passalid beetles in the presence of xylose-fermenting gut yeasts. *Molecular Ecology,* **12**, 3137–3145.

Suh, S.-O., McHugh, J.V., Pollock, D.D. & Blackwell, M. 2005. The beetle gut: a hyperdiverse source of novel yeasts. *Mycological Research,* **109**, 261–265.

Suh, S.-O., Blackwell, M., Kurtzman, C.P. & Lachance, M.-A. 2006. Phylogenetics of Saccharomycetales, the ascomycete yeasts. *Mycologia,* **98**, 1006–1017.

Švácha, P. 1994. Bionomics, behaviour and immature stages of *Pelecotoma fennica* (Paykull) (Coleoptera: Rhipiphoridae). *Journal of Natural History,* **28**, 585–618.

Svenning, J.-C. 2002. A review of natural vegetation openness in north-western Europe. *Biological Conservation,* **104**, 133–148.

Svenning, J.-C., Normand, S. & Kageyama, M. 2008. Glacial refugia of temperate trees in Europe: insights from species distribution modeling. *Journal of Ecology,* **96**, 1117–1127.

Svensson, G.P., Larsson, M.C. & Hedin, J. 2004. Attraction of the larval predator *Elater ferrugineus* to the sex pheromone of its prey, *Osmoderma eremita,* and its implication for conservation biology. *Journal of Chemical Ecology,* **30**, 353–363.

Sverdrup-Thygeson, A. & Birkemoe, T. 2008. What window traps can tell us: effect of placement, forest openness and beetle reproduction in retention trees. *Journal of Insect Conservation,* **13**, 183–191.

Sverdrup-Thygeson, A. & Lindenmayer, D.B. 2003. Ecological continuity and assumed indicator fungi in boreal forest: the importance of the landscape matrix. *Forest Ecology and Management*, **174**, 353–363.

Sverdrup-Thygeson, A., Borg, P. & Bergsaker, E. 2008. A comparison of biodiversity values in boreal forest regeneration areas before and after forest certification. *Scandinavian Journal of Forest Research*, **23**, 236–243.

Swaine, M.D., Lieberman, D. & Putz, F.E. 1987. The dynamics of tree populations in tropical forests: a review. *Journal of Tropical Ecology*, **3**, 359–366.

Swift, M.J. & Boddy, L. 1984. Animal–microbial interactions during wood decomposition. *In:* Anderson, J.M., Rayner, A.D.M. & Walton, D.W.H. (eds.) *Invertebrate–Microbe Interactions*. Cambridge: Cambridge University Press, 89–131.

Swift, M.J., Heal, O.W. & Anderson, J.M. 1979. *Decomposition in Terrestrial Ecosystems*. Oxford: Blackwell.

Syrjänen, K., Kalliola, R., Puolasmaa, A. & Mattsson, J. 1994. Landscape structure and forest dynamics in subcontinental Russian European taiga. *Annales Zoologici Fennici*, **31**, 19–34.

Szymczakowski, V.W. 1975. Unerwarteter Fund einer neuen Eocatops-Art in Schweden und Finnland (Col. Catopidae). *Entomologisk Tidskrift*, **96**, 3–7.

Taberlet, P. & Cheddadi, R. 2002. Quaternary refugia and persistence of biodiversity. *Science*, **297**, 2009–2010.

Takasugi, M., Nagao, S., Masamune, T., Shirata, A. & Takahashi, K. 1979. Structures of moracins E, F, G and H: new phytoalexins from diseased mulberry. *Tetrahedron Letters*, **48**, 4675–4678.

Talbot, P.H.B. 1977. The *Sirex–Amylostereum–Pinus* association. *Annual Review of Phytopathology*, **15**, 41–54.

Talkkari, A., Peltola, H., Kellomäki, S. & Strandman, H. 2000. Integration of component models from the tree, stand and regional levels to assess the risk of wind damage at forest margins. *Forest Ecology and Management*, **135**, 303–313.

Tanhashi, M., Matsushita, N. & Togashi, K. 2009. Are stag beetles fungivorous? *Journal of Insect Physiology*, **55**, 983–988.

Tarasov, P.E., Volkova, V.S., Webb, T., III, Guior, J., Andreev, A.A., Bezunsko, L.G., Bezunsko, T.V., Bykova, G.V., Dorofeyuk, N.I., Kvavandze, E.V., Osipova, I.M., Panova, N.K. & Sevastyanov, D.V. 2000. Last glacial maximum biomes reconstructed from pollen and plant macrofossil data from northern Eurasia. *Journal of Biogeography*, **27**, 609–620.

Tavakilian, G., Berkov, A., Meurer-Grimes, B. & Mori, S. 1997. Neotropical tree species and their faunas of xylophagous longicorns (Coleoptera: Cerambycidae) in French Guiana. *Botanical Review*, **63**, 303–355.

Taylor, E.L. & Taylor, T.N. 2009. Seed ferns from the late Paleozoic and Mesozoic: any angiosperm ancestors lurking there? *American Journal of Botany*, **96**, 237–251.

Taylor, J.W. & Berbee, M.L. 2006. Dating divergences in the fungal tree of life: review and new analyses. *Mycologia*, **98**, 838–849.

Taylor, J. W., Spatafora, J., O'Donnell, K., Lutzoni, F., James, T. Y., Hibbett, D. S., Geiser, D., Bruns, T. D. & Blackwell, M. 2004. The Fungi. *In:* Cracraft, J. & Donoghue, M. J. (eds.) *Assembling the Tree of Life*. New York: Oxford University Press, 171–194.

Taylor, R. L. 1929. The biology of the white pine weevil, *Pissodes strobi* (Peck), and a study of its insect parasites from an economic viewpoint. *Entomologica Americana*, **9**, 166–246; **10**, 1–86.

Taylor, T. N. & Osborn, J. M. 1996. The importance of fungi in shaping the paleoecosystem. *Review of Palaeobotany and Palynology*, **90**, 249–262.

Taylor, T. N., Hass, H. & Kerp, H. 1999. The oldest fossil ascomycetes. *Nature*, **399**, 648.

Taylor, T. N., Klavins, S. D., Krings, M., Taylor, E. L., Kerp, H. & Hass, H. 2004. Fungi from the Rhynie chert: a view from the dark side. *Transactions of the Royal Society of Edinburgh: Earth Sciences*, **94**, 457–473.

Taylor, T. N., Hass, H., Kerp, H., Krings, M. & Hanlin, R. T. 2005. Perithecial ascomycetes from the 400 million-year-old Rhynie chert: an example of ancestral polymorphism. *Mycologia*, **97**, 269–285.

Tedersoo, L., Kõljalg, U., Hallenberg, N. & Larsson, K.-H. 2003. Fine-scale distribution of ectomycorrhizal fungi and roots across substrate layers including coarse woody debris in a mixed forest. *New Phytologist*, **159**, 153–165.

Tenow, O. 1972. The outbreaks of *Oporinia autumnata* Bkh. and *Operophtera* spp. (Lep. Geometridae) in the Scandinavian mountain chain and northern Finland 1862–1968. *Zoologiska Bidrag från Uppsala*, **2**, 1–107.

Terho, M. & Hallaksela, A.-M. 2008. Decay characteristics of hazardous *Tilia*, *Betula*, and *Acer* trees felled by municipal urban tree managers in the Helsinki city area. *Forestry*, **81**, 151–159.

Terho, M., Hantula, J. & Hallaksela, A.-M. 2007. Occurrence and decay patterns of common wood-decay fungi in hazardous trees felled in the Helsinki City. *Forest Pathology*, **37**, 420–432.

Thiel, M. & Gutow, L. 2005. The ecology of rafting in the marine environment. I. The floating substrata. *Oceanography and Marine Biology*, **42**, 181–263.

Thuiller, W., Lavorel, S., Araujo, M. B., Sykes, M. T. & Prentice, I. C. 2005. Climate change threats to plant diversity in Europe. *Proceedings of the National Academy of Sciences of the USA*, **102**, 8245–8250.

Thunes, K. H. 1994. The coleopteran fauna of *Piptoporus betulinus* and *Fomes fomentarius* (Aphyllophorales: Polyporaceae) in western Norway. *Entomologica Fennica*, **5**, 157–168.

Thunes, K. H., Midtgaard, F. & Gjerde, I. 2000. Diversity of coleoptera of the bracket fungus *Fomitopsis pinicola* in a Norwegian spruce forest. *Biodiversity and Conservation*, **9**, 833–852.

Tibell, L. 1997. Anamorphs in mazaediate lichenized fungi and the Mycocaliciaceae ("Caliciales s.lat."). *Symbolae Botanicae Upsalensis*, **32**, 291–322.

Tibell, L. & Wedin, M. 2000. Mycocaliciales, a new order for nonlichenized calicioid fungi. *Mycologia*, **92**, 577–581.

Tien, M. & Kirk, T. K. 1983. Lignin-degrading enzyme from hymenomycete *Phanerochaete chrysosporium* Burds. *Science*, **221**, 661–663.

Tilman, D. 1994. Competition and biodiversity in spatially structured habitats. *Ecology*, **75**, 2–16.

Tilman, D., May, R. M., Lehman, C. L. & Nowak, M. A. 1994. Habitat destruction and the extinction debt. *Nature*, **371**, 65–66.

Timms, L. L., Smith, S. M. & De Groot, P. 2006. Patterns in the within-tree distribution of the emerald ash borer *Agrilus planipennis* (Fairmaire) in young, green-ash plantations of south-western Ontario, Canada. *Agricultural and Forest Entomology*, **8**, 313–321.

Timonen, J., Siitonen, J., Gustafsson, L., Kotiaho, J. S., Stokland, J. N., Sverdrup-Thygeson, A. & Mönkkönen, M. 2010. Woodland key habitats in northern Europe: concepts, inventory and protection. *Scandinavian Journal of Forest Research*, **25**, 309–324.

Tinker, D. B. & Knight, D. H. 2001. Temporal and spatial dynamics of coarse woody debris in harvested and unharvested lodgepole pine forests. *Ecological Modelling*, **141**, 125–149.

Tlalka, M., Bebber, D., Darrah, P. R. & Watkinson, S. C. 2008. Mycelial networks: nutrient uptake, translocation and role in ecosystems. *In:* Boddy, L., Frankland, J. C. & van West, P. (eds.) *Ecology of Saprotrophic Basidiomycetes.* London: Academic Press/Elsevier, 43–62.

Toivanen, T. & Kotiaho, J. S. 2007a. Burning of logged sites to protect beetles in managed boreal forests. *Conservation Biology*, **21**, 1562–1572.

Toivanen, T. & Kotiaho, J. S. 2007b. Mimicking natural disturbances of boreal forest: the effects of controlled burning and creating dead wood on beetle diversity. *Biodiversity Conservation*, **16**, 3193–3211.

Trail, B. J. & Lill, A. 1997. Use of tree hollows by two sympatric gliding possums, the squirrel glider, *Petaurus norfolcensis* and the sugar glider, *P. breviceps*. *Australian Mammalogy*, **20**, 79–88.

Travaglini, D., Barbati, A., Chirici, G., Lombardi, F., Marchetti, M. & Corona, P. 2007. Forest inventory for supporting plant biodiversity assessment: ForestBIOTA data on deadwood monitoring in Europe. *Plant Biosystems*, **141**, 222–230.

Tudge, C. 2005. *The Secret Life of Trees*. London: Penguin Books.

Turner, M. G. & Romme, W. H. 1994. Landscape dynamics in crown fire ecosystems. *Landscape Ecology*, **9**, 59–77.

Turner, R. D. 1973. Wood-boring bivalves, opportunistic species in the deep sea. *Science*, **180**, 1377–1379.

Tyrrell, L. E. & Crow, T. R. 1994. Dynamics of dead wood in old-growth hemlock hardwood forests of northern Wisconsin and northern Michigan. *Canadian Journal of Forest Research*, **24**, 1672–1683.

Tzean, S. S. & Liou, J. Y. 1993. Nematophagous resupinate basidiomycetous fungi. *Phytopathology*, **83**, 1015–1020.

Ulyshen, M. D. & Hanula, J. L. 2010. Patterns of saproxylic beetle succession in loblolly pine. *Agriculture and Forest Entomology*, **12**, 187–194.

Ungerer, M. J., Ayers, M. P. & Lombardero, M. J. 1999. Climate and the northern distribution limits of *Dendroctonus frontalis* Zimmermann (Coleoptera: Scolytidae). *Journal of Biogeography*, **26**, 1133–1145.

Unterseher, M. & Tal, O. 2006. Influence of small-scale conditions on the diversity of wood decay fungi in a temperate, mixed deciduous forest canopy. *Mycological Research*, **110**, 169–178.

Unterseher, M., Otto, P. & Morawetz, W. 2005. Species richness and substrate specificity of lignicolous fungi in the canopy of a temperate, mixed deciduous forest. *Mycological Progress*, **4**, 117–132.

Urcelay, C. & Robledo, G. 2009. Positive relationship between wood size and basidiocarp production of polypore fungi in *Alnus incana* forest. *Fungal Ecology*, **2**, 135–139.

Vallauri, D., André, J. & Blondel, J. 2003. Le bois mort, une lacune des forêts gérées. *Revue Forestière Française*, **55**, 99–112.

van Balen, J.H., Booy, C.J.H., van Franeker, J.A. & Osieck, E.R. 1982. Studies on hole-nesting birds in natural nest sites. I. Availability and occupation of natural nest sites. *Ardea*, **70**, 1–24.

van Mantgem, P.J., Stephenson, N.L., Byrne, J.C., Daniels, L.D., Franklin, J.F., Fulé, P.Z., Harmon, M.E., Larson, A.J., Smith, J.M., Taylor, A.H. & Veblen, T.T. 2009. Widespread increase of tree mortality rates in the western United States. *Science*, **323**, 521–524.

Vanderwel, M.C., Malcolm, J.R., Smith, S.M. & Islam, N. 2006. Insect community composition and trophic guild structure in decaying logs from eastern Canadian pine-dominated forests. *Forest Ecology and Management*, **225**, 190–199.

Vanha-Majamaa, I. & Jalonen, J. 2001. Green-tree retention in Fennoscandian forestry. *Scandinavian Journal of Forest Research*, **16** (Suppl. 3), 79–90.

Vasiliauskas, R., Juska, E., Vasiliauskas, A. & Stenlid, J. 2002. Community of Aphyllophorales and root rot in stumps of *Picea abies* on clear-felled forest sites in Lithuania. *Scandinavian Journal of Forest Research*, **17**, 398–407.

Vasiliauskas, R., Lygis, V., Thor, M. & Stenlid, J. 2004. Impact of biological (Rotstop) and chemical (urea) treatments on fungal community structure in freshly cut *Picea abies* stumps. *Biological Control*, **31**, 405–413.

Veblen, T.T., Hadley, K.S., Nel, E.M., Kitzberger, T., Reid, M. & Villalba, R. 1994. Disturbance regime and disturbance interactions in a Rocky Mountain subalpine forest. *Journal of Ecology*, **82**, 125–135.

Veerkamp, M.T., De Vries, B.W.L. & Kuyper, T.W. 1997. Shifts in species composition of lignicolous macromycetes after application of lime in a pine forest. *Mycological Research*, **101**, 1251–1256.

Venäläinen, M., Harju, A.M., Kainulainen, P., Viitanen, H. & Nikulainen, H. 2003. Variation in the decay resistance and its relationship with other wood characteristics in old Scots pines. *Annals of Forest Science*, **60**, 409–417.

Vera, F.W.M. 2000. *Grazing Ecology and Forest History*. Wallingford, UK: CAB International.

Vernon, P. & Vannier, G. 2001. Freezing susceptibility and freezing tolerance in Palaearctic Cetoniidae (Coleoptera). *Canadian Journal of Zoology*, **79**, 67–74.

Vicuña, R. 2000. Ligninolysis: a very peculiar microbial process. *Molecular Biotechnology*, **14**, 173–176.

Vieira, S., Trumbore, S., Camargo, P.B., Selhorst, D., Chambers, J.Q., Higuchi, N. & Martinelli, L.A. 2005. Slow growth rates of Amazonian trees:

consequences for carbon cycling. *Proceedings of the National Academy of Sciences of the USA*, **102**, 18502–18507.

Villard, M.-A. & Jonsson, B. G. 2009a. *Setting Conservation Targets for Managed Forest Landscapes*. Cambridge: Cambridge University Press.

Villard, M.-A. & Jonsson, B. G. 2009b. Putting conservation target science to work. *In:* Villard, M.-A. & Jonsson, B. G. (eds.) *Setting Conservation Targets for Managed Forest Landscapes*. Cambridge: Cambridge University Press, 393–401.

Vispo, C. & Hume, I. D. 1995. The digestive tract and digestive function in the North American porcupine and beaver. *Canadian Journal of Zoology*, **73**, 967–974.

Vodka, S., Konvicka, M. & Cizek, L. 2009. Habitat preferences of oak-feeding xylophagous beetles in a temperate woodland: implications for forest history and management. *Journal of Insect Conservation*, **13**, 553–562.

Vogt, K. A. & Edmonds, R. L. 1980. Patterns of nutrient concentration in basidiocarps in western Washington. *Canadian Journal of Botany*, **58**, 694–698.

von Sydow, F. 1993. Fungi occurring in the roots and basal parts of one- and two-year-old spruce and pine stumps. *Scandinavian Journal of Forest Research*, **8**, 174–184.

Vonhof, J. M. & Gwilliam, J. C. 2007. Intra- and interspecific patterns of day roost selection by three species of forest-dwelling bats in Southern British Colombia. *Forest Ecology and Management*, **252**, 165–175.

Waddell, K. L. 2002. Sampling coarse woody debris for multiple attributes in extensive resource inventories *Ecological Indicators*, **1**, 139–153.

Wagner, M. R., Clancy, K. M., Lieutier, F. & Paine, T. D. 2002. *Mechanisms and Deployment of Resistance in Trees to Insects*. Dordrecht, The Netherlands: Kluwer Academic Publishers.

Wainhouse, D., Ashburner, R., Ward, E. & Rose, J. 1998. The effect of variation in light and nitrogen on growth and defence in young Sitka spruce. *Functional Ecology*, **12**, 561–572.

Wald, P., Crockatt, M., Gray, V. & Boddy, L. 2004a. Growth and interspecific interactions of the rare oak polypore *Piptoporus quercinus*. *Mycological Research*, **108**, 189–197.

Wald, P., Pitkänen, S. & Boddy, L. 2004b. Interspecific interactions between the rare tooth fungi *Creolophus cirrhatus*, *Hericium erinaceus* and *H. coralloides* and other wood decay species in agar and wood. *Mycological Research*, **108**, 1447–1457.

Walker, L. P. & Wilson, D. B. 1991. Enzymatic hydrolysis of cellulose: an overview. *Bioresource Technology*, **36**, 3–14.

Wallace, A. R. 1878. *Tropical Nature and Other Essays*. London: MacMillan.

Walter, S. T. & Maguire, C. C. 2005. Snags, cavity-nesting birds, and silvicultural treatments in western Oregon. *Journal of Wildlife Management*, **69**, 1578–1591.

Waring, R. H. 1987. Characteristics of trees predisposed to die: stress causes distinctive changes in photosynthate allocation. *BioScience*, **37**, 561–583.

Warren, M. S. & Key, R. S. 1991. Woodlands: past, present and potential for insects. *In:* Collins, N. M. & Thomas, J. A. (eds.) *The Conservation of Insects and their Habitats*. London: Academic Press, 155–212.

Waterbury, J.B., Calloway, C.B. & Turner, R.D. 1983. A cellulolytic nitrogen-fixing bacterium cultured from the gland of Deshayes in shipworms (Bivalvia: Teredinidae). *Science*, **221**, 1401–1403.

Watt, A.S. 1947. Pattern and process in the plant community. *Journal of Ecology*, **35**, 1–22.

Webb, A., Buddle, C.M., Drapeau, P. & Saint-Germain, M. 2008. Use of remnant boreal forest habitats by saproxylic beetle assemblages in even-aged managed landscapes. *Biological Conservation*, **141**, 815–826.

Webb, J.K. & Shine, R. 1997. Out on a limb: conservation implications of tree-hollow use by a threatened snake species (*Hoplocephalus bungaroides*: Serpentes, Elapidae). *Biological Conservation*, **81**, 21–33.

Weber, R.W.S. 2006. On the ecology of fungal consortia of spring sap flows. *Mycologist*, **20**, 140–143.

Weber, R.W.S., Davoli, P. & Anke, H. 2006. A microbial consortium involving the astaxanthin producer *Xanthophyllomyces dendrorhous* on freshly cut birch stumps in Germany. *Mycologist*, **20**, 57–61.

Webster, C.R. & Jenkins, M.A. 2005. Coarse woody debris dynamics in the southern Appalachians as affected by topographic position and anthropogenic disturbance history. *Forest Ecology and Management*, **217**, 319–330.

Wedin, M., Doring, H. & Gilenstam, G. 2004. Saprotrophy and lichenization as options for the same fungal species on different substrata: environmental plasticity and fungal lifestyles in the *Stictis–Conotrema* complex. *New Phytologist*, **164**, 459–465.

Wegensteiner, R., Weiser, J. & Fuhrer, J. 1996. Observations on the occurrence of pathogens in the bark beetle *Ips typographus* L. (Col., Scolytidae). *Journal of Applied Entomology*, **120**, 199–204.

Weier, J. 2009. The mushroom messiah. *Conservation Magazine*, **10**, 13–17.

Weiss, H.B. 1920. The insect enemies of polyporoid fungi. *American Naturalist*, **54**, 443–447.

Weiss, H.B. & West, E. 1920. Fungous insects and their hosts. *Proceedings of the Biological Society of Washington*, **33**, 1–19.

Wells, K. 1994. Jelly fungi, then and now. *Mycologia*, **86**, 18–48.

Wermelinger, B. 2004. Ecology and management of the spruce bark beetle *Ips typographus*: a review of recent research. *Forest Ecology and Management*, **202**, 67–82.

Werner, P.A. & Prior, L.D. 2007. Tree-piping termites and growth and survival of host trees in savanna woodland of north Australia. *Journal of Tropical Ecology*, **23**, 611–622.

Werner, R.A., Holsten, E.H., Matsuoka, S.M. & Burnside, R.E. 2006. Spruce beetles and forest ecosystems in south-central Alaska: a review of 30 years of research. *Forest Ecology and Management*, **227**, 195–206.

Wertheim, B., van Baalen, E.J.A., Dicke, M. & Vet, L.E.M. 2005. Pheromone-mediated aggregation in nonsocial arthropods: an evolutionary ecological perspective. *Annual Review of Entomology*, **50**, 321–346.

Weslien, J. 1992. The arthropod complex associated with *Ips typographus* (L.) (Coleoptera, Scolytidae): species composition, phenology, and impact on bark beetle productivity. *Entomologica Fennica*, **3**, 205–213.

Weslien, J. & Schröter, H. 1996. Natürliche Dynamik des Borkenkäferbefalls nach Windwurf. *AFZ der Wald*, **19**, 1052–1055.

Wesołowski, T. 2005. Virtual conservation: how the European Union is turning a blind eye to its vanishing primaeval forests. *Conservation Biology*, **19**, 1349–1358.

Wesołowski, T. 2007. Lessons from long-term hole-nester studies in a primeval temperate forest. *Journal of Ornithology*, **148** (Suppl. 2), 395–405.

Westoby, M. 1984. The self-thinning rule. *Advances in Ecological Research*, **14**, 167–225.

Wheeler, Q. 1984. Evolution of slime mold feeding in leiodid beetles. *In:* Wheeler, Q. & Blackwell, M. (eds.) *Fungus–Insect Relationships: Perspectives in Ecology and Evolution*. New York: Columbia University Press, 446–477.

Wheeler, Q.D. & Miller, K.B. 2005. Slime-mold beetles of the genus *Agathidium* Panzer in North and Central America. Part I. Coleoptera: Leiodidae. *Bulletin of the American Museum of Natural History*, **290**, 1–95.

Whitehouse, N.J. 2006. The Holocene British and Irish ancient forest fossil beetle fauna: implications for forest history, biodiversity and faunal colonization. *Quaternary Science Reviews*, **25**, 1755–1789.

Whitford, K.R. 2002. Hollows in jarrah (*Eucalyptus marginata*) and marri (*Corymbia calophylla*) trees. I. Hollow sizes, tree attributes and ages. *Forest Ecology and Management*, **160**, 201–214.

Whitney, H.S., Bandoni, R.J. & Oberwinkler, F. 1987. *Entomocorticium dendroctoni* gen. et sp. nov. (Basidiomycotina), a possible nutritional symbiote of the mountain pine beetle in lodgepole pine in British Columbia. *Canadian Journal of Botany*, **65**, 95–102.

Wiebe, K.L. 2001. Microclimate of tree cavity nests: is it important for reproductive success in northern flickers? *Auk*, **118**, 412–421.

Wikars, L.-O. 1992. Forest fires and insects [in Swedish with an English summary]. *Entomologisk Tidskrift*, **113**, 1–11.

Wikars, L.-O. 1997a. Forest disturbance regimes affect the trade-offs between dispersal and reproduction in three species of buprestid beetles. *In:* Wikars, L.-O. (ed.) *Effects of Forest Fire and the Ecology of Fire-adapted Insects*. Comprehensive Summaries of Uppsala Dissertations from the Faculty of Science and Technology 272. Uppsala University.

Wikars, L.-O. 1997b. Pyrophilous insects in Orsa Finnmark, central Sweden: biology, distribution, and conservation [in Swedish with an English summary]. *Entomologisk Tidskrift*, **118**, 155–169.

Wikars, L.-O. 2004. Habitat requirements of the pine wood-living beetle *Tragosoma depsarium* (Coleoptera: Cerambycidae) at log, stand and landscape scale. *Ecological Bulletins*, **51**, 287–294.

Wilhelmsson, L., Arlinger, J., Spångberg, K., Lundquist, S.-E., Grahn, T., Hedenberg, Ö. & Olsson, L. 2002. Models for predicting wood properties in stems of *Picea abies* and *Pinus sylvestris* in Sweden. *Scandinavian Journal of Forest Research*, **17**, 330–350.

Wilhere, G.F. 2003. Simulations of snag dynamics in an industrial Douglas-fir forest. *Forest Ecology and Management*, **174**, 521–539.

Williamson, G.B., Laurance, W.F., Oliveira, A.A., Delamônica, P., Lovejoy, T.E., Gascon, C. & Pohl, L. 2000. Amazonian wet forest resistance to the 1997–98 El Niño drought. *Conservation Biology*, **14**, 1538–1542.

Willis, K.J. & Van Andel, T.H. 2004. Trees or no trees? The environments of central and eastern Europe during the Last Glaciation. *Quaternary Science Reviews*, **23**, 2369–2387.

Wilson, G.F. & Hort, N.D. 1926. Insect visitors to sap exudations of trees. *Transactions of the Royal Entomological Society of London*, **74**, 243–254.

Wimberly, M., Spies, T.A., Long, C.J. & Whitlock, C. 2000. Simulating historical variability in the amount of old forests in the Oregon Coast Range. *Conservation Biology*, **14**, 167–180.

Wingfield, M.J. 1993. *Leptographium* species as anamorphs of *Ophiostoma*: progress in establishing acceptable generic and species concepts. *In:* Wingfield, M.J., Seifert, K.A. & Webber, J.F. (eds.) *Ceratocystis and Ophiostoma: Taxonomy, Ecology and Pathogenicity*. St. Paul, MN: The American Phytopathological Society, 43–51.

Winter, S. & Möller, G.C. 2008. Microhabitats in lowland beech forests as monitoring tool for nature conservation. *Forest Ecology and Management*, **255**, 1251–1261.

Witzell, J. & Martín, J.A. 2008. Phenolic metabolites in the resistance of northern forest trees to pathogens: past experiences and future prospects. *Canadian Journal of Forest Research*, **38**, 2711–2727.

Woldendorp, G., Keenan, R.J., Barry, S. & Spencer, R.D. 2004. Analysis of sampling methods for coarse woody debris. *Forest Ecology and Management*, **198**, 133–148.

Wong, S.T., Servheen, C.W. & Ambu, L. 2004. Home range, movements and activity patterns, and breeding sites of Malayan sun bears *Helarctos malayanus* in the rainforest of Borneo. *Biological Conservation*, **119**, 169–181.

Wood, D.L. 1982. The role of pheromones, kairomones, and allomones in the host selection and colonization behaviour of bark beetles. *Annual Review of Entomology*, **27**, 411–446.

Wood, S.L. & Bright, D.E. 1992. *A Catalog of Scolytidae and Platypodidae (Coleoptera). Part 2: Taxonomic Index*. Great Basin Naturalist Memoirs 13, 1553 pp.

Woodman, J.D., Cooper, P.D. & Haritos, W.S. 2007. Effects of temperature and oxygen availability on water loss and carbon dioxide release in two sympatric saproxylic invertebrates. *Comparative Biochemistry and Physiology, Part A*, **147**, 514–520.

Woodroffe, G.E. 1953. An ecological study of the insects and mites in the nests of certain birds in Britain. *Bulletin of Ecological Research*, **44**, 739–772.

Woodward, S. 1992. Responses of gymnosperm bark tissues to fungal infections. *In:* Blanchette, R.A. & Biggs, A.R. (eds.) *Defense Mechanisms of Woody Plants Against Fungi*. Berlin: Springer-Verlag, 62–75.

Woodward, S., Stenlid, J., Karjalainen, R. & Hüttermann, A. 1998. *Heterobasidion annosum: Biology, Ecology, Impact and Control*. Wallingford, UK: CAB International.

Wormington, K.R., Lamb, D., McCallum, H.I. & Moloney, D.J. 2003. The characteristics of six species of living hollow-bearing trees and their importance for arboreal marsupials in the dry sclerophyll forests of southeast Queensland, Australia. *Forest Ecology and Management*, **182**, 75–92.

Worrall, J.J. 1994. Population structure of *Armillaria* species in several forest types. *Mycologia*, **86**, 401–407.

Worrall, J.J. & Harrington, T. C. 1988. Etiology of canopy gaps in spruce–fir forests at Crawford Notch, New Hampshire. *Canadian Journal of Forest Research*, **18**, 1463–1469.

Worrall, J.J., Anagnost, S.E. & Zabel, R.A. 1997. Comparison of wood decay among diverse lignicolous fungi. *Mycologia*, **89**, 199–219.

Worrall, J.J., Lee, T.D. & Harrington, T.C. 2005. Forest dynamics and agents that initiate and expand canopy gaps in *Picea abies* forests of Crawford Notch, New Hampshire, USA. *Journal of Ecology*, **93**, 178–190.

WRI 2000. *World Resources 2000–2001: People and Ecosystems – The Fraying Web of Life.* Washington DC: World Resources Institute (WRI) in collaboration with United Nations Environment Programme (UNEP), United Nations Development Programme (UNDP) and World Bank.

Wright, D.H. 1983. Species-energy theory: an extension of species-area theory. *Oikos*, **41**, 496–506.

Wu, J., Yu, X.-D. & Zhou, H.-Z. 2008. The saproxylic beetle assemblage associated with different host trees in southwest China. *Insect Science*, **15**, 251–261.

Yakovlev, E.B. 1994. *Palaearctic Diptera Associated with Fungi and Myxomycetes.* [in Russian with an English summary] Petrozavodsk, Russia: Karelian Research Center – Russian Academy of Sciences.

Yamada, T. 2001. Defense mechanisms in the sapwood of living trees against microbial infection. *Journal of Forest Research*, **6**, 127–137.

Yamaoka, Y., Chung, W.-H., Masua, H. & Hizai, M. 2009. Constant association of ophiostomatoid fungi with the bark beetle *Ips subelongatus* invading Japanese larch logs. *Mycoscience*, **50**, 165–172.

Yamazaki, K. 2007. Cicadas 'dig wells' that are used by ants, wasps and beetles. *European Journal of Entomology*, **104**, 347–349.

Yang, H.H., Effland, M.J. & Kirk, T.K. 1980. Factors influencing fungal degradation of lignin in a representative lignocellulosic, thermomechanical pulp. *Biotechnology and Bioengineering*, **22**, 65–77.

Yang, J., Kamdem, D.P., Keathley, D.E. & Han, K.-H. 2004. Seasonal changes in gene expression at the sapwood–heartwood transition zone of black locust (*Robinia pseudoacacia*) revealed by cDNA microarray analysis. *Tree Physiology*, **24**, 461–474.

Yang, Y., Yang, E., An, Z. & Liu, X. 2007. Evolution of nematode-trapping cells of predatory fungi of the Orbiliaceae based on evidence from rRNA-encoding DNA and multiprotein sequences. *PNAS*, **104**, 8379–8384.

Yanoviak, S.P. 2001. The macrofauna of water-filled tree holes on Barro Colorado Island, Panama. *Biotropica*, **33**, 110–120.

Yee, M., Grove, S. & Closs, L.B. 2007. Giant velvet worms (*Tasmanipatus barretti*) and postharvest regeneration burns in Tasmania. *Ecological Management and Restoration*, **8**, 66–71.

Yin, X. 1999. The decay of forest woody debris: numerical modeling and implications based on some 300 data cases from North America. *Oecologia*, **121**, 81–98.

Yoshimoto, J. & Nishida, T. 2007. Boring effect of carpenterworms (Lepidoptera: Cossidae) on sap exudation of the oak, *Quercus acutissima*. *Applied Entomology and Zoology*, **42**, 403–410.

Yoshimoto, J., Kakutani, T. & Nishida, T. 2005. Influence of resource abundance on the structure of the insect community attracted to fermented tree sap. *Ecological Research*, **20**, 405–414.

Yu, Q., Yang, D.-Q., Zhang, S.Y., Beaulieau, J. & Duchesne, I. 2003. Genetic variation in decay resistance and its correlation to wood density and growth in white spruce. *Canadian Journal of Forest Research*, **33**, 2177–2183.

Zabel, R.A. & Morrell, J.J. 1992. *Wood Microbiology: Decay and its Prevention*. San Diego, CA: Academic Press.

Zachariassen, K.E., Li, N.G., Laugsand, A.E., Kristiansen, E. & Pedersen, S.A. 2008. Is the strategy for cold hardiness in insects determined by their water balance? A study on two closely related families of beetles: Cerambycidae and Chrysomelidae. *Journal of Comparative Physiology B*, **178**, 977–984.

Zackrisson, O. 1977. Influence of forest fires on north Swedish boreal forest. *Oikos*, **29**, 22–32.

Zahradnik, P. 1993. New species of the genus *Dorcatoma* from central Europe (Coleoptera: Anobiidae). *Folia Heyrovskiana*, **1**, 80–83.

Zeran, R.M., Andersson, R.S. & Wheeler, T.A. 2006. Effect of small-scale forest management on fungivorous Coleoptera in old-growth forest fragments in southeastern Ontario, Canada. *Canadian Entomologist*, **139**, 118–130.

Zhang, N., Castlebury, L.A., Miller, A.N., Huhndorf, S.M., Schoch, C.L., Seifert, K.A., Rossman, A.M., Rogers, J.D., Kohlmeyer, J., Volkmann-Kohlmeyer, B. & Sung, G.-H. 2006. An overview of the systematics of the Sordariomycetes based on a four-gene phylogeny. *Mycologia*, **98**, 1076–1087.

Zhou, X.D., de Beer, Z.W. & Wingfield, M.J. 2006. DNA sequence comparisons of *Ophiostoma* spp., including *Ophiostoma aurorae* sp. nov., associated with pine bark beetles in South Africa. *Studies in Mycology*, **55**, 269–277.

Zipfel, R.D., de Beer, Z.W., Jacobs, K., Wingfield, B. & Wingfield, M.J. 2006. Multi-gene phylogenies define *Ceratocystiopsis* and *Grosmannia* distinct from *Ophiostoma*. *Studies in Mycology*, **55**, 77–99.

Zjawiony, J.K. 2004. Biologically active compounds from Aphyllophorales (polypore) fungi. *Journal of Natural Products*, **67**, 300–310.

Zugmaier, W., Bauer, R. & Oberwinkler, F. 1994. Mycoparasitism of some *Tremella* species. *Mycologia*, **86**, 49–56.

zur Strassen, R. 1957. Zur Ökologie des *Velleius dilatatus* Fabricius, eines als Raumgast bei *Vespa crabro* Linnaeus lebenden Staphyliniden (Ins. Col.). *Zeitschrift für Morphologie und Ökologie der Tiere*, **46**, 243–292.

Index

Printed in the United States
By Bookmasters

Printed in the United States
By Bookmasters